Microfluidics for Biomedical Applications

Microfluidics for Biomedical Applications

Editors

Nan Xiang
Zhonghua Ni

MDPI • Basel • Beijing • Wuhan • Barcelona • Belgrade • Manchester • Tokyo • Cluj • Tianjin

Editors
Nan Xiang
Southeast University
Nanjing, China

Zhonghua Ni
Southeast University
Nanjing, China

Editorial Office
MDPI
St. Alban-Anlage 66
4052 Basel, Switzerland

This is a reprint of articles from the Special Issue published online in the open access journal *Biosensors* (ISSN 2079-6374) (available at: https://www.mdpi.com/journal/biosensors/special_issues/microflu_med).

For citation purposes, cite each article independently as indicated on the article page online and as indicated below:

LastName, A.A.; LastName, B.B.; LastName, C.C. Article Title. *Journal Name* **Year**, *Volume Number*, Page Range.

ISBN 978-3-0365-7126-3 (Hbk)
ISBN 978-3-0365-7127-0 (PDF)

Cover image courtesy of Nan Xiang

Contents

About the Editors

Nan Xiang

Nan Xiang is a Professor in the School of Mechanical Engineering at Southeast University, Nanjing, China. He is now the PI of the Microfluidics and Soft Robotics Group in Jiangsu Key Laboratory for Design and Manufacture of Micro-Nano Biomedical Instruments. He received his B.E. in Biosystems Engineering from Zhejiang University in 2009, and his Ph.D. in Mechanical Engineering from Southeast University in 2014. His research interests include inertial microfluidics, soft robotics, microfluidic cell separation, viscoelastic microfluidics, point-of-care testing devices, microflow cytometer, and dielectrophoresis. He has published over 100 peer-reviewed articles and applied for over 70 patents. His research has gained increasing recognition through important awards and honors, such as First Prize of Natural Science of the Ministry of Education, the Top Cited paper Award from Institute of Physics, the Jiangsu Science Fund for Excellent Young Scholars, Provincial Excellent Doctoral Dissertation of Jiangsu, the Six Talent Peaks Project of Jiangsu Province, 333 Young Academic Leader of Jiangsu Province, and the Zhishan Young Scholar Fellowship of Southeast University (First class).

Zhonghua Ni

Zhonghua Ni is a Professor in the School of Mechanical Engineering and the director of the Jiangsu Key Laboratory for Design and Manufacture of Micro-Nano Biomedical Instruments, Southeast University, Nanjing, China. He received his B.E. from the China University of Petroleum in 1990 and his Ph.D. from Southeast University in 2001. His research interests include microfluidics, biomedical microdevices, and biosensors. He has published over 300 peer-reviewed articles in well-known SCI journals such as *Lab on a Chip*, *Analytical Chemistry*, and *Applied Physics Letters*.

biosensors

Editorial

Microfluidics for Biomedical Applications

Nan Xiang [1,2,*] and Zhonghua Ni [1,2]

1 School of Mechanical Engineering, Jiangsu Key Laboratory for Design and Manufacture of Micro-Nano Biomedical Instruments, Southeast University, Nanjing 211189, China
2 State Key Laboratory of Bioelectronics, Southeast University, Nanjing 210096, China
* Correspondence: nan.xiang@seu.edu.cn

Microfluidics refers to a technique for controlling and analyzing the fluids or micro-/nano-bioparticles in microscale channels or structures [1]. The developed, integrated systems are also known as micro-total analysis systems (μTAS) or lab-on-a-chip (LOC) systems. The advent of microfluidics has provided important insights into the fields of biomedical research and clinical diagnosis [2]. Compared with conventional techniques, microfluidics offers significant advantages, such as low sample consumption, a high efficiency, a small device footprint, and multifunction integration [3]. To date, the technique of microfluidics has been widely used for a range of biomedical applications, such as efficient sample pretreatment [4], single-cell analysis [5], high-throughput microflow cytometry [6], organ-on-a-chip [7], and biosensing [8]. As a result, great improvements and successes have been achieved in the application of microfluidics for biomedical research. On the basis of these applications, various point-of-care testing (POCT) devices and novel analytical instruments have been invented, some of which have been successfully commercialized [9]. We believe that microfluidics will inevitably lead to the revolution of biomedical research and clinical diagnosis. Our Special Issue is devoted to the most recent technical innovations and developments in the area of microfluidics, particularly in relation to biomedical applications. A total of 15 outstanding papers (including nine research articles and six reviews) were published in our Special Issue.

The detection of rare circulating tumor cells (CTCs) in the peripheral blood has been regarded as a noninvasive liquid biopsy technique and is of great significance for the early diagnosis, therapeutic efficacy monitoring, and personalized treatment of cancers [10]. Due to the rarity of CTCs (typically less than 50 CTCs in 1 mL of blood), the pre-separation of CTCs from the overwhelming majority of blood cells is a prerequisite for downstream detection [11]. Altay et al. [12] proposed a hybrid device coupling the passive spiral inertial microfluidics with the active surface acoustic waves. They numerically simulated the separation of CTCs from red blood cells (RBCs) and white blood cells (WBCs). In their hybrid device, the first-stage spiral channel enabled the differential focusing of the cells under the effect of the balance between the inertial lift force and Dean drag force. Then, these cells were further trapped and lined up on the multiplex nodal lines of surface acoustic waves in the second stage, allowing for the simultaneous separation of the CTCs, RBCs, and WBCs. The hybrid device inherits the significant advantages of both passive and active sorting methods and thus allows for CTC separation with a high efficiency and a high sensitivity.

Inertial microfluidics utilizes the inherent fluid inertial effect to perform the focusing, sorting, and concentration of cells [13]. However, its performance is heavily dependent on the flow rate. Xiang et al. [14] reported a syringe-tip inertial microfluidic centrifuge (i-centrifuge) which consists of a syringe-tip flow stabilizer and a four-channel paralleled inertial microfluidic concentrator. The unstable flow rate generated by hand pushing the syringe was stabilized and regulated by the flow stabilizer and then accurately powered the cell flow in the inertial microfluidics. The key components of i-centrifuge are made of low-cost polymer films and double-sided tapes, which enable the low-cost and disposable

Citation: Xiang, N.; Ni, Z. Microfluidics for Biomedical Applications. *Biosensors* **2023**, *13*, 161. https://doi.org/10.3390/bios13020161

Received: 15 January 2023
Accepted: 18 January 2023
Published: 20 January 2023

use of the i-centrifuge. In the hand-powered mode, the developed i-centrifuge was able to concentrate the cells at a high throughput of 16 mL/min.

The channel cross-sections in inertial microfluidics are commonly rectangles. Mehran et al. [15] developed a spiral inertial microfluidic channel with a unique U-shaped cross-section for the separation of WBCs from whole blood. The utilization of the U-shaped cross-section improved the isolation efficiency through the regulation of the Dean flow in the non-rectangular cross-sections. After optimization, the device was able to isolate over 95% WBCs, with a high purity of 88% at a high throughput of 6 mL/min.

In addition to the inertial effect, the viscoelasticity effect can be enhanced in inertial microfluidics to improve the manipulation accuracy and create a new manipulation capacity. Dai et al. [16] realized the blood plasma extraction in a serpentine channel by adding Poly (ethylene oxide) (PEO) to alter the fluid viscoelasticity of the blood. The focusing of the blood cells could be significantly improved by adding the PEO. After optimizing the critical operational parameters, the authors successfully achieved the blood plasma extraction with less than 1% hemolysis at a throughput of 15 μL/min.

Exosomes are nanoscale extracellular vesicles (EVs) secreted by the cells in the body fluids and are associated with cancer development and metastasis [17]. Due to their small sizes and heterogeneity, the isolation of exosomes is challenging. Zhou et al. [18] developed an aptamer-affinity-based microfluidic device for the rapid and efficient isolation of exosomes with diameters of 30–100 nm. The aptamer-targeting, exosome-carried proteins CD63 and PTK7 could realize a capture efficiency of 10^7–10^8 particles/mL in 20 min. These kinds of aptamer-immobilized microfluidic devices could also be customized to enrich other rare cells for clinical diagnosis.

Meggiolaro et al. [19] summarized the recent advances in microfluidic devices for the isolation of EVs. The current EV isolation methods can be divided into two categories: physical and chemical techniques. The physical isolation techniques are realized using active and passive principles, while the chemical techniques rely on immunocapture. In this paper, the advantages and disadvantages of microfluidics for the isolation of EVs were discussed by comparing this technique with other existing techniques. A perspective on the use of these microfluidic devices for clinical applications was highlighted.

RBC sedimentation is regarded as a promising indicator of hematological diseases and disorders. Kang et al. [20] proposed a new RBC sedimentation index to quantify the RBC sedimentation in syringes filled with blood (hematocrit of 50%) based on the shear stress of the blood flow. Their method showed an over 10-fold increase in sensitivity compared to the conventional methods and could be employed to monitor the RBC sedimentation during the blood delivery over a period of 10 min.

Pathogen infection is the major challenge affecting guided bone regeneration. Shi et al. [21] developed an asymmetric microfluidic/chitosan device for releasing drugs so as to prevent infections and guide bone regeneration. Their device achieved the controllable drug release (the deviation of the minocycline release was only 12.7% over 5 days). An excellent antibacterial performance of over 95% against E. coli and Streptococcus mutans was realized. Their microfluidic/chitosan device offered the advantages of a controllable drug release and low device cost and has potential for clinical use.

Generating the concentration dilutions of diffusible molecules in microfluidic devices is important for high-throughput biochemical analysis. Chen et al. [22] reported a microfluidic device integrated with programmable pneumatic microvalves for adjusting both the concentration in a single channel and the concentration distributions in different channels. They systematically studied the performance of the pneumatic microvalves via experiments and numerical simulations. Then, an empirical formula-supported computational program was used to provide the activation pressures required to generate the specific concentration profiles. Their device showed the ability to dynamically adjust the concentration profiles of microfluidic channels.

The rapid sensing and detection of single molecules and viruses are important functions of microfluidics. Wu et al. [23] proposed an efficient method that can be used to enrich the plasmonic hotspots for surface-enhanced Raman scattering (SERS) detection. In their method, the superhydrophobic concave dome array was used to realize the dynamic enrichment of plasmonic nanoparticles so as to obtains uniform SERS signals in the measurements. The limit of detection of their SERS platform for melamine in milk could reach 5×10^{-7} M, which is lower than the safety value provided by the Food and Drug Administration.

In addition to nanoparticles, two-dimensional (2D) carbon nanomaterials have been used to construct biosensors. Salama et al. [24] made recent advances (2010–2021) in the application of 2D carbon nanomaterial-based fluorescent biosensors for detecting viruses (e.g., Rotavirus, Ebola virus, Influenza virus H3N2, HIV, Hepatitis C virus, and Hepatitis B virus), using the principle of the Förster resonance energy transfer (FRET) mechanism. The application of the FRET–graphene oxide biosensor for virus detection based on multiplexed detection was introduced. The challenges involved in the fluorescent biosensors and the possible solutions required to address these challenges were highlighted.

Electrical impedance is an important technique for single-cell analysis, achieved by acquiring the electrical parameters of single cells in a rapid, non-invasive, and label-free manner [25]. Zhang et al. [26] reviewed the basic principles, analytical models, and design concepts of impedance flow cytometry and electrical impedance spectroscopy. In addition, recent advances in the application of electrical impedance for cell counting, cell recognition, phenotypic assays, viability detection, and cell screening were summarized. Finally, the prospects of impedance sensing were provided.

In addition to the electrical sensors, magnetic biosensors have attracted increasing interest in recent years. Jimenez et al. [27] summarized the history of magnetoimpedance biosensors over the past decade. They also introduced state-of-the-art magnetoimpedance biosensors for healthcare monitoring, including the monitoring of the COVID-19 pandemics. The opportunities and challenges in this field were discussed to guide the future development of this technology.

Droplet microfluidics is a powerful tool for a variety of biomedical applications, including single-cell analysis, micro-robotics, molecular diagnosis, cell manipulation, and so on [28]. Nazari et al. [29] reviewed the recent advances, current challenges, and future trends in droplet microfluidics for advanced regenerative medicine. As an effective tool for encapsulating various biomaterials in picoliter-sized droplets, droplet microfluidics provides the microenvironments necessary for manipulating gametes, fertilization, and embryo cultures in advanced regenerative medicine. The authors' review focused on the latest progress in the application of droplet microfluidics in stem cell therapy, tissue engineering, reproductive biology, and gene therapy.

Zhu et al. [30] reviewed the droplet manipulation technique using magnetofluidic technologies. The deformation, displacement, rotation, levitation, splitting, and fusion of droplets could be realized in a magnetic field, which can enable the remote, wireless, and programmable manipulation of droplets for drug synthesis, biochemistry, and sample preparation. The authors systematically introduced the basic theories, working principles, and functions of different magnetically induced droplet behaviors. They also proposed the challenges involved in the design and fabrication of magnetic droplet manipulation devices.

Our Special Issue, with 15 outstanding papers, is presented as a small step towards the development of new microfluidic devices for various biomedical applications.

Funding: This research was funded by the National Key Research and Development Program of China (2021YFC2103300), the National Natural Science Foundation of China (51875103), and the Natural Science Foundation of Jiangsu Province (BK20190064).

Acknowledgments: The authors are grateful for the valuable opportunity to serve as the guest editors of the Special Issue, "Microfluidics for Biomedical Applications". We appreciate all the authors for their contributions to this Special Issue. We also thank the editorial and publishing staff of *Biosensors* for their support.

Conflicts of Interest: The author declares no conflict of interest.

References

1. Whitesides, G.M. The origins and the future of microfluidics. *Nature* **2006**, *442*, 368–373. [CrossRef] [PubMed]
2. Sackmann, E.K.; Fulton, A.L.; Beebe, D.J. The present and future role of microfluidics in biomedical research. *Nature* **2014**, *507*, 181–189. [CrossRef]
3. Battat, S.; Weitz, D.A.; Whitesides, G.M. An outlook on microfluidics: The promise and the challenge. *Lab. A Chip* **2022**, *22*, 530–536. [CrossRef]
4. Cui, F.; Rhee, M.; Singh, A.; Tripathi, A. Microfluidic Sample Preparation for Medical Diagnostics. *Ann. Rev. Biomed. Eng.* **2015**, *17*, 267–286. [CrossRef] [PubMed]
5. Chen, Y.; Zhou, Z.; Zhu, S.; Ni, Z.; Xiang, N. Label-free microfluidics for single-cell analysis. *Microchem. J.* **2022**, *177*, 107284. [CrossRef]
6. Wang, M.; Liang, H.; Chen, X.; Chen, D.; Wang, J.; Zhang, Y.; Chen, J. Developments of Conventional and Microfluidic Flow Cytometry Enabling High-Throughput Characterization of Single Cells. *Biosensors* **2022**, *12*, 443. [CrossRef]
7. Regmi, S.; Poudel, C.; Adhikari, R.; Luo, K.Q. Applications of Microfluidics and Organ-on-a-Chip in Cancer Research. *Biosensors* **2022**, *12*, 459. [CrossRef]
8. Luka, G.; Ahmadi, A.; Najjaran, H.; Alocilja, E.; DeRosa, M.; Wolthers, K.; Malki, A.; Aziz, H.; Althani, A.; Hoorfar, M. Microfluidics Integrated Biosensors: A Leading Technology towards Lab-on-a-Chip and Sensing Applications. *Sensors* **2015**, *15*, 30011–30031. [CrossRef]
9. Vashist, S.K.; Luppa, P.B.; Yeo, L.Y.; Ozcan, A.; Luong, J.H.T. Emerging Technologies for Next-Generation Point-of-Care Testing. *Trends Biotechnol.* **2015**, *33*, 692–705. [CrossRef]
10. Shen, Z.; Wu, A.; Chen, X. Current detection technologies for circulating tumor cells. *Chem. Soc. Rev.* **2017**, *46*, 2038–2056. [CrossRef]
11. Zhu, S.; Jiang, F.; Han, Y.; Xiang, N.; Ni, Z. Microfluidics for label-free sorting of rare circulating tumor cells. *Analyst* **2020**, *145*, 7103–7124. [CrossRef] [PubMed]
12. Altay, R.; Yapici, M.K.; Koşar, A. A Hybrid Spiral Microfluidic Platform Coupled with Surface Acoustic Waves for Circulating Tumor Cell Sorting and Separation: A Numerical Study. *Biosensors* **2022**, *12*, 171. [CrossRef] [PubMed]
13. Xiang, N.; Ni, Z. Inertial microfluidics: Current status, challenges, and future opportunities. *Lab. A Chip* **2022**, *22*, 4792–4804. [CrossRef] [PubMed]
14. Xiang, N.; Ni, Z. Hand-Powered Inertial Microfluidic Syringe-Tip Centrifuge. *Biosensors* **2022**, *12*, 14. [CrossRef] [PubMed]
15. Mehran, A.; Rostami, P.; Saidi, M.S.; Firoozabadi, B.; Kashaninejad, N. High-Throughput, Label-Free Isolation of White Blood Cells from Whole Blood Using Parallel Spiral Microchannels with U-Shaped Cross-Section. *Biosensors* **2021**, *11*, 406. [CrossRef]
16. Dai, Y.; Cha, H.; Simmonds, M.J.; Fallahi, H.; An, H.; Ta, H.T.; Nguyen, N.-T.; Zhang, J.; McNamee, A.P. Enhanced Blood Plasma Extraction Utilising Viscoelastic Effects in a Serpentine Microchannel. *Biosensors* **2022**, *12*, 120. [CrossRef]
17. Kalluri, R.; LeBleu, V.S. The biology, function, and biomedical applications of exosomes. *Science* **2020**, *367*, eaau6977. [CrossRef]
18. Zhou, Z.; Chen, Y.; Qian, X. Target-Specific Exosome Isolation through Aptamer-Based Microfluidics. *Biosensors* **2022**, *12*, 257. [CrossRef] [PubMed]
19. Meggiolaro, A.; Moccia, V.; Brun, P.; Pierno, M.; Mistura, G.; Zappulli, V.; Ferraro, D. Microfluidic Strategies for Extracellular Vesicle Isolation: Towards Clinical Applications. *Biosensors* **2023**, *13*, 50. [CrossRef]
20. Kang, Y.J. Red Blood Cell Sedimentation Index Using Shear Stress of Blood Flow in Microfluidic Channel. *Biosensors* **2022**, *12*, 547. [CrossRef] [PubMed]
21. Shi, X.; Ma, B.; Chen, H.; Tan, W.; Ma, S.; Zhu, G. An Asymmetric Microfluidic/Chitosan Device for Sustained Drug Release in Guided Bone Regeneration Applications. *Biosensors* **2022**, *12*, 847. [CrossRef] [PubMed]
22. Chen, C.; Li, P.; Guo, T.; Chen, S.; Xu, D.; Chen, H. Generation of Dynamic Concentration Profile Using a Microfluidic Device Integrating Pneumatic Microvalves. *Biosensors* **2022**, *12*, 868. [CrossRef] [PubMed]
23. Wu, J.; Cai, J.; Fan, Y.; Zhang, Y.; Fang, H.; Yan, S. Effective Enrichment of Plasmonic Hotspots for SERS by Spinning Droplets on a Slippery Concave Dome Array. *Biosensors* **2022**, *12*, 270. [CrossRef] [PubMed]
24. Salama, A.M.; Yasin, G.; Zourob, M.; Lu, J. Fluorescent Biosensors for the Detection of Viruses Using Graphene and Two-Dimensional Carbon Nanomaterials. *Biosensors* **2022**, *12*, 460. [CrossRef] [PubMed]
25. Zhu, S.; Zhang, X.; Zhou, Z.; Han, Y.; Xiang, N.; Ni, Z. Microfluidic impedance cytometry for single-cell sensing: Review on electrode configurations. *Talanta* **2021**, *233*, 122571. [CrossRef] [PubMed]
26. Zhang, Z.; Huang, X.; Liu, K.; Lan, T.; Wang, Z.; Zhu, Z. Recent Advances in Electrical Impedance Sensing Technology for Single-Cell Analysis. *Biosensors* **2021**, *11*, 470. [CrossRef]
27. Jimenez, V.O.; Hwang, K.Y.; Nguyen, D.; Rahman, Y.; Albrecht, C.; Senator, B.; Thiabgoh, O.; Devkota, J.; Bui, V.D.A.; Lam, D.S.; et al. Magnetoimpedance Biosensors and Real-Time Healthcare Monitors: Progress, Opportunities, and Challenges. *Biosensors* **2022**, *12*, 517. [CrossRef]
28. Shang, L.; Cheng, Y.; Zhao, Y. Emerging Droplet Microfluidics. *Chem. Rev.* **2017**, *117*, 7964–8040. [CrossRef]

29. Nazari, H.; Heirani-Tabasi, A.; Ghorbani, S.; Eyni, H.; Razavi Bazaz, S.; Khayati, M.; Gheidari, F.; Moradpour, K.; Kehtari, M.; Ahmadi Tafti, S.M.; et al. Microfluidic-Based Droplets for Advanced Regenerative Medicine: Current Challenges and Future Trends. *Biosensors* **2022**, *12*, 20. [CrossRef]
30. Zhu, G.-P.; Wang, Q.-Y.; Ma, Z.-K.; Wu, S.-H.; Guo, Y.-P. Droplet Manipulation under a Magnetic Field: A Review. *Biosensors* **2022**, *12*, 156. [CrossRef]

biosensors

MDPI

Article

A Hybrid Spiral Microfluidic Platform Coupled with Surface Acoustic Waves for Circulating Tumor Cell Sorting and Separation: A Numerical Study

Rana Altay [1], Murat Kaya Yapici [1,2] and Ali Koşar [1,2,*]

1 Faculty of Engineering and Natural Sciences, Sabanci University, Istanbul 34956, Turkey; raltay@sabanciuniv.edu (R.A.); mkyapici@sabanciuniv.edu (M.K.Y.)
2 Center of Excellence for Functional Surfaces and Interfaces for Nano-Diagnostics, Faculty of Engineering and Natural Sciences, Sabanci University, Istanbul 34956, Turkey
* Correspondence: kosara@sabanciuniv.edu

Abstract: The separation of circulating tumor cells (CTCs) from blood samples is crucial for the early diagnosis of cancer. During recent years, hybrid microfluidics platforms, consisting of both passive and active components, have been an emerging means for the label-free enrichment of circulating tumor cells due to their advantages such as multi-target cell processing with high efficiency and high sensitivity. In this study, spiral microchannels with different dimensions were coupled with surface acoustic waves (SAWs). Numerical simulations were conducted at different Reynolds numbers to analyze the performance of hybrid devices in the sorting and separation of CTCs from red blood cells (RBCs) and white blood cells (WBCs). Overall, in the first stage, the two-loop spiral microchannel structure allowed for the utilization of inertial forces for passive separation. In the second stage, SAWs were introduced to the device. Thus, five nodal pressure lines corresponding to the lateral position of the five outlets were generated. According to their physical properties, the cells were trapped and lined up on the corresponding nodal lines. The results showed that three different cell types (CTCs, RBCs, and WBCs) were successfully focused and collected from the different outlets of the microchannels by implementing the proposed multi-stage hybrid system.

Keywords: cell separation; cell sorting; circulating tumor cells (CTCs); inertial microfluidics; surface acoustic wave (SAW); hybrid separation

Citation: Altay, R.; Yapici, M.K.; Koşar, A. A Hybrid Spiral Microfluidic Platform Coupled with Surface Acoustic Waves for Circulating Tumor Cell Sorting and Separation: A Numerical Study. *Biosensors* **2022**, *12*, 171. https://doi.org/10.3390/bios12030171

Received: 31 January 2022
Accepted: 8 March 2022
Published: 11 March 2022

1. Introduction

Cancer is one of the most pervasive and fatal diseases for humanity. Yet, microfluidic platforms can provide early cancer diagnosis, which is vital in the treatment of patients. According to the American Cancer Society, around 1.8 million new cancer cases were diagnosed in the United States of America in 2020, while the number of deaths was reported to be 606,520 [1]. The World Health Organization (WHO) reported that the number of deaths related to cancer can be reduced by up to 30% by diagnosing cancer before the metastasis stage [2].

Cancer involves an abnormal change of the cell's shape and function due to defects in the genetic material or DNA, which differentiates these cells from healthy ones [3,4]. These modified cells experience cell proliferation, multiply rapidly in an uncontrolled manner, gather in groups, and form tumors. Cancer metastasis starts when tumor cells called circulating tumor cells (CTCs) spread through the body via the lymph and blood systems and invade other tissues that they colonize, forming secondary tumors [5]. It has been reported that metastasis is the main reason for at least two-thirds of the deaths of patients with cancer [6]. However, the isolation of CTCs from a blood sample enables further understanding of a patient's health, since these cells carry different genomic and phenotypic properties with respect to primary tumor cells [7]. This has been the main motivation for many valuable studies in the literature conducted during the last decade. Microfluidics has been exploited in this regard for cancer diagnosis and treatment [8–14].

Microfluidic devices provide many advantages such as the consumption of a low amount of samples and reagents, which makes these devices preferable from the perspectives of cost, efficiency, rapidity of the analysis, and flexibility for their integration to lab-on-a-chip or organ-on-a-chip platforms. Since the number of CTCs (1–10/mL) in adult human blood with metastatic disease is very low compared to the number of white blood cells (5–10 $\times$ 10^6/mL) and red blood cells (5–9 $\times$ 10^9/mL) [15], the integration of these devices in cancer research mainly depends on the enrichment and separation of CTCs from a blood sample. For this purpose, several studies have been conducted involving the analysis of different particle sizes and density, which have helped develop methods for the isolation of CTCs from human blood samples [2,16–20].

Based on the need for an external source for operation, both label-free particle and cell separation studies have been mainly categorized into active and passive approaches. Active cell separation approaches utilize several external forces such as magnetic [21,22], electric [23,24], acoustic [25,26], and optical [27,28] forces to identify cells according to their response to applied fields or their physical characteristics such as size and density [4,14]. Passive approaches depend on the physical properties of the cells, which can be manipulated by inertial forces [29], filtration [30], deterministic lateral displacement [31], or adhesion [32]. The operation does not require any external sources, which makes the passive approach preferable.

The utilization of inertial forces in curved microchannels has attracted much attention for a wide range of applications [33–37]. Due to the utilization of secondary flow and emerging Dean Drag force, which provide a redistribution of velocity profiles, the lateral positions of particles vary along the cross section of the channel, which can be near the inner wall, the outer wall, or the centerline [16,38]. For example, Ozbey et al. performed a study on symmetrically curved serpentine microchannels in which they utilized Dean flow physics and inertial forces to obtain parallelizable size-based CTC sorting [7]. Hou et al. presented a study related to the separation of CTCs from blood using inertial-based microfluidic separation techniques in a spiral microchannel in which larger cells migrated towards the inner wall [39]. Sun et al. described the size-dependent separation of tumor cells from diluted peripheral whole blood in double-spiral microchannels using secondary flow and Dean Drag force and reported that 92.28% of blood cells were collected at the inner outlet, while 96.77% of tumor cells were collected at the middle outlet [34]. Recently, hybrid microfluidic platforms consisting of both active and passive components have emerged due to their advantages such as multi-target cell processing, wider operational range, high sensitivity, and efficiency [40]. For example, a system with three microfluidic components called CTC-iChip was introduced [41]. The proposed system utilized hydrodynamic cell sorting, inertial focusing, and magnetophoresis in the same platform. RBCs, platelets, CTCs, and WBCs were successfully collected from different outlets. Later, Zhang et al. [42] presented a novel hybrid microfluidic device, in which they obtained particle sorting in a serpentine microchannel by employing inertial and dielectrophoresis forces. The combination of acoustofluidics and inertial forces was performed by Wu et al. in an asymmetric serpentine microchannel for focusing of 10-, 12-, and 15-micron particles and erythrocytes, leukocytes, and cancer cells [43]. They integrated three- dimensional acoustofluidic tweezers (3D-AFT) to a serpentine microchannel. According to their results, they obtained more than 80% purity at the outlet for each size of beads using 3D-AFT-based separation. Similarly, Ren et al. presented a hybrid study in which they utilized standing surface acoustic waves with fluorescence-activated cell sorting in a serpentine microchannel with two outlets [44]. They obtained purities near 90% for HeLa cells without using sheath flow.

The potential of combined hybrid platforms, which utilize surface acoustic waves and inertial forces for cell separation and CTC enrichment, has been studied in several curved microchannels. Yet, the combination of surface acoustic waves with inertial forces, which occurs in a spiral microchannel, has not been adequately revealed. However, implementing surface acoustic waves in passive separation devices with a spiral structure offers a high potential for multi-target cell processing with high-efficiency and high-sensitivity label-free enrichment of circulating tumor cells. By increasing the outlet number of a microchannel and proposing the sorting of cells using SAWs on the generated pressure nodal lines, which occur in the same lateral position of outlets, multi-target processing capabilities can also be accomplished.

Motivated by the abovementioned studies, in this numerical study, we investigated the separation capabilities of hybrid microfluidic devices consisting of active and passive parts. As a passive separation approach, a spiral microchannel structure was utilized to obtain the separation of CTCs based on inertial forces throughout the device. As an active separation approach, a straight microchannel design with a higher rectangular cross-sectional area was added to the end of the spiral microchannel, where interdigital transducers (IDTs) were embedded in the walls of the microchannel reciprocally. In this study, a 3.3 MHz driving frequency was applied to the sidewall of the microchannel from the straight outlet section to represent the transmission of the selected driving frequency into the working fluid in the microchannel. Surface acoustic waves (SAWs) with 454 μm wavelengths produced five nodal lines corresponding to the lateral position of five outlets. According to the physical parameters of the utilized cells, they were trapped and focused on the nodal lines. The separation of CTCs was obtained without using sheath flow. Small footprint of the presented microfluidic platform with two-loop spiral decreased the time needed for the to reach to the outlet, which consequently increased the viability of the cells during the operation. The microchannel, which is presented in this study, with a height of 70 μm and an initial spiral radius of 5 mm, successfully separated three different cell types (CTCs, WBCs, and RBCs) in 0.28 s and 0.187 s at Reynolds numbers of 40 and 65. Cells that exited the passive separation part (spiral section) were sorted according to their diameter using inertial forces. The additional active separation provided enhanced separation of large cells from small ones in the microfluidic platform. Producing five different nodal lines corresponding to the five different outlets allowed trapping the cells, which increased the separation capability. In the presented microfluidic platform, up to five different cell types with various cell diameters can be separated. By implementing the SAWs and nodal lines corresponding to the five outlets, clogging of the cells at the outlet section, which is an important factor in cell separation studies, can also be prevented. Thus, we propose high-throughput, multi-staged, and hybrid microfluidic devices which allow cell separation and the isolation of different cell types according to their sizes. In addition, we present the effect of the microchannel dimensions on the trajectories of the cells for both passive and active separation.

2. Theory

Considering the fluid domain in a laminar flow regime in the proposed study, the Reynolds number (Re) is represented as:

$$\mathrm{Re} = (\rho \mathrm{U} \mathrm{D_h})/(\mu) \tag{1}$$

where ρ, U, D_h, and μ represent the fluid density (kg/m^3), the average flow velocity (m/s), the microchannel hydraulic diameter (m), and the fluid dynamic viscosity (kg/ms), respectively.

In microchannels having a rectangular cross section, the hydraulic diameter is represented as:

$$\mathrm{D_h} = 2\mathrm{ab}/((\mathrm{a} + \mathrm{b})) \tag{2}$$

The magnitude of the formation of Dean vortices and secondary flow in curved microchannels is represented by the dimensionless Dean number [45]:

$$\mathrm{De} = \mathrm{Re}\sqrt{(\delta)} = \mathrm{Re}\sqrt{(\mathrm{D_h}/2\mathrm{R})} \tag{3}$$

where R and δ represent the radius of curvature of the microchannel and the ratio of the channel hydraulic diameter to the microchannel radius of curvature, respectively.

A particle or cell in a curved microchannel having a Poiseuille flow experiences inertial Lift force, wall-induced Lift force, Dean Drag force, gravitational force, and buoyancy force. The gravitational force and buoyancy force are assumed to be in equilibrium due to the small dimension of the microchannels. Accordingly, the Dean Drag force (F_{De}) [45] and the net inertial force (F_L) [46] are defined as:

$$F_{De} = 3\pi\mu U_{De}\, a_p \tag{4}$$

where a_p is the particle diameter, and U_{De} is the Dean velocity represented as [45]:

$$U_{De} = 1.8 \times 10^{-4}\, De^{1.63} \tag{5}$$

$$F_L = C_L((\rho(U_{max})^2(a_p)^4)/(D_h)^2) \tag{6}$$

where U_{max} is the maximum fluid velocity, and C_L is the Lift coefficient.

The minimum threshold (T), which represents the limitation of successful inertial focusing of the particles, is stated as [43]:

$$T = a_p/D_h \geq 0.07 \tag{7}$$

Accordingly, the ratio of the particle diameter to the channel hydraulic diameter needs to be greater than or equal to 0.07.

In this study, surface acoustic waves (SAWs) were introduced to the microchannel in the straight section before the outlets. Particles experienced an acoustic radiation force (F_r) and a viscous Drag force (F_d) (Stokes law) due to the generated pressure distribution in the straight section, which is expressed as [47]:

$$F_r = -(\pi(p_0)^2 V_p \beta_w)/2\lambda)\ \phi\ (\beta, \rho)\ \sin(2ky) \tag{8}$$

$$\phi\ (\beta, \rho) = ((5\rho_p - 2\rho_m)/(2\rho_p + \rho_m)) - (\beta_p/\beta_m) \tag{9}$$

$$F_d = -6\pi\mu R_{pur} \tag{10}$$

where p_0, λ, V_p ρ_m, ρ_p, β_m, β_p, R_p, and u_r represent the acoustic pressure, the acoustic wavelength, the volume of the particle, the density of the medium, the density of the particle, the compressibility of the medium, the compressibility of the particle, the radius of the particle, and the relative velocity of the particle, respectively. ϕ is the acoustic contrast factor, which is used to determine whether the particles will move to a pressure nodal line or to an antinodal line according to its physical properties (volume, compressibility, density). By introducing SAWs, five different pressure nodal lines corresponding to the five outlets were produced, which promoted the movement of the cells towards to the nodal lines. Even though three cell types with different diameters were utilized in this study, the proposed microchannel platform can be implemented for various applications that require a separation of up to five different particles or cell types.

3. Material and Methods

3.1. Microchannel Design

The microchannel configurations simulated in this study consisted of two-loop spirals with two inlets and five equally distanced 120 µm-wide outlets. The width of each microchannel was 500 µm along the spiral part. The distance between each spiral was fixed at 500 µm. Overall, an aim of the design was to ensure a small footprint.

Active separation methods require low flow rates, while passive separation occurs at high flow rates [48]. To resolve this issue, the channel geometry was modified according to the requirements of the flow rates of the application. In this regard, an expansion was introduced to the microchannel at the end of the last spiral. A short straight rectangular part

(8 mm) decreased the fluid velocity at the active separation component (Figure 1). The width of the microchannel sharply increased to 1 mm at the beginning of this rectangular part.

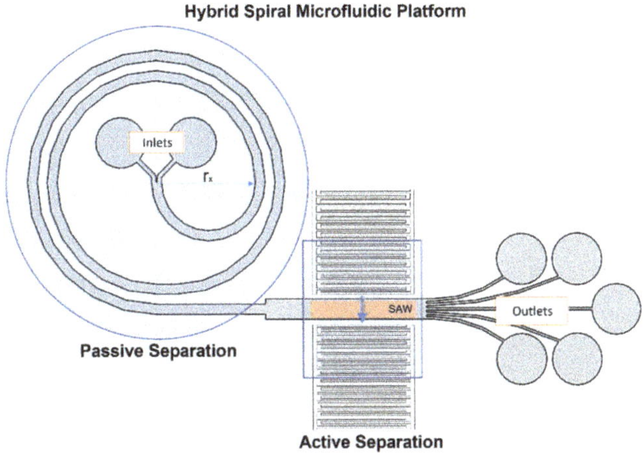

Figure 1. Schematic representation of the hybrid spiral microfluidic platform consisting of passive (spiral) and active (IDTs) components for focusing and separating cells.

The radii of the first spiral and the height of the microchannels differed for each microchannel, as presented in Table 1. The S1 and S2 microchannels had a height of 100 μm, while the S3 and S4 microchannels had a height of 70 μm. The radii of the first spiral (r_x) were set to 10 mm for the S1 and S3 microchannels and to 5 mm for the S2 and S4 microchannels.

Table 1. Common geometrical parameters and dimensions of the microchannels which are named as S1, S2, S3, and S4.

Channel Name	Channel Width at the Spiral Part (μm)	Channel Width at the Straight Part (mm)	Channel Height (μm)	Radius of the First Spiral (r_x) (mm)
S1	500	1	100	10
S2			100	5
S3			70	10
S4			70	5

Figure 1 shows the schematic representation of the hybrid spiral microfluidic platform, which consisted of passive and active components. In this study, the spiral part served as a passive component and allowed focusing and separating the cells by utilizing the Dean and Lift forces. The straight rectangular section constituted the active component, which enhanced the separation of the cells using surface acoustic waves (SAWs). In Figure 1, input and output IDTs represent the locations of the applied surface acoustic waves schematically.

3.2. COMSOL Multiphysics Modeling

Passive separation, which utilized Lift and Drag forces in the spiral microchannel, was performed using the Laminar Flow and Particle Tracing Module for Fluid Flow Physics of the Software COMSOL Multiphysics 5.5 [49]. In the simulations, white blood cells (WBCs), red blood cells (RBCs), and circulating tumor cells (CTCs) were defined by considering them as particles with different diameters and a density (ρ_p) of 1050 kg/m^3. WBCs, RBCs, and CTCs were represented as particles with a diameter of 9 μm, 6 μm, and 20 μm, respectively.

The inlet of the microchannel was arranged in such a way that the particles were released from one common inlet. The physical model for the Laminar Flow Module was chosen as incompressible flow.

To include the effect of the actual shear gradient force which acted on the particles, the discretization was set as P2 + P2. By defining different inlet velocities, the Reynolds number at which efficient separation was obtained was found. With a stationary study, the Laminar Flow Physics was explored for each microchannel presented in Table 1 at Reynolds numbers of 40 and 65. The governing Navier–Stokes equations for an incompressible and steady laminar flow are as follows:

$$\rho(\mathbf{u}\cdot\nabla)\mathbf{u} = \nabla\cdot\left[-p\mathbf{I} + \mu(\nabla\mathbf{u} + (\nabla\mathbf{u})^{\mathrm{T}}\right] \tag{11}$$

$$\rho\nabla\cdot\mathbf{u} = 0 \tag{12}$$

where $\mathbf{u}$ and p represent the velocity vector and the pressure, respectively. The dynamic viscosity and density were implemented from COMSOL Multiphysics 5.5 as a selection of material, i.e., water, liquid.

After finding the velocity field from the stationary study, the Particle Tracing Module for Fluid Flow Physics was utilized to introduce different cell types. The Newtonian formulation for particle release and propagation and bounce wall condition was chosen. Due to low Reynolds numbers, the Drag Law was considered as the Stokes Law, while the Lift Force was set to the wall-induced Lift Law. For both Laminar Flow and Particle Tracing Module for Fluid Flow Physics, a physics-controlled mesh with normal element size was used for the microchannels. With a time-dependent study, particle trajectories were obtained so that the cells were separated according to their diameter by the effects of the Drag and wall-induced Lift forces in the spiral microchannels. The governing equations for particle tracing for fluid flow are as follows:

$$\mathbf{F}_{\mathrm{t}} = \frac{\mathrm{d}(\mathrm{m_p}\mathbf{v})}{\mathrm{dt}} \tag{13}$$

where $\mathbf{F}_{\mathrm{t}}$, $\mathrm{m_p}$, and $\mathbf{v}$ are the total force, the particle mass, and the particle velocity, respectively.

Active separation, which was induced by pressure distribution due to the applied SAWs, was also simulated using the Pressure Acoustics, Frequency Domain, and Particle Tracing Module for Fluid Flow Physics of the Software COMSOL Multiphysics 5.5. Since SAWs were applied in the straight rectangular section of the spiral microchannels before the particles were split into five outlets (Figure 2), the straight rectangular part was isolated and studied separately. In the Pressure Acoustics Module, the fluid model was set to linear elastic. The Drag Law was considered as the Stokes Law. The particle type was chosen as solid and implemented for the radiation force model. The length of the transducer (l) was defined as 5 mm. The driving frequency (f) was set to 3.3 MHz. Thus, assuming the speed of sound in water, c, as 1500 m/s, the wavelength of the SAWs (λ) was 454 µm, which produced five nodal lines in the microchannel and corresponded to the lateral position of the five outlets (Figure 2). After obtaining the acoustic pressure distribution and sound pressure level in the microchannel by using a frequency-domain study, three different particles with different diameters representing RBCs, WBCs, and CTCs were defined. We chose a density of 1050 kg/m^3, based on the density of the cells, and the particles were released from the inlet. The location of the initial position of the particles was mesh-based with zero initial velocity before applying the SAWs. As a result of the generated pressure distribution, the particles experienced the acoustic radiation forces and viscous Drag forces. The particle trajectories were determined using a time-dependent study. The governing equations for Frequency-Domain Pressure Acoustics are as follows:

$$Q_m = \nabla = \left(-\frac{1}{\rho_c}(\nabla p_t - q_d)\right) - \frac{k_{eq}^2 p_t}{\rho_c} \tag{14}$$

$$p_t = p + p_b \tag{15}$$

$$k_{eq}^2 = \left(\frac{\omega}{c_c}\right)^2 - k_z^2 \tag{16}$$

where Q_m, p_c, p_t, q_d, p_b, ω, c_c, and k represent the monopole domain source, the complex density, the total acoustic pressure, the dipole domain source, the background pressure, the angular velocity, the speed of sound, and the wave number, respectively.

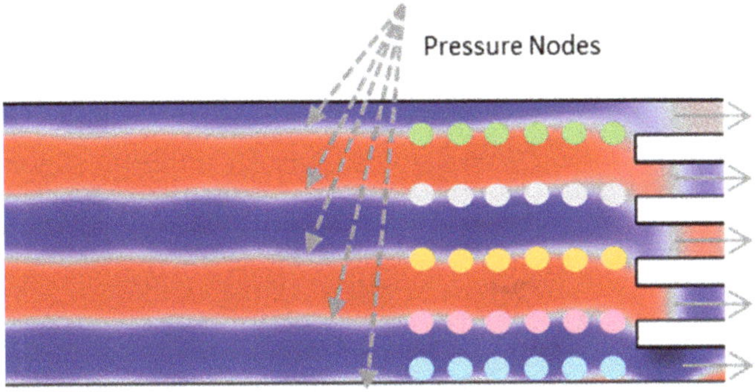

Figure 2. Acoustic pressure distribution in the straight outlet section. Due to the applied SAWs, five nodal pressure lines corresponding to the lateral position of the outlets were generated.

For the Laminar Flow Physics, a no-slip boundary condition was applied on the channel walls. For the inlet boundary condition, normal inflow velocity was applied at the studied Reynolds numbers of 40 and 65 for each microchannel, whereas a zero-pressure boundary condition was imposed for the outlets.

In the Particle Tracing Module for Fluid Flow Physics, a bounce boundary condition was applied on the channel walls. In the passive separation study, a freeze boundary condition was implemented on the outlets of the channel walls to enable the detection of the state of the cells at the outlet.

It should be noted that for the active separation analysis, cells with different diameters and zero initial velocities were simulated in the applied acoustic field to demonstrate the effect of the 3.3 MHz driving frequency on a large number of particles. The active separation analysis was conducted independently of the passive separation analysis.

4. Results and Discussion

4.1. Passive Separation

The results obtained from the numerical simulations of passive separation in the spiral microchannel illustrated that the separation of cells could be obtained from different outlets at different Reynolds numbers. For the analysis, four microchannels with a two-loop spiral structure with a straight rectangular section were used. The presented microchannels had different channel heights and radii of the first spirals, as shown in Table 1. All microchannels were considered at Reynolds numbers of both 40 and 65. According to the behavior of the cells in different velocity fields, the trajectories of RBCs, WBCs, and CTCs were obtained (Figures S1–S8 for velocity field distributions, in Supplementary Materials). Figures 3–6 show the passive cell separation results of the S1, S2, S3, and S4 microchannels, respectively.

Due to the a/D_h ratio represented in Equation (7), cells with a diameter greater than or equal to 11.7 μm were affected by inertial forces in the S1 and S2 microchannels. For the S3 and S4 microchannels, this diameter was 8.6 μm. Particles with larger diameters were exposed to greater Dean Drag forces. CTCs with a diameter of 20 μm were more affected by the Dean Drag force due to their larger diameter compared to the other particles, which resulted in focusing the CTC cell line near the inner wall of each presented microchannel.

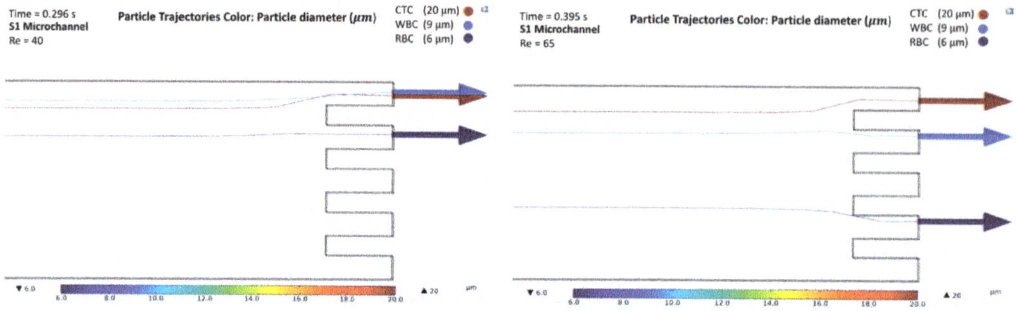

(a) (b)

Figure 3. Numerical results of the cell separation study in the S1 microchannel. CTC, WBC, and RBC cells are represented by dark red, light blue, and dark blue colors, respectively: (**a**) Cell distribution at the outlet of the S1 microchannel at a Reynolds number of 40; (**b**) Cell distribution at the outlet of the S1 microchannel at a Reynolds number of 65.

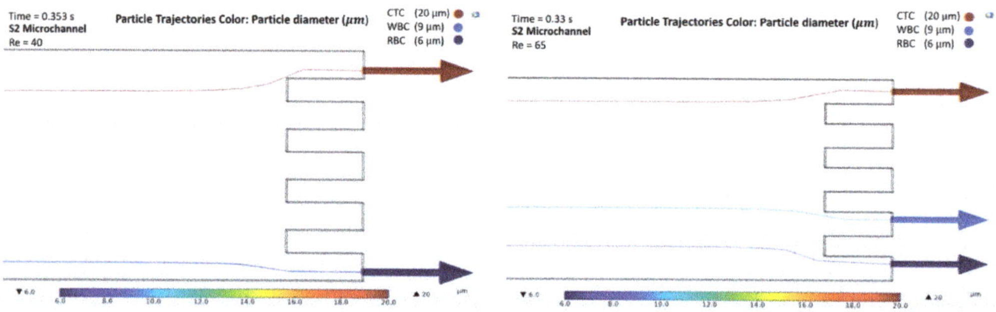

(a) (b)

Figure 4. Numerical results of the cell separation study in the S2 microchannel. CTC, WBC, and RBC cells are represented by dark red, light blue, and dark blue colors, respectively: (**a**) Cell distribution at the outlet of the S2 microchannel at a Reynolds number of 40; (**b**) Cell distribution at the outlet of the S2 microchannel at a Reynolds number of 65.

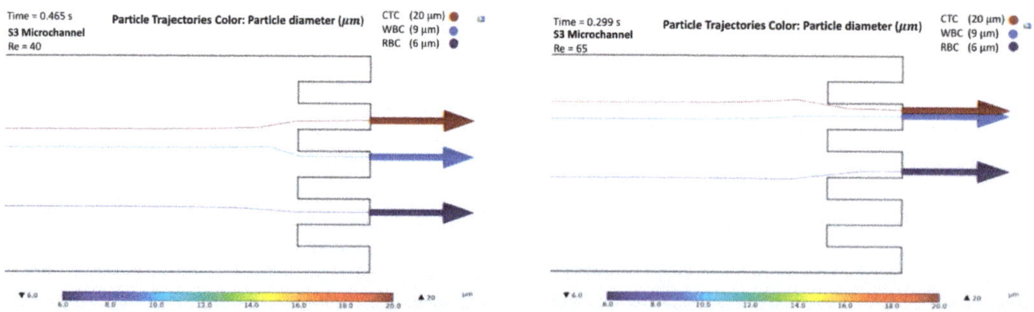

(a) (b)

Figure 5. Numerical results of the cell separation study in the S3 microchannel. CTC, WBC, and RBC cells are represented by dark red, light blue, and dark blue colors, respectively: (**a**) Cell distribution at the outlet of the S3 microchannel at a Reynolds number of 40; (**b**) Cell distribution at the outlet of the S3 microchannel at a Reynolds number of 65.

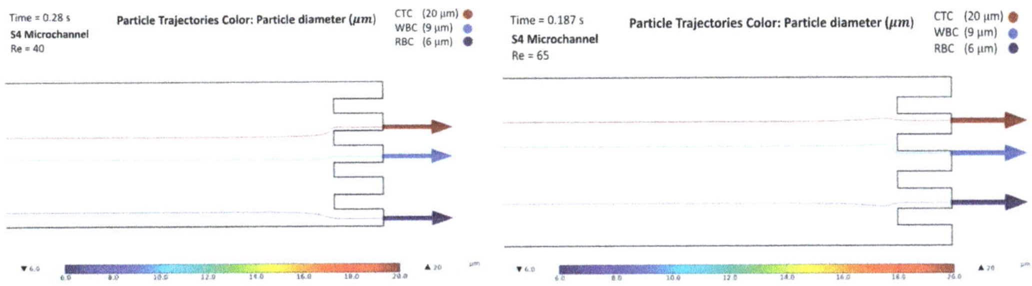

(a) (b)

Figure 6. Numerical results of the cell separation study in the S4 microchannel. CTC, WBC, and RBC cells are represented by dark red, light blue, and dark blue colors, respectively: (**a**) Cell distribution at the outlet of the S4 microchannel at a Reynolds number of 40; (**b**) Cell distribution at the outlet of the S4 microchannel at a Reynolds number of 65.

Figures 3a and 4a show the trajectories of the cells at a Reynolds number of 40 in the S1 and S2 microchannels, respectively. At a Reynolds number of 40, the S1 and S2 microchannels were not capable of providing sufficient separation. However, an increase in the inlet velocity caused the dominance of the Dean Drag and net inertial forces along the microchannels, which consequently affected the trajectories of the cells. Figures 3b and 4b show that CTC, WBC, and RBC cells were collected from the different outlets at a Reynolds number of 65 for the S1 and S2 microchannels, respectively. However, due to the shorter length of the S2 microchannel, the time needed for the cells to be collected from the outlets, 0.33 s, was shorter than the time needed for cell collection from the S1 microchannel, that was 0.395 s.

Figure 5 presents the results of the cell trajectories in the S3 microchannel. At Re = 40, CTC, WBC, and RBC were collected in 0.465 s from the second, third, and fourth outlets, respectively (Figure 5a). However, an increase in the inlet velocity caused a larger Dean Drag force on WBC, which forced them to move towards the second outlet at Re = 65 (Figure 5b). Thus, an efficient separation of the cells was not achieved with the S3 microchannel at a Reynolds number of 65.

In the S4 microchannel, at both Reynolds numbers (Re = 40 and Re = 65), the three kinds of cells (CTC, WBC, and RBC) exited the microchannel from different outlets (Figure 6a,b). CTCs, which had a diameter of 20 μm, were affected more by the Dean Drag force due to their larger diameter compared to those of the other cells (WBCs and RBCs), which resulted in focusing the CTCs near the inner wall of the microchannel. Thus, CTCs were collected from the second outlet at Reynolds numbers of 40 and 65. WBCs, which are smaller than CTCs and larger than RBCs, were less affected by the Dean Drag force and exited the microchannel from the third outlet. RBCs exited the microchannel from the fifth and fourth outlets at Re = 40 and Re = 65, respectively. The duration of the cell collecting times at Reynolds numbers of 40 and 65 were 0.28 and 0.187 s, respectively, for the S4 microchannel. Overall, the S4 microchannel was superior to the other microchannels as it provided a faster separation of the cells in 0.187 s at Re = 65.

4.2. Active Separation

In the hybrid microfluidic platform, we proposed focusing of the cells prior to their exposure to SAWs in acoustofluidic separation (active separation), which was obtained by using the spiral microchannel structure (passive separation). However, to increase the sensitivity and separation efficiency of the device, an active separation technique was realized by introducing SAWs. The combination of inertial focusing with SAWs facilitated the trapping of the particles along the nodal lines leading to the five outlets. The proposed two-step approach is promising as a means of increasing the sensitivity of separation, as indicated by the excellent localization of the 20 µm particles along the nodal lines, whereas the 6 µm and 9 µm particles experienced clear dispersion and mixing.

SAWs with 454 µm wavelengths were introduced from the sidewall of the microchannels. The simulation parameters related to SAW study are represented in Table 2. Five different pressure nodal lines were produced in the same lateral position as the outlets. It has been reported in the literature that these multiple pressure nodal line formations along microchannels have advantages for separation applications since they provide significantly improved efficiency and sensitivity compared to SAW-based separation devices with single pressure nodal or antinodal lines [43].

Table 2. Simulation parameters related to the SAW study.

Description	Expression	Value
Driving frequency	f	3.3 [MHz]
Speed of sound	c	1500 [m/s]
Wavelength	λ	454 [µm]
Transducer length	l	5 [mm]
Particle density	ρ_p	1050 [kg/m^3]

Figure 7a presents the acoustic pressure field (Pa) distribution along the straight rectangular section of the microchannel. A maximum acoustic pressure of 1.8×10^5 Pa (dark red) and a minimum acoustic pressure of -1.8×10^5 Pa (dark blue) were generated. The areas in the Figure 7 indicated in blue and red represent the pressure antinodal lines, whereas the white areas represent the pressure nodal lines. Three kinds of cells with different diameters (CTCs, WBCs, and RBCs) were located with zero initial velocity at t = 0 s (Figure 7b).

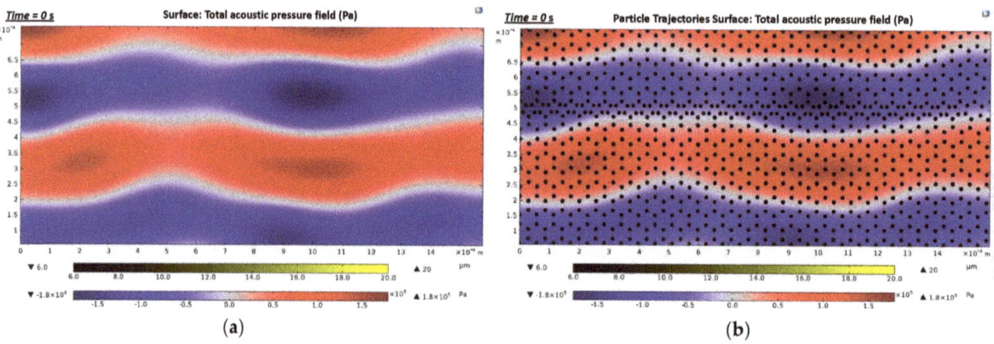

(a) (b)

Figure 7. Total acoustic pressure field (zoom-in images that show three pressure nodal lines occurring in the straight rectangular section): (**a**) Formation of an acoustic pressure field along the straight rectangular section of the microchannel. The blue and red areas represent the pressure antinodal lines. The white areas represent the pressure nodal lines; (**b**) Cells trajectories at t = 0 s. The cells were set to zero initial velocity.

Figure 8a represents the location of the cells when they were exposed to a 3.3 MHz driving frequency for 0.2 s. As SAWs were applied from the sidewall of the microchannel, the cells were exposed to the acoustic radiation force. They lined up on the pressure nodal lines, as described in Figure 7a (white areas). In Figure 8b, the arrows represent the motions of the cells while they moved towards the pressure nodal lines.

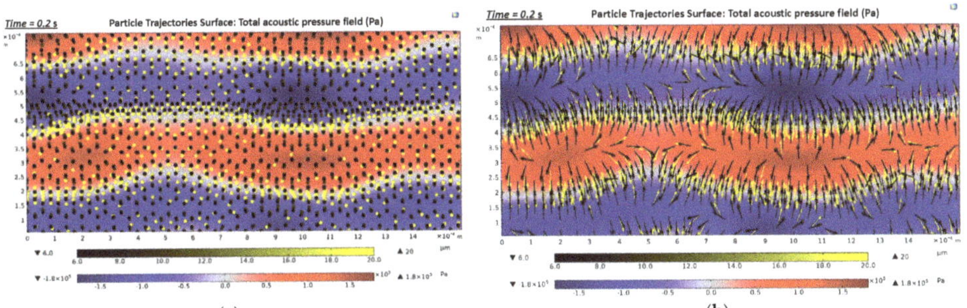

(**a**) (**b**)

Figure 8. Trajectories of the cells during 0.2 s of exposure to a 3.3 MHz driving frequency (zoom-in images that show three pressure nodal lines occurring in the straight rectangular section): (**a**) Location of the cells in t = 0.2 s along the straight rectangular section of the microchannel; (**b**) Motions of the cells while they move towards the pressure nodal lines in t = 0.2 s.

Particles with larger diameters experienced larger acoustic radiation forces and Drag forces. This led to a clear positioning of larger cells (CTCs) along the nodal lines, while smaller cells (WBCs, RBCs) displayed a less accurate distribution (Figure 9). In Figure 9, CTCs with larger diameters (20 μm) are in light yellow and were more affected by SAWs, sorting on the pressure nodal lines. The CTCs moved towards the pressure nodal lines in 0.1 s upon exposure to 3.3 MHz driving frequency (Figure 9a). At t = 0.3 s, the CTCs were trapped on the pressure nodal lines (Figure 9b). However, smaller particles (WBCs and RBCs) required more time to be sorted in the pressure nodal lines (Figure 9c,d).

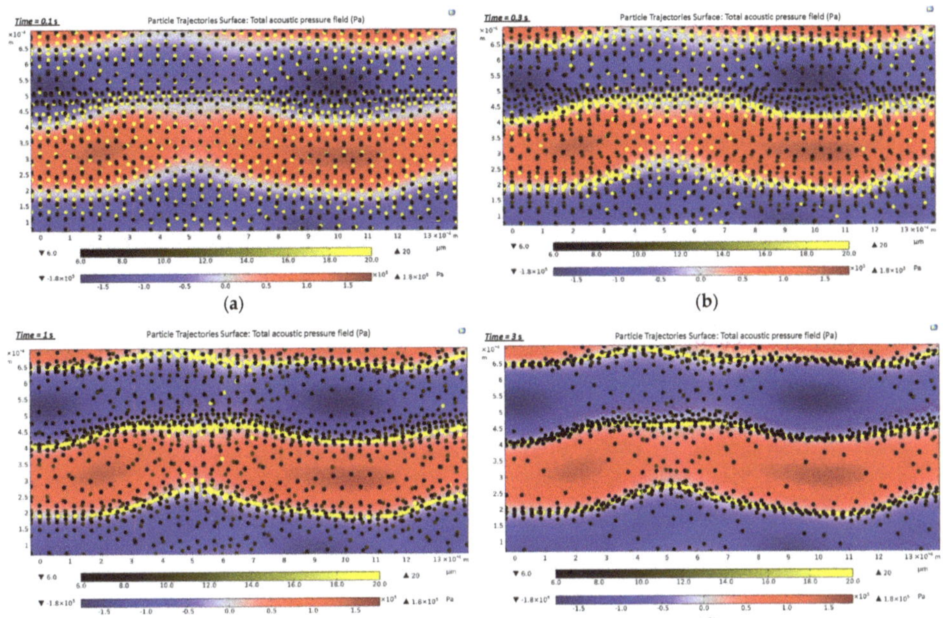

(**c**) (**d**)

Figure 9. Distribution of the cells subjected to the acoustic pressure field along the straight rectangular section of the microchannel serving as an active separation in the hybrid device (zoom-in images that show three pressure nodal lines occurring in the straight rectangular section): (**a**) Cell trajectories at t = 0.1 s; (**b**) Cell trajectories at t = 0.3 s; (**c**) Cell trajectories at t = 1 s; (**d**) Cell trajectories at t = 3 s.

5. Conclusions

This study presents a multi-stage hybrid microfluidics platform that utilizes inertial and acoustic radiation forces for particle/CTC separation studies. The spiral microchannel structure allowed focusing different kinds of cells using the Dean Drag force and Lift forces. These cells later experienced acoustic radiation forces that were applied along the straight rectangular section of the microchannel. The SAWs wavelength was set to 454 µm. Five pressure nodal lines, which were in the same lateral position of the outlets, were produced. Thus, the cells, which were not lined up after exiting the spiral microchannel, were able to be captured and trapped on the pressure nodal lines depending on their diameters.

A microchannel, with a height of 70 µm and an initial spiral radius of 5 mm, successfully separated three different cell types (CTCs, WBCs, and RBCs) in 0.28 s and 0.187 s at Reynolds numbers of 40 and 65, respectively.

In the light of the findings of this study, we propose that microchannels with these features, which are capable of separating cells at both studied Reynolds numbers and provided the fastest cell separation in this study, can be fabricated as polydimethylsiloxane (PDMS) microchannels, and IDT structures (length, finger distance, thickness) can be designed and coupled with such PDMS microchannels.

Supplementary Materials: The following supporting information can be downloaded at: https://www.mdpi.com/article/10.3390/bios12030171/s1, Figure S1: The results of the cell trajectories and velocity field in the S1 microchannel at Reynolds number of 40. CTC, WBC, and RBC were represented by dark red, light blue, and dark blue colors, respectively; Figure S2: The results of the cell trajectories and velocity field in the S1 microchannel at Reynolds number of 65. CTC, WBC, and RBC were represented by dark red, light blue, and dark blue colors, respectively; Figure S3: The results of the cell trajectories and velocity field in the S2 microchannel at Reynolds number of 40. CTC, WBC, and RBC were represented by dark red, light blue, and dark blue colors, respectively; Figure S4: The results of the cell trajectories and velocity field in the S2 microchannel at Reynolds number of 65. CTC, WBC, and RBC were represented by dark red, light blue, and dark blue colors, respectively; Figure S5: The results of the cell trajectories and velocity field in the S3 microchannel at Reynolds number of 40. CTC, WBC, and RBC were represented by dark red, light blue, and dark blue colors, respectively; Figure S6: The results of the cell trajectories and velocity field in the S3 microchannel at Reynolds number of 65. CTC, WBC, and RBC were represented by dark red, light blue, and dark blue colors, respectively; Figure S7: The results of the cell trajectories and velocity field in the S4 microchannel at Reynolds number of 40. CTC, WBC, and RBC were represented by dark red, light blue, and dark blue colors, respectively; Figure S8: The results of the cell trajectories and velocity field in the S4 microchannel at Reynolds number of 65. CTC, WBC, and RBC were represented by dark red, light blue, and dark blue colors, respectively; Video S1: The video shows the particle trajectories in the velocity field in the S4 microchannel at a Reynolds number of 65. CTC, WBC, and RBC cells are identified by color. Accordingly, CTC, WBC, and RBC cells are shown in dark red, light blue, and dark blue color, respectively. Video S2: The video shows the particle trajectories in the applied acoustic pressure fields for 3 s. The maximum and minimum acoustic pressures that were formed in the microchannel are shown in the legend. The maximum and minimum of approximately 1.8×10^5 Pa and -1.8×10^5 Pa acoustic pressure fields are shown in dark red and dark blue colors, respectively. CTC, WBC, and RBC cells are shown in yellow shades. Accordingly, CTC, WBC, and RBC cells are ranged from light yellow to dark yellow, respectively.

Author Contributions: Conceptualization, R.A. and A.K.; methodology, R.A., M.K.Y. and A.K.; software, R.A.; validation, R.A. and A.K.; formal analysis, R.A., M.K.Y. and A.K.; investigation, R.A., M.K.Y. and A.K.; resources, R.A.; data curation, R.A. and A.K.; writing—original draft preparation, R.A.; writing—review and editing, R.A., M.K.Y. and A.K.; visualization, R.A.; supervision, A.K.; project administration, A.K.; funding acquisition, A.K. All authors have read and agreed to the published version of the manuscript.

Funding: This research was funded by TUBITAK (The Scientific and Technological Research Council of Turkey) 1004 NANOSIS Programme, Grant Number 20A6012.

Institutional Review Board Statement: Not applicable.

Data Availability Statement: Not applicable.

Acknowledgments: The author would like to thank Osman Şahin for useful advice and valuable discussion. Graduate financial support from TUBITAK (The Scientific and Technological Research Council of Turkey) 1004 NANOSIS Programme Project Grant Number 20A6012 and graduate student support provided by the Faculty of Engineering and Natural Sciences of Sabanci University are greatly appreciated. The authors also thank Daniel L. Calvey from Sabanci University for his kind help in editing the manuscript.

Conflicts of Interest: The authors declare no conflict of interest.

References

1. American Cancer Society. *Cancer Facts & Figures 2020*; American Cancer Society: Atlanta, GA, USA, 2020; pp. 1–52.
2. Tabatabaei, S.A.; Targhi, M.Z. Design and experimental investigation of a novel spiral microfluidic chip to separate wide size range of micro-particles aimed at cell separation. *Proc. Inst. Mech. Eng. H J. Eng. Med.* **2021**, *235*, 1315–1328. [CrossRef] [PubMed]
3. Cho, H.; Kim, J.; Song, H.; Sohn, K.Y.; Jeon, M.; Han, K.H. Microfluidic technologies for circulating tumor cell isolation. *Analyst* **2018**, *143*, 2936–2970. [CrossRef] [PubMed]
4. Farahinia, A.; Zhang, W.J.; Badea, I. Novel microfluidic approaches to circulating tumor cell separation and sorting of blood cells: A review. *J. Sci. Adv. Mater. Devices* **2021**, *6*, 303–320. [CrossRef]
5. Chaffer, C.L.; Weinberg, R.A. A perspective on cancer cell metastasis. *Science* **2011**, *331*, 1559–1564. [CrossRef] [PubMed]
6. Dillekås, H.; Rogers, M.S.; Straume, O. Are 90% of deaths from cancer caused by metastases? *Cancer Med.* **2019**, *8*, 5574–5576. [CrossRef] [PubMed]
7. Ozbey, A.; Karimzadehkhouei, M.; Kocaturk, N.M.; Bilir, S.E.; Kutlu, O.; Gozuacik, D.; Kosar, A. Inertial focusing of cancer cell lines in curvilinear microchannels. *Micro Nano Eng.* **2019**, *2*, 53–63. [CrossRef]
8. Jackson, E.L.; Lu, H. Advances in microfluidic cell separation and manipulation. *Curr. Opin. Chem. Eng.* **2013**, *2*, 398–404. [CrossRef] [PubMed]
9. Manzoor, A.A.; Romita, L.; Hwang, D.K. A review on microwell and microfluidic geometric array fabrication techniques and its potential applications in cellular studies. *Can. J. Chem. Eng.* **2021**, *99*, 61–96. [CrossRef]
10. Maghsoudi, S.; Rabiee, N.; Ahmadi, S.; Rabiee, M.; Bagherzadeh, M.; Karimi, M. *An Overview of Microfluidic Devices*; Elsevier Inc.: Amsterdam, The Netherlands, 2021; ISBN 9780128187913.
11. Al-Faqheri, W.; Thio, T.H.G.; Qasaimeh, M.A.; Dietzel, A.; Madou, M.; Al-Halhouli, A. Particle/cell separation on microfluidic platforms based on centrifugation effect: A review. *Microfluid. Nanofluidics* **2017**, *21*, 102. [CrossRef]
12. Bhagat, A.A.S.; Bow, H.; Hou, H.W.; Tan, S.J.; Han, J.; Lim, C.T. Microfluidics for cell separation. *Med. Biol. Eng. Comput.* **2010**, *48*, 999–1014. [CrossRef]
13. Wu, Z.; Hjort, K. Microfluidic Hydrodynamic Cell Separation: A Review. *Micro Nanosyst.* **2010**, *1*, 181–192. [CrossRef]
14. Shields, C.W., IV; Reyes, C.D.; López, G.P. Microfluidic cell sorting: A review of the advances in the separation of cells from debulking to rare cell isolation. *Lab Chip* **2015**, *15*, 1230–1249. [CrossRef] [PubMed]
15. Alvarez Cubero, M.J.; Lorente, J.A.; Robles-Fernandez, I.; Rodriguez-Martinez, A.; Puche, J.L.; Serrano, M.J. Circulating tumor cells: Markers and methodologies for enrichment and detection. In *Circulating Tumor Cells*; Humana Press: New York, NY, USA, 2017.
16. Özbey, A.; Karimzadehkhouei, M.; Akgönül, S.; Gozuacik, D.; Koşar, A. Inertial focusing of microparticles in curvilinear microchannels. *Sci. Rep.* **2016**, *6*, 38809. [CrossRef] [PubMed]
17. Mitra, P.; Dutta, S.; Hens, N.A. Separation of particles in spiral micro-channel using Dean' s flow fractionation. *J. Braz. Soc. Mech. Sci. Eng.* **2020**, *42*, 405. [CrossRef]
18. Yeh, P.Y.; Dai, Z.; Yang, X.; Bergeron, M.; Zhang, Z.; Lin, M.; Cao, X. An efficient spiral microchannel for continuous small particle separations. *Sens. Actuators B Chem.* **2017**, *252*, 606–615. [CrossRef]
19. García-Sánchez, P.; Ramos, A. Continuous Particle Separation in Microfluidics: Deterministic Lateral Displacement Assisted by Electric Fields. *Micromachines* **2021**, *12*, 66. [CrossRef]
20. Bhagat, A.A.S.; Kuntaegowdanahalli, S.S.; Papautsky, I. Continuous particle separation in spiral microchannels using dean flows and differential migration. *Lab Chip* **2008**, *8*, 1906–1914. [CrossRef]
21. Robert, D.; Pamme, N.; Conjeaud, H.; Gazeau, F.; Iles, A.; Wilhelm, C. Cell sorting by endocytotic capacity in a microfluidic magnetophoresis device. *Lab Chip* **2011**, *11*, 1902–1910. [CrossRef]
22. Zhu, T.; Cheng, R.; Lee, S.A.; Rajaraman, E.; Eiteman, M.A.; Querec, T.D.; Unger, E.R.; Mao, L. Continuous-flow ferrohydrodynamic sorting of particles and cells in microfluidic devices. *Microfluid. Nanofluid.* **2012**, *13*, 645–654. [CrossRef]
23. Sahin, O.; Elitas, M.; Yapici, M.K. Simulation of Dielectrophoresis based Separation of Red Blood Cells (RBC) from Bacteria Cells. In Proceedings of the 2020 21st International Conference of Thermal, Mechanical and Multi-Physics Simulation and Experments in Microelectronics and Microsystems, EuroSimE 2020, Cracow, Poland, 5–8 July 2020; pp. 8–11. [CrossRef]
24. Zhang, X.; Xu, X.; Ren, Y.; Yan, Y.; Wu, A. Numerical simulation of circulating tumor cell separation in a dielectrophoresis based Y-Y shaped microfluidic device. *Sep. Purif. Technol.* **2021**, *255*, 117343. [CrossRef]

25. Wu, F.; Shen, M.H.; Yang, J.; Wang, H.; Mikhaylov, R.; Clayton, A.; Qin, X.; Sun, C.; Xie, Z.; Cai, M.; et al. An Enhanced Tilted-Angle Acoustofluidic Chip for Cancer Cell Manipulation. *IEEE Electron Device Lett.* **2021**, *42*, 577–580. [CrossRef]
26. Jo, M.C.; Guldiken, R. A label-free cell separation using surface acoustic waves. In Proceedings of the 2011 Annual International Conference of the IEEE Engineering in Medicine and Biology Society, Boston, MA, USA, 30 August–3 September 2011; pp. 7691–7694. [CrossRef]
27. MacDonald, M.P.; Spalding, G.C.; Dholakia, K. Microfluidic sorting in an optical lattice. *Nature* **2003**, *426*, 421–424. [CrossRef] [PubMed]
28. Lau, A.Y.; Lee, L.P.; Chan, J.W. An integrated optofluidic platform for Raman-activated cell sorting. *Lab Chip* **2008**, *8*, 1116–1120. [CrossRef] [PubMed]
29. Shiriny, A.; Bayareh, M. Inertial focusing of CTCs in a novel spiral microchannel. *Chem. Eng. Sci.* **2021**, *229*, 116102. [CrossRef]
30. Choi, S.; Song, S.; Choi, C.; Park, J.K. Continuous blood cell separation by hydrophoretic filtration. *Lab Chip* **2007**, *7*, 1532–1538. [CrossRef]
31. Beech, J.P.; Holm, S.H.; Adolfsson, K.; Tegenfeldt, J.O. Sorting cells by size, shape and deformability. *Lab Chip* **2012**, *12*, 1048–1051. [CrossRef]
32. Lee, C.H.; Bose, S.; Van Vliet, K.J.; Karp, J.M.; Karnik, R. Examining the lateral displacement of HL60 cells rolling on asymmetric P-selectin patterns. *Langmuir* **2011**, *27*, 240–249. [CrossRef]
33. Rzhevskiy, A.S.; Bazaz, S.R.; Ding, L.; Kapitannikova, A.; Sayyadi, N.; Campbell, D.; Walsh, B.; Gillatt, D.; Warkiani, M.E.; Zvyagin, A.V. Rapid and label-free isolation of tumour cells from the urine of patients with localised prostate cancer using inertial microfluidics. *Cancers* **2020**, *12*, 81. [CrossRef]
34. Sun, J.; Li, M.; Liu, C.; Zhang, Y.; Liu, D.; Liu, W.; Hu, G.; Jiang, X. Double spiral microchannel for label-free tumor cell separation and enrichment. *Lab Chip* **2012**, *12*, 3952–3960. [CrossRef]
35. Altay, R.; Yetisgin, A.A.; Erdem, K.; Koşar, A. The effect of varying radius of curvature on mixing in elliptical spiral microchannels. *Chem. Eng. Process.—Process Intensif.* **2021**, *164*, 108401. [CrossRef]
36. Ahmadi, V.E.; Butun, I.; Altay, R.; Bazaz, S.R.; Alijani, H.; Celik, S.; Warkiani, M.E.; Koşar, A. The effects of baffle configuration and number on inertial mixing in a curved serpentine micromixer: Experimental and numerical study. *Chem. Eng. Res. Des.* **2021**, *168*, 490–498. [CrossRef]
37. Guzniczak, E.; Otto, O.; Whyte, G.; Willoughby, N.; Jimenez, M.; Bridle, H. Deformability-induced lift force in spiral microchannels for cell separation. *Lab Chip* **2020**, *20*, 614–625. [CrossRef] [PubMed]
38. Martel, J.M.; Toner, M. Inertial focusing dynamics in spiral microchannels. *Phys. Fluids* **2012**, *24*, 032001. [CrossRef] [PubMed]
39. Hou, H.W.; Warkiani, M.E.; Khoo, B.L.; Li, Z.R.; Soo, R.A.; Tan, D.S.W.; Lim, W.T.; Han, J.; Bhagat, A.A.S.; Lim, C.T. Isolation and retrieval of circulating tumor cells using centrifugal forces. *Sci. Rep.* **2013**, *3*, 1259. [CrossRef] [PubMed]
40. Yan, S.; Zhang, J.; Yuan, D.; Li, W. Hybrid microfluidics combined with active and passive approaches for continuous cell separation. *Electrophoresis* **2017**, *38*, 238–249. [CrossRef]
41. Ozkumur, E.; Shah, A.M.; Ciciliano, J.C.; Emmink, B.L.; Miyamoto, D.T.; Brachtel, E.; Yu, M.; Chen, P.I.; Morgan, B.; Trautwein, J.; et al. Inertial focusing for tumor antigen-dependent and -independent sorting of rare circulating tumor cells. *Sci. Transl. Med.* **2013**, *5*, 179ra47. [CrossRef]
42. Zhang, J.; Yan, S.; Alici, G.; Nguyen, N.T.; Di Carlo, D.; Li, W. Real-time control of inertial focusing in microfluidics using dielectrophoresis (DEP). *RSC Adv.* **2014**, *4*, 62076–62085. [CrossRef]
43. Wu, M.; Chen, K.; Yang, S.; Wang, Z.; Huang, P.H.; Mai, J.; Li, Z.Y.; Huang, T.J. High-throughput cell focusing and separation: Via acoustofluidic tweezers. *Lab Chip* **2018**, *18*, 3003–3010. [CrossRef]
44. Ren, L.; Yang, S.; Zhang, P.; Qu, Z. Standing Surface Acoustic Wave SSAW-Based Fluorescence-Activated Cell Sorter. *Small* **2018**, *14*, 1801996. [CrossRef]
45. Kuntaegowdanahalli, S.S.; Bhagat, A.A.S.; Kumar, G.; Papautsky, I. Inertial microfluidics for continuous particle separation in spiral microchannels. *Lab Chip* **2009**, *9*, 2973–2980. [CrossRef]
46. Ying, Y.; Lin, Y. Inertial Focusing and Separation of Particles in Similar Curved Channels. *Sci. Rep.* **2019**, *9*, 16575. [CrossRef] [PubMed]
47. Ding, X.; Peng, Z.; Lin, S.C.S.; Geri, M.; Li, S.; Li, P.; Chen, Y.; Dao, M.; Suresh, S.; Huang, T.J. Cell separation using tilted-angle standing surface acoustic waves. *Proc. Natl. Acad. Sci. USA* **2014**, *111*, 12992–12997. [CrossRef] [PubMed]
48. Rostami, Z.; Rajabi, F.; Shamloo, A. Cell separation by using active and passive methods together. In Proceedings of the 4th International Conference on Innovative Technologies in Science, Engineering and Technology, Istanbul, Turkey, 12 November 2020.
49. *COMSOL Multiphysics® v.5.6*; COMSOL AB: Stockholm, Sweden, 2020. Available online: www.comsol.com (accessed on 20 December 2021).

biosensors

MDPI

Article

Hand-Powered Inertial Microfluidic Syringe-Tip Centrifuge

Nan Xiang [1,2,*] and Zhonghua Ni [1,2]

1 School of Mechanical Engineering, Jiangsu Key Laboratory for Design and Manufacture of Micro-Nano Biomedical Instruments, Southeast University, Nanjing 211189, China; nzh2003@seu.edu.cn
2 State Key Laboratory of Bioelectronics, Southeast University, Nanjing 210096, China
* Correspondence: nan.xiang@seu.edu.cn

Abstract: Conventional sample preparation techniques require bulky and expensive instruments and are not compatible with next-generation point-of-care diagnostic testing. Here, we report a manually operated syringe-tip inertial microfluidic centrifuge (named i-centrifuge) for high-flow-rate (up to 16 mL/min) cell concentration and experimentally demonstrate its working mechanism and performance. Low-cost polymer films and double-sided tape were used through a rapid nonclean-room process of laser cutting and lamination bonding to construct the key components of the i-centrifuge, which consists of a syringe-tip flow stabilizer and a four-channel paralleled inertial microfluidic concentrator. The unstable liquid flow generated by the manual syringe was regulated and stabilized with the flow stabilizer to power inertial focusing in a four-channel paralleled concentrator. Finally, we successfully used our i-centrifuge for manually operated cell concentration. This i-centrifuge offers the advantages of low device cost, simple hand-powered operation, high-flow-rate processing, and portable device volume. Therefore, it holds potential as a low-cost, portable sample preparation tool for point-of-care diagnostic testing.

Keywords: inertial microfluidics; cell concentration; hand-powered; point-of-care diagnostic testing

Citation: Xiang, N.; Ni, Z. Hand-Powered Inertial Microfluidic Syringe-Tip Centrifuge. *Biosensors* **2022**, *12*, 14. https://doi.org/10.3390/bios12010014

Received: 27 November 2021
Accepted: 28 December 2021
Published: 29 December 2021

1. Introduction

Sample preparation is the first critical but a most time-consuming step in medical diagnostics or biochemical analysis [1–3]. For example, the isolation of rare cells, such as circulating tumor cells and fetal nucleated red blood cells, from complex cell populations [4–6] or the concentration of specific pathogens, parasites, or microorganisms from large-volume biological or environmental sample fluids [7–9] can significantly improve the detection sensitivity and accuracy. Currently, point-of-care testing and on-site rapid analysis pose a new challenge to sample preparation [10]. Conventional sample preparation techniques, commonly based on centrifugation [11] and fluorescence-activated cell sorting [12], rely heavily on expensive and bulky equipment, external power sources, time-consuming procedures, and experienced technicians, making them not compatible with newly emerging point-of-care diagnostic testing. In contrast, ideal sample preparation approaches for point-of-care diagnostic testing should be simple, rapid, low-cost, and portable, and thus easily applied outside the laboratory by a non-specialist without training.

As a novel approach to precisely control and manipulate cells and fluids in the microscale space, microfluidics provides new insights for on-site sample preparation owing to the advantages of miniaturization, a low device cost, small-sample consumption, and high integration [13,14]. To date, various novel microfluidic devices have been developed to realize separation [15,16], ordering [17,18], concentration [19,20], trapping [21], enrichment [22], filtration [23,24], and lysis [25,26] of cells on a single chip. In addition to the diversified application functions, continuous efforts have been made toward the development of simple, low-cost, and portable microfluidic devices for achieving effective sample preparation in resource-constrained environments at remote sites.

One option is to simplify and reduce the cost of the device fabrication process, allowing the device to be massively produced and disposed of in small clinics, homes, and field settings. In addition to well-established soft lithography [27], many new fabrication techniques have been proposed to simplify the fabrication process or to create complex non-planar structures. For example, three-dimensional (3D) printing [28,29] and laser direct writing [30,31] are new methods for directly creating innovative 3D structures in bulk materials. However, the devices fabricated by these approaches are commonly in the millifluidic size scale and require an additional step to remove the sacrificial or fused materials. Other mass production techniques include injection molding [32], roll-to-roll hot embossing [33], and xurography [34]. In addition to rapid prototyping techniques, the materials for fabricating microfluidic devices have evolved from polydimethylsiloxane [35] to low-cost paper [36], off-the-shelf tubing [37,38], and commercially available polymer films [39,40].

Alternatively, the straightforward principle of microfluidic sample preparation can be used so that the required supporting peripheral equipment can be simplified. Currently, microfluidic sample preparation can be categorized into active and passive microfluidics. Active microfluidics commonly employ external force fields, including electric [41,42], magnetic [43,44], optical [45], and acoustic [46,47] forces, to manipulate fluids or microscale objects. These methods have a high manipulation resolution but limited throughput. More importantly, complex, expensive microstructures (such as microelectrodes and interdigital transducers) and bulk field generators (such as signal generators) are still required, which prevents the miniaturization of these systems. The ideal principle for microfluidic sample preparation should be simple, low-cost, portable, and external field-free. In contrast, passive microfluidics solely applies hydrodynamic effects or specially designed microstructures to engineer fluids and cells [4,16,48] and thus are more suitable for point-of-care diagnostic testing. Among the reported passive microfluidics, inertial microfluidics has attracted increasing interest in recent years owing to its advantages of a high processing throughput, simple channel geometry, and easy operation [49–53]. Thus, many novel inertial microfluidic devices have been developed for various sample preparation functions, such as cell single-line ordering [54], selective trapping [55], solution exchange [56], differential separation [39,57], and efficient mixing [58].

Although passive microfluidics can work without the use of external field generators, the operation of most of these devices still relies heavily on bulk external fluid-driven systems (such as syringe pumps, peristaltic pumps, and gas-driven fluid pumping systems), which require electricity as the power source and are difficult to miniaturize. To address this limitation, various on-chip fluid pumping systems using capillary force [59], surface energy gradient [60], electroosmotic flow [61], and acoustic streaming [62] have been explored. However, the flow rates provided by these pumping systems are very low and thus are incapable of driving flows in high-flow-rate systems (such as inertial microfluidics). In turn, human power and finger actuation may be the ideal power source for driving sample fluids for point-of-care diagnostics [63–65], but the precise control of fluid flow generated by these low-cost power sources remains a challenge.

Great success has been achieved in simplifying device fabrication, working principle, and fluid pumping system; nonetheless, a simple, low-cost, and portable device that allows rapid sample preparation is still rarely reported. Here, we developed a hand-operated syringe-tip inertial microfluidic centrifuge (named i-centrifuge) for high-flow-rate and continuous-flow cell concentrations. The i-centrifuge consists of a syringe-tip flow stabilizer for regulating the flow generated by hand power and a four-channel paralleled inertial microfluidic device for high-flow-rate (up to 16 mL/min) cell concentrations. The integration of a flow stabilizer enables the concentration performance to be entirely independent on the operations and experiences of the user. We demonstrated the design concept, experimentally characterized the performance, and applied this novel i-centrifuge to hand-operated cell concentration. The developed i-centrifuge offers the advantages of low device cost, hand-powered simple operation, high-flow-rate processing, and portable device volume, thereby holding potential for sample preparation in resource-constrained environments.

2. Materials and Methods

2.1. Device Fabrication

For the four-channel paralleled inertial microfluidic concentrator, each channel unit was fabricated by enclosing a patterned 95 μm thick polyvinyl chloride (PVC) film with two commercially available laminating films with the assistance of a desktop laminator (LM8-330, Rayson, Foshan, China). The bonding of different channel units was achieved using double-sided tape (180 μm thick, 3M, Shanghai, China). The patterns in the PVC film, laminating films, and double-sided tapes were cut using a laser cutting system (TH-UV200A, Tianhong, Suzhou, China) equipped with a UV laser source (Awave 355-10W-30K, Advanced Optowave, New York, NY, USA).

To fabricate the elastic membrane in the syringe-tip flow stabilizer, the mixed and degassed PDMS liquid (Sylgard 184, Dow Corning, MI, USA) with a base to curing agent ratio of 10:1 was spin-coated onto a polyethylene terephthalate (PET) film and then cured at 100 °C for over 100 min. After being transferred onto the double-sided tape, the PET film was peeled off from the PDMS membrane (measured thickness of 60 μm). The holes or grooves in the PDMS membrane and double-sided tapes were cut using the abovementioned laser cutting system. The PDMS membranes with different thicknesses are also commercially available. In addition to the PDMS membrane, other commercially available elastic films could be used in the syringe-tip flow stabilizer.

The housings of the concentrator and syringe-tip flow stabilizer were directly printed in the photocurable resin (SZUV-W8001, DigitalManu, Shanghai, China) using a laser-based stereolithography 3D printer (3DSL-450S, DigitalManu, Shanghai, China).

2.2. Preparation of Particle/Cell Suspensions

Fluorescent particles with diameters of 10 μm (G1000B, 1% solid content, Thermal Fisher Scientific, Waltham, MA, USA) were diluted to low concentrations with phosphate-buffered saline (Sigma-Aldrich, Burlington, MA, USA). Before performing the experiments, the particle suspensions were uniformly dispersed using a vortex mixer (Thermal Fisher Scientific, Waltham, MA, USA).

Unicellular green microalgal cells (GY-H1 *Platymonas helgolandica tsingtaoensis*, Shanghai Guangyu Biological Technology, Shanghai, China) were used to characterize the concentration performance. The microalgal cells had a polydisperse size of 5–20 μm (average size: 12 μm) and a non-spherical flat shape with an average circularity of approximately 0.5. The microalgal cells were cultured in the F/2+Si medium according to the manufacturer's instructions. After harvesting, the microalgal cells were diluted with phosphate-buffered saline to specific concentrations. In addition to the microalgal cells, human breast cancer MCF-7 cells were cultured in the high-glucose Dulbecco's modified Eagle's medium (DMEM, Thermo Fisher Scientific, Waltham, MA, USA) containing 10% fetal bovine serum (Thermo Fisher Scientific, Waltham, MA, USA) and 1% penicillin-streptomycin (Thermo Fisher Scientific, Waltham, MA, USA). After harvesting, MCF-7 cells were dispensed in phosphate-buffered saline (PBS, Sigma-Aldrich, Burlington, MA, USA) at specific concentrations.

2.3. Experimental Setup

As the 3D printed housing of our concentrator was not transparent, the device was clamped by two transparent poly(methylmethacrylate) plates to characterize the particle distribution in the four-channel paralleled concentrator. The entire device was fixed on the observation platform of an inverted fluorescence microscope (IX 71, Olympus, Tokyo, Japan). The inlet and outlets of the concentrator were connected to the syringe and centrifuge tubes using tubing. The prepared samples were loaded into a plastic syringe, which was driven by a precise syringe pump (Legato 270, KD Scientific, Holliston, MA, USA) to generate the desired flow rate. The particle distribution in the device was observed and captured using a high-speed camera (Exi Blue, Qimaging, British Columbia, Canada) under a long exposure time of 500 ms. To avoid random errors, over 100 image frames were vertically overlaid using ImageJ software (https://imagej.nih.gov/ij/ (accessed on 26 December 2021)) to create a composite image illustrating the statistical particle distribution across the channel width. When inertial focusing was achieved, a bright fluorescent stream could be clearly observed. The fluorescence intensity profile across the channel width was measured using this software. Gaussian fitting of the intensity profile was performed to obtain the full width at half maximum (FWHM) for evaluating the focusing performance.

A gas-driven flow system (Figure S1) was established to characterize the flow-stabilizing performance of the integrated syringe-tip flow stabilizer. First, a pressure controller (OB1 Base MkIII, Elveflow, Paris, France) was employed to regulate the input compressed air at a specific pressure to push the liquid out of the hermetic sample reservoir into the syringe-tip flow stabilizer. The mass of the liquid output by the syringe-tip flow stabilizer was continuously monitored using an electronic balance. Based on these data, the output volumetric flow rate of the syringe-tip flow stabilizer at a specific input pressure was calculated. Finally, the flow rate-pressure curves could be plotted.

For cell concentration application, the syringe was manually pushed, and a syringe pump was not required. During the cell concentration process, the liquids from both outlets were collected separately. The volumes and concentrations of the initial sample and the collected target samples were measured. Cell concentrations were counted using a Countess II FL automated cell counter (Thermo Fisher Scientific, Waltham, MA, USA).

3. Results and Discussion

3.1. Conceptual Design and Working Principle of i-Centrifuge

The i-centrifuge (Figure 1a), consisting of a syringe-tip flow stabilizer and a syringe filter-like inertial microfluidic concentrator, can be quickly mounted onto the syringe tip via a simple press-fit connection. The sample liquid in the syringe is first injected into the flow stabilizer under the hand-pushing operation. However, fluid flow generated by manually pushing the syringe is heavily dependent on the experience of the operator and may be highly unstable and uncontrollable. In our hand-operated system, this unstable liquid flow can be regulated to be stable at a specific flow rate using an integrated flow stabilizer. As illustrated in Figure 1b, when the liquid is injected into the flow stabilizer, the liquid will flow through the hole in the suspended membrane toward the outlet of the flow stabilizer. As the hole in the suspended membrane is small, the fluid will accumulate above the suspended membrane and apply positive pressure to the suspended membrane. The suspended membrane deforms toward the bottom wall when a positive flow pressure (ΔP) is applied to the top of the suspended membrane, resulting in the increase in the flow resistance (ΔR) of the formed entire flow path. The deformation degree of the elastic membrane and the resulting flow resistance of the flow path varies with the pressure applied to the membrane. Therefore, by dynamically adjusting the flow resistance of the flow path according to the input pressure, a constant output flow rate ($Q = \frac{P}{R} = \frac{P+\Delta P}{R+\Delta R}$) can be achieved under varied pressures by using the flow stabilizer. The flow regulation mechanism is passive and electricity-free, which makes this flow stabilizer especially suitable for hand-powered operations. The only requirement for actuating the flow stabilizer is to apply a pressure larger than the threshold value to induce sufficient membrane deformation for flow autoregulation. The threshold pressure is the minimum pressure when the flow-rate variation is within 5%.

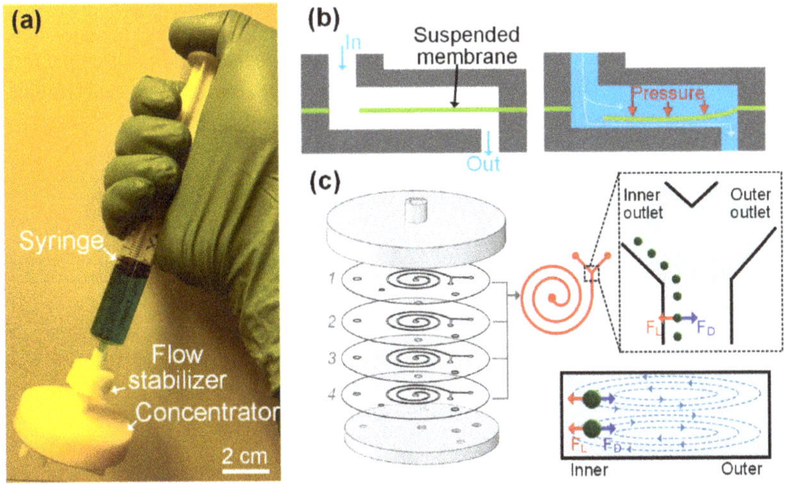

Figure 1. (**a**) Photograph illustrating the operation of the hand-operated syringe-tip inertial microfluidic centrifuge (i-centrifuge) for continuous-flow cell concentration. The i-centrifuge consists of a syringe-tip flow stabilizer and a syringe filter-like inertial microfluidic concentrator and can be quickly mounted onto the syringe tip via a simple press-fit connection. (**b**) Working principle of the flow stabilizer for regulating varied input liquid flow to be at a desired stable flow rate. (**c**) Structure and working principle of the syringe filter-like inertial microfluidic concentrator. Four spiral inertial microfluidic channels were integrated for achieving ultra-high throughout processing. The right part showed the cell inertial focusing principle at the position before the Y-shaped outlet system and in the cross-section of each spiral channel.

After passing through the flow stabilizer, the unstable liquid flow generated by manually pushing the syringe can be automatically regulated to be at the desired value and driven into the downstream four-channel paralleled inertial microfluidic concentrator. In our concentrator, four-channel paralleled spiral inertial microfluidic channels were designed to achieve a passive cell concentration in an ultra-high-flow-rate manner. As illustrated in Figure 1c, the sample liquid flowed into each channel unit via the central inlet hole. The inlet and outlet housings (Figure S2) were used as the world-to-chip interface to quickly clamp the microfluidic concentrator. When flowing along the spiral channel at finite Reynolds numbers, the cells simultaneously suffer from the coupled effects of inertial migration and cross-sectional Dean flow [66,67], which resulted in lateral cell migration perpendicular to the main flow stream. The mechanics for inertial migration is the inertial lift force (F_L) caused by the inherent inertia of microfluids. The equation of F_L can be expressed as [67]: $F_L = \frac{U_m^2 a_p^4 \rho}{D_h^2} f_L$, where U_m is the maximum velocity, a_p is the cell diameter, ρ is the fluid density, D_h is the hydraulic diameter, and f_L is the lift coefficient. The F_L is the net force of a shear-induced inertial lift force (F_{LS}) and a wall-induced inertial lift force (F_{LW}) [68]. The parabolic flow profile induces an F_{LS} to push the cell down the shear gradient toward the channel wall. The wall, in turn, induces a repulsive F_{LW} to push the cell away from the channel wall. In addition to F_L, the cross-sectional Dean flow induces a lateral Dean drag force (F_D) on cells [53]. A scaling of F_D can be expressed as [68]: $F_D \propto \rho U_m^2 a_p D_h^2 R^{-1}$, where the R is the radius of curved channels. In spiral channels, the cells are focused into a cell train at a lateral focusing position near the inner channel wall under specific flow rates. The inertial focusing position is actually the equilibration position, where the net force acting on the cells equals zero. To achieve the inertial focusing, cells need to satisfy the criterion ($a_p/H \geq 0.07$, where H is the channel height) [67]. In this work, the channel height H and channel width W were designed to be 90 μm and 500 μm, respectively. The loop number was controlled to be two. More details on the channel dimensions can be found in Table S1. According to the focusing criterion, the designed channel is able to focus the cells with diameters larger than 6.3 μm. By utilizing a Y-shaped outlet system, the focused cell train can be collected via the inner outlet, whereas the blank cell-free fluid in the outer half channel can be removed via the outer outlet. The removal of cell-free fluids can significantly reduce the volume of the target samples and thus increase the cell concentration. The passive cell concentration based on the principle of inertial focusing does not rely on external force fields and can work in a high-flow-rate, continuous-flow manner.

3.2. Low-Cost Four-Channel Paralleled Spiral Inertial Microfluidics for Ultra-High-Flow-Rate Processing

To lower the device cost and enable disposable use, a spiral inertial microfluidic channel designed to achieve an ultra-high-flow-rate and continuous-flow cell concentration was fabricated in low-cost polymer films using a rapid process of laser cutting and lamination bonding. Specifically, the channel geometries were patterned by cutting through grooves in a PVC film using a laser. Then, the patterned PVC film was sandwiched between two laminating films, and the through channel was enclosed using lamination (Figure 2a). This single device is highly transparent, which enables the clear observation of cell-focusing dynamics in the channels. The material cost for each device was only $0.01, which allowed its disposable use. To achieve ultra-high-flow-rate processing, four identical channel layers were precisely aligned and vertically stacked with the assistance of locating holes in each channel layer and a corresponding fixture (Figure 2b). The bonding of the different channel layers was accomplished using a patterned double-sided adhesive. The fabrication process of the four-channel paralleled device was completed within 15 min. Although four layers were stacked, the total thickness of the four-channel paralleled device was only 2 mm.

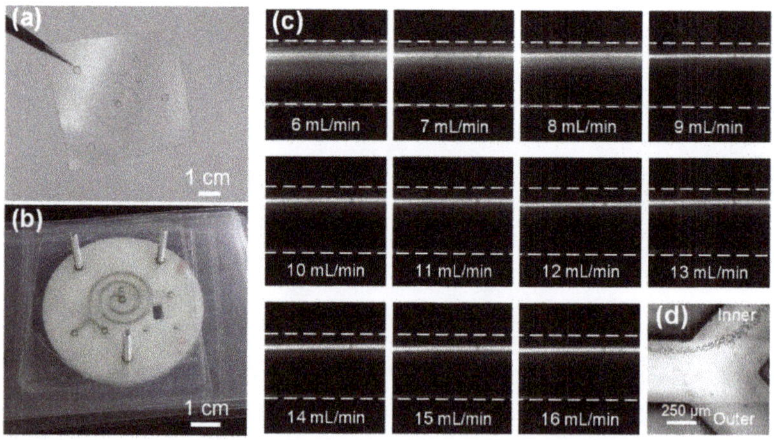

Figure 2. (**a**) Photograph of the single device that was fabricated in polymer films using a rapid process of laser cutting and lamination bonding. (**b**) Photograph of the four-channel paralleled device for ultra-high-flow-rate processing. Four identical channel layers were precisely aligned and vertically stacked with the assistance of locating holes in each channel layer and a corresponding fixture. (**c**) Composite images illustrating the distributions of 10 μm particles across the channel width near the outlet at the flow rates of 6–16 mL/min with an interval of 1 mL/min. White dotted lines indicate the channel walls. The upper wall is the inner wall. (**d**) Stacked bright field image illustrating the particle focusing at the Y-shaped outlet.

The cell concentration process using this four-channel paralleled device was based on the principle of inertial focusing. Therefore, we characterized the focusing performance of our device to better understand device physics. The 10 μm particle suspensions were pumped into the device via the inlet at the flow rates of 6–16 mL/min with an interval of 1 mL/min. Figure 2c illustrates the stacked composite images illustrating the particle distributions across the channel width before the outlet at various flow rates. It was observed that the fluorescent band gradually narrowed down and moved into the inner half channel at low flow rates of 6–8 mL/min. With a further increase in the flow rate, a clear fluorescent focusing stream was observed, indicating the formation of a focused particle train near the inner channel wall. Given the focusing of particles near the inner channel wall, the entire focused particle train could be completely removed via the inner outlet, whereas the cell-free fluids could be removed via the outer outlet (Figure 2d and Supplementary Video S1). From the abovementioned focusing phenomena, we concluded that our device could be successfully applied for cell concentration over a wide flow rate range of 9–16 mL/min. To the best of our knowledge, an operational flow rate of up to 16 mL/min is the highest value among previously reported microfluidic concentrators.

To quantitatively illustrate the effect of the flow rate on particle focusing, the fluorescence intensity profiles across the channel width at various flow rates were measured (Figure 3a,b). At flow rates of 6–9 mL/min, the fluorescence intensity in the outer half channel gradually decreased with the narrowing of the fluorescent peak. At high flow rates of 10–16 mL/min, a narrow fluorescent stream was clearly observed near the inner channel wall. The lateral position of the focusing stream remained nearly unchanged during this flow rate range.

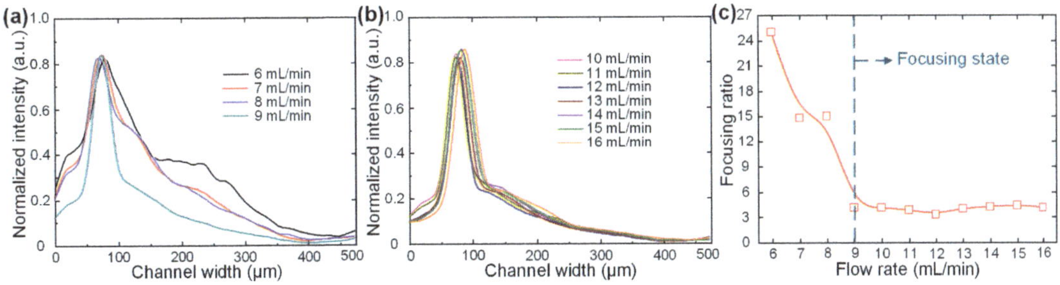

Figure 3. (**a**,**b**) Normalized fluorescence intensity profiles across the channel width at flow rates of (**a**) 6–9 and (**b**) 10–16 mL/min. (**c**) Focusing ratios at increasing flow rates. Particles were regarded as in focusing state in the device after the flow rate reached 9 mL/min.

In addition to the lateral focusing position, we defined a dimensionless focusing ratio to evaluate the focusing quality (Figure 3c). The focusing ratio was defined by dividing the FWHM of the fluorescent profile by the particle diameter. It was clearly observed that the focusing ratio rapidly decreased with the increasing flow rate and then became stable after the flow rate was greater than 9 mL/min. Therefore, we concluded that the particles were in the focusing state in our device after the flow rate reached 9 mL/min. Taken together, these results demonstrate that our device is capable of focusing particles into a train with stable lateral positions and focusing qualities over a board flow rate range of 9–16 mL/min. In addition to standard-sized particles, we tested the focusing performances of microalgal cells. Supplementary Video S2 illustrates the distribution of microalgal cells before the outlet at the representative flow rate of 10 mL/min. It was found that the focusing of microalgal cells was worse than that of standard-sized particles due to the polydisperse size and irregular shape of microalgal cells. However, nearly all the microalgal cells could still be collected via the inner outlet over a board flow rate range of 9–16 mL/min. The stable performance over the board flow rate range and the sheathless and external field-free operation make this device a good choice for hand-operated applications. We next characterized the flow stabilization in different channel layers. Figure S3 illustrated the focusing performances of 10 μm particles in the top and bottom channel layers (layers 1 and

4) at the flow rate of 10 mL/min. It was found that the particles in these two layers could be focused into tight streams and completely exported by the inner outlet. The focusing positions in these two layers are slightly different due to the small flow-rate variation across different layers. We further characterized the concentration performances of devices with different layer numbers. For devices with 1–4 layers, the 100% recovery of particles could be achieved, which validates the effectiveness of our four-layer design. Further increasing the channel number would deteriorate the concentration performance (a 7.36% decrease in particle recovery was observed in devices with five layers).

3.3. *Integration of Flow Stabilizer for Enabling Hand-Powered Operation*

When pushing the syringe by hand, the generated flow rates may inevitably vary during the pushing process. Although the four-channel paralleled concentrator can operate over a board flow rate range of 9–16 mL/min, it is still challenging to manually push the syringe to generate the desired flow rate within this range. Noteworthy, a flow rate that is too high or too low will deteriorate the concentration performance. To enable the concentration performance to be totally independent of the experiences and operations of the users, a syringe-tip flow stabilizer (Figure 4a–c) was integrated with the four-channel paralleled inertial microfluidic concentrator to stabilize and regulate the flow generated by hand pushing the syringe. The flow-stabilizing actuator was fabricated by stacking two layers of double-sided tapes (i and iii) and one layer of PDMS membrane (ii) in the order of tape–PDMS–tape (Figure 4a). The locating holes in each layer were used to precisely align the different layers. The two layers of double-sided tape were patterned with three branching channels. After transferring the PDMS membrane onto one surface of the double-sided tape, a triangular through hole was cut on the PDMS membrane in the central region of the three branching channels (Figures 4c and S4).

By stacking the tape–PDMS–tape, a portion of the PDMS membrane was suspended over the three branching channels, forming three parallel flow regulators. In the current design, three flow regulators were radially arrayed to increase the output stable flow rate, and these flow regulators shared the same inlet. The working principle of each parallel flow regulator is described in Figure 1b. When the sample is injected from the inlet, apart from flowing through the central hole, part of the fluids will be distributed into the upper branching channels with dead ends. The fluids in the upper branching channels above the membrane apply positive pressure on the suspended PDMS membrane, resulting in the deformation of the PDMS membrane toward the cavities of the lower three branching channels. As the fluids will flow along the lower branching channels to the outlets, the deformation of the PDMS membrane into the lower branching channels significantly increases the resistance of the flow paths. By dynamically regulating the flow resistance of the flow path according to the input flow pressure, a constant output flow rate can be achieved under a hand-powered operation. The only requirement for actuating the flow stabilizer is to apply a pressure larger than the threshold value. The integration of three parallel flow regulators enables the syringe-tip flow stabilizer to provide an ultra-high driving flow rate with small deviations.

Next, we fabricated a prototype of the syringe-tip flow stabilizer (Figure 4d) that could output a flow rate within the optimal working flow rate range of the four-channel paralleled inertial microfluidic concentrator. The detailed design parameters for the syringe-tip flow stabilizer are listed in Table S2. The detailed mechanisms behind these design parameters can be found in our previous study [69]. In the current work, we combined three parallel flow regulators (each with a constant flow rate of 3.50 mL/min) in a syringe-tip flow stabilizer for outputting an ultra-high driving flow rate and then integrated the syringe-tip flow stabilizer with the syringe filter-like inertial microfluidic concentrator for concentrating the cells from large-volume samples.

After fabrication, we set up a gas-driven flow system (Figure S1) to characterize the flow-stabilizing performance of the syringe-tip flow stabilizer. The analysis of the output flow rates of the flow stabilizer with increasing applied pressure showed that the pressure first increased with increasing pressure and then became constant at 9.95 mL/min after the pressure increased to more than 50 kPa (Figure 4e). A small deviation in the flow rate after achieving a stable output was only 3.6%. Under a flow rate of about 10 mL/min, the injection of 10 mL of the sample could be completed within 1 min, which is affordable for hand-pushing operations. The low threshold pressure for outputting a constant flow rate could be easily provided by hand pushing the syringe (pressure of 63–150 kPa measured by different adult operators, $N > 5$). Therefore, the operators only need to continuously push the syringe at their own comfortable speed.

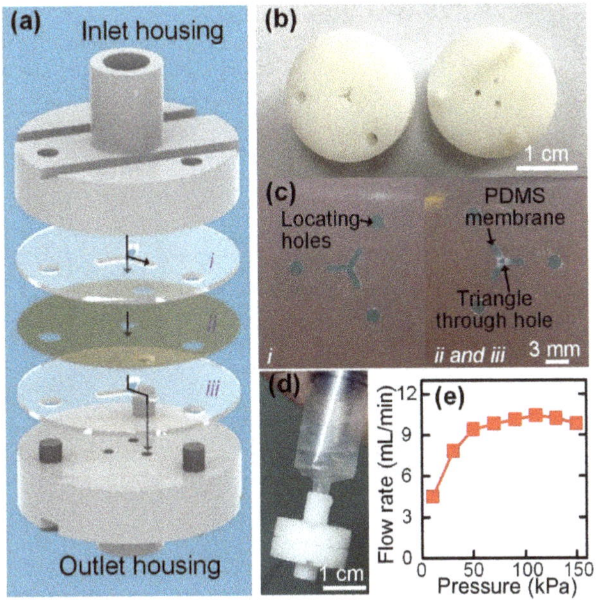

Figure 4. (**a**) Schematic diagram of the syringe-tip flow stabilizer containing an inlet housing, a flow stabilizing actuator (i, ii, and iii), and an outlet housing. The arrows in this figure illustrated the flow paths of one parallel flow regulator. (**b**) Photographs of the inlet and outlet housings that were fabricated by 3D printing. (**c**) Fabricated parts (i, ii, and iii) for assembling the flow-stabilizing actuator. Three parallel flow regulators were integrated in the syringe-tip flow stabilizer. (**d**) Photograph showing the prototype of the syringe-tip flow stabilizer. (**e**) Output flow rates of the syringe-tip flow stabilizer under different applied pressures.

3.4. Hand-Operated Cell Concentration

Finally, we integrated the syringe-tip flow stabilizer with our four-channel paralleled inertial microfluidic concentrator for hand-operated cell concentrations. To set up the integrated device, a syringe was first plugged into the inlet of the flow stabilizer, while the outlet of the flow stabilizer was directly connected to the central inlet of the concentrator. The press-fit connection enables the quick assembly or disassembly of these components without leakage, and the integrated device can be easily operated using a single hand. The suspensions of microalgal cells, 10 μm standard-sized particles, and MCF-7 tumor cells with initial concentrations of about 4×10^5 counts/mL were prepared and employed in this experiment. Five volunteers, none of whom underwent any operation training before the test, were invited to operate the syringe to inject the samples into the integrated device. The unstable flow rate generated by manually operating the syringe could be regulated to be the desired stable value after passing through the syringe-tip flow stabilizer. Then, the stable sample flow could power inertial focusing of cells in the concentrator, achieving the cell concentration at the optimal performance. To quantitatively characterize the performance of the integrated device for cell concentration under hand-powered operation, the liquid volumes and cell concentrations of the initial samples and those collected from both outlets were analyzed. We defined two dimensionless parameters: recovery efficiency (RE) and concentration factor (CF). RE was calculated by dividing the cell number (n_{target}) in the target sample collected from the inner outlet by the total cell number in all outlets (n_{total}), and CF was calculated by dividing the cell concentration (c_{target}) of the target sample collected from the inner outlet by the cell concentration (c_{initial}) of the initial sample.

Under the hand-operated mode, an RE of 100% and a CF of 2.06 $\pm$ 0.02 was achieved for 10 μm particles (T1 in Figure 5). For microalgal cells, an RE of 86.1 $\pm$ 2.9% and a CF of 1.74 $\pm$ 0.07 was obtained (T2 in Figure 5). Moreover, for MCF-7 tumor cells, an RE of 89.3 $\pm$ 1.8% and a CF of 1.82 $\pm$ 0.06 was obtained (T4 in Figure 5). The concentration performances of microalgal cells and MCF-7 tumor cells were found to be worse than of standard-sized particles because of the highly polydisperse cell sizes.

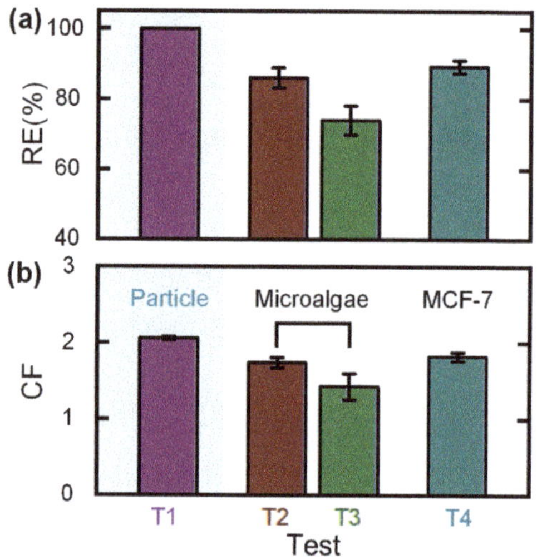

Figure 5. (**a**) Recovery efficiency (RE) and (**b**) concentration factor (CF) of 10 μm standard-sized particles, microalgal cells, and MCF-7 tumor cells under different experimental tests (T1−T4). Tests T1, T2, and T4 were performed with the integrated device using 10 μm standard-sized particles, microalgal cells, and MCF-7 cells, respectively. Test T3 was a control experiment on the concentration of microalgal cells using a device without a flow stabilizer. All the experiments were repeated five times under the hand-operated mode. The error bar in this figure denoted the standard deviation.

For comparison, we performed a control experiment on the concentration of microalgal cells using a device without a flow stabilizer (T3 in Figure 5). It was noticeable that the RE decreased to 74.1 $\pm$ 4.1%, and the CF decreased to 1.43 $\pm$ 0.17 due to the unstable flow rate generated by varied pushing operations (such as pushing at non-uniform speeds). As the syringe-tip flow stabilizer can provide a constant flow rate, the use of our flow stabilizer can make the concentration performance of the downstream concentrator totally independent of the user-pushing operations. The low threshold pressure (50 kPa for the current device) required to drive the flow stabilizer to output a constant flow rate can be easily provided by any adult operator.

To deal with samples with low initial cell concentrations at the level of 10^4 counts/mL, a multistep serial concentration was performed by reinjecting the collected target sample into the integrated device. With the 10 mL sample being reduced to less than 1 mL, the concentrations of microalgal cells in the target samples gradually increased from 0.586×10^5 to 6.96×10^5 counts/mL. After the multistep serial concentration, a total CF of 11.9 was achieved. The entire concentration process (including the sample reloading time) was completed within 5 min. Through observation of the migration of microalgal cells under a microscope, it was found that the cells remained alive and active after running through the integrated device. For MCF-7 cells, the cell viability was evaluated by Trypan blue exclusion. As illustrated in Figure S5, the cells remained alive after being processed with the integrated device. We next explored the effect of initial cell concentration on concentration performance using the microalgal cells. When the initial cell concentrations were below 2.4×10^5 counts/mL, an RE approaching 95% and a CF of about 2 could be achieved. Further increasing the initial cell concentration to be 5×10^5 counts/mL, the RE and CF decreased to be 70.8% and 1.35, respectively. The deterioration of concentration performance was caused by the heavy cell interactions at an increased cell concentration. In addition to microalgal cells and MCF-7 cells, the i-centrifuge can be applied to concentrate various other cells (e.g., the pre-sorted rare circulating tumor cells in clinical samples and the pathogenic bacterium in environmental samples). Through adjusting the outlet system, our i-centrifuge is possible to separate different-sized particles/cells according to their differential focusing positions. Centrifugation is the gold standard for cell concentration but requires electricity and an expensive centrifuge. As compared with the centrifuge, our i-centrifuge offers the advantages of electricity-free hand-operated operation, a low device cost, and a small footprint, which makes our device especially suitable for cell concentration in the field or other resource-poor settings.

4. Conclusions

In this study, we describe a novel integrated syringe-tip inertial microfluidic centrifuge (i-centrifuge) consisting of a syringe-tip flow stabilizer and a four-channel paralleled inertial microfluidic concentrator. The key components of the i-centrifuge were fabricated with low-cost polymer films and double-sided tape using a rapid nonclean-room process of laser cutting and lamination bonding, which enables their disposable use. Moreover, the i-centrifuge can be directly mounted onto the syringe tip via a quick press-fit connection and can be operated by hand power independent of the experiences and operations of the user. Noteworthy, the unstable and undesired liquid flow generated by the hand-pushing syringe can be regulated to be at a specific flow rate with the assistance of an integrated syringe-tip flow stabilizer. Then, the regulated liquid flow was used to drive the four-channel paralleled inertial microfluidic concentrator to enable high-flow-rate (up to 16 mL/min) cell concentration. Overall, we experimentally demonstrated the working mechanism and performance characterization of the i-centrifuge, successfully applying it to the hand-operated concentration of particles and cells. In summary, the i-centrifuge offers the advantages of a low device cost, simple hand-powered operation, high-flow-rate processing, and portable device volume. Therefore, it holds potential as a low-cost, portable sample preparation tool for point-of-care diagnostic testing.

Supplementary Materials: The following are available online at https://www.mdpi.com/article/10.3390/bios12010014/s1, Figure S1: The gas-driven flow system for characterizing the performance of our syringe-tip flow stabilizer. The compressed air was regulated via a computer-controlled pressure controller to generate a specific pressure for driving the liquid in the hermetic sample reservoir. The values of pressures were monitored and recorded using the software. The liquid then flowed through our syringe-tip flow stabilizer, and the mass of the output fluid was continuously monitored using an electronic balance. On the basis of these data, the stable output flow rates at specific pressures could be calculated. Figure S2: Images illustrating the designs and structures of inlet and outlet housings. Figure S3: Focusing performances of 10 μm particles in the top and bottom channel layers. Figure S4: Assembly and fabrication process of the PDMS membrane. The PDMS membrane was transferred onto one side of the patterned double-sided tape using a polyethylene terephthalate (PET) film. After being transferred onto the double-sided tape, the PET film was carefully peeled off, and a triangle through hole was cut on the PDMS membrane at the central region of the three branching channels. Figure S5: Image illustrating the cell viability. Table S1: Dimensions of spiral channels. Table S2: Dimensions of the three parallel flow regulators in the syringe-tip flow stabilizer. Video S1: A video illustrating the particle focusing at the Y-shaped outlet. Video S2: A video illustrating the distribution of microalgal cells at the Y-shaped outlet.

Author Contributions: Conceptualization, N.X. and Z.N.; methodology, N.X.; validation, N.X.; investigation, N.X.; writing—original draft preparation, N.X.; writing—review and editing, N.X. and Z.N.; project administration, N.X. and Z.N.; funding acquisition, N.X. and Z.N. All authors have read and agreed to the published version of the manuscript.

Funding: This research work is supported by the National Natural Science Foundation of China (51875103 and 81727801), the Natural Science Foundation of Jiangsu Province (BK20190064), the Six Talent Peaks Project of Jiangsu Province (SWYY-005) and the Zhishan Youth Scholar Program of SEU.

Institutional Review Board Statement: Not applicable.

Informed Consent Statement: Not applicable.

Data Availability Statement: The data presented in this study are available on request from the corresponding author.

Conflicts of Interest: The authors declare no conflict of interest.

References

1. Cui, F.; Rhee, M.; Singh, A.; Tripathi, A. Microfluidic Sample Preparation for Medical Diagnostics. *Annu. Rev. Biomed. Eng.* **2015**, *17*, 267–286. [CrossRef] [PubMed]
2. Xia, L.; Yang, J.; Su, R.; Zhou, W.; Zhang, Y.; Zhong, Y.; Huang, S.; Chen, Y.; Li, G. Recent Progress in Fast Sample Preparation Techniques. *Anal. Chem.* **2020**, *92*, 34–48. [CrossRef] [PubMed]
3. Mach, A.J.; Adeyiga, O.B.; Di Carlo, D. Microfluidic sample preparation for diagnostic cytopathology. *Lab Chip* **2013**, *13*, 1011–1026. [CrossRef] [PubMed]
4. Zhu, S.; Jiang, F.; Han, Y.; Xiang, N.; Ni, Z. Microfluidics for label-free sorting of rare circulating tumor cells. *Analyst* **2020**, *145*, 7103–7124. [CrossRef]
5. Wei, X.; Chen, K.; Guo, S.; Liu, W.; Zhao, X.-Z. Emerging Microfluidic Technologies for the Detection of Circulating Tumor Cells and Fetal Nucleated Red Blood Cells. *ACS Appl. Bio Mater.* **2021**, *4*, 1140–1155. [CrossRef]
6. Vaidyanathan, R.; Soon, R.H.; Zhang, P.; Jiang, K.; Lim, C.T. Cancer diagnosis: From tumor to liquid biopsy and be-yond. *Lab Chip* **2019**, *19*, 11–34.
7. Kant, K.; Shahbazi, M.-A.; Dave, V.P.; Ngo, T.A.; Chidambara, V.A.; Than, L.Q.; Bang, D.D.; Wolff, A. Microfluidic devices for sample preparation and rapid detection of foodborne pathogens. *Biotechnol. Adv.* **2018**, *36*, 1003–1024. [CrossRef]
8. Bridle, H.; Miller, B.; Desmulliez, M.P. Application of microfluidics in waterborne pathogen monitoring: A review. *Water Res.* **2014**, *55*, 256–271. [CrossRef]
9. Rajapaksha, P.; Elbourne, A.; Gangadoo, S.; Brown, R.; Cozzolino, D.; Chapman, J. A review of methods for the detec-tion of pathogenic microorganisms. *Analyst* **2019**, *144*, 396–411. [CrossRef]
10. Vashist, S.K.; Luppa, P.B.; Yeo, L.Y.; Ozcan, A.; Luong, J.H.T. Emerging Technologies for Next-Generation Point-of-Care Testing. *Trends Biotechnol.* **2015**, *33*, 692–705. [CrossRef]
11. Mrema, J.E.; Campbell, G.H.; Miranda, R.; Jaramillo, A.L.; Rieckmann, K.H. Concentration and separation of eryth-rocytes infected with Plasmodium falciparum by gradient centrifugation. *Bull. World Health Organ.* **1979**, *57*, 133–138.
12. Bonner, W.A.; Hulett, H.R.; Sweet, R.G.; Herzenberg, L.A. Fluorescence Activated Cell Sorting. *Rev. Sci. Instrum.* **1972**, *43*, 404–409. [CrossRef]
13. Sackmann, A.F.; Fulton, A.L.; Beebe, D.J. The present and future role of microfluidics in biomedical research. *Nature* **2014**, *507*, 181–189. [CrossRef] [PubMed]
14. Chen, L.; Yang, C.; Xiao, Y.; Yan, X.; Hu, L.; Eggersdorfer, M.; Chen, D.; Weitz, D.; Ye, F. Millifluidics, microfluidics, and nanofluidics: Manipulating fluids at varying length scales. *Mater. Today Nano* **2021**, *16*, 100136. [CrossRef]
15. Witek, M.A.; Freed, I.M.; Soper, S.A. Cell Separations and Sorting. *Anal. Chem.* **2019**, *92*, 105–131. [CrossRef] [PubMed]
16. Nasiri, R.; Shamloo, A.; Ahadian, S.; Amirifar, L.; Akbari, J.; Goudie, M.J.; Lee, K.; Ashammakhi, N.; Dokmeci, M.R.; Di Carlo, D.; et al. Microfluidic-Based Approaches in Targeted Cell/Particle Separation Based on Physical Properties: Fundamentals and Applications. *Small* **2020**, *16*, e2000171. [CrossRef] [PubMed]
17. Yan, S.; Yuan, D. Continuous microfluidic 3D focusing enabling microflow cytometry for single-cell analysis. *Talanta* **2021**, *221*, 121401. [CrossRef]
18. Daniele, M.A.; Boyd, D.A.; Mott, D.; Ligler, F.S. 3D hydrodynamic focusing microfluidics for emerging sensing technologies. *Biosens. Bioelectron.* **2015**, *67*, 25–34. [CrossRef]
19. Xiang, N.; Li, Q.; Ni, Z. Combining Inertial Microfluidics with Cross-Flow Filtration for High-Fold and High flow-rate Passive Volume Reduction. *Anal. Chem.* **2020**, *92*, 6770–6776. [CrossRef]

20. Giordano, B.C.; Burgi, D.S.; Hart, S.J.; Terray, A. On-line sample pre-concentration in microfluidic devices: A review. *Anal. Chim. Acta* **2012**, *718*, 11–24. [CrossRef]
21. Luan, Q.; Macaraniag, C.; Zhou, J.; Papautsky, I. Microfluidic systems for hydrodynamic trapping of cells and clusters. *Biomicrofluidics* **2020**, *14*, 031502. [CrossRef]
22. Xie, Y.; Rufo, J.; Zhong, R.; Rich, J.; Li, P.; Leong, K.W.; Huang, T.J. Microfluidic Isolation and Enrichment of Nanopar-ticles. *ACS Nano* **2020**, *14*, 16220–16240. [CrossRef]
23. Liu, Y.; Xu, H.; Li, T.; Wang, W. Microtechnology-enabled filtration-based liquid biopsy: Challenges and practical con-siderations. *Lab Chip* **2021**, *21*, 994–1015. [CrossRef]
24. Quan, Y.; Chen, K.; Xiang, N.; Ni, Z. A single-view field filter device for rare tumor cell filtration and enumeration. *Electrophoresis* **2020**, *41*, 2000–2006. [CrossRef]
25. Grigorov, E.; Kirov, B.; Marinov, M.; Galabov, V. Review of Microfluidic Methods for Cellular Lysis. *Micromachines* **2021**, *12*, 498. [CrossRef]
26. Zhu, S.; Wu, D.; Han, Y.; Wang, C.; Xiang, N.; Ni, Z. Inertial microfluidic cube for automatic and fast extraction of white blood cells from whole blood. *Lab Chip* **2020**, *20*, 244–252. [CrossRef]
27. Whitesides, G.M.; Ostuni, E.; Takayama, S.; Jiang, X.; Ingber, D.E. Soft Lithography in Biology and Biochemistry. *Annu. Rev. Biomed. Eng.* **2001**, *3*, 335–373. [CrossRef]
28. Balakrishnan, H.K.; Badar, F.; Doeven, E.H.; Novak, J.I.; Merenda, A.; Dumée, L.F.; Loy, J.; Guijt, R.M. 3D Printing: An Alternative Microfabrication Approach with Unprecedented Opportunities in Design. *Anal. Chem.* **2021**, *93*, 350–366. [CrossRef] [PubMed]
29. Nielsen, A.V.; Beauchamp, M.J.; Nordin, G.P.; Woolley, A.T. 3D Printed Microfluidics. *Annu. Rev. Anal. Chem.* **2020**, *13*, 45–65. [CrossRef] [PubMed]
30. Xu, B.-B.; Zhang, Y.-L.; Xia, H.; Dong, W.-F.; Ding, H.; Sun, H.-B. Fabrication and multifunction integration of microflu-idic chips by femtosecond laser direct writing. *Lab Chip* **2013**, *13*, 1677–1690. [CrossRef] [PubMed]
31. Paiè, P.; Bragheri, F.; Di Carlo, D.; Osellame, R. Particle focusing by 3D inertial microfluidics. *Microsyst. Nanoeng.* **2017**, *3*, 17027. [CrossRef] [PubMed]
32. Lee, U.N.; Su, X.; Guckenberger, D.J.; Dostie, A.M.; Zhang, T.; Berthier, E.; Theberge, A.B. Fundamentals of rapid injec-tion molding for microfluidic cell-based assays. *Lab Chip* **2018**, *18*, 496–504. [CrossRef] [PubMed]
33. Peng, L.; Deng, Y.; Yi, P.; Lai, X. Micro hot embossing of thermoplastic polymers: A review. *J. Micromech. Microeng.* **2014**, *24*, 013001. [CrossRef]
34. Speller, N.C.; Morbioli, G.G.; Cato, M.E.; Cantrell, T.P.; Leydon, E.M.; Schmidt, B.E.; Stockton, A.M. Cutting edge microfluidics: Xurography and a microwave. *Sens. Actuators B Chem.* **2019**, *291*, 250–256. [CrossRef]
35. McDonald, J.C.; Whitesides, G.M. Poly(dimethylsiloxane) as a Material for Fabricating Microfluidic Devices. *Acc. Chem. Res.* **2002**, *35*, 491–499. [CrossRef]
36. Ma, J.; Yan, S.; Miao, C.; Li, L.; Shi, W.; Liu, X.; Luo, Y.; Liu, T.; Lin, B.; Wu, W.; et al. Paper Microfluidics for Cell Analysis. *Adv. Healthc. Mater.* **2019**, *8*, 1801084. [CrossRef]
37. Pasitka, L.; van Noort, D.; Lim, W.; Park, S.; Mandenius, C.-F. A Microbore Tubing Based Spiral for Multistep Cell Frac-tionation. *Anal. Chem.* **2018**, *90*, 12909–12916. [CrossRef]
38. Wang, X.; Gao, H.; Dindic, N.; Kaval, N.; Papautsky, I. A low-cost, plug-and-play inertial microfluidic helical capillary device for high flow-rate flow cytometry. *Biomicrofluidics* **2017**, *11*, 014107. [CrossRef]
39. Zhu, Z.; Wu, D.; Li, S.; Han, Y.; Xiang, N.; Wang, C.; Ni, Z. A polymer-film inertial microfluidic sorter fabricated by jigsaw puzzle method for precise size-based cell separation. *Anal. Chim. Acta* **2021**, *1143*, 306–314. [CrossRef]
40. Xiang, N.; Zhang, R.; Han, Y.; Ni, Z. A Multilayer Polymer-Film Inertial Microfluidic Device for High flow-rate Cell Concentration. *Anal. Chem.* **2019**, *91*, 5461–5468. [CrossRef]
41. Xuan, X. Recent advances in direct current electrokinetic manipulation of particles for microfluidic applications. *Electrophoresis* **2019**, *40*, 2484–2513. [CrossRef] [PubMed]
42. Liu, L.; Chen, K.; Xiang, N.; Ni, Z. Dielectrophoretic manipulation of nanomaterials: A review. *Electrophoresis* **2019**, *40*, 873–889. [CrossRef] [PubMed]
43. Pamme, N. Magnetism and microfluidics. *Lab Chip* **2006**, *6*, 24–38. [CrossRef] [PubMed]
44. Huang, D.; Xiang, N. Rapid and precise tumor cell separation using the combination of size-dependent inertial and size-independent magnetic methods. *Lab Chip* **2021**, *21*, 1409–1417. [CrossRef] [PubMed]
45. Paiè, P.; Zandrini, T.; Vázquez, R.M.; Osellame, R.; Bragheri, F. Particle Manipulation by Optical Forces in Microfluidic Devices. *Micromachines* **2018**, *9*, 200. [CrossRef]
46. Zhang, P.; Bachman, H.; Ozcelik, A.; Huang, T.J. Acoustic Microfluidics. *Annu. Rev. Anal. Chem.* **2020**, *13*, 17–43. [CrossRef]
47. Wu, M.; Ozcelik, A.; Rufo, J.; Wang, Z.; Fang, R.; Huang, T.J. Acoustofluidic separation of cells and particles. *Microsyst. Nanoeng.* **2019**, *5*, 1–18. [CrossRef]
48. Yan, S.; Zhang, J.; Yuan, D.; Li, W. Hybrid microfluidics combined with active and passive approaches for continuous cell separation. *Electrophoresis* **2017**, *38*, 238–249. [CrossRef]
49. Stoecklein, D.; Di Carlo, D. Nonlinear Microfluidics. *Anal. Chem.* **2019**, *91*, 296–314. [CrossRef]
50. Jiang, D.; Ni, C.; Tang, W.; Huang, D.; Xiang, N. Inertial microfluidics in contraction–expansion microchannels: A review. *Biomicrofluidics* **2021**, *15*, 041501. [CrossRef]

51. Tang, W.; Zhu, S.; Jiang, D.; Zhu, L.; Yang, J.; Xiang, N. Channel innovations for inertial microfluidics. *Lab Chip* **2020**, *20*, 3485–3502. [CrossRef] [PubMed]
52. Zhang, J.; Yan, S.; Yuan, D.; Alici, G.; Nguyen, N.-T.; Warkiani, M.E.; Li, W. Fundamentals and applications of inertial microfluidics: A review. *Lab Chip* **2016**, *16*, 10–34. [CrossRef] [PubMed]
53. Amini, H.; Lee, W.; Di Carlo, D. Inertial microfluidic physics. *Lab Chip* **2014**, *14*, 2739–2761. [CrossRef] [PubMed]
54. Martel, J.M.; Toner, M. Inertial Focusing in Microfluidics. *Annu. Rev. Biomed. Eng.* **2014**, *16*, 371–396. [CrossRef]
55. Khojah, R.; Stoutamore, R.; Di Carlo, D. Size-tunable microvortex capture of rare cells. *Lab Chip* **2017**, *17*, 2542–2549. [CrossRef]
56. Dudani, J.; Gossett, D.R.; Tse, H.T.K.; Lamm, R.J.; Kulkarni, R.P.; Di Carlo, D. Rapid inertial solution exchange for enrichment and flow cytometric detection of microvesicles. *Biomicrofluidics* **2015**, *9*, 014112. [CrossRef]
57. Warkiani, M.E.; Khoo, B.L.; Wu, L.; Tay, A.K.P.; Bhagat, A.A.; Han, J.; Lim, C.T. Ultra-fast, label-free isolation of circulating tumor cells from blood using spiral microfluidics. *Nat. Protoc.* **2016**, *11*, 134–148. [CrossRef]
58. Sudarsan, A.P.; Ugaz, V.M. Multivortex micromixing. *Proc. Natl. Acad. Sci. USA* **2006**, *103*, 7228–7233. [CrossRef]
59. Jang, I.; Kang, H.; Song, S.; Dandy, D.S.; Geiss, B.J.; Henry, C.S. Flow control in a laminate capillary-driven microfluidic device. *Analyst* **2021**, *146*, 1932–1939. [CrossRef]
60. Walker, G.M.; Beebe, D.J. A passive pumping method for microfluidic devices. *Lab Chip* **2002**, *2*, 131–134. [CrossRef]
61. Bengtsson, K.; Robinson, N.D. A large-area, all-plastic, flexible electroosmotic pump. *Microfluid. Nanofluidics* **2017**, *21*, 178. [CrossRef]
62. Du, X.Y.; Swanwick, M.E.; Fu, Y.Q.; Luo, J.K.; Flewitt, A.J.; Lee, D.S.; Maeng, S.; Milne, W.I. Surface acoustic wave induced streaming and pumping in 128° Y-cut LiNbO3for microfluidic applications. *J. Micromech. Microeng.* **2009**, *19*, 035016. [CrossRef]
63. Iwai, K.; Shih, K.C.; Lin, X.; Brubaker, T.A.; Sochol, R.D.; Lin, L. Finger-powered microfluidic systems using multi-layer soft lithography and injection molding processes. *Lab Chip* **2014**, *14*, 3790–3799. [CrossRef] [PubMed]
64. Park, J.; Han, D.H.; Park, J.-K. Towards practical sample preparation in point-of-care testing: User-friendly microfluidic devices. *Lab Chip* **2020**, *20*, 1191–1203. [CrossRef] [PubMed]
65. Bhamla, M.S.; Benson, B.; Chai, C.; Katsikis, G.; Johri, A.; Prakash, M. Hand-powered ultralow-cost paper centrifuge. *Nat. Biomed. Eng.* **2017**, *1*, 0009. [CrossRef]
66. Kuntaegowdanahalli, S.S.; Bhagat, A.A.S.; Kumar, G.; Papautsky, I. Inertial microfluidics for continuous particle sepa-ration in spiral microchannels. *Lab Chip* **2009**, *9*, 2973–2980. [CrossRef] [PubMed]
67. Xiang, N.; Shi, Z.; Tang, W.; Huang, D.; Zhang, X.; Ni, Z. Improved understanding of particle migration modes in spiral inertial microfluidic devices. *RSC Adv.* **2015**, *5*, 77264–77273. [CrossRef]
68. Di Carlo, D.; Irimia, D.; Tompkins, R.G.; Toner, M. Continuous inertial focusing, ordering, and separation of particles in microchannels. *Proc. Natl. Acad. Sci. USA* **2007**, *104*, 18892–18897. [CrossRef] [PubMed]
69. Xiang, N.; Han, Y.; Jia, Y.; Shi, Z.; Yi, H.; Ni, Z. Flow stabilizer on a syringe tip for hand-powered microfluidic sample injection. *Lab Chip* **2019**, *19*, 214–222. [CrossRef]

biosensors

MDPI

Article

High-Throughput, Label-Free Isolation of White Blood Cells from Whole Blood Using Parallel Spiral Microchannels with U-Shaped Cross-Section

Amirhossein Mehran [1], Peyman Rostami [1], Mohammad Said Saidi [1], Bahar Firoozabadi [1,*] and Navid Kashaninejad [2,*]

[1] School of Mechanical Engineering, Sharif University of Technology, Tehran 11155, Iran; amirhossein.mehran@alum.sharif.edu (A.M.); peyman.rmi@gmail.com (P.R.); mssaidi@sharif.edu (M.S.S.)
[2] Queensland Micro- and Nanotechnology Centre, Nathan Campus, Griffith University, 170 Kessels Road, Brisbane, QLD 4111, Australia
* Correspondence: firoozabadi@sharif.edu (B.F.); n.kashaninejad@griffith.edu.au (N.K.); Tel.: +98-2166165521 (B.F.); +61-7373-55391 (N.K.)

Abstract: Rapid isolation of white blood cells (WBCs) from whole blood is an essential part of any WBC examination platform. However, most conventional cell separation techniques are labor-intensive and low throughput, require large volumes of samples, need extensive cell manipulation, and have low purity. To address these challenges, we report the design and fabrication of a passive, label-free microfluidic device with a unique U-shaped cross-section to separate WBCs from whole blood using hydrodynamic forces that exist in a microchannel with curvilinear geometry. It is shown that the spiral microchannel with a U-shaped cross-section concentrates larger blood cells (e.g., WBCs) in the inner cross-section of the microchannel by moving smaller blood cells (e.g., RBCs and platelets) to the outer microchannel section and preventing them from returning to the inner microchannel section. Therefore, it overcomes the major limitation of a rectangular cross-section where secondary Dean vortices constantly enforce particles throughout the entire cross-section and decrease its isolation efficiency. Under optimal settings, we managed to isolate more than 95% of WBCs from whole blood under high-throughput (6 mL/min), high-purity (88%), and high-capacity (360 mL of sample in 1 h) conditions. High efficiency, fast processing time, and non-invasive WBC isolation from large blood samples without centrifugation, RBC lysis, cell biomarkers, and chemical pre-treatments make this method an ideal choice for downstream cell study platforms.

Keywords: WBC isolation; spiral microchannels; inertial microfluidics; passive cell separation; high-throughput separation

Citation: Mehran, A.; Rostami, P.; Saidi, M.S.; Firoozabadi, B.; Kashaninejad, N. High-Throughput, Label-Free Isolation of White Blood Cells from Whole Blood Using Parallel Spiral Microchannels with U-Shaped Cross-Section. *Biosensors* **2021**, *11*, 406. https://doi.org/10.3390/bios11110406

Received: 28 September 2021
Accepted: 17 October 2021
Published: 20 October 2021

1. Introduction

In medical science, the study of characteristics and mechanisms of cell functions, identifying external damaging factors, and finding methods to prevent and treat cellular disorder diseases in the human body requires the preparation of cell samples with suitable purity. Blood is the most important component of the human body, containing various types of essential cells, including red blood cells (RBCs), white blood cells (WBCs), and platelets which are produced in the bone marrow and released into the bloodstream [1]. WBCs comprise about 1% of whole blood and have a significant role in the body's immune system. The change in the total number of WBCs in the blood is the harbinger of infections, autoimmune reactions, and other malignancies [2]. For example, in blood cancer diseases, such as leukemia and myeloma, the life cycle of normal blood cells is interrupted by the abnormal growth in the number of WBCs in the bone marrow and bloodstream. Hence, enumeration and monitoring of WBCs are critical in diagnosing various kinds of diseases [3]. Since RBCs comprise the majority of whole blood, complete or fractional separation of RBCs from a blood sample is usually required for subsequent cellular and molecular examination of WBCs. However, conventional cell sorting techniques, such as density gradient centrifugation, RBC lysis, porous filtration, and other methods are often labor-intensive, require a large volume of sample, and sometimes use expensive particular biomarkers or labels to identify the target cells [4].

In recent years, various microfluidic cell separation devices have been developed to minimize the amount of required sample, processing time, and operator-based errors while maintaining high-throughput and isolation efficiency [5,6]. Although they have been successful, there is always a trade-off between the efficiency and throughput of these platforms. Based on the separation mechanism, microfluidic cell isolation techniques can be categorized into active and passive methods [6].

Active methods require an externally applied force so that cells are separated due to their different optical [7], electrical [8], or magnetic [9] properties. Some examples of active methods are dielectrophoresis (DEP) [10], acoustophoresis [11], and magnetophoresis [12]. Although active methods have been proven to be useful in various biological applications, the low-throughput, complex mechanism, and high fabrication costs have limited their widespread applications in most cases.

Passive methods rely on label-free separation of cells using their physical properties without any externally applied force field. Some passive separation methods include deterministic lateral displacement (DLD) [13], pinch flow fractionation (PFF) [14], hydrodynamic filtration [15], and inertial migration [16].

Among passive cell separation techniques, inertial microfluidics has attracted much attention in recent years. The efficiency of inertial microfluidic for cell separation depends on inertial migration and hydrodynamic forces. Inertial migration is a function of the geometrical parameters of the device, while hydrodynamic forces separate particles based on their physical properties, such as density [17], size [18], or deformability [19].

Segre and Silberberg [20,21] were the first to report the inertial migration effect by showing that randomly incoming dispersed particles in a circular tube with a radius of R, can laterally migrate toward the channel walls and form a ring-shaped annulus with a radius of 0.6 R at the outlet. This observation led many researchers to investigate the physics of this phenomenon [22,23]. Particle lateral migration is affected by the shear gradient lift force, which moves the particles toward the tube walls. The wall induces a lift force that pushes the particles through the tube centerline and prevents them from getting close to the walls. As a result, particles reach a certain equilibrium position in the tube's cross-section depending on the magnitude of these forces, flow velocity, and particle's physical properties [24]. Due to difficulties with particle separation at the outlet of a circular tube, several studies have investigated particle inertial migration in non-circular channels [25–27]. In a square microchannel, eight stable particle equilibrium positions (corners and midline of channel edge) exist in relatively low Reynolds number flows ($Re < 100$) [25]. However, increasing the Reynolds number ($Re > 500$) will reduce the equilibrium positions to four, in which particles are focused at microchannel corners [25]. In a straight rectangular microchannel, randomly dispersed particles with $a_p/D_h \sim 0.1$ (where a_p is particle diameter and D_h is hydraulic diameter of the channel) align in four equilibrium positions where shear gradient lift force and wall induced lift force balance each other [27]. According to recent studies, in a spiral microchannel, particles are focused in one single equilibrium position close to the inner wall of the microchannel while smaller sized particles continue to recirculate by the effect of the secondary Dean flow [16,28].

Recently, many studies have been performed to optimize separation by increasing throughput, separation efficiency, and resolution in spiral microchannels [29,30]. Some studies have developed numerical calculations for a better understanding of the inertial focusing mechanism [31,32]. Some studies have been investigated to modify the microchip geometry to optimize the flow rate and isolation efficiency. Label-free tumor cell separation from whole blood has been conducted using a double spiral microchannel device with an 88.5% tumor recovery rate as well as 92.28% recovery for blood cells [33]. Higher particle isolation efficiency was reached by introducing a spiral microchannel with trapezoidal cross-section [34,35] and ordered micro-obstacles [36]. However, the complex fabrication of trapezoidal cross-section and micro-obstacles has limited the wide application of this kind of geometry in microfluidic platforms. Our group proposed a spiral microchannel with a stair-like cross-section for size-based particle separation. Using equilibrated vortices present in the spiral microchannel, it was observed that there is a size-dependent threshold for flow rate for separating each specific particle [37].

This work overcomes the limitations of previous methods by presenting a passive microfluidic approach for high-throughput continuous isolation of WBCs from whole blood in a spiral microchannel with U-shape cross-section using size-dependent inertial migration. The proposed method uses secondary Dean drag force to move smaller cells away and inertial forces to equilibrate larger cells through the spiral microchannel. One of the advantages of the U-shaped cross-section is that target cells can be collected from separate outlets in discrete cell streams without any restriction on the channel length. The U-shaped cross-section also avoids the re-circulation of smaller particles through the entire cross-section. Hence, it removes certain drops in isolation efficiency due to the mixing problems of conventional rectangular cross-sections. Unlike previous microfluidic devices with rather complex geometries and cross-sections, e.g., microchannels with trapezoidal cross-section, or spiral microchannels with ordered micro-bars, our proposed spiral microchannels with the U-shaped cross-section are easy to fabricate. Also, we have developed a novel numerical algorithm to calculate the inertial forces exerting on particles to optimize the proposed device's cell separation capability that can be used for similar inertial focusing methods for future works. It also further eliminates the time-consuming process of trial and error to find the optimized design of the device. Based on numerical results, we select an optimized geometry regarding the proper placement of the secondary vortices and their primary role in carrying the target cells to proper microchip outlets. Utilizing this method, high separation efficiency for WBCs and high removing ratios for RBCs and platelets were achieved using the fabricated parallel spiral microchannels with a U-shaped cross-section. It was reached under a high-throughput (as high as 6 mL/min), high-capacity (360 mL sample in 1 h) label-free cell sorting without the need for any significant pre-processing or post-processing. Single-layer spiral microchannel with the U-shaped cross-section, which is a modified version of a spiral microchannel with rectangular cross-section, has the following advantages: (i) Strong commercialization potentials due to simplified fabrication methods; (ii) Low fabrication costs and efforts; (iii) A wide range for the separation of cells/particles with different sizes. Such an optimized microfluidic device has great potential for biological cell separation in clinical applications.

2. Design Principle

In straight channels, shear-induced lift force resulting from the parabolic nature of the velocity profile in a Poiseuille flow tends to move the dispersed particles away from the center of the microchannel toward the channel walls. As the particle gets closer to the walls, an asymmetric wake around the particle forms a wall-induced lift force that prevents the particles from getting close to the walls and directs them away from the walls toward the channel center [38]. Hence, the particles are focused in equilibrium positions in narrow bands where these two opposite forces are equal [20,21,25,28,39,40]. The overall magnitude of the inertial forces which cause the lateral migration of the particles in a straight microchannel can be estimated by Equation (1) [41]:

$$F_L = \rho G^2 C_L(Re,\ x_p)\ {a_p}^4 \tag{1}$$

where ρ is the density of the fluid, G is the fluid shear rate ($G = 2U_f/D_h$), U_f is the average flow velocity, C_L is a non-dimensional lift coefficient which is a function of Reynolds number and particle position in the channel's cross-section, U is the maximum velocity in the microchannel, D_h is the hydraulic diameter of the microchannel and a_p is the particle diameter. The magnitude of C_L, thus, F_L starts from zero in the channel centerline reaches a maximum value, and then goes back to zero again around $0.2D_h$ away from the channel wall, which is considered as the equilibrium position of particles [41]. Beyond this distance, C_L becomes negative in sign showing the dominant effect of wall-induced lift force and further increases in magnitude by moving toward the walls [27].

Fluid flow in curvilinear channels experiences radially outward centrifugal acceleration resulting in the shift of maximum velocity toward the outer wall of the channel, leading to the formation of two counter-rotating vortices known as Dean vortices in the top and bottom halves of the channel cross-section plane [42,43]. The strength of these two identical vortices can be described using the dimensionless Dean number (De):

$$De = \frac{\rho U_f D_h}{\mu}\sqrt{\frac{D_h}{2R}} = Re\sqrt{\frac{D_h}{2R}} \tag{2}$$

where ρ is the fluid density, U_f is the average flow velocity, μ is the fluid viscosity, D_h is the hydraulic diameter, R is the flow path curvature, and Re is the flow Reynolds number. According to Stokes law, the applied drag force on a single smooth spherical particle (larger than 1 μm in diameter) in a laminar flow can be calculated using Equation (3) [44]:

$$F_D = 6\pi\mu\, r_p V \tag{3}$$

where μ is the fluid viscosity, r_p is the particle radius, and V is the particle velocity.

Assuming Stokes law, particles are affected by the drag force due to the presence of the secondary flow in the microchannel, which can be described as:

$$F_D = 3\pi\mu\, U_{Dean}\, a_p \tag{4}$$

where a_p is the particle diameter. Ookawara et al. [45] numerically investigated Dean flows in a rectangular curved microchannel and presented a correlation, in the form of a power function, for estimating the average dean flow velocity. They conducted their simulations based on using the SIMPLE algorithm with a second-order upwind scheme in the Fluent software with boundary conditions of uniform velocity inlet and pressure outlet for the microchannel inlet and outlet, respectively. Based on their formulation, the average Dean velocity can be expressed by Equation (4):

$$U_{Dean} = 1.8 \times 10^{-4}\, De^{1.63} \tag{5}$$

Thus, the drag force caused by the secondary Dean vortices exerting on the particles can be estimated by:

$$F_D = 5.4 \times 10^{-4}\, \pi\mu\, De^{1.63}\, a_p \tag{6}$$

The inertial forces and the Dean drag force act in the same direction on the particles near the outer wall while particles near the inner wall experience these two forces in opposite directions. According to Equations (1) and (5), the ratio of the inertial lift force to the Dean drag force (F_L/F_D) is proportional to ${a_p}^3$, therefore, as particles become larger, the inertial lift force exerting on them becomes greater than the Dean drag force and vice versa for small particles. Hence, in a spiral microchannel, larger particles tend to stay in equilibrium positions near the microchannel inner wall, while smaller ones are constantly circulated by the secondary flow.

In this paper, we take advantage of the simultaneous acting of inertial forces and Dean drag force on the blood cells in a spiral microchannel to develop a high-throughput device that can isolate WBCs from whole blood without the needs of blood lysis, cell labeling, centrifugation, and other chemical or physical pre-processing techniques on the sample.

3. Materials and Methods

3.1. Device Fabrication

Micromilling is a sub-branch of micromachining which has been widely used to fabricate a wide range of microfluidic devices. The advantages of using this technique are faster fabrication, straightforward process, easier user interface, and the ability to manufacture complex geometries, which makes this method an ideal choice in rapid prototyping microfluidic platforms for testing, validation, and research. Unlike other means of fabrication such as PDMS casting, etching, etc., micromilling can be employed on various types of materials and can support the fabrication of multi-level and complex structures. Moreover, micromilling devices, such as CNC machines can be found nearly in any university or workshop and are commonly used in manufacturing processes. PMMA (polymethyl methacrylate) material is mainly used in microfluidics due to its excellent characteristics, such as biocompatibility, good strength, low cost, and optical properties. In this study, the master mold was manufactured by micromilling on a PMMA substrate by Dahlih MCV-1020A milling machine which has an accuracy of 2 μm in the 3 moving directions. Since the microchannel geometry consists of two sets of levels with different heights, several layer cuts are needed, including a couple of rough cuts, and two final cuts with the depths of 1 mm, 120 μm, and 200 μm, respectively. In order to cut the microchannels on the PMMA substrate, 3 different micromilling bits with different sizes were used. A 6 mm diameter 4-flute endmill cutting with a spindle speed of 3000 rpm and cutting feed rate of 100 mm/min was used for starting rough cuts, a 0.8 mm diameter 2-flute endmill cutting with a spindle speed of 6000 rpm and cutting feed rate of 100 mm/min was used for cutting the spaces between the spiral loops, and a 0.1 mm diameter 2-flute endmill cutting with a spindle speed of 7000 rpm and cutting feed rate of 50 mm/min was used for carving the space above the cross-section passway. Once the PMMA mold was ready, we proceeded to the next step. The microchip was fabricated using the standard soft lithography method. Briefly, degassed PDMS (polydimethylsiloxane), previously mixed in a 10:1 pre-polymer base to the curing agent, was cast onto the master mold and baked in a vacuum oven for 4 h at 75 °C. Cured PDMS layer with embedded channels was peeled off, and inlets and outlets holes were punched using a 2 mm biopsy punch. Finally, the PDMS layer was irreversibly bonded to a thick standard glass slab with an oxygen plasma and subsequently baked for 2 h at 85 °C to further improve the bonding. Silicone tubes with an outer diameter of 2 mm were press-fitted into the channel inlets and outlets. The device consists of two parallel 4-loop spiral geometry with 4 inlets, and 2 shared inner and outer outlets. The cross-section of the channels is a U-shape geometry with 700 μm width and 200 μm height, and the two sections of the channel are connected by a 200 μm × 80 μm passway as shown in Figure 1B.

Each spiral microchannel has an initial radius of curvature of 6 mm and a distance of 1.9 mm between two successive loops. The total length of each spiral is approximately 17.6 cm (Figure 1A).

3.2. Numerical Simulation

The main idea of simulation is to calculate the forces exerting on the cells in the fluid domain. Since the cells' size is relatively comparable to microchannel dimensions (14 μm vs. 200 μm), the effect of cells on the fluid domain cannot be ignored. In terms of investigations on the particle lateral migration, most theoretical examinations involve using simplified models, such as circular tubes, parallel plates, and negligible particle diameters ($a/H << 1$). These models manage to approximate the fluid domain by treating the particle as a dimensionless point with a point-force applying to it, hence failing to take the effect of the particle on the fluid domain into the account. Originally, Di Carlo et al. [31] developed an algorithm to overcome these limitations and calculate the inertial forces exerting on particles with diameters comparable to that of the microchannel dimensions in a straight rectangular microchannel. The main drawback was that it was limited to conventional rectangular cross-section microchannels. However, in this study, we further developed the algorithm so that we could assess the performance of arbitrary complex microchannel cross-sections, hence extending it to investigate the viability of our U-shaped cross-section microchannel on isolating particles with different sizes. Several other useful information can be collected while using this algorithm as a simulation approach, such as the threshold flow rate estimation in which smaller particles begin to migrate toward the outer microchannel section, equilibrium positions of different-sized particles throughout microchannels with complex cross-sections, the amount of shear stress on the surface of particles to evaluate the possibility of cell damage, etc. Herein, we represent our modified version of this algorithm which is developed for a U-shaped cross-section microchannel in the present work as well as supporting more complex cross-sections alongside treating dispersed particles as objects affecting the flow field domain. Also, in terms of simulations, although blood cells don't exactly match that of a circular shape, they were treated as smooth solid spherical particles for the sake of simplicity.

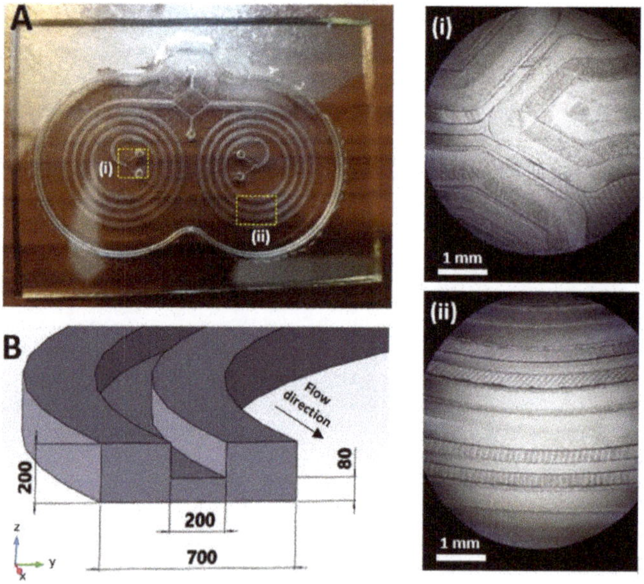

Figure 1. (**A**) The 4-loop parallel spiral microchannels fabricated in PDMS with the corresponding mold preview. (**i**) The magnified view of section (i). (**ii**) The magnified view of section (ii). (**B**) Schematic of the cross-section of the spiral microchannel (The dimensions are in μm).

Inertial lift force is a function of particle location within the cross-section of a microchannel and flow shear rate around the particle. We managed to calculate this force within the cross-section and investigate the equilibrium positions of particles through the cross-section plane. We took the x-direction as the main flow direction in any arbitrary cross-section throughout the spiral microchannel. The simulation consists of three steps:

(a) A 3D steady laminar flow is solved through the whole spiral microchannel, and the velocity profiles are extracted using COMSOL Multiphysics® software.

(b) A straight channel with a proposed cross-section with a length of 20 times the particle diameter is created.

The velocity profile solved at Step (a) is set as the inlet and outlet boundary conditions, and the flow field is calculated through the entire channel. Linear and rotational velocities of flow on some cross-section points will be used in the next step.

(c) Particle modeled as a solid sphere is put in the middle of the microchannel alongside the x-direction, and its location is set to change in the yz-plane in each solution step.

Steady-state linear and rotational velocities of the particle are unknown and should be calculated using an iterative trial-and-error process.

The flowchart of the proposed algorithm is shown in Figure 2.

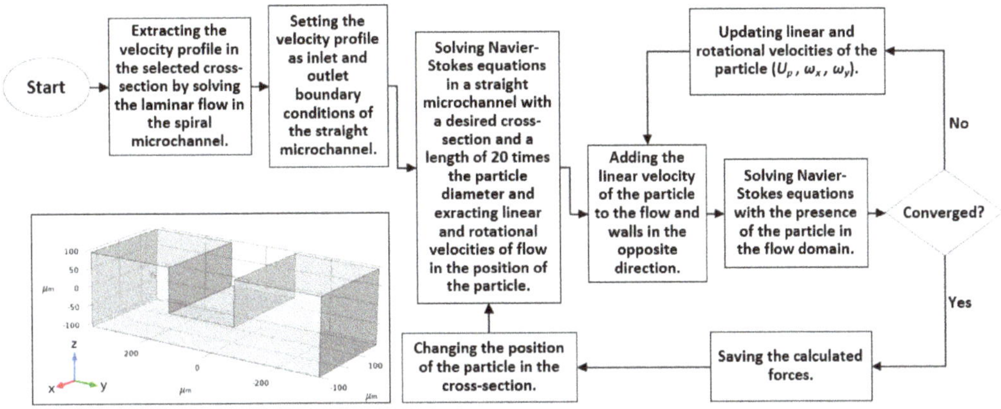

Figure 2. Flowchart of the numerical algorithm used in this study for the calculation of the forces acting on particles. The original algorithm was proposed to investigate particle inertial migration in a straight microchannel [31]. It was later modified to be used in a rectangular spiral microchannel [32]. We extended the original algorithm to run support the corresponding calculations of any other desired spiral microchannel cross-sections.

The governing equations for fluid flow within the microchannel are continuity and Navier-Stokes equations as follows [46,47]:

$$\rho \nabla \cdot u = 0 \tag{7}$$

$$\rho(\boldsymbol{u} \cdot \nabla)\boldsymbol{u} = \nabla \cdot \left[-p\boldsymbol{I} + \mu\left(\nabla \boldsymbol{u} + (\nabla \boldsymbol{u})^T\right)\right] \tag{8}$$

where ρ is the fluid density, ∇ is the differential (Del) operator, $\boldsymbol{u}$ is the flow velocity vector, $\boldsymbol{I}$ is the identity tensor and p is the pressure. The Navier-Stokes equations are solved using the finite element method. We used piecewise-polynomial approximations of equal order to spatially approximate both velocity and pressure [48]. Specifically, piece-wise quadratic (P2 + P2) scheme is chosen to discretize the equations. P2 + P2 scheme uses second-order elements for the velocity field and second-order elements for pressure [48].

To start the algorithm (Step c), we set the initial linear and rotational velocity of the particle equal to the fluid velocities at the exact location of the particle, which was solved in step (b). In fact, these initially estimated velocities are from the that no particle is present in the flow field.

Moving wall boundary condition is applied to channel walls with velocities equal to particle's linear velocity (U_p) in the opposite direction (–x):

$$U_{wall} = -U_p \tag{9}$$

Channel inlet and outlet are set to the same velocity profile given in step (b) and are updated in every iteration using the particle's axial velocity as follows:

$$U_{in} = U_{fluid,\ old} - U_p \tag{10}$$

Particle is treated as a solid sphere with rotating walls:

$$\vec{U} = \vec{\Omega}_p \times \left(\vec{r} - \vec{r}_p\right) \tag{11}$$

where $\left(\vec{r} - \vec{r_p}\right)$ represents the location of a point on the particle surface relative to its center with a flow velocity vector of $\vec{U}$, and $\vec{\Omega_p}$ represents the particle rotational velocity vector.

Particle rotational and axial linear velocities are updated during the calculations until the axial force, and cross-sectional components of momentum exerting on the particle become less than 1×10^{-14} N and 1×10^{-20} N·m, respectively, hence, to be considered as a steady-state condition for the particle. Once convergence occurs, calculations will start over with the particle being placed at another location in the yz-plane. The following formulas are used to update the velocities:

$$U_p'' = U_p' + a_z \times \Delta t \tag{12}$$

$$\Omega_y'' = \Omega_y' + \alpha_y \times \Delta t \tag{13}$$

$$\Omega_z'' = \Omega_z' + \alpha_z \times \Delta t \tag{14}$$

where superscripts indices ($''$) and ($'$) refer to old and new calculated parameters, respectively, U_p is the axial linear velocity of the particle, and a_x, α_y and α_z are the linear and cross-sectional rotational accelerations of the particle, respectively, which can be found using Newton's second law of motion.

After the inertial forces are calculated for a specific particle in the whole cross-section, we can investigate whether the particle can be forced to move through the middle passway by the secondary flow field or not, hence characterizing the ability of the microchannel to separate the target cells.

3.3. Mesh Independency

A mixed structured-unstructured type of mesh was used with the presence of the particle in the microchannel (Figure 3A). The microchannel was split into two distinct parts: (i) a block containing the particle with a cross-section similar to the cross-section of the microchannel in which triangular and tetrahedral grids were used for the surface of the particle and the rest of the zone, respectively, (ii) other zones excluding the particle in which structured grids were used. The inertial lift force exerting on the particle in the y-direction throughout the microchannel cross-section was selected as the mesh independence criteria and various mesh resolutions were considered to evaluate grid independence. The results show that the maximum value for different parameters, including the width of the block containing the particle, size of the tetrahedral grids, and size of the triangular grids on the surface of the particle are $5 \times D_p$, $0.1 \times D_p$, and $0.05 \times D_p$, respectively (Figure 3B), in order to preserve less than 0.1% change in the y-component of the lift force with D_p being the particle diameter.

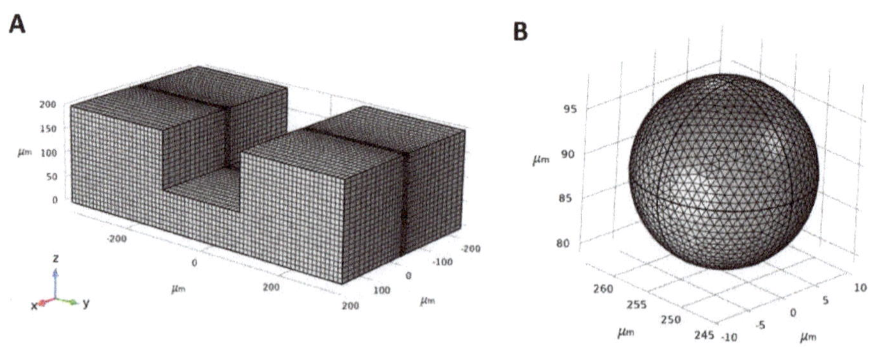

Figure 3. Illustration of the mesh in the fluid domain. (**A**) The domain was separated into two zones in which tetrahedral mesh was used between the particle surface and the microchannel walls with a maximum size of $0.1 \times D_p$, and the structured mesh was used outside the block, including the particle. (**B**) Triangular mesh was used on the surface of the particle with a maximum size of $0.05 \times D_p$.

3.4. Sample Preparation

Human blood samples were collected using a blood collection tube. 10 mL of blood was mixed with 35 μL ethylenediaminetetraacetic acid (EDTA) solution (7 μL per 2 mL of blood) to prevent blood clotting. This study was approved by the Research Ethics Committees of the Iran University of Medical Sciences (ID: IR.IUMS.REC.1400.285), and informed consent was obtained from the subject involved in the study. In addition, all clinical experiments were conducted in accordance with the ethical principles and related guidelines. Due to the non-Newtonian behavior of blood and high concentration of RBCs in the blood, we diluted the starting sample with DI water. The donor's blood was measured to be at 45% hematocrit upon donation. Three tubes of the sample were prepared by mixing the whole blood with DI water with the ratios of 1:45, 1:22.5, and 1:9, therefore, the yielding blood samples were measured to be at 1%, 2%, and 5% hematocrits, respectively.

Although in this case, minor RBC lysis may occur due to hemolysis which is caused by osmotic effects of blood dilution, since the main aim of this research is to prepare WBC samples, this effect is ignored. However, in cases of RBC isolation and RBC sample preparation, the usage of saline solutions (i.e., PBS) is recommended in order to boost and preserve RBC recovery.

3.5. Experimental Approach

Throughout the tests, the prepared blood samples were pumped through the spirals' inner inlets as well as an equal flow of DI water as the sheath flow through the outer inlets using two 10 mL syringes with a syringe pump with flows varying from 2 mL/min up to 8 mL/min. Processed samples were collected from inner and outer outlets in silicone tubes, and cell counting was done using a Neubauer chamber.

A total number of 4 distinct tests were done for each flow rate and sample hematocrit to reduce uncertainty and errors caused by devices and cell enumeration. To get well-balanced results, DI water was set to be pumped into the spiral microchannel to wash it out from the remaining blood cells from previous use before running further tests.

Since this research aims to isolate WBCs from whole blood and remove RBCs and platelets from the target outlet (inner outlet), we introduce the terms 'isolation efficiency' and 'removing ratio' for WBCs and RBCs + platelets, respectively, to evaluate the device performance.

The ratio of the number of WBCs in the inner outlet to the whole number of WBCs in both outlets is reported as the 'WBC isolation efficiency' of the spiral microchannel and the ratio of the number of WBCs to the whole number of blood cells in the inner outlet is reported as 'WBC purity':

$$\%\text{WBC} = \frac{\#\text{ of WBC in the inner outlet}}{\#\text{ of WBC in (inner} + \text{outer) outlets}} \tag{15}$$

$$\%\text{Purity} = \frac{\#\text{ of WBC in the inner outlet}}{\#\text{ of total cells in the inner outlet}} \tag{16}$$

Moreover, the removing ratio of RBCs and platelets are obtained using the equations below:

$$\%\text{RBC} = \frac{\#\text{ of RBC in the outer outlet}}{\#\text{ of RBC in (inner} + \text{outer) outlets}} \tag{17}$$

$$\%\text{Platelets} = \frac{\#\text{ of Platelets in the outer outlet}}{\#\text{ of Platelet in (inner} + \text{outer) outlets}} \tag{18}$$

4. Results and Discussion

4.1. Device Design for Cell Separation

The concept of cell separation in this method is illustrated in Figure 4. The U-shaped cross-section design proposed in this study can keep large particles, such as WBCs, in the inner microchannel section, while forcing smaller cells, like RBCs and platelets, to migrate to the outer microchannel section through the middle passway without recirculating and coming back to the inner section of the microchannel. Additionally, the proposed U-shaped cross-section eliminates the re-circulation and mixing problems of conventional rectangular cross-sections. In order to reach a decent isolation flow rate, two identical spiral microchannels are set as parallel to each other, with inner outlets and outer outlets being merged into a single inner and outer outlet, respectively. At the end of the experiment, WBCs ($a_p \sim 14$ µm) are collected from the inner outlets and RBCs ($a_p \sim 7$ µm) and platelets ($a_p \sim 3$ µm) are collected from outer outlets. Due to the size and flow dependence of the forces exerting on the blood cells, we managed to investigate the effect of flow rate and cell concentration on the WBC isolation efficiency and RBC and platelets removing ratios in the spiral microchannel to find the optimum flow rate.

4.2. Validation of the Numerical Simulations

The microfluidic device proposed by Di Carlo et al. [31] was a straight channel with a 50 µm × 40 µm rectangular cross-section, and the diameter of the particle was 10 µm in a flow with Re = 80. In order to validate our algorithm, inertial lift forces acting on a 10 µm particle were calculated in the same straight microchannel (Figure 5A). Particle equilibrium positions are located in the areas where the force vector approaches zero value, approximately 0.6 times of half channel width. As can be seen in Figure 5B, our simulations are quite consistent with Di Carlo's results [31] and demonstrate the same equilibrium regions. Next, we investigated the ability of the developed algorithm in this study by performing the calculations on the U-shaped cross-section microchannel.

4.3. Simulation of the Secondary Flow in the Spiral Microchannel

In a conventional rectangular cross-section spiral microchannel, pressure gradient results in the formation of the secondary flow consisting of two similar counter-rotating Dean vortices. However, in our geometry, the secondary flow consists of four Dean vortices locating at the top and bottom of the inner and outer channel sections. Dean vortices' strength is increased by further increasing the flow rate and the Dean number and, therefore, the vortex cores start to appear (Figure 6).

Contours of velocity magnitude and the schematic of the flow velocity profile in this study are shown in Figure 6A,B. As can be seen, there are two locations of maximum velocity in the inner and outer section of the microchannel, with the magnitude of maximum velocity in the inner section being higher than the outer one. This difference generates a pressure gradient that causes the secondary flow's streamline to be directed from the inner section toward the outer microchannel section. Thus, the secondary flow prevents the migrated particles from returning to the inner microchannel section through the middle passway. This option removes the possibilities of cell circulation through the entire cross-section and increases the isolation efficiency.

The calculated inertial force field acting on a 10 µm particle in the U-shape channel is shown in Figure 7C. According to the simulation results, 10 µm particles migrate from the center of the microchannel's inner section toward the microchannel walls. As the particles get closer to the walls, a rather large inertial force stops them from further approaching the walls. The particles tend to reach equilibrium positions where the total amount of inertial forces exerting on them is zero; these positions are shown as blue dashed lines.

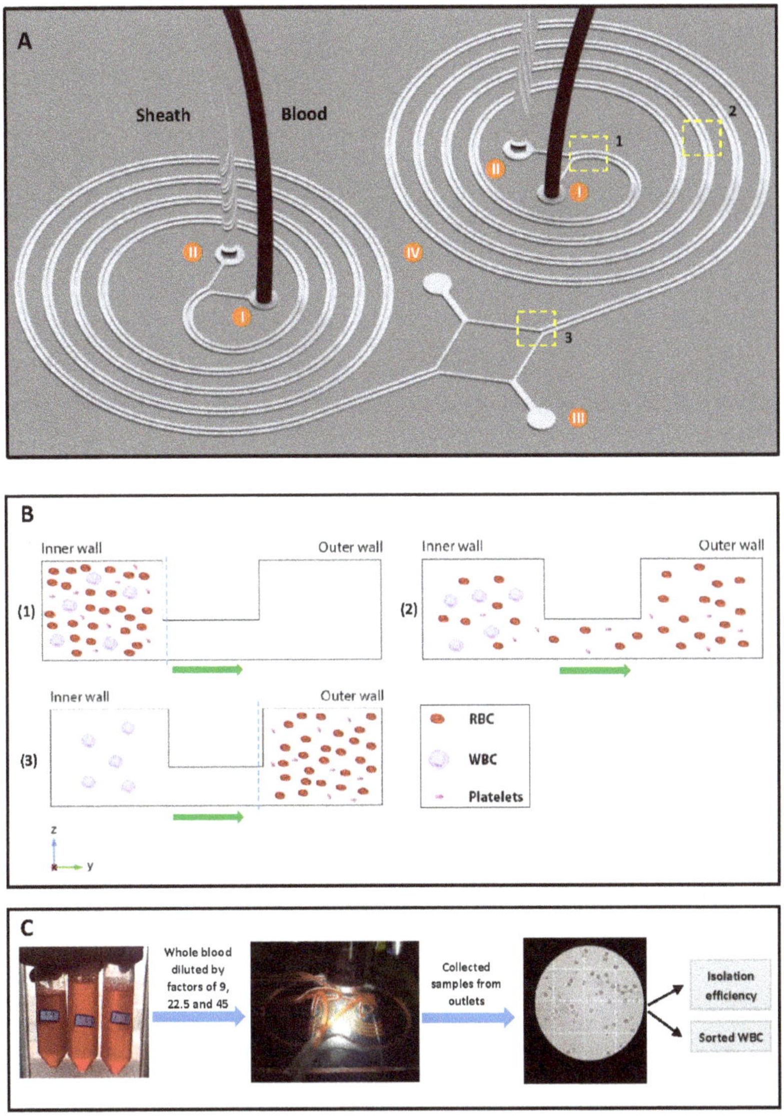

Figure 4. (**A**) 3D schematic of the proposed spiral microchip in this study. The design includes two parallel U-shaped cross-section spiral microchannels that consist of: (I) two inner inlets for the injection of blood samples, (II) two outer inlets for the injection of sheath fluid, (III) shared outer outlet for the extraction of RBCs and platelets, and (IV) shared inner outlet for the extraction of WBCs. (**B**) Larger blood cells, i.e., WBCs, affected by dominant inertial forces are focused along the inner microchannel wall. In comparison, smaller blood cells, i.e., RBCs and platelets, are forced to migrate toward the outer microchannel section by the effect of the secondary flow (flow direction is shown in green arrows), which in turn stops the smaller cells from getting back to the inner microchannel section. (**C**) The overall process of WBC sorting and enumeration. Blood cell count is done using a Neubauer chamber, and the resulting WBC isolation efficiency is calculated using the fractional ratio of blood cells in specific outlets to the total count of blood cells in both outlets.

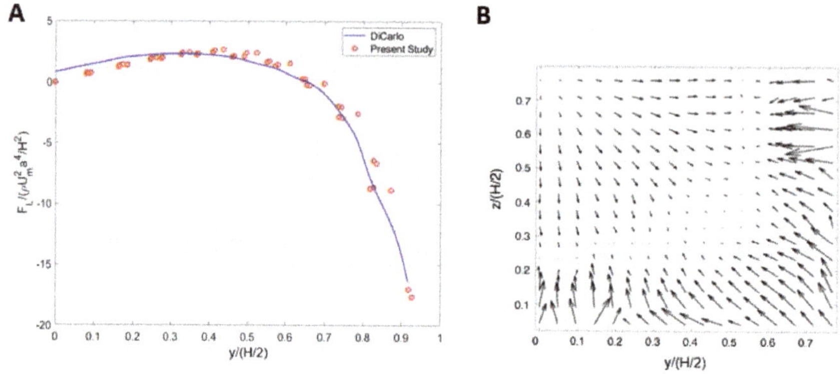

Figure 5. (**A**) Calculated inertial force in a straight channel at z = 0 using the algorithm in the present study versus inertial force reported by Ref. [31]. (**B**) Calculated inertial force field acting on a 10 μm particle in the present study.

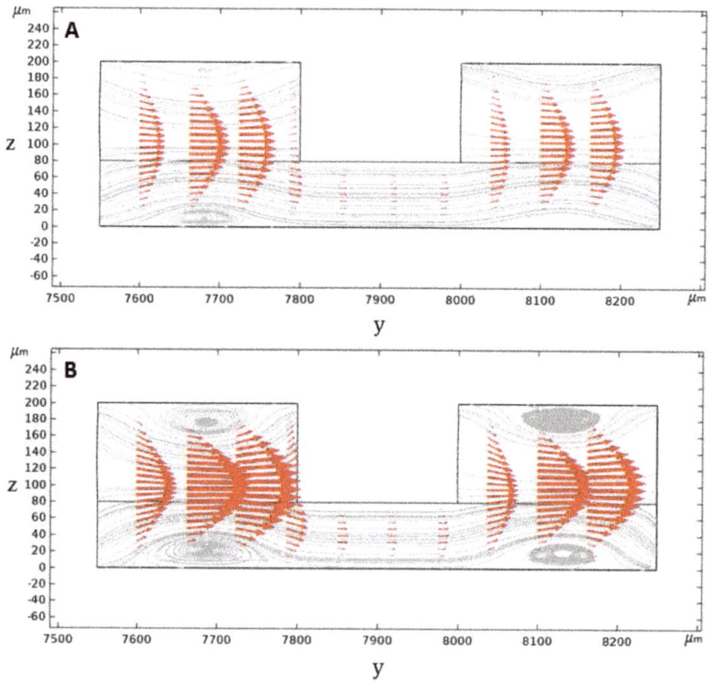

Figure 6. Calculated secondary flow streamlines and velocity vectors in the second loop of the designed spiral microchannel cross-section at x = 0. Velocity vectors are being directed from the inner wall to the outer wall in the middle passway with: (**A**) 2 mL/min, (**B**) 3 mL/min inlet flow rate (Note that in this figure the origin of y-direction lies on the origin of the spiral microchannel).

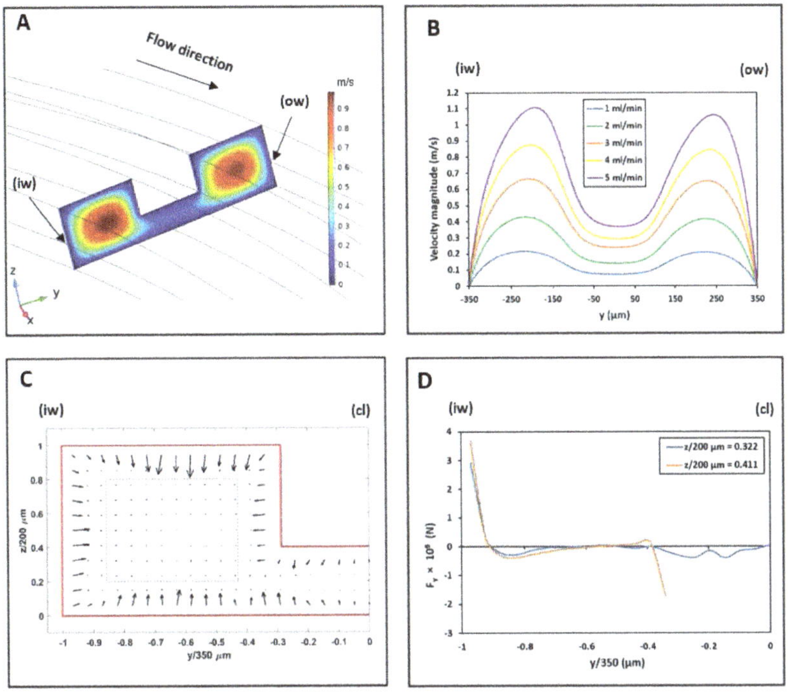

Abbreviations: iw (inner wall), ow (outer wall), cl (centerline).

Figure 7. (**A**) Contour of velocity magnitude in the spiral microchannel at the second loop at x = 0 with an inlet flow rate of 3 mL/min. (**B**) Velocity magnitude profile with the change in inlet flow rates at z = 40 μm (z = 0 indicates the bottom wall of the microchannel and y = 0 indicates the centerline of the microchannel cross-section (cl), this difference in the maximum velocity magnitude results in the streamlines being directed from the inner microchannel wall (iw) toward the outer microchannel wall (ow). (**C**) The calculated inertial force field acting on a 10 μm particle in the inner section of the spiral microchannel with an inlet flow rate of 5 mL/min. Equilibrium positions are shown as blue dashed lines in the microchannel cross-section. (**D**) The inertial force applying to a 10 μm particle alongside y-direction at z = 64.4 μm (z/200 μm = 0.322) and z = 82.2 μm (z/200 μm = 0.411). As can be seen, the inertial force at the beginning of the passway is much lower than the wall-induced inertial force. The Dean drag force can overcome this inertial force, therefore, allowing the particles smaller than 10 μm to move toward the outer section of the spiral microchannel.

The U-shaped microchannel's capability to isolate the particles depends on the value of the exerting the inertial forces and the Dean drag force on the particles near the middle passway. Suppose the Dean drag force overcomes the inertial forces. In that case, the particles migrate to the outer microchannel section through the middle passway and, due to the direction of the streamlines, they cannot come back to the inner section. The inertial lift force profile applying to a 10 μm particle alongside the y-direction is given in two different microchannel heights at the beginning of the middle passway in Figure 7D. According to the calculation results for 5 mL/min flowrate, the Dean drag force is strong enough to overcome the inertial forces for particles smaller than 10 μm, hence moving them to the outer microchannel section. However, for particles larger than 10 μm, the inertial lift force near the passway walls is strong enough to overcome the Dean drag force to keep these particles in the inner section. We did our calculations based on a 10 μm particle to ensure a safe particle diameter margin in which WBCs ~ 14 μm are isolated from smaller particles (RBCs ~ 7 μm and platelets ~ 3 μm) in the designed spiral microchannel.

The optimum flow rate for the isolation of blood cells will be investigated experimentally, which will be discussed in the next sections.

Contours of the total shear stress exerting on a 14 μm particle in the 3 directions can be seen in Figure 8. According to the simulations, the particle is prone to maximum stress in the flow rate of 8 mL/min at the top corner of the cross-section near the microchannel inner wall shortly after the microchip inlet. In the worst-case scenario, the magnitude of the total stress applying to a 14 μm is calculated to be less than 45 Pa. Cell loss and damage due to fluid shear stress is a function of both stress magnitude and exposure time. The less exposure time to a particular shear stress, the less the chances of having cells damaged [49]. The critical magnitude of shear stress in this study for a 14 μm particle which is set to represent a WBC, remains relatively below the known threshold of WBC damaging shear stress [49]. Additionally, the low exposure time of WBCs to these amounts of shear stress, which equals to the amount of time that a single cell manages to pass through the microchannel (about 0.306 s) further lessons the possibility of losing WBCs in the process.

4.4. Presence of the Secondary Flow in the Spiral Microchip

In order to confirm the presence of the secondary flow in the spiral microchannel, two syringes of fluid were prepared. One syringe was filled with a colored fluid, while the other one was filled with DI water. The colored fluid was introduced into the spiral microchannels through the inner inlets, and DI water was set to enter through the outer inlets. The two syringes with 10 mL capacity were driven at 5 mL/min (2.5 mL/min per spiral) ($Re = 81.7$, $De = 11.2$) using a syringe pump, and images of the flow streams were captured at the inlets and outlets of the left side spiral. As shown in Figure 9, the colored fluid filled the entire section of the microchannel at the outlet bifurcation, showing the Dean vortices' effect to mix the flow in both the inner and outer sections of the spiral microchannel.

4.5. The Effect of the Flow Rate

To investigate the effect of the flow rate on the separation of blood cells, blood sample with the hematocrit of 1% was injected into the microchip through the inner inlets with 2–8 mL/min flowrate (with 1 mL/min step), and DI water was injected through the outer inlets as the sheath flow with the same flow rate.

The corresponding Reynolds and Dean number for various testing flow rates for the microchip are listed in Table 1.

As shown in Figure 10A, RBCs' removing ratio reaches a maximum value of 95.7% by increasing the blood sample flow rate up to 6 mL/min and then decreases with further flowrate increment.

The initial expectation would be that for sample flow rates higher than 6 mL/min, small cells do not have enough time (length) to migrate from the inner to outer section due to higher linear velocity. However, further investigations show that increasing the flow rate increases the inertial force more than the Dean drag force. The resultant force stops the majority of the RBCs from migrating through the middle passway, and, hence, the overall efficiency decreases. Additionally, local perturbations in the place of the outlet bifurcation negatively affect the isolation efficiency and removing ratios, which causes the cells to exit through the wrong outlets and further reducing the device's performance.

Unlike RBCs and WBCs in which inertial and Dean drag forces acting on them compete with each other all the time, platelets are always affected by Dean drag force due to their very small diameter. The removing ratio for platelets reaches a maximum value of 94.6% at a flow rate of 4 mL/min and remains almost constant up to 8 mL/min. Since the inertial forces and the Dean drag force are proportional to the 4th and 1st power of the particle diameter, respectively, the value of the inertial forces applying to the platelets are negligible, and they accumulate in the outer section of the spiral microchannel with a high average removing ratio (Figure 10B).

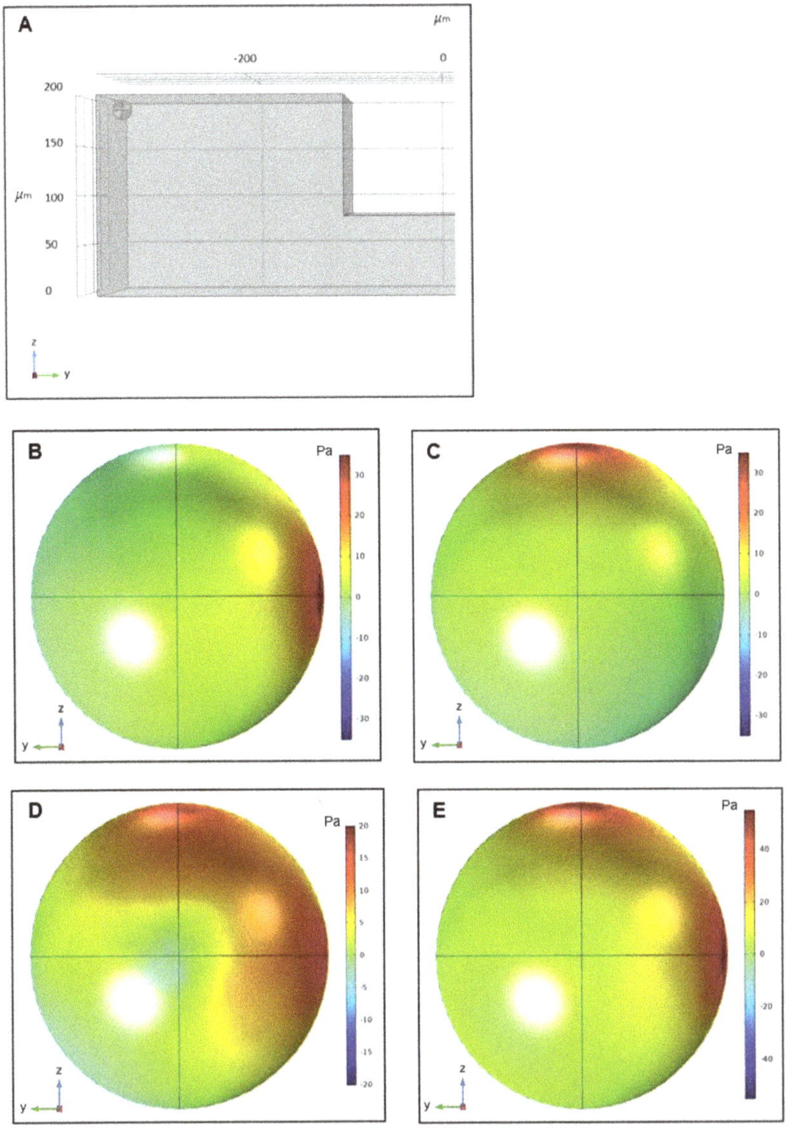

Figure 8. Schematic of (**A**) the critical location for a 14 μm particle in terms of experiencing the maximum shear stress throughout the spiral microchannel, (**B**) the total shear stress exerting on the particle in the x-direction, (**C**) the total shear stress exerting on the particle in the y-direction, (**D**) the total shear stress exerting on the particle in the z-direction, (**E**) the total magnitude of shear stress exerting on the particle. The resulting maximum stress in the x, y, and z-direction was found to be 35 Pa, 35 Pa, and 20 Pa, respectively. Moreover, the maximum value for the magnitude of total shear stress is 45 Pa.

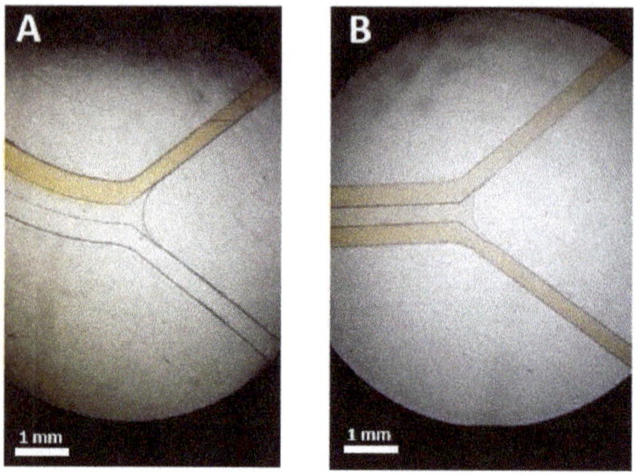

Figure 9. Images of the left side spiral at (**A**) Inlets, (**B**) outlets showing the effect of the secondary flow on mixing the colored fluid throughout the entire microchannel. The colored fluid and DI water were pumped through the microchannel from the inner and outer inlets, respectively.

Table 1. The microchip's specifications in terms of sample flow rate, microchannel Reynolds number, and Dean number.

Sample Flow Rate (mL/min)	Sample Flow Rate Per Spiral (mL/min)	*Re*	*De*
2	1	32.68	4.50
3	1.5	49.02	6.75
4	2	63.36	9.00
5	2.5	81.70	7.47
6	3	98.04	13.50
7	3.5	114.38	15.75
8	4	130.72	18.00

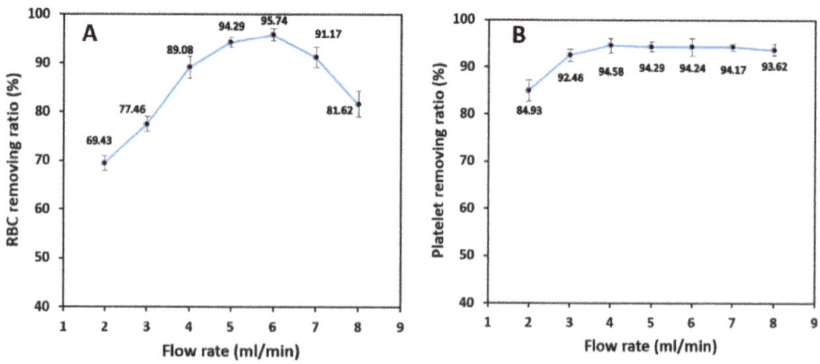

Figure 10. The effect of the sample inlet flow rate on the removing ratio of (**A**) RBCs, (**B**) platelets for 1% blood hematocrit sample. RBC removing ratio is the most sensitive to changes in inlet flow rate with a maximum value at 6 mL/min. The removing ratio of platelets is the least dependent on the flow rate and they can be extracted from the shared outer outlets with high removing ratios.

Test results show that the isolation efficiency of WBCs from whole blood is almost stable up to a flow rate of 6 mL/min with 96.8% maximum efficiency at 5 mL/min. Increasing the sample flow rate has virtually no significant effect on the isolation of WBCs because the inertial forces have the leading role in keeping the cells in the inner section of the spiral microchannel. However, further increase in flow rate from 6 mL/min increases the Dean vortices' strength, which tends to move the white blood cells to the outer section and decreases the WBC isolation efficiency (Figure 11A). According to Figure 11B, WBC purity increases significantly from 44.3% at 2 mL/min to 88.3% at 6 mL/min by increasing the flow rate since the removing ratio of RBCs has also the same trend with respect to flow rate. A further increase in the flow rate results in a fall in purity since RBCs and platelets manage to remain in the inner section within the strong vortices.

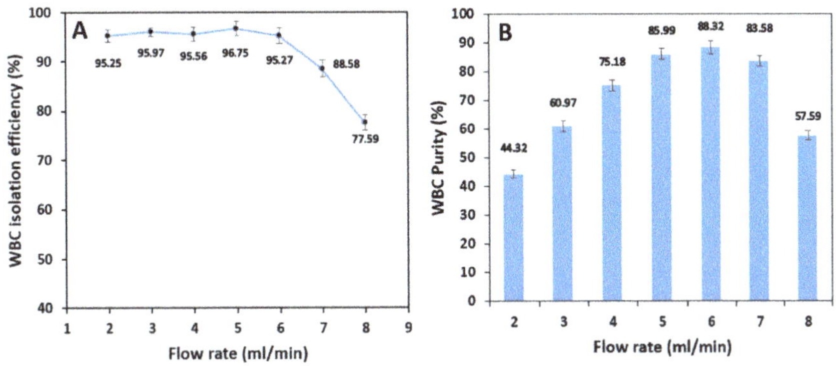

Figure 11. The effect of the sample inlet flow rate on WBC: (**A**) isolation efficiency, and (**B**) purity for a 1% blood hematocrit sample. The WBC isolation efficiency experiences a massive drop in higher flow rates due to the existence of high-strength dean vortices which force the WBCs to migrate to the outer microchannel wall. The outlet samples are the purest with flow rates of 4, 5, and 6 mL/min. The higher the removing ratio of RBCs and platelets, the higher WBC purity.

Since RBC removing ratio is highly dependent on the inlet flow rate, the overall criteria of the success of the spiral microchip on isolating the WBCs from whole blood must be selected based on the best isolation efficiency of WBCs and removing ratio of RBCs at the same time. As a result, for a 1% hematocrit blood sample, our spiral microchip can isolate WBCs from whole blood with 96.8% efficiency at the inlet flow rate of 6 mL/min with a WBC sample purity of 88.3%. Moreover, the removing ratios of RBCs and platelets, in this case, are 95.8% and 94.2%, respectively at this flow rate.

Using flow rates of more than 8 mL/min is not applicable due to cell damage possibilities. Additionally, higher flow rates may break up the PDMS-glass slide bonding and are dangerous for test operators, so continuing the tests with flow rates of more than 8 mL/min was not possible.

4.6. The Effect of Hematocrit

The same methods and tests were applied to blood samples with 2% and 5% hematocrit to find the effect of the hematocrit on the efficiency of cell separation.

As shown in Figure 12A, the general trend of the removing ratio of RBCs remains the same for three tested blood hematocrits. RBC removing ratio peaks at 88.4% and 82.5% at 4 mL/min and 5 mL/min flow rates for 2% and 5% hematocrit, respectively, which is considered as a significant drop in the RBC removing ratio compared with 95.7% of 1% hematocrit blood sample at 6 mL/min flow rate. At relatively low flow rates, the difference in removing ratios is negligible for three hematocrits. However, in the 3–7 mL/min flow rate band, the ratio drops with the increase in hematocrit, and this difference tends to grow by increasing the flow rate up to 7 mL/min. The decrease in removing ratio with an increase in hematocrit is due to the higher concentrations of RBCs in higher blood hematocrits that raise the effect of cell-cell interaction force alongside the inertial forces and Dean drag force, which further degrades the overall removing ratio. Platelets typically focus in the outer microchannel section and can be extracted from the outer outlet at relatively high removing ratios Figure 12B. The effect of hematocrit on the isolation is relatively low in this case. It peaks at a value of 93.8% at 5 mL/min and 91.6% at 4 mL/min for 2% and 5% blood hematocrit, respectively, which is pretty much close to 94.2% extraction at 6 mL/min for 1% blood hematocrit. The reason is that the platelets are quite small, and cell-cell interactions are negligible. Hence, the primary force acting on the platelets is the secondary Dean drag force which carries them to the outer section of the spiral microchannel through the middle passway.

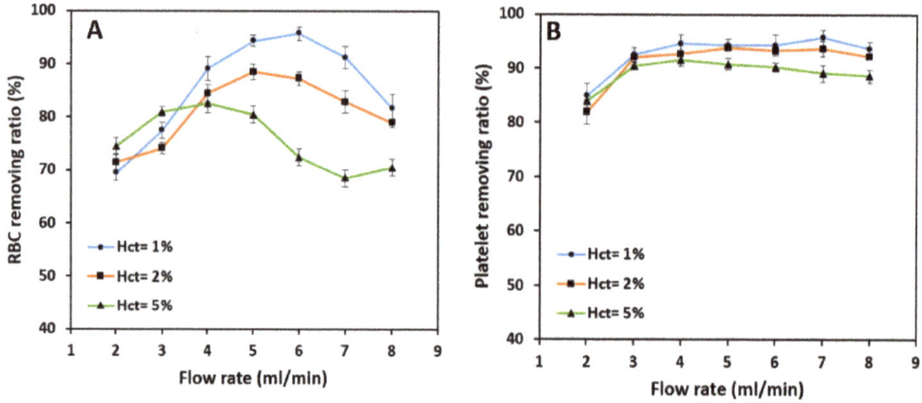

Figure 12. The effect of blood hematocrit on the removing ratio of (**A**) RBC, (**B**) Platelets. The removing ratio mainly decreases with an increase in blood hematocrit due to a rise in the effect of cell-cell interaction. While RBC removing ratio is highly dependent on blood hematocrit, the removing ratio of platelets does not seem to change very much with a change in blood hematocrit.

According to Figure 13A, the overall isolation efficiency of WBCs decreases as hematocrit increases. The efficiency starts at relatively high amounts at the flow rate of 2 mL/min; i.e., 92% and 90.2% for 2% and 5% hematocrits which are lower than the 95.2% efficiency value reached by 1% blood hematocrit.

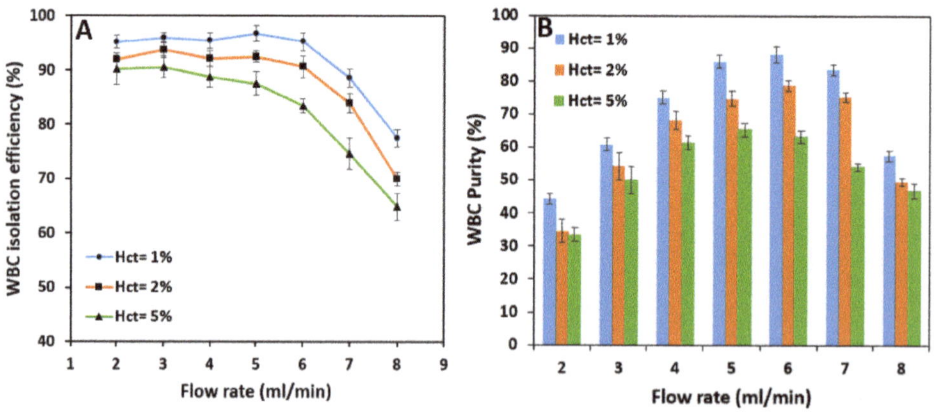

Figure 13. The effect of blood hematocrit on WBCs': (**A**) isolation efficiency, (**B**) purity. Separation efficiency mainly decreases with an increase in blood hematocrit due to the rise in the effect of cell-cell interaction in higher flow rates. Sample purities for lower hematocrits are mainly higher since to amount of RBCs in the diluted sample is lower. WBC purity is higher at the inlet flow rates of 5, 6, and 7 mL/min due to an increase in RBCs removing ratio.

As the flow rate increases, the reduction in isolation efficiency becomes noticeable for higher hematocrits. In terms of purity, higher sample hematocrits are less pure than the sample with 1% hematocrit due to the fact that more RBCs and platelets are present in these samples. The WBC outlet sample has a maximum purity of 78.8% at 6 mL/min and 65.5% at 5 mL/min flow rate for 2% and 5% hematocrits, respectively (Figure 13B). These figures are pretty much lower than the maximum purity for 1% hematocrit which is 88.3%. Therefore, increasing the sample hematocrit has a negative effect on WBC purity.

The three best overall performances of the microchip tested at different flow rates and hematocrits are listed and summarized in Table 2.

Table 2. Summary of the spiral microchip best performance results in the isolation of WBC, RBC, and Platelets based on the inlet sample flow rate and hematocrit.

		Sample Hematocrit		
		1%	**2%**	**5%**
	Flow rate (mL/min)	6	5	4
WBC	**Purity (%)**	88.3	74.8	61.5
	Isolation efficiency (%)	95.3	92.5	88.8
RBC	**Removing ratio (%)**	95.7	88.4	82.5
Platelet	**Removing ratio (%)**	94.2	93.8	91.6

With all things considered, optimum results are achieved under a flow rate of 6 mL/min for a 1% hematocrit blood sample. The illustration of flow and collected blood samples from the microchip outlets are shown in Figure 14.

The isolation of WBCs from whole blood has been the main purpose of our spiral microchip. Since the RBC removing ratio is highly sensitive to inlet flow rates, even a slight improvement in removing the red blood cells from the target outlet is a success. Because any sample contains numerous RBCs compared to other blood components (about 1 WBC to every 600 RBC), we managed to select the best performance based on the highest value for RBC removing ratio. Yet, we had to sacrifice a small amount of WBC isolation efficiency.

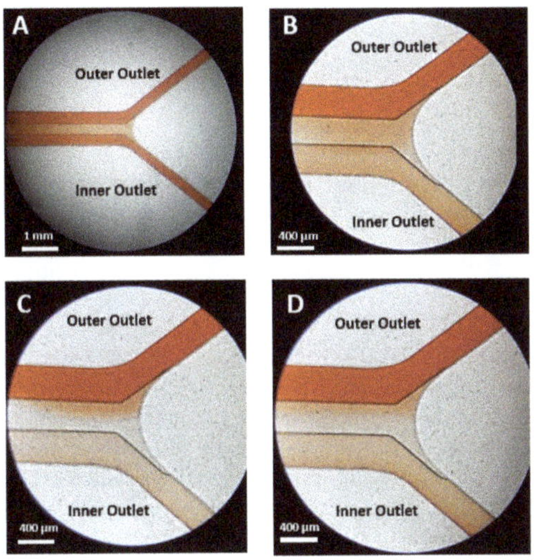

Figure 14. *Cont.*

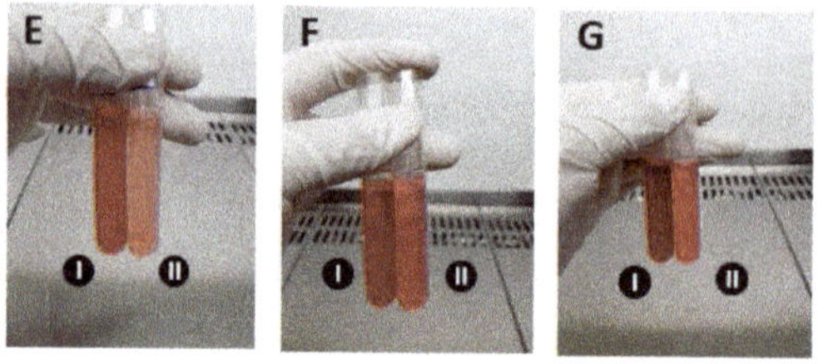

Figure 14. Illustration of the flow in outlet bifurcation in the left spiral for 1% hematocrit blood sample with (**A**) 2 mL/min (**B**) 4 mL/min (**C**) 6 mL/min and (**D**) 7 mL/min flow rates and collected blood samples with (**E**) 1%, (**F**) 2% and (**G**) 5% hematocrit for 6 mL/min flow rate from microchip's: (I) outer outlets and (II) inner outlets. As can be seen, for constant sample hematocrit, increasing the flow rate reduces the concentration of RBCs in the inner outlets, which results in an increase in WBC WBC and isolation efficiency and RBC removing ratio. For constant flow rates, reducing the sample hematocrit (i.e., higher dilution factors) further reduces the concentration of RBCs in the inner outlets, thus, the collected sample from the inner outlet looks more transparent in comparison to higher sample hematocrits.

The numerical simulations showed that a 5 mL/min flow rate could force particles smaller than 10 μm to migrate to the outer microchannel section while holding the particles larger than 10 μm in the inner section. Although the simulation settings are for the case of a single solid particle dispersed in a Newtonian fluid without the interference of other particles, the experimental reported results are reasonably acceptable for 1% hematocrit at 5–6 mL/min flow rate.

5. Conclusions

This paper proposed a passive cell separating method using two parallel U-shaped spiral microchannels for size-based, label-free and continuous separation of WBCs from whole blood based on cell inertial migration. The spiral microchannel with U-shaped cross-section increases WBC isolation efficiency by avoiding the re-circulation of smaller cells (RBCs and platelets) through the entire microchip, thus eliminating the major drawback of the conventional rectangular spiral microchannel. Numerical simulations were performed to study the flow behavior, secondary flow distribution, and particle migration mechanism in the proposed design. An algorithm was developed to calculate the hydrodynamic forces exerting on particles present in the U-shaped spiral microchannel to further predict the design's performance before proceeding to the fabrication of the microchip as well as checking the possibility of WBC damage due to the applying shear stress to them. Subsequently, we successfully isolated WBCs from whole blood in our experiments and investigated the effect of flow rate and hematocrit on the WBC isolation efficiency and RBC and platelet removing ratios. The best performance can be achieved for a 6 mL/min flow rate with 95.3% and 88.3% isolation efficiency and purity for WBCs, respectively, and removing ratios of 95.7%, and 94.2% for RBCs, and platelets, respectively, for a 1% hematocrit blood sample. Also, our device can process an amount of 360 mL of 1% hematocrit blood sample in 1 h under its best condition which is considered to be higher than most of the conventional cell separating microfluidics platforms. Additionally, the proposed spiral microchip allows a high-throughput, high-efficiency, non-invasive, size-based label-free WBC separation from whole blood without the assistance of RBC lysis, density gradient centrifugation, cell biomarkers, and chemical treatments on the sample. Most importantly, it can be integrated into more complex microfluidic platforms.

Moreover, we successfully tested the double parallel spiral microchannels. We reached decent isolation results that are already superior to most of the microfluidic cell sorting platforms in case of fast processing time. There is still room for improvement in microchip throughput, which can be done by parallelizing more spiral microchannels while maintaining high performance.

Author Contributions: Conceptualization, A.M.; methodology, A.M.; software, A.M. and P.R.; validation, P.R.; investigation and resources, A.M. and M.S.S.; writing—original draft preparation, A.M. and P.R.; writing—review and editing, A.M., N.K. and B.F.; visualization, A.M. and N.K.; supervision, M.S.S., B.F. and N.K.; funding acquisition, M.S.S. All authors have read and agreed to the published version of the manuscript.

Funding: This research received no external funding.

Institutional Review Board Statement: The study was conducted according to the guidelines of the Declaration of Helsinki and approved by the Research Ethics Committees of Iran University of Medical Sciences (ID: IR.IUMS.REC.1400.285).

Informed Consent Statement: Informed consent was obtained from all subjects involved in the study.

Data Availability Statement: The datasets used and/or analyzed during the current study are available from the corresponding author on reasonable request.

Conflicts of Interest: The authors declare no conflict of interest.

References

1. Kumar, D.; Jain, N.; Khurana, A.; Mittal, S.; Satapathy, S.C.; Senkerik, R.; Hemanth, J.D. Automatic Detection of White Blood Cancer From Bone Marrow Microscopic Images Using Convolutional Neural Networks. *IEEE Access* **2020**, *8*, 142521–142531. [CrossRef]
2. Braun, D.P.; Harris, J.E. Relationship of Leukocyte Numbers, Immunoregulatory Cell Function, and Phytohemagglutinin Responsiveness in Cancer Patients23. *JNCI J. Natl. Cancer Inst.* **1981**, *67*, 809–814. [CrossRef]
3. Nairn, J.; Hodge, G.; Henning, P. Changes in leukocyte subsets: Clinical implications for children with chronic renal failure. *Pediatr. Nephrol.* **2005**, *20*, 190–196. [CrossRef]
4. Gossett, D.R.; Weaver, W.M.; Mach, A.J.; Hur, S.C.; Tse, H.T.; Lee, W.; Amini, H.; Di Carlo, D. Label-free cell separation and sorting in microfluidic systems. *Anal. Bioanal. Chem.* **2010**, *397*, 3249–3267. [CrossRef]
5. Rostami, P.; Kashaninejad, N.; Moshksayan, K.; Saidi, M.S.; Firoozabadi, B.; Nguyen, N.-T. Novel approaches in cancer management with circulating tumor cell clusters. *J. Sci. Adv. Mater. Devices* **2019**, *4*, 1–18. [CrossRef]
6. Tajik, P.; Saidi, M.S.; Kashaninejad, N.; Nguyen, N.-T. Simple, Cost-Effective, and Continuous 3D Dielectrophoretic Microchip for Concentration and Separation of Bioparticles. *Ind. Eng. Chem. Res.* **2020**, *59*, 3772–3783. [CrossRef]
7. Lee, G.-H.; Kim, S.-H.; Ahn, K.; Lee, S.-H.; Park, J.Y. Separation and sorting of cells in microsystems using physical principles. *J. Micromech. Microeng.* **2015**, *26*, 013003. [CrossRef]
8. Jeon, H.; Kim, Y.; Lim, G. Continuous particle separation using pressure-driven flow-induced miniaturizing free-flow electrophoresis (PDF-induced μ-FFE). *Sci. Rep.* **2016**, *6*, 19911. [CrossRef]
9. Kang, J.H.; Krause, S.; Tobin, H.; Mammoto, A.; Kanapathipillai, M.; Ingber, D.E. A combined micromagnetic-microfluidic device for rapid capture and culture of rare circulating tumor cells. *Lab Chip* **2012**, *12*, 2175–2181. [CrossRef]
10. Moon, H.-S.; Kwon, K.; Kim, S.-I.; Han, H.; Sohn, J.; Lee, S.; Jung, H.-I. Continuous separation of breast cancer cells from blood samples using multi-orifice flow fractionation (MOFF) and dielectrophoresis (DEP). *Lab Chip* **2011**, *11*, 1118–1125. [CrossRef]
11. Evander, M.; Johansson, L.; Lilliehorn, T.; Piskur, J.; Lindvall, M.; Johansson, S.; Almqvist, M.; Laurell, T.; Nilsson, J. Noninvasive Acoustic Cell Trapping in a Microfluidic Perfusion System for Online Bioassays. *Anal. Chem.* **2007**, *79*, 2984–2991. [CrossRef]
12. Hejazian, M.; Li, W.; Nguyen, N.T. Lab on a chip for continuous-flow magnetic cell separation. *Lab Chip* **2015**, *15*, 959–970. [CrossRef]
13. McGrath, J.; Jimenez, M.; Bridle, H. Deterministic lateral displacement for particle separation: A review. *Lab Chip* **2014**, *14*, 4139–4158. [CrossRef]
14. Yamada, M.; Nakashima, M.; Seki, M. Pinched Flow Fractionation: Continuous Size Separation of Particles Utilizing a Laminar Flow Profile in a Pinched Microchannel. *Anal. Chem.* **2004**, *76*, 5465–5471. [CrossRef]
15. Oakey, J.; Allely, J.; Marr, D.W.M. Laminar-Flow-Based Separations at the Microscale. *Biotechnol. Prog.* **2002**, *18*, 1439–1442. [CrossRef]
16. Di Carlo, D.; Irimia, D.; Tompkins, R.G.; Toner, M. Continuous inertial focusing, ordering, and separation of particles in microchannels. *Proc. Natl. Acad. Sci. USA* **2007**, *104*, 18892. [CrossRef]
17. Sugiyama, D.; Teshima, Y.; Yamanaka, K.; Briones-Nagata, M.P.; Maeki, M.; Yamashita, K.; Takahashi, M.; Miyazaki, M. Simple density-based particle separation in a microfluidic chip. *Anal. Methods* **2014**, *6*, 308–311. [CrossRef]
18. Yoon, Y.; Kim, S.; Lee, J.; Choi, J.; Kim, R.K.; Lee, S.J.; Sul, O.; Lee, S.B. Clogging-free microfluidics for continuous size-based separation of microparticles. *Sci. Rep.* **2016**, *6*, 26531. [CrossRef] [PubMed]
19. Holmes, D.; Whyte, G.; Bailey, J.; Vergara-Irigaray, N.; Ekpenyong, A.; Guck, J.; Duke, T. Separation of blood cells with differing deformability using deterministic lateral displacement†. *Interface Focus* **2014**, *4*, 20140011. [CrossRef] [PubMed]
20. Segré, G.; Silberberg, A. Radial Particle Displacements in Poiseuille Flow of Suspensions. *Nature* **1961**, *189*, 209–210. [CrossRef]

21. Segré, G.; Silberberg, A. Behaviour of macroscopic rigid spheres in Poiseuille flow Part 1. Determination of local concentration by statistical analysis of particle passages through crossed light beams. *J. Fluid Mech.* **1962**, *14*, 115–135. [CrossRef]
22. Vasseur, P.; Cox, R.G. The lateral migration of a spherical particle in two-dimensional shear flows. *J. Fluid Mech.* **1976**, *78*, 385–413. [CrossRef]
23. Feuillebois, F. Some theoretical results for the motion of solid spherical particles in a viscous fluid. *Multiph. Sci. Technol.* **1989**, *4*, 583–789. [CrossRef]
24. Dijkshoorn, J.P.; Schutyser, M.A.I.; Wagterveld, R.M.; Schroën, C.G.P.H.; Boom, R.M. A comparison of microfiltration and inertia-based microfluidics for large scale suspension separation. *Sep. Purif. Technol.* **2017**, *173*, 86–92. [CrossRef]
25. Chun, B.; Ladd, A.J.C. Inertial migration of neutrally buoyant particles in a square duct: An investigation of multiple equilibrium positions. *Phys. Fluids* **2006**, *18*, 031704. [CrossRef]
26. Kim, Y.W.; Yoo, J.Y. The lateral migration of neutrally-buoyant spheres transported through square microchannels. *J. Micromech. Microeng.* **2008**, *18*, 065015. [CrossRef]
27. Bhagat, A.A.S.; Kuntaegowdanahalli, S.S.; Papautsky, I. Enhanced particle filtration in straight microchannels using shear-modulated inertial migration. *Phys. Fluids* **2008**, *20*, 101702. [CrossRef]
28. Bhagat, A.A.S.; Kuntaegowdanahalli, S.S.; Papautsky, I. Continuous particle separation in spiral microchannels using dean flows and differential migration. *Lab Chip* **2008**, *8*, 1906–1914. [CrossRef]
29. Warkiani, M.E.; Khoo, B.L.; Wu, L.; Tay, A.K.P.; Bhagat, A.A.S.; Han, J.; Lim, C.T. Ultra-fast, label-free isolation of circulating tumor cells from blood using spiral microfluidics. *Nat. Protoc.* **2016**, *11*, 134–148. [CrossRef] [PubMed]
30. Rafeie, M.; Zhang, J.; Asadnia, M.; Li, W.; Warkiani, M.E. Multiplexing slanted spiral microchannels for ultra-fast blood plasma separation. *Lab Chip* **2016**, *16*, 2791–2802. [CrossRef]
31. Di Carlo, D.; Edd, J.F.; Humphry, K.J.; Stone, H.A.; Toner, M. Particle Segregation and Dynamics in Confined Flows. *Phys. Rev. Lett.* **2009**, *102*, 094503. [CrossRef]
32. Martel, J.M. Particle Focusing in Microchannels. Ph.D. Thesis, Harvard University, Cambridge, MA, USA, 2014.
33. Sun, J.; Li, M.; Liu, C.; Zhang, Y.; Liu, D.; Liu, W.; Hu, G.; Jiang, X. Double spiral microchannel for label-free tumor cell separation and enrichment. *Lab Chip* **2012**, *12*, 3952–3960. [CrossRef]
34. Guan, G.; Wu, L.; Bhagat, A.A.; Li, Z.; Chen, P.C.Y.; Chao, S.; Ong, C.J.; Han, J. Spiral microchannel with rectangular and trapezoidal cross-sections for size based particle separation. *Sci. Rep.* **2013**, *3*, 1475. [CrossRef]
35. Syed, M.S.; Marquis, C.; Taylor, R.; Warkiani, M.E. A two-step microengineered system for high-density cell retention from bioreactors. *Sep. Purif. Technol.* **2021**, *254*, 117610. [CrossRef]
36. Shen, S.; Tian, C.; Li, T.; Xu, J.; Chen, S.-W.; Tu, Q.; Yuan, M.-S.; Liu, W.; Wang, J. Spiral microchannel with ordered micro-obstacles for continuous and highly-efficient particle separation. *Lab Chip* **2017**, *17*, 3578–3591. [CrossRef]
37. Ghadami, S.; Kowsari-Esfahan, R.; Saidi, M.S.; Firoozbakhsh, K. Spiral microchannel with stair-like cross section for size-based particle separation. *Microfluid. Nanofluid.* **2017**, *21*, 115. [CrossRef]
38. Di Carlo, D. Inertial microfluidics. *Lab Chip* **2009**, *9*, 3038–3046. [CrossRef]
39. Matas, J.-P.; Morris, J.F.; Guazzelli, É. Inertial migration of rigid spherical particles in Poiseuille flow. *J. Fluid Mech.* **2004**, *515*, 171–195. [CrossRef]
40. Matas, J.-P.; Morris, J.; Guazzelli, E. Lateral Forces on a Sphere. *Oil Gas Sci. Technol.* **2004**, *59*, 59. [CrossRef]
41. Asmolov, E.S. The inertial lift on a spherical particle in a plane Poiseuille flow at large channel Reynolds number. *J. Fluid Mech.* **1999**, *381*, 63–87. [CrossRef]
42. Dean, W.R. XVI. Note on the motion of fluid in a curved pipe. *Lond. Edinb. Dublin Philos. Mag. J. Sci.* **1927**, *4*, 208–223. [CrossRef]
43. Dean, W.R. The streamline motion of fluid in a curved pipe. *Phil. Mag.* **1928**, *5*, 673–693. [CrossRef]
44. Bird, R.B.; Edwin, W.E.S.; Lightfoot, N. *Transport Phenomena*, 2nd ed.; John Wiley & Sons: Hoboken, NJ, USA, 2001; p. 920.
45. Ookawara, S.; Higashi, R.; Street, D.; Ogawa, K. Feasibility study on concentration of slurry and classification of contained particles by microchannel. *Chem. Eng. J.* **2004**, *101*, 171–178. [CrossRef]
46. Barisam, M.; Saidi, M.S.; Kashaninejad, N.; Vadivelu, R.; Nguyen, N.-T. Numerical Simulation of the Behavior of Toroidal and Spheroidal Multicellular Aggregates in Microfluidic Devices with Microwell and U-Shaped Barrier. *Micromachines* **2017**, *8*, 358. [CrossRef] [PubMed]
47. Sheidaei, Z.; Akbarzadeh, P.; Kashaninejad, N. Advances in numerical approaches for microfluidic cell analysis platforms. *J. Sci. Adv. Mater. Devices* **2020**, *5*, 295–307. [CrossRef]
48. Liu, J.-G.; Liu, J.; Pego, R.L. Stable and accurate pressure approximation for unsteady incompressible viscous flow. *J. Comput. Phys.* **2010**, *229*, 3428–3453. [CrossRef]
49. Suwannaphan, T.; Srituravanich, W.; Sailasuta, A.; Piyaviriyakul, P.; Bhanpattanakul, S.; Jeamsaksiri, W.; Sripumkhai, W.; Pimpin, A. Investigation of Leukocyte Viability and Damage in Spiral Microchannel and Contraction-Expansion Array. *Micromachines* **2019**, *10*, 772. [CrossRef]

biosensors

Article

Enhanced Blood Plasma Extraction Utilising Viscoelastic Effects in a Serpentine Microchannel

Yuchen Dai [1], Haotian Cha [1], Michael J. Simmonds [2], Hedieh Fallahi [1], Hongjie An [1], Hang T. Ta [3], Nam-Trung Nguyen [1], Jun Zhang [1,*] and Antony P. McNamee [2,*]

1 Queensland Micro-Nanotechnology Centre, Griffith University, Nathan, QLD 4111, Australia; y.dai@griffith.edu.au (Y.D.); haotian.cha@griffithuni.edu.au (H.C.); hedieh.fallahi@griffithuni.edu.au (H.F.); hongjie.an@griffith.edu.au (H.A.); nam-trung.nguyen@griffith.edu.au (N.-T.N.)
2 Biorheology Research Laboratory, Menzies Health Institute Queensland, Griffith University, Gold Coast, QLD 4222, Australia; m.simmonds@griffith.edu.au
3 School of Environment and Science, Griffith University, Nathan, QLD 4111, Australia; h.ta@griffith.edu.au
* Correspondence: jun.zhang@griffith.edu.au (J.Z.); a.mcnamee@griffith.edu.au (A.P.M.)

Abstract: Plasma extraction from blood is essential for diagnosis of many diseases. The critical process of plasma extraction requires removal of blood cells from whole blood. Fluid viscoelasticity promotes cell migration towards the central axis of flow due to differences in normal stress and physical properties of cells. We investigated the effects of altering fluid viscoelasticity on blood plasma extraction in a serpentine microchannel. Poly (ethylene oxide) (PEO) was dissolved into blood to increase its viscoelasticity. The influences of PEO concentration, blood dilution, and flow rate on the performance of cell focusing were examined. We found that focusing performance can be significantly enhanced by adding PEO into blood. The optimal PEO concentration ranged from 100 to 200 ppm with respect to effective blood cell focusing. An optimal flow rate from 1 to 15 μL/min was determined, at least for our experimental setup. Given less than 1% haemolysis was detected at the outlets in all experimental combinations, the proposed microfluidic methodology appears suitable for applications sensitive to haemocompatibility.

Keywords: blood plasma separation; microfluidics; viscoelasticity; particle migration

Citation: Dai, Y.; Cha, H.; Simmonds, M.J.; Fallahi, H.; An, H.; Ta, H.T.; Nguyen, N.-T.; Zhang, J.; McNamee, A.P. Enhanced Blood Plasma Extraction Utilising Viscoelastic Effects in a Serpentine Microchannel. *Biosensors* **2022**, *12*, 120. https://doi.org/10.3390/bios12020120

Received: 22 December 2021
Accepted: 12 February 2022
Published: 14 February 2022

1. Introduction

Blood is a two-phase suspension of red blood cells (RBCs), white blood cells (WBCs), and platelets suspended in plasma. Given the high protein content, including serum albumin, globulins, and fibrinogen, within plasma, this fluid is often separated from whole blood and used for analytical purposes, including in vitro diagnostics. Blood plasma also contains glucose, electrolytes, hormones, carbon dioxide, and oxygen, so it is crucial for the intravascular osmotic effect which maintains the balance of electrolyte concentration and protects the body from infection and other blood diseases. However, the presence of cellular components inhibits diagnostic sensitivity, accuracy, and reproducibility. Hence, it is of great importance to remove blood cells from the whole blood and extract plasma for medical applications. Two conventional methods, namely centrifugal separation and membrane filtration, have become standard practice over the last century. Although the centrifugal technique is widely used in biological labs, such a device is bulky and involves a batch-based separation technique, thus it is limited for integration with other units to form a fully automated system for point of care diagnosis. Moreover, inappropriate centrifugation, despite the low likelihood, may cause haemolysis of RBCs, leading to contamination risks and misdiagnosis due to handling errors [1]. On the other hand, membrane filtration technique has finer structural details, but clogging and fouling issues inherent in this mode severely deteriorate filtration performance, especially for samples with high cellular content [2].

Microfluidic techniques refer to the systems that process fluids in channels with typical dimensions (width and height) in the order of 10 to 100 μm. The corresponding characteristic length scale enables precise manipulation of fluid and particles, such as cellular components of blood, and thus facilitates the purity and efficiency requirements for plasma separation. Microfluidic technology has consequently become a promising candidate for plasma separation in recent years [3–5].

Based on the origin of manipulating forces, microfluidics can be classified as either active or passive techniques. The former utilises external active forces, such as dielectrophoresis [6], magnetophoresis [7], or acoustophoresis [8], while passive methods exploit the intrinsic fluid dynamics of a sample [9,10]. Of particular relevance for passive microfluidic techniques are inertial microfluidics, which permit simple fabrication, low cost of development, and high throughput [11,12]. Note that within passive systems, the suspending medium is commonly a Newtonian fluid with a constant viscosity.

More recently, the migration of particles immersed in viscoelastic fluids has attracted increased attention due to the simple focusing equilibrium position caused by the intrinsic nonlinear elastic properties of viscoelastic fluids [13–17]. Leshansky et al. [18] observed that, when particles were suspended in viscoelastic fluids, flow within straight circular channels with negligible inertial conditions led to particle migration towards the centreline by the normal stress difference between the axis and the wall. In contrast, Yang et al. [19] reported that flow within straight rectangular channels of a viscoelastic suspension led to particles with different sizes migrating towards the corners, in addition to the centreline. The latter finding is more practical since microfluidic channels are mainly made in planar geometries, i.e., rectangular cross sections, due to standard photolithography and soft lithography. Based on the size-dependent migration in rectangular channels, viscoelastic fluids have been previously used to separate platelets from diluted whole blood [20], *E. coli* bacteria from RBCs [21], malaria parasites from WBCs [22], Jurkat cells from yeast cells [23], and Chlorella microalgae from *B. subtilis* bacteria [24]. In addition, Lu et al. [25,26] experimentally found that particles can be sorted by their shape/morphology in viscoelastic microfluidics [27–29]. Later, based on the simple shape-dependent migration principle, viscoelastic fluids were adopted to separate abnormal-shaped yeast cells from those that were regular shaped [30], incubated *Candida albicans* with germ tube formations from spherical candida cells [31], and cyanobacterial Anabaena with different rod aspect ratios [32]. Moreover, Yang et al. [33] demonstrated that elasto-inertial particles can also be sorted by their deformability, successfully isolating rigidified RBCs from fresh RBCs.

In contrast to sorting different cells into different locations, it is of more interest to focus all the cells into a single equilibrium position in terms of blood plasma separation. In Yang et al. [19], it was experimentally demonstrated that the number of equilibrium positions can be reduced to one (centreline) even in rectangular channels by increasing the fluid flow inertial force without altering any other condition. Additionally, secondary flows created in curved channels or straight channels with disturbance obstacles were found to facilitate migration and modify the equilibrium positions depending on the balance between the inertial lift and Dean drag forces [34,35]. Yuan et al. [36] then combined both Dean-flow and elasto-inertial effects, and successfully demonstrated the separation of blood plasma in a straight channel with asymmetrical expansion–contraction cavity arrays (ECCA channel); the purity of separated plasma was as high as 99.99%. However, the blood was highly diluted (~20 times) and the effect of blood dilution on plasma separation performance was not studied. Moreover, the haemocompatibility of the added viscoelasticity on blood processing (e.g., the intactness of RBCs) remains unknown.

In this work, we systematically investigated viscoelastic blood plasma extraction in a serpentine microchannel. Whole blood was diluted using a viscoelastic phosphate-buffered saline (PBS) solution dissolved with poly (ethylene oxide) (PEO). The influences of PEO solution concentration, blood dilution, and the inertial force on the blood plasma separation performance were investigated. Finally, the haemocompatibility of viscoelastic microfluidic blood plasma separation was examined by measuring levels of free haemoglobin within the separated plasma.

2. Theoretical Background

Among the theories of particle migration in Newtonian fluids, Asmolov's solution [37] has been widely used to determine the lateral forces (F_L) exerted on particles immersed within Newtonian fluids in inertial flow [38], as follows:

$$F_L = \frac{\rho_f U_m^2 a^4}{D_h^2} f_L(Re_c, x_c)$$

where ρ_f, U_m, a, D_h denote the fluid density, mean velocity, spherical diameter of the particles, and hydraulic diameter of the channel, respectively. The lift coefficient $f_L(Re_c, x_c)$ is a function of the particle position within the cross section x_c and the channel Reynolds number Re_c, which is expressed as:

$$Re_c = \frac{\rho_f U_m D_h}{\mu_f}$$

where μ_f represents the dynamic viscosity of the fluid.

In the presence of non-Newtonian viscoelastic fluids, an extra elastic force is exerted on particles. The elasticity can be characterised by Weissenberg number (Wi) in rectangular channel flows as [39]:

$$Wi = \lambda \dot{\gamma}_c = \frac{2\lambda Q}{hw^2}$$

where λ, $\dot{\gamma}_c$, Q denote the relaxation time of the fluid, characteristic shear rate and volumetric flow rate, respectively. For elasto-inertial flows in rectangular channels, both the first ($N_1 = \tau_{xx} - \tau_{yy}$) and second ($N_2 = \tau_{yy} - \tau_{zz}$) normal stress differences contribute to the particle migration. τ_{xx}, τ_{yy}, and τ_{zz} denote normal stresses along the main flow, the velocity gradient, and vorticity direction, respectively. It is known that the second normal stress difference N_2 leads to the development of the secondary motions in viscoelastic flows and it becomes negligibly small compared with N_1 in diluted PEO solutions [40,41]. Thus, the lateral force F_E exerted on a particle by the normal stress difference is expressed as [42]

$$F_E \propto a^3 \frac{\partial N_1}{\partial x}$$

3. Materials and Methods

3.1. Design and Fabrication of Microfluidic Devices

The microfluidic channel used in this study is illustrated in Figure 1. The channel consisted of a 90 mm serpentine section and a 6 mm straight section with a uniform depth of 70 μm. The width of each U-turn in the serpentine section was 120 μm. The width of the serpentine and straight channel were 60 μm and 200 μm, respectively. The middle outlet is marked as Outlet A, while the two side outlets are labelled as Outlet B1 and B2. All devices were fabricated by standard photolithography and soft lithography techniques [43], including silicon master fabrication, and poly-dimethylsiloxane (PDMS) replica molding as well as bonding through plasma oxidation.

3.2. Preparation of Viscoelastic Fluids

A 0.5% *w*/*v* PEO (average Mw~2 million Da, Sigma–Aldrich, St. Louis, MO, USA) in phosphate-buffered saline (PBS) solution was prepared by rapid mixing to form a uniform PEO solution with a concentration of 5000 parts per million (ppm). PEO was chosen as the viscoelastic agent because it is reported to be biocompatible [44] and is widely used in viscoelastic microfluidics [13]. The 5000 ppm PEO solution was subsequently mixed with whole blood and PBS at different ratios to achieve the following: blood dilutions (1/2,

1/3, 1/5, and 1/10); and PEO concentrations (100, 200, 500, 1000, and 2000 ppm). The viscosity of the varied blood–PEO suspensions was measured through a Brookfield DV3T rheometer with a CPA-40Z spindle. The results were assessed across a range of shear rates (75~1500 s^{-1}) and is presented in the Supplementary Figure S1.

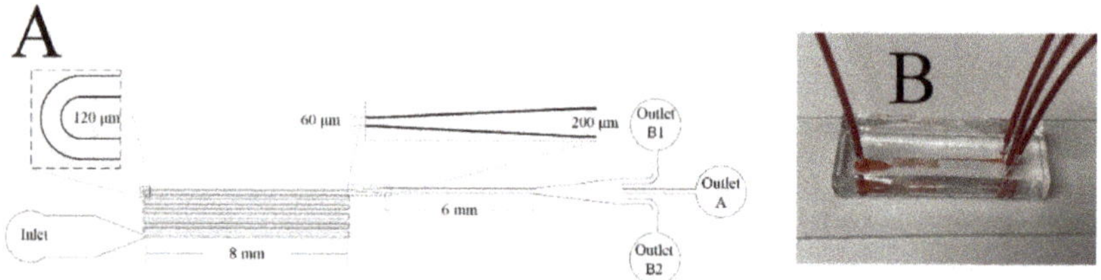

Figure 1. The microfluidic channel used in this study: (**A**) a schematic diagram; (**B**) an image of the fabricated microfluidic device during use with whole blood.

3.3. Preparation of Blood Cells

Blood was collected from healthy humans via venepuncture of a prominent vein in the antecubital region. Blood was collected into vacutainers containing 1.8 mg/mL of the anticoagulant K_2EDTA. Experimental procedures were completed within 6 h of initial blood collection. All protocols were reviewed and approved by the Griffith University Human Research Ethics Committee (protocol number: 2021/773), which conforms with the Declaration of Helsinki.

3.4. Flow Cytometry and Haemoglobin Analysis

Cell counts and filtration efficiency was determined for samples collected from outlets A and B using flow cytometry and spectrophotometric haemoglobin analysis for each concentration of PEO. Flow cytometry was performed using a FC500 flow cytometer (Beckman Coulter, CA, USA), with RBCs gates identified using initial antibody staining of glycophorin A (CD35a) conjugated to fluorescein isothiocyanate. To obtain accurate cell counts, absolute cell counting beads (123count eBeads, Invitrogen, Waltham, MA, USA) were added to each sample prior to analysis. Absolute total events were used for analysis to compare cell counts between outlet A (centre) and outlet B (side) to gain a measure of filtration efficiency.

To ensure the microfluidic device did not damage RBCs in the process of cell focusing and phase separation, samples collected from all outlets were assessed for haemolysis using the Harboe spectrophotometric method (PMID: 13646603) [45]. Briefly, the supernatant was isolated via centrifugation at 3000× *g* for 10 min, before being diluted 1:10 in 0.01% Na_2CO_3 solution. Each sample was thoroughly mixed and subsequently loaded into a microplate reader (FLUOstar Omega, BMG Labtech, Mornington, VIC, Australia) and absorbance was recorded at 380, 415, and 450 nm. Free haemoglobin concentration (mg/dL) was then calculated using the following formula: $[Hb] = (167.2A_{415} - 83.6A_{450} - 83.6A_{380})$. Haemolysis was determined for central (A) and side (B) outlets relative to each respective total haemoglobin concentration.

3.5. Experimental Setup and Data Analysis

Blood samples diluted in PEO PBS were infused into the microfluidic device at specific flow rates by a syringe pump (neMESYS, Centoni GmbH, Korbußen, Germany). To avoid the presence of air bubbles inside the microchannel, which could adversely interrupt the flow characteristics, the microchannel as well as the connected tubing were fully filled with PBS before each experiment. In experiments, to reduce sedimentation of blood cells (which drastically influences the blood concentration within the microchannel), the syringe pump was regularly arranged from vertical to horizontal orientation, and vice versa. However, a mixing device should be integrated inline in practical applications, such as placing a magnetic stirrer bar in the sample chamber and activating it by a rotating magnetic field. The outputs from Outlet A and Outlet B were collected into two 2 mL tubes for each case. An inverted microscope (Nikon Eclipse Ti) equipped with a high-speed camera (Phantom Miro3, Vision Research) at an ultra-short (~20 µs) exposure time was utilised to monitor and record the flow of samples inside the microchannels. The open-source software ImageJ (National Institutes of Health, Bethesda, MD, USA) was adopted to analyse the captured videos. The images of each video were stacked so that the statistical distribution of blood cells could be visualised. The polynomial curve fitting method was adopted to fit the data of focusing performance through MATLAB.

4. Results and Discussions

4.1. Blood Plasma Extraction Mechanism

When blood samples diluted in PEO PBS solution flow through the serpentine channel, three forces exert on the cellular components, including an inertial lift force induced by the flow inertia [37], an elastic force caused by the viscoelastic fluid [18], and a Dean drag force resulting from the curved structure at each turn [34]. The Dean drag force is absent in the straight regions of each channel. In general, the inertial lift force focuses the particles at a certain distance away from the channel walls. Meanwhile, the viscoelastic force causes a lateral force on the blood cells by the first normal stress difference, leading to cell movement away from the wall. In addition, the serpentine channel contains curved regions with two opposite directions in a periodic pattern, resulting in the alternate directions of corresponding Dean drag forces. This process aids the focusing of cells towards the centreline of flow. Hence, under the synergistic effect of the three forces, the cellular components of the diluted blood are expected to be focused on the central area of the channel after a certain flowing distance. Finally, the blood plasma can be extracted from cell-free regions nearby two side walls and collected at the side outlets B1 and B2.

4.2. Fluid Viscoelastic Enhanced Focusing of Blood Cells

In this section, we studied whether the added fluid viscoelasticity facilitated the focusing of blood cells. The blood sample diluted in only PBS was selected to be the control, which is labelled as $C_{PEO} = 0$ ppm. We first compare it with whole blood diluted in 500 ppm PEO PBS solution. Figure 2 presents the distribution of blood cells after the serpentine region of the channel for the two cases, at the same blood concentration (1/10) and volume flow rates ranging 1 to 50 µL/min. The side walls of the microfluidic channel are identified in each test with a dashed line overlay. As observed, the added viscoelasticity from PEO facilitated the focusing of blood cells. For blood sample of $C_{PEO} = 0$ ppm, even with 10 times dilution, the cell-free layers are very narrow. However, the cell-free area is increased by the viscoelastic effect at certain range of flow rates for blood samples of $C_{PEO} = 500$ ppm. Furthermore, the total width of free cell layers first increases and then decreases with the increase of flow rate (Figure 2B), which indicates that the enhancement of cells focusing by viscoelasticity is also sensitive to flow rate.

To quantify the focusing performance, a cell-free area ratio δ is introduced as follows:

$$\delta = \frac{W_f}{W_c} = \frac{W_{f1} + W_{f2}}{W_c}$$

where W_f denotes the total averaged cell-free layer width, which is the sum of the two cell-free layer widths W_{f1} and W_{f2} nearby channel walls. W_c represents the channel width. To calculate W_f, we first stacked each frame of the micrographs into a single figure. A rectangular region of interest was then selected at a discrete point before the expansion. Finally, the profile of grey values along the channel's width was obtained to determine the cell-free area with the error areas considered. Figure 3 illustrates the process using the data from the 1/10 blood concentration and 500 ppm PEO at flow rate of 10 µL/min as an example. Since the greyscale value of 0 represents black, it is not difficult to determine the widths of the error areas (err_1 and err_2), and then the total width of cell-free layers on two sides W_f can be estimated, so as δ.

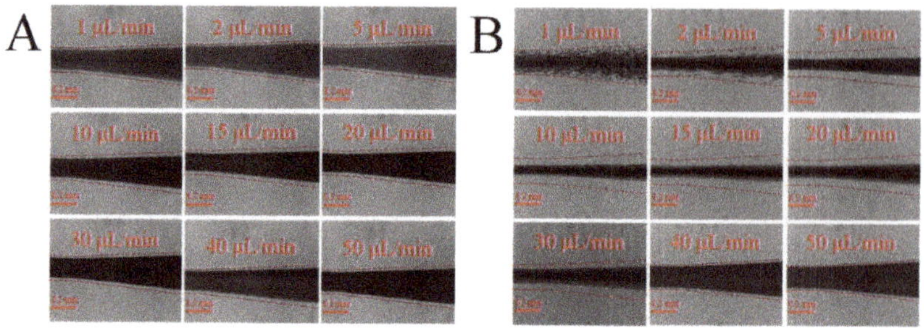

Figure 2. Blood cell distribution at the end of the serpentine channel at various flow rates: (**A**) whole blood diluted (×1/10) in PBS; (**B**) blood diluted (×1/10) in 500 ppm PEO PBS solution. The added viscoelasticity from PEO can facilitate the focusing of blood cells at channel centre, and the expanded cell-free area is beneficial for blood plasma extraction.

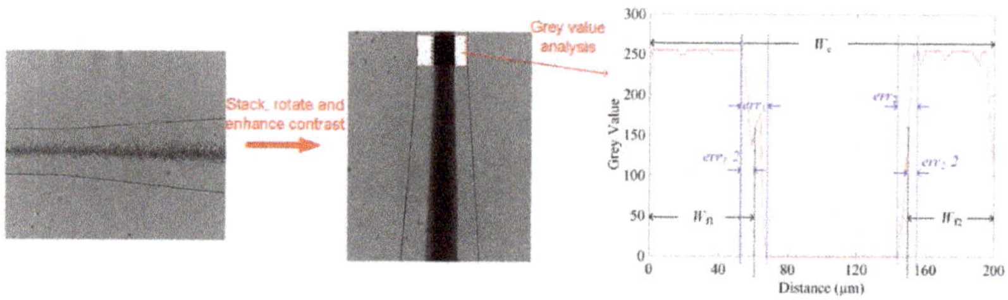

Figure 3. Image process and averaged greyscale value analysis to determine the boundaries of the cell-free areas. 50 frames of images captured by a high-speed camera were stacked. The greyscale values along the channel lateral position were obtained at the end of serpentine channel. The boundaries of the cell-free layers on two sides were determined by half of the error area width.

By repeating the procedures above, the effect of flow rate on the cell-free area ratio δ can be obtained (Figure 4). As presented, the cell-free area ratio δ of 10 times diluted blood in 500 ppm PEO PBS is significantly higher than that in only PBS at low to moderate flow rates (1~30 μL/min), while a narrowed cell-free area ratio δ was found at high flow rates (40 and 50 μL/min). Importantly, a peak value of δ was identified for the focusing performance of 10 times diluted blood in 500 ppm PEO PBS, as demonstrated by the fitting curve. Specifically, when the flow rate was 10 μL/min, δ of 10 times diluted blood in 500 ppm PEO PBS was at its highest value of approximately 0.56, which is more than two times of the peak δ value (around 0.23) of PBS.

4.3. *Effects of PEO Concentration and Blood Dilution*

We further investigated the effects of PEO concentration and blood dilution on blood cells' focusing performance. Various PEO concentrations and blood dilutions demonstrated a similarly peaked response (Figure 5); the cell-free area ratio δ generally shows an initial increase and subsequent decrease when flow rate increased, except for the case with the highest PEO concentration (C_{PEO} = 2000 ppm) which demonstrates a consistently decreasing trend. Additionally, low PEO concentrations (C_{PEO} = 100, 200, and 500 ppm) demonstrated superior performance compared to counterparts with high PEO concentration (C_{PEO} = 1000, 2000 ppm). Specifically, the highest δ was approximately 0.44 of C_{PEO} = 100 ppm at 15 μL/min, 0.58 of C_{PEO} = 200 ppm at 15 μL/min, and 0.56 of C_{PEO} = 500 ppm at 10 μL/min. In general, optimal δ were found within low to moderate flow rates, and this optimised flow rate decreased from 15 to 1 μL/min when C_{PEO} increased from 100 to 2000 ppm (Figure 5A). Furthermore, among the various blood dilution cases, the 10 times diluted blood consistently shows a superior effect to others, as presented in Figure 5B. This is not unexpected, since larger dilution has fewer cells, thus leading to larger cell-free layers. Similarly, optimal focusing performances were found within low to moderate flow rates ranging from 5 to 15 μL/min. Among all the cases, the highest cell-free area ratio δ was approximately 0.58 with 200 ppm PEO concentration and 10 times diluted blood at 15 μL/min.

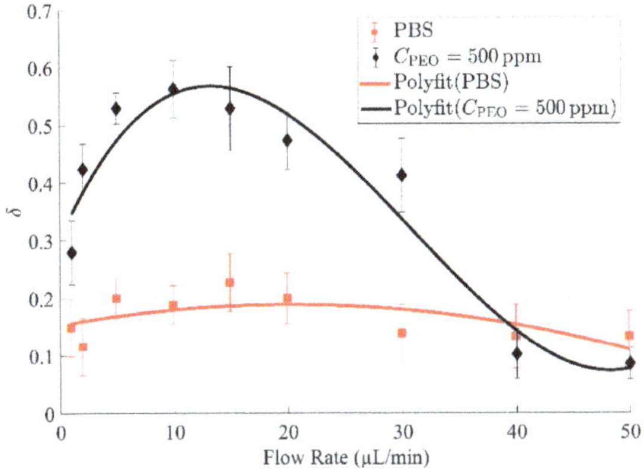

Figure 4. Comparison of cell-free area ratio δ of diluted ($\times 1/10$) whole blood samples in PBS and 500 ppm PEO PBS, respectively. δ of diluted blood in 500 ppm PEO PBS is significantly higher than that in only PBS at low to moderate flow rates (1~30 μL/min).

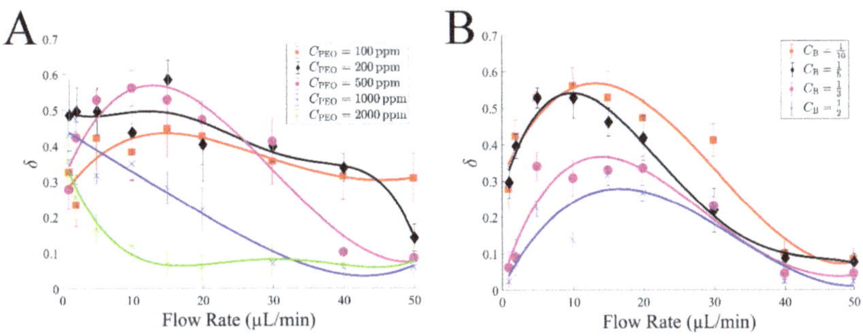

Figure 5. The effects of PEO concentration and blood dilution number on cell-free area ratio δ: (**A**) 10 times diluted blood with various PEO concentrations (100, 200, 500, 1000, and 2000 ppm); (**B**) various blood dilutions (1/2, 1/3, 1/5, and 1/10) at 500 ppm PEO concentration.

4.4. Blood Plasma Extraction and Haemoglobin Analysis

Figure 6 presents the focusing of blood cells and plasma separation at the bifurcation outlets at 15 μL/min for several typical cases. As observed, at a moderate flow rate, either increasing the PEO concentration or decreasing blood dilution results in wider cell layers. Consequently, the collection of cell-free plasma from blood becomes less effective. Indeed, in the present "best case scenario" (10 times dilution of blood in 500 ppm PEO), several blood cells may still be observed at the two-sided outlets, likely due to a suboptimal flow resistance ratio between the outlets; thus, opportunities that exist to optimise the outlets should improve performance of plasma separation.

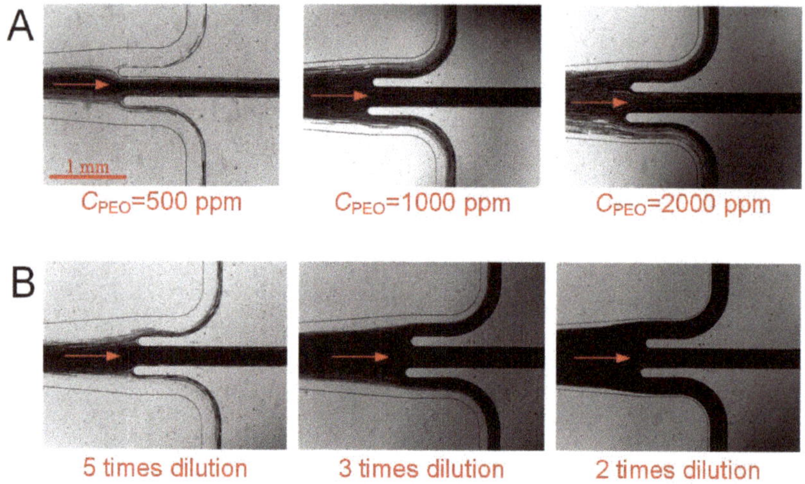

Figure 6. Focusing of blood cells and plasma extraction at the bifurcation outlets: (**A**) 10 times diluted blood with various PEO concentrations; (**B**) various blood dilutions with 200 ppm PEO concentration.

After determining the optimal viscoelastic focusing at 15 μL/min, the blood plasma separation was conducted for diluted blood samples (1/10) with different PEO concentrations and the collected plasma from the different outlets was examined. Figure 7 displays the flow cytometric data for central and side outlets for increasing PEO concentration, where gated regions identify RBCs populations and eBeads. Absolute cell counts (n) in the RBCs region increased for samples collected from the side outlet for the blood sample with PEO concentration ≥500 ppm.

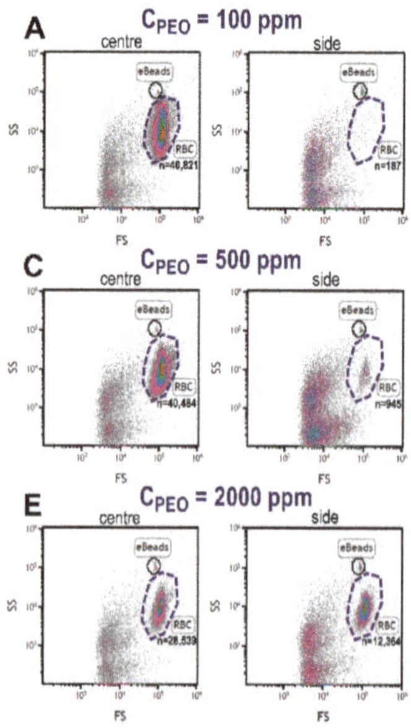

Figure 7. Flow cytometric cell counts in central and side outlets of the microfluidic device for blood samples suspended at 1:10 dilution of increasing concentrations of PEO; 100 (**A**), 200 (**B**), 500 (**C**), 1000 (**D**), and 2000 ppm (**E**). RBCs: red blood cells; FS: forward scatter; SS: side scatter.

To determine an index of filtration efficiency for plasma extraction, absolute cell counts in the side outlet were presented relative to total cell counts in the central channel for each PEO concentration (Figure 8A). The filtration efficiency into the side channel decreased at PEO concentrations $\geq$500 ppm (Figure 8A), which can be supported by the fact that the total RBCs haemoglobin content in the side channel increased (Figure 8B). Although total measured haemoglobin content increased at PEO concentrations of 500, 1000, and 2000 ppm, no substantial haemolysis (i.e., cell-free haemoglobin; <1%) was detected in the outlet of any condition (Figure 8C), indicating that the microfluidic device does not rupture RBCs in the process of viscoelastic focusing and plasma extraction. Based on the data of filtration efficiency and haemolysis in Figure 8, the blood samples with PEO concentrations of 100 and 200 ppm are optimal for blood plasma extraction.

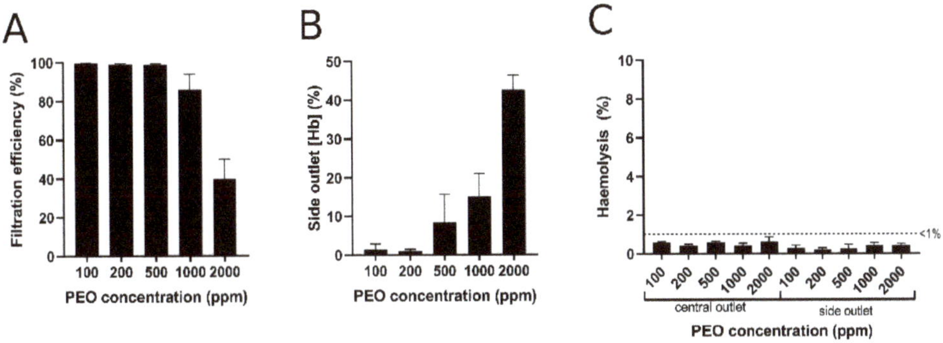

Figure 8. Filtration efficiency as determined by comparative RBCs counts (**A**) and relative total haemoglobin concentration (**B**) of the side outlet versus central outlet of the microfluidic device for each PEO concentration. (**C**) Measured cell-free haemoglobin (i.e., haemolysis) at the central and side outlets of the microfluidic device for each PEO concentration. (Hb): haemoglobin concentration. Error bars represent standard error of the mean (n = 3).

While this study mainly focused on RBCs movement and blood plasma extraction for haemoglobin detection, movement of other blood cells (WBCs and platelets) must also be considered. Since particle size dictates cell centreline focusing or margination, WBCs (which are generally larger than RBCs) should focus towards the centreline, and platelets (which are smaller than RBCs) should have less alignment. Therefore, there are likely fewer WBCs and more platelets in the separated plasma, evidenced by the flow cytometric data in Figure 7. In our data collection, platelets were not qualitatively identified due to their size; however, the spectrophotometric analyses for haemoglobin detection are not sensitive to platelet concentration or the presence of WBCs [46].

The processing throughput of the proposed device is far less than standard centrifugation. However, many analytical methods only require small volumes of isolated plasma (less than 30 μL; e.g., the Harboe spectrophotometric method for haemoglobin assessment). While increasing the throughput of the device would not be required for future microfluidic devices with embedded inline analytics, one potential solution for increasing throughput would be to implement multiple parallel channels to amplify the output. The proposed plasma extraction microfluidic technology in the present study is part of a staged theme of works seeking to develop a fully-automated bedside point-of-care free-haemoglobin quantification system. Given the developments towards inline analytics, we believe that the relatively low throughput may not be a significant issue. The advantages of microfluidic plasma extraction technology and potential integration with other functional units and automation will likely increase clinical efficiency and reduce human intervention.

Finally, it is plausible that the current microfluidic device could be deployed to ascertain blood phase separation for cell focusing and plasma isolation, with prospects to integrate inline plasma analytics at side outlets for development into a future stand-alone device for blood analytics.

5. Conclusions

The current study investigated viscoelastic blood plasma extraction in a sheathless serpentine microchannel. The influences of PEO solution concentration, blood dilution, and flow rate on the cell focusing performance were tested. The experimental data demonstrated that the cell focusing performance, in general, shows a biphasic trend (first increasing and then decreasing) when flow rate increases, at least for the measured PEO concentrations and blood dilutions used. Hence, the optimal cell focusing performance for each case was determined with the flow rate ranging from 1 to 15 μL/min. Furthermore, the highest cell focusing performance among all the cases was found to be approximately 0.58 with 200 ppm PEO concentration and 10 times diluted blood at 15 μL/min flow rate. Finally, the haemocompatibility results showed that only <1% haemolysis was detected in the outlet region of any condition, indicating that the microfluidic device does not rupture RBCs in the process of viscoelastic focusing and plasma extraction.

Supplementary Materials: The following supporting information can be downloaded at: https://www.mdpi.com/article/10.3390/bios12020120/s1, Figure S1: (A) Blood viscosity at different PEO concentrations. Blood dilution is 1/10. (B) Blood viscosity at different dilutions. PEO concentration is 500 ppm.

Author Contributions: Conceptualisation, J.Z. and A.P.M.; methodology, J.Z.; software, Y.D.; formal analysis, Y.D. and A.P.M.; investigation, Y.D. and H.C.; resources, Y.D. and H.C.; data curation, Y.D., J.Z. and A.P.M.; writing—original draft preparation, Y.D.; writing—review and editing, J.Z., A.P.M., M.J.S., H.A., H.T.T. and N.-T.N.; visualisation, Y.D. and H.F.; supervision, J.Z. and A.P.M.; project administration, A.P.M. and M.J.S.; funding acquisition, A.P.M., M.J.S., N.-T.N. and J.Z. All authors have read and agreed to the published version of the manuscript.

Funding: This research was funded by MHIQ Collaborative Interdisciplinary Grants 2021, Discovery Project (Grant No. DP180100055), ARC DECRA fellowship (Grant No. DE210100692), ARC Future Fellowships (FT180100361), and Griffith University Postdoctoral Fellowship Scheme. This work was performed in part at the Queensland node—Griffith—of the Australian National Fabrication Facility, a company established under the National Collaborative Research Infrastructure Strategy to provide nano and microfabrication facilities for Australian researchers.

Institutional Review Board Statement: The study was conducted according to the guidelines of the Declaration of Helsinki, and approved by the Institutional Review Board (or Ethics Committee) of Griffith University (protocol number: 2021/773).

Informed Consent Statement: Informed consent was obtained from all subjects involved in the study. Written informed consent has been obtained from the patient(s) to publish this paper.

Data Availability Statement: The data presented in this study are available on request from the corresponding authors.

Conflicts of Interest: The authors declare no conflict of interest.

References

1. Lippi, G.; Plebani, M.; di Somma, S.; Cervellin, G. Hemolyzed specimens: A major challenge for emergency departments and clinical laboratories. *Crit. Rev. Clin. Lab. Sci.* **2011**, *48*, 143–153. [CrossRef] [PubMed]
2. Yu, Z.T.F.; Yong, K.M.A.; Fu, J. Microfluidic blood cell sorting: Now and beyond. *Small* **2014**, *10*, 1687–1703. [CrossRef] [PubMed]
3. Tripathi, S.; Kumar, Y.B.V.; Prabhakar, A.; Joshi, S.S.; Agrawal, A. Passive blood plasma separation at the microscale: A review of design principles and microdevices. *J. Micromech. Microeng.* **2015**, *25*, 083001. [CrossRef]
4. Zhang, J.; Yuan, D.; Sluyter, R.; Yan, S.; Zhao, Q.; Xia, H.; Tan, S.H.; Nguyen, N.-T.; Li, W. High-throughput separation of white blood cells from whole blood using inertial microfluidics. *IEEE Trans. Biomed. Circuits Syst.* **2017**, *11*, 1422–1430. [CrossRef]
5. Kuan, D.-H.; Wu, C.-C.; Su, W.-Y.; Huang, N.-T. A microfluidic device for simultaneous extraction of plasma, red blood cells, and on-chip white blood cell trapping. *Sci. Rep.* **2018**, *8*, 1–9. [CrossRef]
6. Çetin, B.; Li, D. Dielectrophoresis in microfluidics technology. *Electrophoresis* **2011**, *32*, 2410–2427. [CrossRef]

7. Forbes, T.P.; Forry, S.P. Microfluidic magnetophoretic separations of immunomagnetically labeled rare mammalian cells. *Lab Chip* **2012**, *12*, 1471–1479. [CrossRef]
8. Kersaudy-Kerhoas, M.; Sollier, E. Micro-scale blood plasma separation: From acoustophoresis to egg-beaters. *Lab Chip* **2013**, *13*, 3323–3346. [CrossRef]
9. Crowley, T.A.; Pizziconi, V. Isolation of plasma from whole blood using planar microfilters for lab-on-a-chip applications. *Lab Chip* **2005**, *5*, 922–929. [CrossRef]
10. di Carlo, D.; Irimia, D.; Tompkins, R.G.; Toner, M. Continuous inertial focusing, ordering, and separation of particles in microchannels. *Proc. Natl. Acad. Sci. USA* **2007**, *104*, 18892–18897. [CrossRef]
11. Zhang, J.; Yan, S.; Yuan, D.; Alici, G.; Nguyen, N.-T.; Warkiani, M.E.; Li, W. Fundamentals and applications of inertial microfluidics: A review. *Lab Chip* **2016**, *16*, 10–34. [CrossRef] [PubMed]
12. Fallahi, H.; Yadav, S.; Phan, H.-P.; Ta, H.; Zhang, J.; Nguyen, N.-T. Size-tuneable isolation of cancer cells using stretchable inertial microfluidics. *Lab Chip* **2021**, *21*, 2008–2018. [CrossRef] [PubMed]
13. Yuan, D.; Zhao, Q.; Yan, S.; Tang, S.-Y.; Alici, G.; Zhang, J.; Li, W. Recent progress of particle migration in viscoelastic fluids. *Lab Chip* **2018**, *18*, 551–567. [CrossRef] [PubMed]
14. Lu, X.; Liu, C.; Hu, G.; Xuan, X. Particle manipulations in non-Newtonian microfluidics: A review. *J. Colloid Interface Sci.* **2017**, *500*, 182–201. [CrossRef]
15. Zhou, J.; Papautsky, I. Viscoelastic microfluidics: Progress and challenges. *Microsyst. Nanoeng.* **2020**, *6*, 1–24. [CrossRef]
16. D'Avino, G.; Greco, F.; Maffettone, P.L. Particle migration due to viscoelasticity of the suspending liquid and its relevance in microfluidic devices. *Annu. Rev. Fluid Mech.* **2017**, *49*, 341–360. [CrossRef]
17. Yuan, D.; Yadav, S.; Ta, H.T.; Fallahi, H.; An, H.; Kashaninejad, N.; Ooi, C.H.; Nguyen, N.T.; Zhang, J. Investigation of viscoelastic focusing of particles and cells in a zigzag microchannel. *Electrophoresis* **2021**, *42*, 2230–2237. [CrossRef]
18. Leshansky, A.M.; Bransky, A.; Korin, N.; Dinnar, U. Tunable nonlinear viscoelastic "focusing" in a microfluidic device. *Phys. Rev. Lett.* **2007**, *98*, 234501. [CrossRef] [PubMed]
19. Yang, S.; Kim, J.Y.; Lee, S.J.; Lee, S.S.; Kim, J.M. Sheathless elasto-inertial particle focusing and continuous separation in a straight rectangular microchannel. *Lab Chip* **2011**, *11*, 266–273. [CrossRef]
20. Nam, J.; Lim, H.; Kim, D.; Jung, H.; Shin, S. Continuous separation of microparticles in a microfluidic channel via the elasto-inertial effect of non-Newtonian fluid. *Lab Chip* **2012**, *12*, 1347–1354. [CrossRef]
21. Liu, C.; Xue, C.; Chen, X.; Shan, L.; Tian, Y.; Hu, G. Size-based separation of particles and cells utilizing viscoelastic effects in straight microchannels. *Anal. Chem.* **2015**, *87*, 6041–6048. [CrossRef] [PubMed]
22. Nam, J.; Shin, Y.; Tan, J.K.S.; Lim, Y.B.; Lim, C.T.; Kim, S. High-throughput malaria parasite separation using a viscoelastic fluid for ultrasensitive PCR detection. *Lab Chip* **2016**, *16*, 2086–2092. [CrossRef] [PubMed]
23. Yuan, D.; Tan, S.H.; Zhao, Q.; Yan, S.; Sluyter, R.; Nguyen, N.-T.; Zhang, J.; Li, W. Sheathless Dean-flow-coupled elasto-inertial particle focusing and separation in viscoelastic fluid. *RSC Adv.* **2017**, *7*, 3461–3469. [CrossRef]
24. Yuan, D.; Zhao, Q.; Yan, S.; Tang, S.-Y.; Zhang, Y.; Yun, G.; Nguyen, N.-T.; Zhang, J.; Li, M.; Li, W. Sheathless separation of microalgae from bacteria using a simple straight channel based on viscoelastic microfluidics. *Lab Chip* **2019**, *19*, 2811–2821. [CrossRef] [PubMed]
25. Lu, X.; Xuan, X. Elasto-inertial pinched flow fractionation for continuous shape-based particle separation. *Anal. Chem.* **2015**, *87*, 11523–11530. [CrossRef] [PubMed]
26. Lu, X.; Zhu, L.; Hua, R.-M.; Xuan, X. Continuous sheath-free separation of particles by shape in viscoelastic fluids. *Appl. Phys. Lett.* **2015**, *107*, 264102. [CrossRef]
27. Valero, A.; Braschler, T.; Rauch, A.; Demierre, N.; Barral, Y.; Renaud, P. Tracking and synchronization of the yeast cell cycle using dielectrophoretic opacity. *Lab Chip* **2011**, *11*, 1754–1760. [CrossRef]
28. Sugaya, S.; Yamada, M.; Seki, M. Observation of nonspherical particle behaviors for continuous shape-based separation using hydrodynamic filtration. *Biomicrofluidics* **2011**, *5*, 024103. [CrossRef]
29. Beech, J.P.; Holm, S.H.; Adolfsson, K.; Tegenfeldt, J.O. Sorting cells by size, shape and deformability. *Lab Chip* **2012**, *12*, 1048–1051. [CrossRef]
30. Li, D.; Zielinski, J.; Kozubowski, L.; Xuan, X. Continuous sheath-free separation of drug-treated human fungal pathogen Cryptococcus neoformans by morphology in biocompatible polymer solutions. *Electrophoresis* **2018**, *39*, 2362–2369. [CrossRef]
31. Nam, J.; Jee, H.; Jang, W.S.; Yoon, J.; Park, B.G.; Lee, S.J.; Lim, C.S. Sheathless shape-based separation of Candida albicans using a Viscoelastic non-Newtonian fluid. *Micromachines* **2019**, *10*, 817. [CrossRef] [PubMed]
32. Yuan, D.; Yan, S.; Zhang, J.; Guijt, R.M.; Zhao, Q.; Li, W. Sheathless Separation of Cyanobacterial Anabaena by Shape Using Viscoelastic Microfluidics. *Anal. Chem.* **2021**, *93*, 12648–12654. [CrossRef] [PubMed]
33. Yang, S.; Lee, S.S.; Ahn, S.W.; Kang, K.; Shim, W.; Lee, G.; Hyun, K.; Kim, J.M. Deformability-selective particle entrainment and separation in a rectangular microchannel using medium viscoelasticity. *Soft Matter* **2012**, *8*, 5011–5019. [CrossRef]
34. di Carlo, D.; Edd, J.F.; Humphry, K.J.; Stone, H.A.; Toner, M. Particle segregation and dynamics in confined flows. *Phys. Rev. Lett.* **2009**, *102*, 094503. [CrossRef]
35. Gossett, D.R.; Carlo, D.D. Particle focusing mechanisms in curving confined flows. *Anal. Chem.* **2009**, *81*, 8459–8465. [CrossRef]
36. Yuan, D.; Zhang, J.; Sluyter, R.; Zhao, Q.; Yan, S.; Alici, G.; Li, W. Continuous plasma extraction under viscoelastic fluid in a straight channel with asymmetrical expansion–contraction cavity arrays. *Lab Chip* **2016**, *16*, 3919–3928. [CrossRef]

37. Asmolov, E.S. The inertial lift on a spherical particle in a plane Poiseuille flow at large channel Reynolds number. *J. Fluid Mech.* **1999**, *381*, 63–87. [CrossRef]
38. di Carlo, D. Inertial microfluidics. *Lab Chip* **2009**, *9*, 3038–3046. [CrossRef]
39. Rodd, L.; Cooper-White, J.; Boger, D.; McKinley, G.H. Role of the elasticity number in the entry flow of dilute polymer solutions in micro-fabricated contraction geometries. *J. Non-Newton. Fluid Mech.* **2007**, *143*, 170–191. [CrossRef]
40. Bird, R.B.; Armstrong, R.C.; Hassager, O. Fluid mechanics. In *Dynamics of Polymeric Liquids*; Wiley: Hoboken, NJ, USA, 1987; Volume 1.
41. Magda, J.; Lou, J.; Baek, S.; DeVries, K. Second normal stress difference of a Boger fluid. *Polymer* **1991**, *32*, 2000–2009. [CrossRef]
42. Tehrani, M. An experimental study of particle migration in pipe flow of viscoelastic fluids. *J. Rheol.* **1996**, *40*, 1057–1077. [CrossRef]
43. Duffy, D.C.; McDonald, J.C.; Schueller, O.J.; Whitesides, G.M. Rapid prototyping of microfluidic systems in poly (dimethylsiloxane). *Anal. Chem.* **1998**, *70*, 4974–4984. [CrossRef] [PubMed]
44. Yuan, D.; Sluyter, R.; Zhao, Q.; Tang, S.; Yan, S.; Yun, G.; Li, M.; Zhang, J.; Li, W. Dean-flow-coupled elasto-inertial particle and cell focusing in symmetric serpentine microchannels. *Microfluid. Nanofluid.* **2019**, *23*, 41. [CrossRef]
45. Harboe, M. A method for determination of hemoglobin in plasma by near-ultraviolet spectrophotometry. *Scand. J. Clin. Lab. Investig.* **1959**, *11*, 66–70. [CrossRef] [PubMed]
46. Cookson, P.; Sutherland, J.; Cardigan, R. A simple spectrophotometric method for the quantification of residual haemoglobin in platelet concentrates. *Vox Sang.* **2004**, *87*, 264–271. [CrossRef] [PubMed]

biosensors

Article

Target-Specific Exosome Isolation through Aptamer-Based Microfluidics

Zixuan Zhou [1], Yan Chen [2] and Xiang Qian [1,*]

[1] Tsinghua-Shenzhen International Graduate School, Tsinghua University, Shenzhen 518055, China; zhouzx19@mails.tsinghua.edu.cn
[2] Shenzhen Institutes of Advanced Technology, Chinese Academy of Sciences, Shenzhen 518055, China; yan.chen@siat.ac.cn
* Correspondence: qian.xiang@sz.tsinghua.edu.cn

Abstract: Exosomes (30–100 nm in diameter) are a group of cell-derived membrane vesicles, packaged as valuable cargo with lipid, proteins, and genetic materials from their parent cells. With the increasing interest in exosomes for diagnostic and therapeutic applications, the rapid isolation of pure exosome populations has become a hot topic. In this paper, we propose modified microchannels with aptamer in a microfluidics system for rapid and efficient isolation of exosomes by targeting exosome-carrying CD63 and PTK 7. The capture efficiency in surface-modified channels reaches around 10^7–10^8 particles/mL in 20 min, and purified exosomes with reliable size can be achieved.

Keywords: aptamer; exosome; isolation; CD63; PTK7; microfluidics

Citation: Zhou, Z.; Chen, Y.; Qian, X. Target-Specific Exosome Isolation through Aptamer-Based Microfluidics. *Biosensors* **2022**, *12*, 257. https://doi.org/10.3390/bios12040257

Received: 20 December 2021
Accepted: 23 February 2022
Published: 18 April 2022

1. Introduction

Different cell types are capable of secreting extracellular vesicles in vitro [1]. Extracellular vesicles (EVs) (30–1000 nm in diameter) are mixed populations of cell-derived membrane structures. They more than just carry wastes, but play an important role in transcellular communication [2,3]. According to biogenesis, or in other words, intracellular origin, EVs have been sorted into two classes, exosomes and other EVs [4]. Exosomes generally refer to 30–100 nm EVs formed by the inward budding of the endosomal membrane during maturation of multi-vesicular endosomes. Endosomes are generated by inward budding from the plasma membrane, and they release exosomes through the fusion with plasma membrane [5]. An exosome contains a specific composition of proteins, mRNA, miRNA, and DNA that carries coding information to communicate between the exosome-producing cell and the target cell, making them attractive candidates for circulating disease biomarkers for clinical application [6,7]. Despite these promising attributes, a major source of ongoing confusion is how to build solid methods for isolating exosomes from mixed EV populations, since all EVs share a similar appearance, overlapping range of size, and often common protein composition, making it difficult to make a clear distinction [8]. Fortunately, however, exosomes bear specific surface molecules such as tetraspanins CD63 and protein tyrosine kinase 7 (PTK7), which can be considered as genuine markers of interest [9]. CD63 is a kind of cell surface-associated membrane protein which is abundantly presented on the exosome surface [10]. PTK7 is a transmembrane receptor that has been reported as present on exosomes and plays an important role in cell-to-cell communication [11].

Reported exosome isolation techniques mainly include gold standard–differential ultracentrifugation [12], sequential centrifugal ultrafiltration [13], immuno-beads, polymeric precipitation [14], and size-exclusion chromatography (SEC) [15], requiring dedicated laboratory instruments and time-consuming operations which challenge the experimental convenience [16]. Successful isolation requires reliable characterization to prove it. Transmission electron microscopy (TEM) and scanning electron microscopy (SEM) have been well documented to observe the morphology of exosome. Nanoparticle tracking analysis (NTA) offers the additional ability to measure size distribution and concentration of exosomes [17].

Microfluidics, with the advantage of being a portable, cheap alternative, is employed as a platform in the isolation of exosomes. Generally, the strategies applied for the isolation of exosomes in a microfluidics system include the immunoaffinity-based method, dynamic-based method, and size-based methods [18]. In 2021, Yu et al. developed the ExoSD microfluidic chip endowed with immunomagnetic capture beads to achieve exosomes isolation [19]. In 2014, Kanwar et al. fabricated an ExoChip functionalized with antibodies to isolate exosomes from serum samples [20]. In 2018, Smith et al. created nanoscale deterministic lateral displacement (DLD) arrays on microfluidic channel to achieve collection of EVs without the further step of isolating exosomes [21]. In 2013, filtration-based microfluidics fabricated by Wand et al. achieved the trapping of EVs on nanowire-on-micropillars [22].

Aptamer, a single-stranded DNA or RNA oligonucleotide with complex three-dimensional shape, often contains 15–60 nucleotides. It is generated from a large random sequence pool via in vitro selection based on its specific recognition for targeting molecules; moreover, it can be tailored for a specific target. Chemical modification can enhance activity or stability of aptamers [23,24]. Aptamers, also called chemical antibodies, are usually compared with antibodies, since both molecules function as affinity reagents [25]. The binding domain between the surface protein and oligonucleotide aptamers is electrostatic attraction, hydrogen bond, and van der Waals force. Since the phosphate backbone of DNA is negatively charged, it can contribute more oxygen atoms, in other words, more electrons, when binding to positively charged amino acids, making specific adsorption more stable. Aptamers provide great promise for isolating and detecting EVs in affinity applications [26].

There is not much literature concerning microchannels modified with aptamer in microfluidic chips to achieve the purpose of separating exosomes. Aptamer related exosome applications focus mainly on exosome detection [27–29]. Recently, studies have usually applied aptamers to target EVs based on magnetic beads. These works are developed from mature immunomagnetic magnetic beads' (IMB) separation. Aptamers act as the role of antibodies in IMB. In 2019, Zhang et al. developed DNA aptamer-based magnetic isolation process for EV's isolation. Cell culture medium or plasma is mixed with biotin labelled CD63 aptamer and then separated using streptavidin modified magnetic beads. NTA characterization of isolated EVs showed 1.4×10^7 particles/mL capture efficiency in 90 min [30]. However, in aptamer-IMB there exists the problem of not eluting independent EVs from beads. In 2020, Song et al. developed a magnetic bead-based exosome immunoaffinity separation system by using CD63 targeting aptamer. NTA results showed that this method captures 8.37×10^7 exosomes particles from 10 mL MDA-MB-231 cell culture mediate in 60 min. The native state exosomes were eluted using NaCl elution [31].

In this paper, we present aptamer-affinity microfluidics to endow the isolation platform with targeting capability. The chip, which is made from polydimethylsiloxane (PDMS) and glass, implements exosomes isolation based on immobilized DNA-aptamers in microchannels. Two exosomes-carried biomarkers, CD63 and PTK7, which are ubiquitously presented on most exosomes, serve as targets for specific capture by Apt-CD63 [32] and Apt-PTK7 [9], respectively. These aptamers are immobilized via the biotin–avidin–desthiobiotin link on a glass slab. The chip benefits from a combination of filtration-based apparatus and aptamer-affinity based apparatus. The microfilter in the inlet achieves the ability of starting from cell culture supernatant without previous treatment, since it in attempts to filtrate cell debris, bacteria, or PDMS residues before aptamer-affinity reaction. The cell culture supernatant sample is composed of large-sized impurities, such as cell debris and proteins, and small-sized particles, such as exosomes and other EVs. The aim of the aptamer-based apparatus is to capture exosomes from mixed EVs populations. According to reviews, exosomes contain CD63 and PTK7 distinct from other EVs, such that we chose these two surface biomarkers as aptamer targeting proteins and also chose them as ELISA proteins in the following exosome characterization experiment.

2. Materials and Methods

2.1. Cancer Cells Culture

The lung cancer cell lines A549 were obtained from the Shanghai cell bank of the Chinese Academy of Science. The cells were cultured at 37 °C in a humidified atmosphere of 5% CO_2 in cell incubator (Thermo Fisher Scientific, Waltham, MA, USA). The cell culture medium contains 90% RPMI Medium 1640 supplemented with 10% fetal bovine serum (FBS) depleted of EVs and 1% penicillin-streptomycin (PS). The freezing medium contains 90% FBS and 10% dimethylsulfoxide (DMSO). Phosphate buffered saline (PBS), FBS, PS, and 0.25%Trypsin-EDTA, RPMI Medium 1640 were purchased from Thermo Fisher Scientific, Waltham, MA, USA. For repeated experiments using the same batch of the cell lines, the slow-freezing and rapid-thawing method was utilized. When the cells reach 80% to 90% coverage, passaging should be operated. Passage 6–8 cell culture supernatant was prepared for experiments.

2.2. Working Principle and Fabrication Processes of the Aptamer-Based Microfluidics

Our aptamer-based microfluidic chip applies the principal of using the biotin–avidin–desthiobiotin binding link to immobilize aptamer on the surface of the channel. DSPE–PEG–Biotin are embedded in the PDMS prepolymer on glass slab by mixing with chloroform, as reported by Huang et al. [33]. Aptamers modified with desthiobiotin, an analogue of biotin, are immobilized by the lower binding affinity with avidin. Desthiobiotin instead of biotin binds with avidin to facilitate easier release of aptamer-conjugated exosomes [34]. Figure 1A displays fabrication steps of functionalized glass slab. A filtration-based apparatus is conducted in order to hinder cell debris and PDMS residues to enter into the downstream. The affinity-based apparatus possesses the ability for discrimination between the exosomes of interest and other EVs in the fluids. Figures 1B and 1D show a photo of PDMS/glass microfluidics and a schematic diagram, respectively. In order to reduce the flow resistance, the microchannel is gradually widened from 100 µm to 1000 µm at the end. Figure 1C shows a SEM image of microfilter in inlet, correspondingly labeled in Figure 1D. The gap between micropillars is around 50 µm.

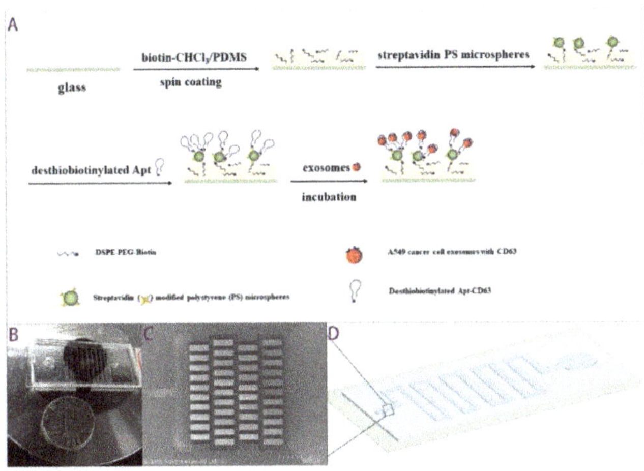

Figure 1. Aptamer-based exosomes isolation microfluidics. (**A**) Immobilization of aptamer onto glass surface for EVs capture. (**B**) Prototype of PDMS/glass chip in comparison with a 1 yuan RMB coin. (**C**) SEM image of micropillars in inlet. The scale bar is 500 µm. (**D**) Schematic diagram of PDMS/glass microfluidics.

The chip was composed of PDMS top slab and glass bottom slab. The PDMS slab was fabricated by soft lithography techniques. The layout of the channel structure was designed in AutoCAD (Autodesk Inc., San Rafael, CA, USA) and printed it on a film (Shenzhen Qingyi Photomask, Shenzhen, China). The pattern on the film was transferred to a silicon wafer with photoresist via lithography using UV-365 alignment (URE-2000/35, Institute of Optics and Electronics, Chinese Academy of Science, Chengdu, China). We adopted an ultrathick photoresist, SU8-2050 (Nippon Kayaku, Tokyo, Japan), and the depth was 100 μm. The silicon wafer with patterned photoresist was utilized as a master mold in molding process. PDMS prepolymer and curing agent mixture (10:1 volume ratio) (Sylgard 184 reagent, Dow Corning, Midland, MI, USA) was poured into master mold and cured in an oven at 80 °C for 1 h. The elastic PDMS slab was obtained by peeling. The bottom slab was a functionalized cover glass. A 1 mL 5 mg/mL biotin-chloroform solution was added to 3 g PDMS prepolymer and the curing agent mixture (10:1 volume ratio). The mixture covered, as thinly as possible, over the entire glass via spin coating step. Otherwise, a thicker layer would cause blocking during the subsequent bonding step. With the limitation of experimental conditions, the maximum speed we used was 1800 rpm. The top slab of the flow channel was then put on the mixture-coated cover glass before the latter was cured at 37 °C overnight.

DSPE–PEG–Biotin and Streptavidin modified green fluorescent polystyrene microspheres (5 μm) were purchased from Xi'an ruixi Biological Technology, Shanghai, China. Apt-CD63: 3′ Desthiobiotin-CAC CCC ACC TCG CTC CCG TGA CAC TAA TGC TA-5′ [32] and Apt-PTK7: 3′ Desthiobiotin-TTT TTT TAT CTA ACT GCT GCG CCG CCG GGA AAA TAC TGT ACG GTT AGA-5′ [9] were synthesized by Sangon Biotech, Shanghai, China with reversed-phase HPLC purification. Streptavidin and biotinylated aptamer were introduced into device one after another by connecting flow control kit (MFCS-EZ, Fluigent, Okabé, France) from the inlet. After the channel surface was modified with aptamers, cell culture supernatant was introduced into the device for 10 min incubation for exosomes isolation, and captured exosomes were expected to be eluted into PBS washing buffer.

2.3. Exosomes Isolation

First, 4 mL A549 Cell culture supernatant was run and incubated through our device for exosomes capture for approximately 10 min and discharged with a pressure pump. Trapped lipid vesicles were recovered intact in 4 mL PBS buffer for 10 min. The whole experiment setup is shown in Figure 2. By pumping nitrogen gas into the flow control kit, a pressure-diven flow was created in microchannel. The flow velocity could be controlled directly by manipulated gas pressure on software. For comparison, another 4 mL A549 Cell culture supernatant was purified by using exoEasy Maxi Kit (Qiagen, Hilden, Germany) and exosomes were eluted into 4 mL buffer.

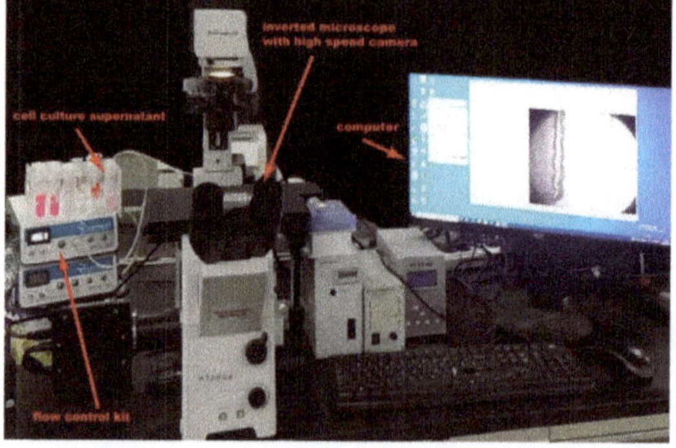

Figure 2. The experiment setup.

2.4. Exosomes Characterization

2.4.1. Morphological Characterization of Chip-Isolated Exosomes

For TEM analysis, 10 μL exosomes/PBS solution were fixed with 10 μL 2% Paraformaldehyde (PFA) (Dalian Meilun Biological Technology, Dalian, China) on a formvar–carbon coated EM grid (Beijing XXBR Technology, Beijing, China) for 30 min at room temperature [35]. The grid was floated on a drop of 2% Phosphotungstic acid hydrate (PAH) (Shenzhen Ziker Biological Technology, Shenzhen, China) three times for several seconds and used a filter paper strip to absorb excess liquid. After washing twice with PBS, the fixed exosomes were dehydrated with a descending sequence of ethanol (40%, 60%, 80%, 96–98%) [36]. Exosomes were observed under low beam energies as 1.5 kV at 1 mA by using TEM (Tecnai G2 F30).

For AFM analysis, 10μL exosomes/PBS were fixed with 10 μL 2% PFA on a freshly cleaved mica sheet for 30 min. After washing with PBS, a filter paper strip was used to absorb excess liquid from the edge of the mica sheet [37]. The sample was tested under Peak Force Modulation at a scan rate of 0.9 Hz, and the vibration amplitude of the oscillating cantilever was 300 kHz by using AFM (Bruker Dimension Icon).

2.4.2. Size Distribution and Concentration Characterization of Isolated Exosomes

The concentration and size distribution of particles in exosomes collected solutions were analyzed by using NanoSight NS300 (Malvern Instruments, Malvern, UK) equipped with video capture and particle-tracking software via measuring the rate of Brownian motion of particles.

2.4.3. Exosomes-Carried Proteins CD63 and PTK7 ELISA Characterization of Isolated Exosomes

100 μL conjugate reagent was added to each exosome sample and then incubated at 37 °C for 1 h. The exosome sample contains 10 μL exosomes recovery and 40 μL sample dilution, which means that the sample's final dilution factor is five-fold. After clearing, the samples were incubated with chromogen solution at 37 °C for 15 min and, subsequently, absorbance was read spectrophotometrically at a wavelength of 450 nm. Human Cluster of differentiation (CD63; TSPAN30) ELISA Kit and Human Protein Tyrosine Kinase 7(PTK7) ELISA Kit were purchased from China Jiangsu Meibiao Biotechnology, Yancheng, China.

3. Results and Discussion

3.1. Morphological Characterization of Chip-Isolated Exosomes

Intuitive characterization of exosomes relies on microscopy techniques with high resolution. Chip-isolated exosomes are observed under TEM and AFM after specialized staining protocols. In Figure 3, microscopic morphological diagnosis demonstrates exosomes captured by chip show intact cupped-croissant morphology in 100–200 nm that are consistent with those previously reported [36,37]. We did not use biological sample-specific probes, so the edges of exosomes are not smooth, which is due to the limited experimental conditions (Figure 3B). Although this AFM image is not particularly ideal, it is still meaningful because it provides further evidence that the exosomes are cup-shaped. Since the AFM image of exosomes captured by using exoEasy Maxi Kit has not been successfully obtained, it failed to be compared with the morphology of exosomes isolated from two methods here.

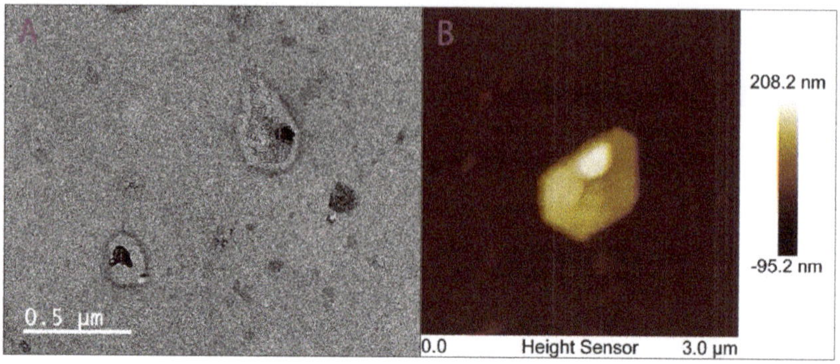

Figure 3. TEM and AFM imaging of exosomes isolated by aptamer-based microfluidics. (**A**) TEM image of EVs isolated by microfluidics presents a distorted cup-shaped morphology and uniform unimodal size distribution following 200 nm filtration. (**B**) AFM image EVs isolated by microfluidics presents a distorted cup-shaped morphology.

3.2. Concentration and Size Distribution Characterization of Isolated Exosomes

TEM and AFM can illustrate the morphological appearance of single exosome particle clearly. However, according to the limited field of view, we cannot acquire the size distribution of exosome populations by enumerating all the TEM pictures. Thus, to demonstrate the ability and efficiency of our aptamer-based microfluidics for isolation of exosomes, we used cell culture media of different passages as a single source of a specific exosome population. Passage 6–8 cell culture supernatant was prepared for NTA characterization. Isolated exosomes using exoEasy Maxi Kit served as a reference for verifying the performance of our device. The NTA results are illustrated in Figure 4. There are six samples in Figure 4, and the number "6" in Figure 4A,B represents NTA results of kit and our chip by using passage 6 cell culture supernatant, respectively. By the analogy, "kit-7" and "chip-7" means NTA result of kit and our chip by using passage 7 cell culture supernatant, respectively, and the same is true for "kit-8" and "chip-8" in Figure 4. Each sample was run three times and the standard error is calculated by the following formula: Standard error = STDEV.S/ SQRT (3). The particle population from NTA shows that enrichment of exosomes by using the kit and our chip is almost identical. It should be noted that chips' peaks show a smaller mean diameter compared to those of the kit under the circumstance of the same passage of cell culture supernatant. This indicates that exosomes isolated by using aptamer-based microfluidics are closer to the theoretical size range of exosomes, since many reviews have reported that exosomes bear smaller diameters (30–100 nm) than other EVs (100–1000 nm), which was pointed out in the introduction section. Exosomes isolated by using our chip show smaller size than those from the exoEasy Maxi Kit, so our aptamer-affinity microfluidics indeed can isolate reliable exosomes compared with the exoEasy Maxi Kit, while kit-isolated exosomes are mixed with other larger EVs.

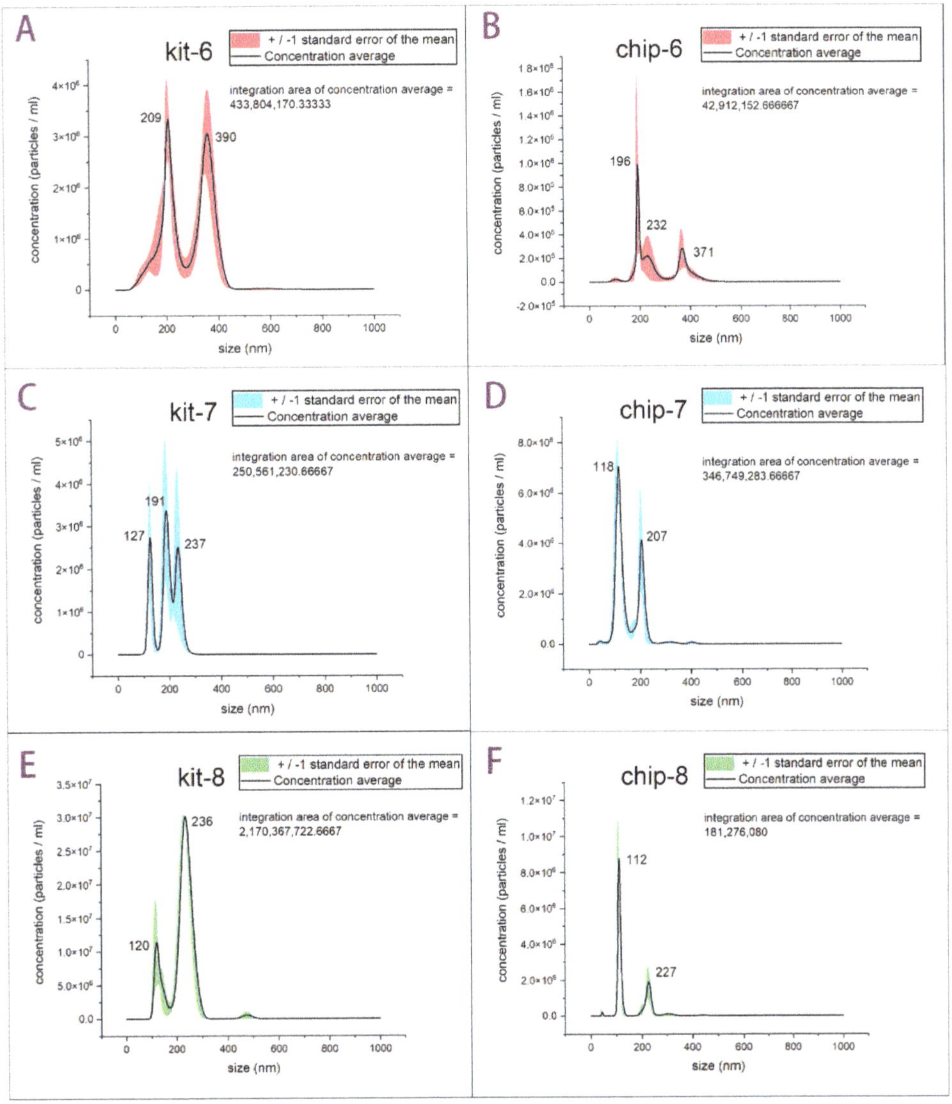

Figure 4. NTA of exosomes by using aptamer-based microfluidics and exoEasy Maxi Kit. (**A**) Averaged concentration/size distribution of exosomes collected from cell culture supernatant (passage-6) by using exoEasy Maxi Kit. (**B**) Averaged concentration/size distribution of exosomes collected from cell culture supernatant (passage-6) by using our device. Error bars indicated +/− standard error of the mean. (**C**,**D**) are the data graphs from cell culture supernatant (passage-7). (**E**,**F**) are the data graphs from cell culture supernatant (passage-8).

Combining the results of morphology and size distribution, we confirmed that our aptamer-based chip can conduct a better approach to isolate exosomes with ideal size distribution and perfect shape. Integrated results show the capture efficiency of around 10^8–10^9 particles/mL in 20 min by using exoEasy Maxi Kit. Our procedure is satisfactory with regard to capture efficiency of 10^7–10^8 particles/mL in 20 min (Figure 4). We estimated the times of each experiment were both around 20 min.

3.3. *Exosomes-Carried Proteins CD63 and PTK7 ELISA Characterization of Isolated Exosomes*

Additionally, to fully demonstrate the target-specific capturing of our aptamer-based microfluidic chip, quantitative enzyme-linked immunosorbent assay (ELISA) for exosome-carried marker CD63 and PTK7 was adopted. Considering that exosomes contain CD63 and PTK7 distinct from other EVs, they were chosen to quantitatively confirm how many particles recovered were true exosomes. We have mentioned above that these two proteins are also our aptamer targeting proteins. According to the ELISA kit use instruction, the standard density is used as the horizontal and the O.D value for the vertical. Linear regression analysis shows the positive correlation between O.D value and standard CD63 concentration, with intercept value ranging in 0.173 $\pm$ 0.078 and slope value ranging in 0.375 $\pm$ 0.027 (Figure S1A). There is a positive correlation between O.D value and standard PTK7 concentration, with intercept value ranging in 0.103 $\pm$ 0.022 and slope value ranging in 0.218 $\pm$ 0.005 (Figure S1B). Thus, the concentration of CD63 and PTK7 in the previously mentioned six samples can be determined by comparing the O.D value of the samples to the standard curve, respectively (Table S1).

In Table 1, we found that each exosome's recovery by using our aptamer-based microfluidics has a higher concentration of CD63 and PTK7 compared with its counterpart (kit-sample). The concentration difference is greater than 1 pg/mL. The comparison results are reliable due to the limit of detection being less than 0.1 pg/mL. The higher protein concentration, the more determinate exosomes in the chip recovery sample, because these two proteins are carried on the exosome's surface. The exoEasy Maxi kit claimed to adopt chromatographic spin columns for purifying exosomes and other EVs, which seems unreliable for a purification of exosomes [38]. Although the kit has a superior EVs yield (10^8–10^9 particles/mL) to our chip (10^7–10^8 particles/mL), our device endows properties of targeting and specificity that will allow one to enrich more exosomes from a complex EVs population. Further work is necessary to understand the mechanism by which our device endows properties of targeting and specificity to enrich more exosomes from complex EVs population. Table 1 summaries particle populations of "kit-6" and "kit-8" is an order of magnitude higher than those data coming from "chip-6"and "chip-8", respectively. Nevertheless, chip samples contain more exosomes of interest even when counted by relatively moderate particles. These imply that aptamer has a great potential in affinity application, such as capturing pollutant organic macromolecules, and not only in its use for targeting specific exosomes. Unfortunately, it should be pointed out that we cannot determine that only exosomes are present in chip recovery, because other EVs may also contain CD63 or PTK7, since the endosomes (the origin of exosomes) and EVs are both budding from the plasma membrane. CD 63 and PTK7 are highly enriched on the plasma membrane.

Table 1. Protein concentration of EVs recovery.

	Sample	Kit-6	Chip-6	Kit-7	Chip-7	Kit-8	Chip-8
CD63	O.D value (dilution 5-fold)	0.5106	0.581	0.5107	0.5316	0.5025	0.5839
	CD63 concentration (pg/mL)	80.39	93.48	80.41	84.29	78.88	94.01
PTK7	O.D value (dilution 5-fold)	0.541	0.548	0.518	0.561	0.568	0.587
	PTK7 concentration (pg/mL)	10.032	10.19	9.51	10.49	10.65	11.09
	Particles/mL	4.3×10^8	4.3×10^7	2.5×10^8	3.4×10^8	2.1×10^9	1.8×10^8

In summary, via aptamer-based microfluidics combined with microfilter apparatuses in our chip, we isolated intact exosomes from A549 cell culture supernatant, displaying not only high capture efficiency but also convenient operation. Our chip is satisfactory, with a capture efficiency of 10^7–10^8 particles/mL in 20 min compared with that of 10^8–10^9 particles/mL in 20 min by using a commercial kit. The cell culture supernatant rapidly contacts with aptamer modified in microchannels as it flows through, which greatly reduces the time by replacing the method of mixing cell culture supernatant with aptamer–IMB. The time is shortened without affecting the performance of aptamer, because this exosome's isolation is based on the microfluidics. Furthermore, before supernatants enter into the aptamer-based apparatus, the microfilter apparatus in our chip can filtrate impurities such as cell debris. However, when we use the exoEasy Maxi Kit, supernatants require extra operations to filter particles larger than 0.8 μm by using syringe filters. Importantly, the aptamer-based apparatus has the ability to capture exosomes from mixed EVs populations. In theory, exosomes show smaller diameters (30–100 nm) than other EVs (100–1000 nm). According to NTA, exosomes isolated by using our chip are closer to the theoretical size of exosomes than those from the exoEasy Maxi Kit. Therefore, our aptamer-affinity microfluidics can indeed isolate reliable exosomes compared with the exoEasy Maxi Kit, while there are other larger EVs mixed in with the kit recoveries. The results of ELISA showed that CD63 and PTK7 carried by exosomes are more abundant in chip recoveries, which further confirmed that the exosomes isolated by our aptamer-based microfluidics are more authentic.

4. Conclusions

Purifying the exosome populations from other EVs or other interferences from biological samples is still an open problem. In this paper, we proposed an aptamer-based microfluidics for exosome isolation, which possesses the advantages of being easy-to-use, high efficiency, and target specific. We have demonstrated the capture ability of exosomes in such aptamer-affinity based microfluidic chips from cell culture supernatants as high as 10^7–10^8 particles/mL with reliable size. In this way, our chip combining an aptamer-based apparatus with a filtration-based apparatus achieves the ability of starting from cell culture supernatant without pre-treatment, potentiating a convenient experiment. The specific aptamer targeting exosome-carried proteins CD63 and PTK7 achieved reliable exosomes isolation. It should be pointed out that we cannot qualify how many aptamers are modified on the microchannel, which is something we need to improve. We simply employed aptamers combined with the microfluidics platform, demonstrating that the aptamer has great potential to specifically capture exosomes. Furthermore, such microfluidics immobilized with aptamers can be customized to different targeted sets of the user's requirements, which will contribute to the development of enriching particles for clinical diagnosis or drug delivery in the future.

Supplementary Materials: The following supporting information can be downloaded at: https://www.mdpi.com/article/10.3390/bios12040257/s1, Figure S1: Standard curve of ELISA, Table S1: Protein concentration of EVs recovery.

Author Contributions: Methodology, analysis, and writing Z.Z.; Conceptualization, supervision, review, and editing, X.Q.; Comments and editing, Y.C. All authors have read and agreed to the published version of the manuscript.

Funding: This research was supported by the Shenzhen Research Funding, grant number JCYJ20200109142824889 and the National Natural Science Foundation of China, grant number 61875221.

Institutional Review Board Statement: Not applicable.

Data Availability Statement: Not applicable.

Conflicts of Interest: The authors declare no conflict of interest. The funders had no role in the design of the study; in the collection, analyses, or interpretation of data; in the writing of the manuscript, or in the decision to publish the results.

References

1. Kowal, J.; Tkach, M.; Thery, C. Biogenesis and secretion of exosomes. *Curr. Opin. Cell Biol.* **2014**, *29*, 116–125. [CrossRef] [PubMed]
2. van Niel, G.; D'Angelo, G.; Raposo, G. Shedding light on the cell biology of extracellular vesicles. *Nat. Rev. Mol. Cell Biol.* **2018**, *19*, 213–228. [CrossRef] [PubMed]
3. El Andaloussi, S.; Breakefield, X.O.; Wood, M.J. Extracellular vesicles: Biology and emerging therapeutic opportunities. *Nat. Rev. Drug Discov.* **2013**, *12*, 347–357. [CrossRef] [PubMed]
4. Tkach, M.; Thery, C. Communication by Extracellular Vesicles: Where We Are and Where We Need to Go. *Cell* **2016**, *164*, 1226–1232. [CrossRef] [PubMed]
5. Kourembanas, S. Exosomes: Vehicles of intercellular signaling, biomarkers, and vectors of cell therapy. *Annu. Rev. Physiol.* **2015**, *77*, 13–27. [CrossRef]
6. Yanez-Mo, M.; Siljander, P.R.; Andreu, Z.; Zavec, A.B.; Borras, F.E.; Buzas, E.I.; Buzas, K.; Casal, E.; Cappello, F.; Carvalho, J.; et al. Biological properties of extracellular vesicles and their physiological functions. *J. Extracell Vesicles* **2015**, *4*, 27066. [CrossRef]
7. Colombo, M.; Raposo, G.; Thery, C. Biogenesis, secretion, and intercellular interactions of exosomes and other extracellular vesicles. *Annu. Rev. Cell Dev. Biol.* **2014**, *30*, 255–289. [CrossRef]
8. Raposo, G.; Stoorvogel, W. Extracellular vesicles: Exosomes, microvesicles, and friends. *J. Cell Biol.* **2013**, *200*, 373–383. [CrossRef]
9. Zou, J.; Shi, M.; Liu, X.; Jin, C.; Xing, X.; Qiu, L.; Tan, W. Aptamer-Functionalized Exosomes: Elucidating the Cellular Uptake Mechanism and the Potential for Cancer-Targeted Chemotherapy. *Anal. Chem.* **2019**, *91*, 2425–2430. [CrossRef]
10. Pols, M.S.; Klumperman, J. Trafficking and function of the tetraspanin CD63. *Exp. Cell Res.* **2009**, *315*, 1584–1592. [CrossRef]
11. Peradziryi, H.; Tolwinski, N.S.; Borchers, A. The many roles of PTK7: A versatile regulator of cell-cell communication. *Arch. Biochem. Biophys.* **2012**, *524*, 71–76. [CrossRef] [PubMed]
12. Bobrie, A.; Colombo, M.; Krumeich, S.; Raposo, G.; Thery, C. Diverse subpopulations of vesicles secreted by different intracellular mechanisms are present in exosome preparations obtained by differential ultracentrifugation. *J. Extracell Vesicles* **2012**, *1*, 18397. [CrossRef] [PubMed]
13. Xu, R.; Simpson, R.J.; Greening, D.W. A Protocol for Isolation and Proteomic Characterization of Distinct Extracellular Vesicle Subtypes by Sequential Centrifugal Ultrafiltration. *Methods Mol. Biol.* **2017**, *1545*, 91–116. [PubMed]
14. Contreras-Naranjo, J.C.; Wu, H.J.; Ugaz, V.M. Microfluidics for exosome isolation and analysis: Enabling liquid biopsy for personalized medicine. *Lab Chip* **2017**, *17*, 3558–3577. [CrossRef]
15. Gurunathan, S.; Kang, M.H.; Jeyaraj, M.; Qasim, M.; Kim, J.H. Review of the Isolation, Characterization, Biological Function, and Multifarious Therapeutic Approaches of Exosomes. *Cells* **2019**, *8*, 307. [CrossRef]
16. Liga, A.; Vliegenthart, A.D.; Oosthuyzen, W.; Dear, J.W.; Kersaudy-Kerhoas, M. Exosome isolation: A microfluidic road-map. *Lab Chip* **2015**, *15*, 2388–2394. [CrossRef]
17. van der Pol, E.; Hoekstra, A.G.; Sturk, A.; Otto, C.; van Leeuwen, T.G.; Nieuwland, R. Optical and non-optical methods for detection and characterization of microparticles and exosomes. *J. Thromb. Haemost.* **2010**, *8*, 2596–2607. [CrossRef]
18. Lin, B.; Lei, Y.; Wang, J.; Zhu, L.; Wu, Y.; Zhang, H.; Wu, L.; Zhang, P.; Yang, C. Microfluidic-Based Exosome Analysis for Liquid Biopsy. *Small Methods* **2021**, *5*, e2001131. [CrossRef]
19. Yu, Z.; Lin, S.; Xia, F.; Liu, Y.; Zhang, D.; Wang, F.; Wang, Y.; Li, Q.; Niu, J.; Cao, C.; et al. ExoSD chips for high-purity immunomagnetic separation and high-sensitivity detection of gastric cancer cell-derived exosomes. *Biosens. Bioelectron.* **2021**, *194*, 113594. [CrossRef]
20. Kanwar, S.S.; Dunlay, C.J.; Simeone, D.M.; Nagrath, S. Microfluidic device (ExoChip) for on-chip isolation, quantification and characterization of circulating exosomes. *Lab Chip* **2014**, *14*, 1891–1900. [CrossRef]
21. Smith, J.T.; Wunsch, B.H.; Dogra, N.; Ahsen, M.E.; Lee, K.; Yadav, K.K.; Weil, R.; Pereira, M.A.; Patel, J.V.; Duch, E.A.; et al. Integrated nanoscale deterministic lateral displacement arrays for separation of extracellular vesicles from clinically-relevant volumes of biological samples. *Lab Chip* **2018**, *18*, 3913–3925. [CrossRef]
22. Wang, Z.; Wu, H.J.; Fine, D.; Schmulen, J.; Hu, Y.; Godin, B.; Zhang, J.X.; Liu, X. Ciliated micropillars for the microfluidic-based isolation of nanoscale lipid vesicles. *Lab Chip* **2013**, *13*, 2879–2882. [CrossRef] [PubMed]
23. Omid, C.; Farokhzad, J.C.; Teply, B.A.; Sherifi, I.; Jon, S.; Kantoff, P.W.; Richie, J.P.; Langer, R. Targeted nanoparticle-aptamer bioconjugates for cancer chemotherapy in vivo. *Proc. Natl. Acad. Sci. USA* **2006**, *103*, 6315–6320.
24. Dunn, M.R.; Jimenez, R.M.; Chaput, J.C. Analysis of aptamer discovery and technology. *Nat. Rev. Chem.* **2017**, *1*, 1–16. [CrossRef]
25. Iliuk, A.B.; Hu, L.; Tao, W.A. Aptamer in bioanalytical applications. *Anal. Chem.* **2011**, *83*, 4440–4452. [CrossRef] [PubMed]
26. Song, S.; Wang, L.; Li, J.; Fan, C.; Zhao, J. Aptamer-based biosensors. *TrAC Trends Anal. Chem.* **2008**, *27*, 108–117. [CrossRef]
27. Zhao, X.; Luo, C.; Mei, Q.; Zhang, H.; Zhang, W.; Su, D.; Fu, W.; Luo, Y. Aptamer-Cholesterol-Mediated Proximity Ligation Assay for Accurate Identification of Exosomes. *Anal. Chem.* **2020**, *92*, 5411–5418. [CrossRef]
28. Yu, X.; He, L.; Pentok, M.; Yang, H.; Yang, Y.; Li, Z.; He, N.; Deng, Y.; Li, S.; Liu, T.; et al. An aptamer-based new method for competitive fluorescence detection of exosomes. *Nanoscale* **2019**, *11*, 15589–15595. [CrossRef]
29. Liu, C.; Zhao, J.; Tian, F.; Chang, J.; Zhang, W.; Sun, J. lambda-DNA- and Aptamer-Mediated Sorting and Analysis of Extracellular Vesicles. *J. Am. Chem. Soc.* **2019**, *141*, 3817–3821. [CrossRef]
30. Zhang, K.; Yue, Y.; Wu, S.; Liu, W.; Shi, J.; Zhang, Z. Rapid Capture and Nondestructive Release of Extracellular Vesicles Using Aptamer-Based Magnetic Isolation. *ACS Sens.* **2019**, *4*, 1245–1251. [CrossRef]

31. Song, Z.; Mao, J.; Barrero, R.A.; Wang, P.; Zhang, F.; Wang, T. Development of a CD63 Aptamer for Efficient Cancer Immunochemistry and Immunoaffinity-Based Exosome Isolation. *Molecules* **2020**, *25*, 5585. [CrossRef] [PubMed]
32. Jiang, Y.; Shi, M.; Liu, Y.; Wan, S.; Cui, C.; Zhang, L.; Tan, W. Aptamer/AuNP Biosensor for Colorimetric Profiling of Exosomal Proteins. *Angew. Chem. Int. Ed. Engl.* **2017**, *56*, 11916–11920. [CrossRef]
33. Huang, B.; Wu, H.; Kim, S.; Kobilka, B.K.; Zare, R.N. Phospholipid biotinylation of polydimethylsiloxane (PDMS) for protein immobilization. *Lab Chip* **2006**, *6*, 369–373. [CrossRef] [PubMed]
34. Lo, T.W.; Zhu, Z.; Purcell, E.; Watza, D.; Wang, J.; Kang, Y.T.; Jolly, S.; Nagrath, D.; Nagrath, S. Microfluidic device for high-throughput affinity-based isolation of extracellular vesicles. *Lab Chip* **2020**, *20*, 1762–1770. [CrossRef]
35. Muller, L.; Hong, C.S.; Stolz, D.B.; Watkins, S.C.; Whiteside, T.L. Isolation of biologically-active exosomes from human plasma. *J. Immunol. Methods* **2014**, *411*, 55–65. [CrossRef] [PubMed]
36. Sokolova, V.; Ludwig, A.K.; Hornung, S.; Rotan, O.; Horn, P.A.; Epple, M.; Giebel, B. Characterisation of exosomes derived from human cells by nanoparticle tracking analysis and scanning electron microscopy. *Colloids Surf. B Biointerfaces* **2011**, *87*, 146–150. [CrossRef]
37. Sharma, S.; Rasool, H.I.; Palanisamy, V.; Mathisen, C.; Schmidt, M.; Wong, D.T.; Gimzewski, J.K. Structural-Mechanical Characterization ofNanoparticle Exosomes in Human Saliva, Using Correlative AFM, FESEM, and Force Spectroscopy. *ACS Nano* **2010**, *4*, 1921–1926. [CrossRef]
38. Bruce, T.F.; Slonecki, T.J.; Wang, L.; Huang, S.; Powell, R.R.; Marcus, R.K. Exosome isolation and purification via hydrophobic interaction chromatography using a polyester, capillary-channeled polymer fiber phase. *Electrophoresis* **2019**, *40*, 571–581. [CrossRef]

biosensors

MDPI

Review

Microfluidic Strategies for Extracellular Vesicle Isolation: Towards Clinical Applications

Alessio Meggiolaro [1], Valentina Moccia [2], Paola Brun [3], Matteo Pierno [1], Giampaolo Mistura [1], Valentina Zappulli [2] and Davide Ferraro [1,*]

[1] Department of Physics and Astronomy, University of Padua, Via Marzolo 8, 35131 Padua, Italy
[2] Department of Comparative Biomedicine and Food Science, University of Padua, Viale dell'Università 16, 35020 Legnaro, Italy
[3] Department of Molecular Medicine, University of Padua, Via Gabelli 63, 35121 Padua, Italy
* Correspondence: davide.ferraro@unipd.it

Abstract: Extracellular vesicles (EVs) are double-layered lipid membrane vesicles released by cells. Currently, EVs are attracting a lot of attention in the biological and medical fields due to their role as natural carriers of proteins, lipids, and nucleic acids. Thus, they can transport useful genomic information from their parental cell through body fluids, promoting cell-to-cell communication even between different organs. Due to their functionality as cargo carriers and their protein expression, they can play an important role as possible diagnostic and prognostic biomarkers in various types of diseases, e.g., cancers, neurodegenerative, and autoimmune diseases. Today, given the invaluable importance of EVs, there are some pivotal challenges to overcome in terms of their isolation. Conventional methods have some limitations: they are influenced by the starting sample, might present low throughput and low purity, and sometimes a lack of reproducibility, being operator dependent. During the past few years, several microfluidic approaches have been proposed to address these issues. In this review, we summarize the most important microfluidic-based devices for EV isolation, highlighting their advantages and disadvantages compared to existing technology, as well as the current state of the art from the perspective of the use of these devices in clinical applications.

Keywords: extracellular vesicles; microfluidics; purification; liquid biopsy; microfabrication; clinics

Citation: Meggiolaro, A.; Moccia, V.; Brun, P.; Pierno, M.; Mistura, G.; Zappulli, V.; Ferraro, D. Microfluidic Strategies for Extracellular Vesicle Isolation: Towards Clinical Applications. *Biosensors* **2023**, *13*, 50. https://doi.org/10.3390/bios13010050

Received: 18 November 2022
Revised: 23 December 2022
Accepted: 24 December 2022
Published: 29 December 2022

1. Introduction

Discoveries in genomics are leading to important outcomes in medicine, improving knowledge of many diseases and leading to the concept of "precision medicine", which is defined as the tailoring of medical treatment to individual characteristics [1]. For example, after a cancer diagnosis, the first approach is often a surgical biopsy to identify the type of tumor by specific marker expression or by genomic analysis [2]. Unfortunately, the latter is an invasive and time-consuming procedure, may not be representative of the entire tumor, and may cause cancer seeding [3]. To face these issues, much attention has been paid to a less invasive procedure called *liquid biopsy*: body fluids (e.g., blood, urine, saliva) are screened for tracers released by cancer tissues, which can provide more rapid and complete information about the original tumor (e.g., type, stage, progression, etc.) and could be used as prognostic and/or diagnostic tools [4]. The most well-known tracers are circulating tumor cells (CTCs) and circulating tumor DNA (ctDNA). The first are cells that are spontaneously released from the cancer tissue and travel in the patient's blood [5]. Similarly, ctDNA are nucleic acid fragments presenting specific tumor mutations that are released from cancer cells and travel in body fluids [6].

Another type of *tracer* that has been discovered in recent decades as potentially useful for liquid biopsy are extracellular vesicles (EVs) [7]. These are double-layered phospholipid membrane structures, released by most cell types, which travel in body fluids carrying various biological molecules of the parental cell (i.e., proteins, lipids, and nucleic acids).

The biogenesis of EVs is mainly related to two pathways: (i) the direct outward budding of the cell membrane and ii) the inward budding of multivesicular bodies that fuse with the cell surface to then be released. In the former case, EVs are known as ectosomes (or microvesicles (MVs), or microparticles), with a size of 100 nm to 1000 nm, and in the latter case, they are known as small extracellular vesicles (sEVs) or exosomes, with a size ranging between 30 and 200 nm [8,9]. Although initially considered cell debris or cell waste, it is now recognized that EVs play a role in cell-to-cell communication, acting as cargo ships between cells by transporting genetic information [10], and therefore participating in a variety of physiological and pathological processes [11]. Therefore, EVs are perfectly suitable for liquid biopsy and are now considered promising diagnostic, predictive, and prognostic biomarkers for many types of diseases. In fact, unlike CTC and cDNA, EVs can provide a variety of information, e.g., either on cardiovascular [12], autoimmune [13], and neurodegenerative [14] diseases, or on various types of cancer [15].

Today, given the invaluable importance of EVs for liquid biopsy, there are some key challenges to overcome regarding their isolation [16]. The current most frequently used approaches, described in Section 2, are based on differential ultracentrifugation (DU), size-exclusion chromatography (SEC), density-gradient separation (DGS), filtration, and immunoaffinity strategies. Additionally, EVs can be collected from various fluids (e.g., cell culture, blood, urine, etc.); thus, the same isolation strategy may present different efficiencies depending on the starting sample [17]. Finally, most of the methods require at least several hours for EV isolation, and thus more rapid isolation protocols are also demanding.

Microfluidic devices have recently been proposed for addressing these issues, as demonstrated by the increasing number of published papers that have appeared over the past ten years on this topic. Figure 1 compares the publications per year obtained using the terms *extracellular vesicles* (or *exosomes*) (Figure 1a) and together with *microfluidics* (Figure 1b) as keywords. Both trends are similarly increasing; however, the ratio between the two numbers (Figure 1b, inset) reports how microfluidics has gained slightly more visibility during the last five years. Notably, considering the low number of articles per year in the microfluidic case, this trend must be monitored in the near future. Microfluidics is commonly defined as the science and technology of systems that manipulate small amounts of fluids (pL and nL ranges), using channels with dimensions that typically range from tens to hundreds of microns [18]. This leads to several advantages, including the development of a portable system for point-of-care analysis, the reduction in sample and reagent volume, down to a million times more than conventional approaches, and the ability to perform parallelized assays that can drastically increase analysis throughput [19]. Given these benefits, it is clear that microfluidics can contribute to simplifying and speeding up the EV isolation process from biofluids, representing a good alternative to conventional protocols. Additionally, microfluidic devices can also be exploited for EV analysis and detection, being embedded within the same microfluidic system or based on other instruments (e.g., a fluorescence microscope). In the latter case, microfluidic devices can be seen as passive tools for EV storage.

In this review, we aim to address the most relevant microfluidic systems devoted to EV isolation, underlining both advantages and disadvantages compared to the conventional existing methods. After reviewing the most common isolation methods, microfluidic approaches are discussed, with particular emphasis on those that seem more promising for future clinical applications. In this context, EVs must be isolated by a microfluidic device and ready for further analysis (see the workflow in Figure 2). Given the variety of microfluidic devices in terms of microfabrication and functionality, they are divided into two main categories: physical and chemical approaches. Whereas the former can be distinguished in active and passive methods, the latter are mainly based on immunocapture on fixed and non-fixed (beads) substrates. In addition, a quantitative analysis of the diffusion of the various microfluidic methods, as well as their capabilities of being used in real clinical studies, is presented. A short description of EV detection methods based on microfluidics is also introduced; however, for a deeper understanding, a dedicated review can already be found in the literature [20].

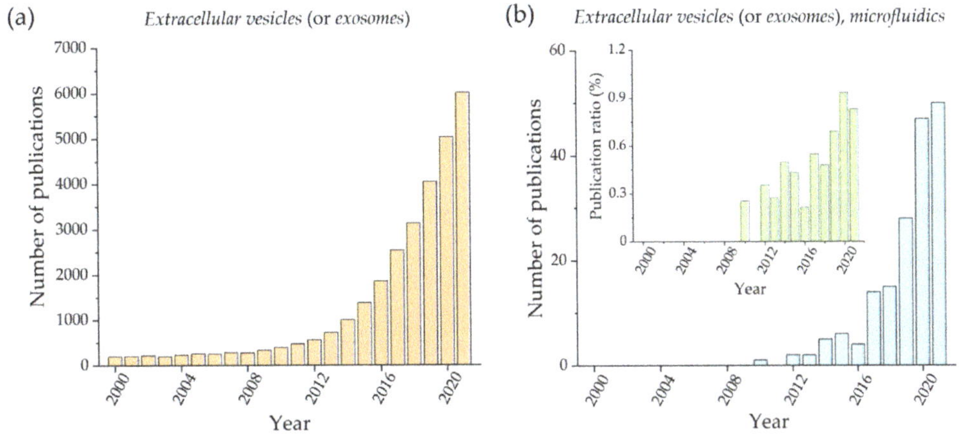

Figure 1. Number of publications per year that contain the keywords: *extracellular vesicles* (or *exosomes*) (**a**) and *extracellular vesicles* (or *exosomes*) and *microfluidics* (**b**). An inset with the ratio between these two numbers is also reported. (Source: Web of Science database, excluding review, meeting abstract, and retracted papers).

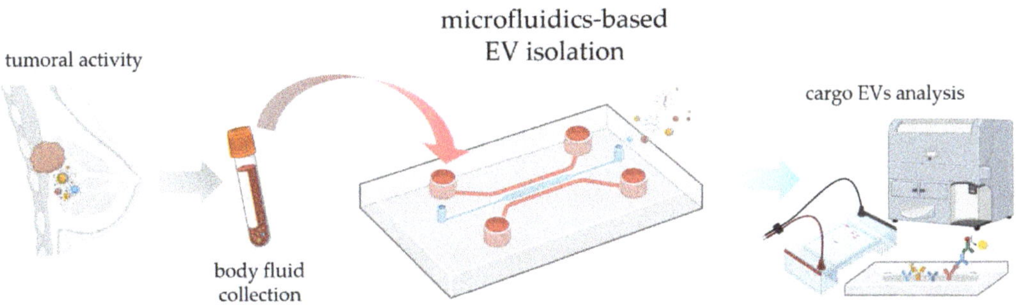

Figure 2. Possible workflow of tumor diagnosis using the microfluidic EV isolation strategy, from sample collection by liquid biopsy to analysis. The elements of the figure are created by BioRender.com.

Importantly, in recent years, different terminology has been used in the literature to classify EVs based on their size or function, since their biogenesis was not easily assessed; however, in 2018, the International Society for Extracellular Vesicles (ISEV) indicated using the word *Extracellular Vesicle* (or EV) as a broad generic term that includes all subtypes of vesicles to avoid confusion in the literature. This recommendation is followed in this review, using the term small extracellular vesicles (sEVs) when the cited articles refer to exosomes or EVs smaller than <200 nm [9].

2. Conventional EV Isolation Strategies

Conventional methods for the isolation and purification of extracellular vesicles can be classified into methods based on the morphological properties (i.e., size, density) and based on their interaction with specific components (solubility, protein reaction). The preferential strategy must be chosen as a function of the initial sample (e.g., cell culture, blood, urine) and the scope of the analysis (e.g., quantity evaluation, diagnosis of specific diseases) [9]. In the following paragraphs, conventional methods will be briefly introduced, reporting their working principle and focusing on their advantages and disadvantages in terms of EV purification. More details on these methods can be found in a specific review [17].

2.1. Differential Ultracentrifugation and Density Gradient Ultracentrifugation

Differential ultracentrifugation (DU) was the first approach used for EV isolation. In general, centrifugation is a label-free method that allows accelerating the natural sedimentation rate of suspended objects that are denser than the surrounding medium [21]. In the case of EV isolation, protocols are typically based on increasing the centrifugal force to progressively remove first cell debris (approximately 1500× *g*), then large EVs (between 10,000 and 20,000× *g*), and finally, to collect small EVs (100,000–200,000× *g*) appearing as a small pellet at the bottom of the centrifugal tube [21]. The limitations of this method are related to the need for expensive equipment (i.e., an ultracentrifuge) and the variable recovery rate (between 5 and 80%), which can often be operator related, preventing comparisons between different studies. In addition, isolated objects are pelleted according to their density, and thus collected materials also contain protein complexes and non-EV nanoparticles (e.g., apoptotic bodies, viruses), leading to an incomplete separation [22]. Additionally, EVs may also cauterize together due to strong centrifugal force. Many different DU protocols can be found in the literature that may puzzle those who are in the research field for the first time. Despite these issues, probably due to its simplicity and recovery rates that can be rather high, DU remains the most widely used approach in research laboratories and is typically combined with other filtration-based EV isolation methods.

Density gradient ultracentrifugation (DGU) is a variation of DU that consists of the addition of specific components (e.g., sucrose) within the suspending medium in order to match the EV density, while allowing the other components to precipitate [23]. Although DGU allows for gaining higher EV purity than DU, some limitations are also reported: the process is time consuming, and molecules with similar EV density (e.g., high-density lipoproteins) can be co-isolated.

2.2. Filtration Methods (Ultrafiltration and Size-Exclusion Chromatography)

Filtration methods are based on the use of a porous membrane to filter objects larger than the porous size [24]. Since EVs typically have sub-micrometric size, membranes with pore sizes between 0.001 μm and 0.01 μm are used combined with ultracentrifugation, in the so-called ultrafiltration (UF) technique; UF allows for faster protocols and better sample quality than DU in terms of purity from protein co-isolation. However, the recovery yield can be biased by the pore size [25].

A highly used filtration-based technique for EV isolation is size exclusion chromatography (SEC), consisting of the elution of EVs in a column composed of packed porous polymeric beads [26,27]. This simple strategy allows for isolating intact EVs from various biological fluids, preserving their biophysical properties, sharp-peaked distribution in size, and high functionality. Thus, among isolation strategies based on the physical properties of EVs, in particular their size, SEC is considered the least invasive in terms of EV integrity [28]. On the other hand, an important disadvantage is the low recovery rate: SEC can be applied to concentrate EV fluids (such as plasma), but in the case of low initial EV concentration (e.g., cell culture media), a pre-concentration step by UF is required. Despite the similarity to microfluidic technologies based on separation by size through pores, SEC relies on the passive motion of particles in a stationary phase, and it is highly time-consuming. Importantly, SEC can also co-isolate other components, such as viruses, protein aggregates, large proteins, and low-density lipoproteins; however, these contaminations are typically less compared to other EV-isolation methods. Today, several commercially available SEC kits can be found specifically designed for EV isolation, depending on the volume and quality of the input sample [29]. It should also be noted that recent studies are trying to improve the capabilities of a standard SEC column, for example, by using a bead size gradient as in particle purification liquid chromatography (PPLC) [30,31]. In any case, despite the good purity of the final sample, SEC requires high costs for disposable filtering columns, in addition to long isolation times.

2.3. Precipitation and Immunoaffinity Methods

Physical–chemical interactions between EVs and solid support are also exploited for their isolation. These approaches are based on EV precipitation [32], adjusting their solubility by chemical compounds, and immunochemistry reactions that exploit the protein present on their surface [33]. EV precipitation can be achieved by properly adjusting the concentration of specific polymers within the starting sample, leading to very simple protocols (e.g., ExoQuick® [34]). However, the final samples are contaminated by the polymer used, which may compromise the downstream analysis. In contrast, immunochemistry reactions are based on the chemical binding between proteins on the EV membrane and specific antibodies, typically grafted onto surfaces or beads. In more detail, some specific tetraspanin molecules are present on most EV membranes (e.g., CD63, CD81, CD9, and others) and are typically used for this purpose [35]. It is noteworthy that the same approach can be applied to isolate a subpopulation of EVs that presents specific membrane proteins associated with a specific EV subtype. The advantages of this approach are its simplicity, the fact that it does not require specific training by the user, and its reproducibility. However, to avoid nonspecific interaction, pre-purification steps are typically required (e.g., differential ultracentrifugation). In addition, isolation kits for specific immunocapturing are usually costly. Another drawback is intrinsic in the approach itself: by selecting the EVs from their surface markers, a subpopulation is always collected, and this could eventually bias the downstream analysis. For this reason, it is preferable to use multimarker antibody cocktails to recover vesicles characterized by different antigens or secreted from heterogeneous cells [36,37].

2.4. Comments

In summary, it is clear that all the conventional approaches listed above have both advantages and disadvantages, as highlighted in Table 1, and the choice between them must be made according to the scope of the study. However, it is important to note that although the chosen approach is the same, the specific parameters for the isolation of EVs are adapted differently from time to time in different laboratories, leading to a lack of standardization and important inconveniencies in comparing the data. On the contrary, microfluidic devices have the potential to overcome some issues, such as the need for expensive facilities and consumables and large sample volumes, which are peculiar to conventional isolation techniques.

Table 1. Summary of the comparison of conventional techniques for EV isolation, based on different isolation strategies. Retention time, quality, and quantity of processed sample and protocol simplicity are evaluated according to the following scale: (–) difficult/very bad, (-) non-trivial/mediocre, (+) easy/good, (++) very easy/excellent.

	Working Principle	Retention Time	Output Sample (Quality and Quantity)	Simplicity
Differential ultracentrifugation	particle size	-	-	++
Density gradient ultracentrifugation	particle density	–	–	++
Ultrafiltration	particle size	–	-	+
Size-exclusion chromatography	particle size	–	-	+
Field-flow fractionation	particle size	-	+	+
Precipitation-based	particle–polymer interaction	-	+	+
Immunoaffinity-based	antigen–antibody binding	-	+	+

3. EV Isolation Methods Based on Microfluidic Devices

Microfluidics is typically applied to bioanalytical protocols by following two different approaches: (i) the miniaturization of existing methods or (ii) the development of new methods that cannot be performed without miniaturization. The first way allows for automating processes while reducing starting volumes, sometimes leading to increased throughput of the analysis and quality of the output sample. In contrast, the second approach tackles the conventional limitations from a completely different angle, leading to results that cannot be compared with existing large-scale methods.

Among the possible ways to classify microfluidic devices, they can be distinguished between "physical" or "chemical" methods according to the nature of forces that regulate the EV isolation process.

3.1. *Physical Methods*

Physical isolation methods are label-free and exploit physical properties to discriminate vesicles (i.e.,: size or density). They can respond to external physical forces or be based on passive EV collection. In the following, physical approaches are divided into passive or active methods, depending on the presence or absence of driving forces that trigger the physical characteristics of the EVs or of the medium in which they are dispersed.

3.1.1. Passive Approaches

Passive separation methods are label-free isolation strategies that do not require external forces or stimuli. They are intended to enrich EVs by filtering processes through membranes integrated in microfluidic channels or by exploiting hydrodynamic flow properties.

Filtration. A simple method to separate EVs from the initial biological sample based on size requires the use of filtration systems, such as a nanoporous membrane, that allow the passage of vesicles having a dimension smaller than the pore size by exploiting pressure provided by external syringe pumps or by pressure controllers. Inspired by UF and SEC, microfluidic protocols have important advantages in terms of cost, required sample volume, and automation. Filters such as polycarbonate track-etched membranes were integrated within microfluidic devices during the fabrication process by the authors of [38–44]. In this way, the final device is simple to use: it does not require labels or surface treatments, allowing for the processing of very large sample volumes. Nevertheless, the production of such devices can be very complicated because of the strong microfabrication skills required, sometimes preventing mass production. As in the case of conventional filters, these devices can be prone to clogging and are typically disposable; additionally, isolation through filters lacks specificity, except for size. Liang et al. presented a prototyping example, developing a polycarbonate-based double-filtration system to isolate vesicles within a range of 30–200 nm starting from the urine of patients with bladder cancer [38]. An isolation chamber is devoted to collect EVs smaller than the pore size of 200 nm of the first membrane, and particles smaller than 30 nm are trapped through a second filter in a waste chamber (Figure 3a). This method allowed for the isolation of EVs from 8 mL of the pre-centrifuged initial sample in approximately 3.5 h, using a flow rate of 40 μL/min. Further multiple-step filters are arranged in the Exosome Total Isolation Chip (ExoTIC), in which five membranes in a series (pores of 200, 100, 80, 50, and 30 nm) allow for a strict differentiation in the size of vesicles from various types of samples, such as plasma, urine, and lavage [39]. The starting samples have an upper volume limit that can range from 20 mL for cell culture to 500 μL for plasma. Another device called Exodisc takes advantage of centrifugal force with double-step filtering through membranes of different pore sizes [44]. Woo et al. were able to isolate exosome from a 1 mL starting sample of cell culture or urine in 30 min with high purity (95% yield). Sunkara et al. optimized the same platform for processing whole blood, despite using smaller samples (30 μL) and reaching lower recovery rates (exceeding 75%) [45]. Other types of filtering devices rely on cross-flow (tangential-flow) processes, in which the feed flows across the surface of the membrane that acts to concentrate EVs larger than the pore size (30–50 nm). These devices can pair the cross-flow strategy with conventional (dead-end) filtration [41,42] or with other isolation techniques, such as immunoaffinity-based capture [43], to improve purification efficiency (see also Section 3.2.1).

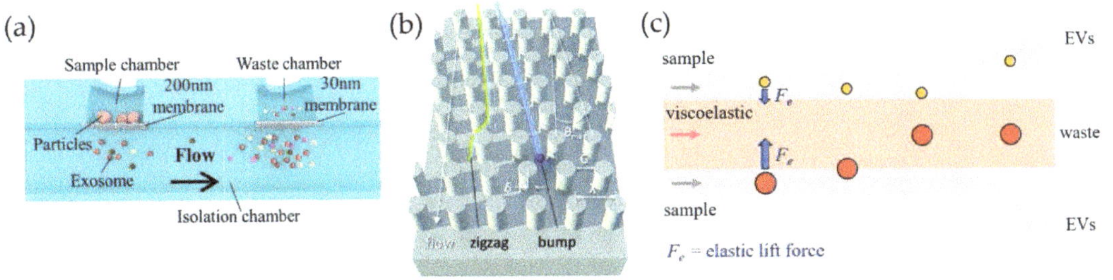

Figure 3. EV isolation methods based on passive approaches. (**a**) Double-membrane system to enrich sEVs (exosomes) of sizes between 30 and 200 nm [38]; (**b**) silicon array of pillars placed inside the chip to induce different zigzag pathways based on particle dimension [46]; (**c**) microfluidic device allowing collection of extracellular vesicles from initial sample mediated by elastic force from viscoelastic fluid.

Inertial force. EV separation can also be performed by exploiting the inertial lift force $\boldsymbol{F_i}$ that particles experience while flowing in a microchannel due to the Poiseuille flow profile [47–49]. In fact, $\boldsymbol{F_i}$ acts to drive the particles orthogonally to the flow direction inside the microchannel in a manner that strongly depends on particle dimension $\boldsymbol{D}$ ($\boldsymbol{F_i} \propto \boldsymbol{D^4}$). Therefore, by properly tuning the channel size and flow rates, it is necessary to focus the particles according to their size [50–55]. A particular configuration consists of spiral channels that strongly favor the lateral migration of particles with different velocities depending on size [56], as used by Tay et al. for rapid isolation from a whole blood sample at 80 μL/min, despite reaching a poor recovery efficiency (20–60%) [57].

Deterministic lateral displacement (DLD). A different approach exploits inertial microfluidics and hydrodynamic interactions of particles with structured channels: particles flowing in microchannels, other than the main force that drives them along the channel itself, also experience lateral forces depending on their size. This effect can be combined with specific and ordered patterns of pillars inside microfluidic channels (see Figure 3b), leading to the so-called deterministic lateral displacement (DLD), which allows for the generation of streamlines that the particles follow depending on the distance of centers $\boldsymbol{\lambda}$ and the gap of the pillars $\boldsymbol{G}$, as well as on the offset angle $\boldsymbol{\theta}$. Their separation can occur whenever the particle size exceeds a critical diameter, which for circular pillars equals $\boldsymbol{D_C = 1.4\ G\tan\theta^{0.48}}$, which acts as a cut-off [58]. Specific details about the physical principles of DLD are discussed in devoted articles [59]. DLD has been widely applied for hydrodynamic cell separation depending on the cells' size, and it has also recently been applied to nanometric objects, such as EVs. For example, the pioneering work of Wunsch et al. provided a sharp resolution of particles between 20 and 110 nm separated with a silicon-based pillar array [60]. The production of these pillars, having gaps from 25 to 235 nm, required a sequence of complex micro- and nano-fabrication steps including photolithography followed by reactive-ion etching, then an electron beam process, and finally, deep-UV lithography. Later, Smith et al. integrated 1024 arrays in parallel in another nanoDLD device (Figure 3b) to isolate EVs from serum and plasma, using a similar fabrication approach [46]. This improved design allows for faster isolation processing (up to 900 μL/h), despite an EV recovery of approximately 50%. Other devices, reporting pillar dimensions and gaps of the order of microns, are instead replicated from silicon wafers produced by standard photolithographic and etching techniques, allowing one to reach a high purity of the final sample, but working at low throughput (of the order of μL/h) [61,62].

Viscoelastic force. Most bodily fluids (such as blood, saliva, semen, etc.) exhibit a non-Newtonian behavior when flowing through channels [63]. This viscoelastic property can be leveraged to separate particles by size by driving a lateral migration owing to the elastic lift forces, without external fields. Specifically, the trajectory of the particles is regulated by the first normal stress difference (N_1), inducing the lateral motion towards the points of minimum shear rate, with relaxation time dependent on medium properties and channel width [64–66]. The resulting elastic lift force F_e depends on the cube of the particle size ($F_e \sim D^3$), and therefore, taking into account a device presenting several outlets (see Figure 3c), larger particles migrate faster to the center line of the channel, whereas smaller EVs are collected at the two sides [67–69]. Unlike DLD-based devices, particles immersed in viscoelastic media can be focused simply by adjusting the rate and width of microchannels, without requiring additional micro- or nanofabricated structures [67,68]. However, to enhance the elastic effect and guarantee good hydrodynamic focusing, specific polymers can be added to the starting samples. As examples, Liu et al. (2017) added a low concentration (0.1 wt%) of a biocompatible polymer to the cell culture medium or serum sample, namely, poly(oxyethylene) (or PEO), to enhance these effects and better control the separation of EVs, achieving high purity and recovery rates greater than 80% and 90%, respectively [70]. Then, in 2019, a similar approach was used to simultaneously separate particles by size and based on membrane protein EVs from breast cancer cell lines and from serum, by using double-stranded λ-DNA molecules in TBE buffer to increase the non-Newtonian effect [71]. In this case, the extracellular vesicles are subjected to the centerline-directed elastic lift force F_e; additionally, larger microvesicles and apoptotic bodies are repelled by the elastic force, competing with the and drag forces F_d ($F_d \propto D$). Asghari et al. exploited oscillatory flows to separate micrometer and sub-micrometer constituents from HEK293T cell lines and was able to focus both λ-DNA strands and vesicles in a sheathless flow [72]. For this purpose, a more complex setup is needed to perform the EV separation, including a pressure-driven chip coupled with an electronic device to actuate valves and generate controlled flow oscillations.

Flow fractionation methods. Another possible way to separate microparticles by size by exploiting hydrodynamic forces is provided by asymmetric flow field-flow fractionation (AF4) [73]. This method requires the implementation of thin microchannels (dozens of μm) having one side made of a membrane that allows the generation of a flow perpendicular to the main stream [74]. Thus, the injected sample under laminar flow conditions is subjected to both the cross-flow field and Brownian diffusion. The accumulation of particles is regulated by the competition of these two counteracting forces, which induce large particles to move in proximity to the membrane, and the smaller particles are easily conveyed along the stream. Typically applied for polymer and protein fractionation, AF4 has been used to isolate EVs, being capable of separating two different subpopulations of vesicles by size (60–80 nm and 90–120 nm) from several tumor cell lines [75]. Shin et al. employed a similar fractionation approach, known as EV separation pinched-flow fractionation (PFF) [76], to isolate EVs from apoptotic bodies [77]. Here, a sheath fluid is applied to achieve EVs by focusing within the microchannel.

3.1.2. Active Approaches

Whenever physical forces are applied to fluids that contain suspended particles, they can respond to the stimulus by changing their motion. Active separation exploits applied fields, such as acoustic or electrical ones, without resorting to channel functionalization or patterning, being label-free and contact-free.

Acoustofluidics. Acoustofluidic devices combine the ability of microfluidics to handle small volumes in confined channels with the ability to trigger particle motion with acoustic waves [78–80]. This label-free and contact-free method employs ultrasound waves to induce differential forces on particles according to their size [81]. Particles can be trapped, separated, focused, or transported by regulating the properties of acoustic waves, which can propagate within the bulk material (bulk acoustic waves, BAWs) [82,83] or along the surface of the medium (surface acoustic waves, SAWs) [84]. Importantly, to emit acoustic waves, electrodes or piezoelectric substrates must be included into the microfluidic devices and properly engineered during their production. In the case of BAWs, the entire piezoelectric material driven by an alternating current (AC) vibrates at the same frequency of the AC signal (100 kHz–10 MHz). In contrast, SAWs are generated by applying an AC signal to interdigitated transducers (IDTs) patterned on a piezoelectric material, which are excited at higher frequencies than BAWs (up to GHz) (Figure 4a). In order to confine small particles such as EVs, high frequency, of the order of dozens of MHz, is generally required, and thus SAW-based devices are employed [85]. As an example, Lee et al. used a $LiNbO_3$ wafer to imprint interdigitated electrodes that can discriminate sEVs and larger microvesicles from red blood cells according to their size; their cutoff value can be set by tuning the acoustic power and flow velocity [86]. Although fabrication is somewhat complex due to the presence of acoustic actuators, acoustofluidic devices can guarantee high isolation efficiency [87–89]. Wu et al. developed a device based on SAWs consisting of two modules to remove larger blood cells and debris, showing separation of EVs with an 82.4% recovery rate and 98.4% purity using flow rates in the order of few μL/min [90]. Other SAW-based platforms have been coupled with commercial acoustic transducers that lead to automated processes [91–93], and others have been implemented together with modules devoted to the detection of sEVs [94–96].

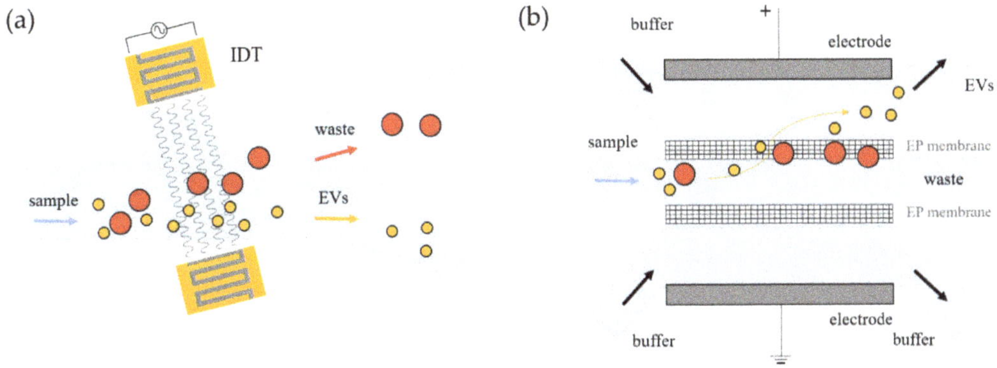

Figure 4. Examples of physical forces to separate vesicles within microfluidic chips. (**a**) Device with interdigitated electrodes generating acoustic waves to separate EVs from the initial sample; (**b**) differentiation of EVs through membranes mediated by electrophoretic (EP) forces.

Electrokinetic force. Electric fields applied to fluids allow for the manipulation of polarizable particles, giving rise to a variety of electrokinetic phenomena: electrophoresis, dielectrophoresis, electroosmosis, etc. These effects provide forces whose magnitude acting on the particles is strictly dependent on their dimension, the dielectric constant, and the charge density of both particles and the surrounding medium. More precisely, the electrophoretic effect (EP) works on monopoles, requiring high forces ($F_{EP} \propto E$) to induce particle manipulation. Dielectrophoresis (DEP) instead allows for controlling the trapping of particles by electric field gradients, to which the DEP force is proportional ($F_{DEP} \propto \nabla|E|^2$) and thus depends on the electrode geometry, rather than the intensity of electric pulses [97]. The latter method happens to be the most widely employed for EV manipulation due to its simplicity with respect to the other electrokinetic phenomena. Ibsen et al. used an alternate current electrokinetic microelectrode to concentrate sEVs from plasma on the edge of the microelectrodes where high field gradients were exerted to process relatively small aliquots (30–50 μL) in less than 30 min, including on-chip fluorescent detection [98]. To improve isolation performance, a device based on dielectrophoretic interactions can be mediated by other types of substrates, such as polystyrene microspheres [99], coupled with automated parts [100], or pneumatically driven components [101]. The latter has been exploited by Davies et al. in a device that takes advantage of the electrophoretic interaction with a pressure-driven filtration stage of porous polymer monolithic membranes (PPMs), which have variable size pores, to isolate vesicles from 240 μL of whole blood in two hours. In addition to DEP, other examples of electrokinetic phenomena already exploited to trap and concentrate vesicles are electrophoresis [102,103] or electro-osmosis [104,105]. For example, Cho et al. developed a device to enrich plasma EVs by coupling a porous membrane with a dedicated electrode (Figure 4b), in order to remove free proteins and debris subjected to electrophoretic migration through 30 nm pores, with an efficiency of approximately 85% [106].

A recent work by Tayebi et al. combined both dielectrophoretic and acoustophoretic forces to sort extracellular vesicles (<200 nm) and microvesicles (>300 nm) from cell cultures that reached high levels of purity (95%) and recovery (81%) [107]. This kind of virtual DLD (vDLD) permits tuning the balance of the two counteracting forces by adjusting properties of the medium and channel sizes, given a fixed electrode geometry.

3.2. *Chemical Methods*

Unlike physical methods, approaches based on the chemical affinity between specific antibodies and antigens allow for the recovery of vesicles in a more selective manner. Immunoaffinity-based capture can occur on flat or patterned substrates, as well as on micrometric solid beads or nanoparticles.

3.2.1. Immunocapture on Fix Support

The selective separation of EVs can be achieved by properly engineering the internal microchannel surfaces by adding a specific antibody that can anchor a specific EV membrane protein. The method provides an extremely good specificity and reproducibility and, in the best cases, allows for processing of samples with very high throughput, even of tens of μL/min.

The simplest strategy is to functionalize unstructured channels. However, by using a straight channel with a typical lateral side of dozen to hundreds of μm and considering the typical size of EVs, the binding area available for vesicle capture is relatively low, causing a poor probability of contact. Moreover, the laminar flow prevents the correct mixing of the solution containing EVs, limiting their accessibility to the molecules anchored on the channel walls. This issue has been faced in two ways: i) improving mixing by patterning channels with specific patterns and ii) including micro- and nanostructures within the channel by creating a sort of filter through which the solution is forced to pass. The first method is well known in the microfluidic community, having already been applied to promote chemical reactions or for isolation purposes [108]. In contrast, the second mimics standard filtration methods by integrating specificity, since these 'filters' are coated to capture EVs showing the desired markers [109]. However, as for filtration methods, this approach suffers from clogging and highly complex microfabrication protocols. To increase the surface-to-volume ratio of channel walls, the inner surfaces of microfluidic chips with micro- and nano- structures are also chemically functionalized with antibodies to ensure the EV chemical affinity [110,111]. The most common microstructures are ordered rows of pillars [112–114], herringbone patterns [108,111,115–118], and properly shaped microposts [119,120]. Meanwhile, in the case of nanopatterning, nanorods, nanowires, and more complex 3D structures [121] are typically used. In 2010, a pioneering work by Chen et al. described a way in which to modify PDMS microfluidic channels presenting herringbone grooves with specific surface treatment [122]. The authors flushed inside the chip a solution of 3-mercaptopropyl trimethoxysilane and incubated it with Neutravidin solution before functionalizing it with biotinylated anti-CD63 antibodies, allowing for the isolation of vesicles from 400 μL of serum within one hour. In another work, Chen et al. used an array of ZnO nanowires (Figure 5a) with interconnected macropores to expand the trapping area [109]. The latter approach was validated with small EVs spiked in saline solutions, showing a trapping rate of up to 30 μL/min, and then with both serum and plasma, from which trapped vesicles were detected with horseradish peroxidase (HRP)-

labeled antibody to allow for colorimetric sensing using 3,3′,5,5′-tetramethylbenzidine (TMB). Zhang et al. developed a chip with graphene oxide/polydopamine (GO-PDA) interfaces that provides specific EV absorption from human plasma [119]. The same group then compared a herringbone pattern with solid structures with colloidal silica nanorods (Figure 5b) that showed an improvement in the detection limit of plasma samples at 10 sEVs/μL [123]. Wang et al. fabricated a microfluidic chip structured with a 3D array of ciliated silicon pillars for multiscale filtering of EV-like vesicles or liposomes, mixing filtration properties and immunocapture [112]. Tests with prototyped 83 nm liposomes revealed a retention rate of approximately 60% from a 30 μL of starting volume. This arrangement has been optimized by Qi et al. to improve capture efficiency and preserve EV integrity for drug delivery (Figure 5c) [113]. In this work, the retention rate of sEVs from MDA-MB-231 (breast cancer) cell culture could be increased to 70%, mainly due to the anti-CD63 functionalization of micropillars.

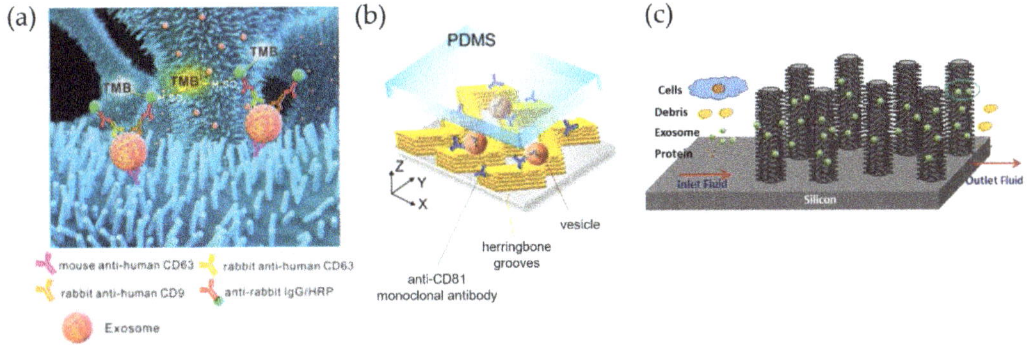

Figure 5. Immunoaffinity capture inside microfluidic devices. (**a**) ZnO nanowires fabricated inside channels for the specific capture of CD63-positive sEVs, modified from [109]; (**b**) colloidal structures arranged in a herringbone configuration inside the microfluidic channel, modified from [123]; (**c**) ciliated silicon nanorods capable of discriminating sEVs from cell debris [113].

3.2.2. Immunocapture on Beads and Nanoparticles

Another strategy to chemically trap EVs requires the use of beads of a size between 0.5 and 20 μm functionalized with the target antibody. In fact, floating beads of micrometer size present a larger surface area. Beads can be directly injected and mixed within the initial sample, enhancing the EV contact probability, without introducing complicated micro- or nano-structures into the microfluidic chip, which typically require costly fabrication approaches and a highly trained operator. Therefore, this technique is one of the most efficient in terms of specificity, but especially for higher recovery rates and analysis sensitivity. On the other hand, the flow rates applied in the microchannel to transport liquids cannot be too low, in order to prevent bead sedimentation, nor too high, to ensure good mixing between vesicles and EVs, even though some works tried to process the sample at a throughput of up to approximately 9 mL/h [124]. There are different possible beads that can vary in terms of material or size. The most commonly used are micrometer-sized commercial immunomagnetic beads that have a paramagnetic core that can be easily handled using external magnets [125–135]. Here, unlike in functionalized channels for immunocapture, an external magnetic force must be applied to manipulate particles and favor isolation. A key aspect in the choice and use of the floating substrate for EV capture is that the time must be sufficient for the substrate to settle in the channel. The sedimentation speed v for a single object dispersed in a viscous fluid can be calculated by the Stokes law: $v = \left(d^2(\rho_d - \rho_f)g\right)/(18\mu_f)$, where d and ρ_d are the diameter and the density of the object, ρ_f and μ_f are the density and the viscosity of the fluid, and g is the gravity acceleration [136]. Thus, considering a polystyrene bead of 1 μm (ρ~1.05 g/cm^3)

and an EV of 200 nm (ρ~1.1 g/cm^3) dispersed in water (μ_f~1 cP), it is possible to notice that the first sediments are more than ten times faster than the second. This occurrence becomes even more critical when using magnetic beads (ρ~1.8 g/cm^3), as their sedimentation speed is two orders of magnitude higher than for EVs. Therefore, the flow rates applied to the liquid within the channel must be fast enough to prevent particle sedimentation, but slow enough to ensure a sufficiently long incubation time for EV capture. Thus, the working range of this type of device is limited.

A highly cited example is given by He et al., who used immunomagnetic beads coated with specific antibodies (e.g., anti-EpCAM) to capture and lyse sEVs inside a unique device to analyze the protein content by chemifluorescent ELISA [125]. Plasma samples of 30 µL volume are processed in about 100 min. Zhao et al. fabricated a device called ExoSearch to enrich sEVs from plasma and measure multiple marker fluorescent signals (Figure 6a) [126].

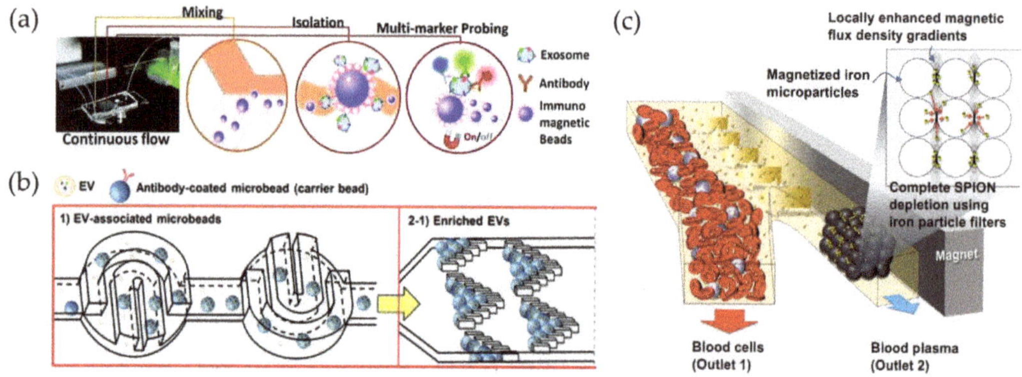

Figure 6. Functionalized beads or nanoparticles used for vesicle capture inside microfluidic devices. (**a**) Immunomagnetic beads coated for the enrichment of vesicles from blood plasma inside ExoSearch chips [126]; (**b**) streptavidin-coated polystyrene beads used as substrate to trap vesicles in herringbone filters and redisperse them [137]; (**c**) microfluidic systems used to flow blood samples and isolate vesicles by means of superparamagnetic nanoparticles (SPIONs), modified from [138].

In addition to magnetic beads, polystyrene beads (PS) can be used for the isolation of EVs [80,99,124,137,139–141] using centrifugation and redispersion instead of magnetic forces. One of the main drawbacks of immunocapturing with beads is the difficulty of breaking the bond with antibodies, preventing EV damage. However, Tayebi et al. used specifically coated PS beads to capture EVs from MCF-7 (a breast cancer cell line) in constrictions along the microfluidic circuit with an aperture of 30 µm to trap a single bead, by exploiting hydrodynamic resistance in channels having different shapes [140]. The trapped EVs were then removed from the beads by rising with a low pH IgG elution buffer (0.1 M glycine-HCl) for 1 min and then waiting 10 min for antibody–antigen dissociation. Finally, neutralization is provided by a solution of pH 7.4 (1 M Tris-HCl). Despite a good purity, this approach limits the amount of EV–bead complexes to the number of trapping sites and the maximum flow rate (50 µL/min). Gwak et al., instead, promoted chaotic stirring of coated PS beads inside horseshoe-shaped mixers, and the work [124], was able to enrich plasma EVs using fish trap-shaped filters (Figure 6b) [137]. Then, using the same elution buffer, different flow rates were tested: under the best conditions, the whole isolation process for the 100 µL sample was completed in 5 min, showing a capture efficiency greater than 97%.

Unlike micrometric beads that are typically larger in size than EVs, an alternative solid substrate is represented by nanoparticles (NPs), which are on the order of a few nanometers, either with magnetic properties (superparamagnetic or ferromagnetic) [134,138,142–147] or simply functionalized with selective markers [148]. In particular, they have a similar or even smaller size than small EVs, so EVs can be used to capture a single EV on their surface, rather than being encapsulated inside, as used in drug delivery [149]. Notably, these NPs can be an active tool for the capture and manipulation of vesicles, rather than a passive substrate. A pivotal work by Shao et al. showed a strategy for labeling and isolating blood glioblastoma microvesicles, implementing properly functionalized magnetic nanoparticles (core of 7 nm) and using a two-step protein targeting to maximize binding. This approach allowed for better detection of CD63 + vesicles with a micronuclear magnetic resonance (μNMR) system [142]. In another work, Ko et al. exploited magnetic NiFe nanopores (600 nm diameter) to trap EVs labeled with 50 nm coated NPs, allowing for the processing of serum and plasma with flow rates of up to 10 mL/h [144]. Recently, increasing attention has also been paid to superparamagnetic iron oxide nanoparticles (SPIONs) to isolate EVs due to the nanoparticles' reversible magnetic property and easy manipulation, as exploited by Kwon et al. for the purification of blood samples [138]. Here, SPIONs and EVs create a complex cluster that can be isolated by exploiting the magnetic force applied by an external magnet (Figure 6c).

In the following, Table 2 reports a summary of the main EV isolation methods exploiting microfluidic strategies, distinguished between physical and chemical, together with their working principle.

Table 2. Summary of relevant published articles dealing with EV isolation from different microfluidic approaches.

	Methods	Working Principle
Physical: Passive	Filtration [38–45,150,151]	Micro-/nano- filtration process by porous membranes inside chip
	Deterministic lateral displacement [46,60–62]	Particle distribution in size by lateral forces conveyed by ordered array of posts
	Inertial force [50–55,77] Viscoelastic force [67–72]	Imbalance of inertial forces or of shear forces in non-Newtonian viscoelastic fluid
Physical: Active	Acoustofluidics [86–96,107]	Acoustic trapping by ultrasound waves
	Electrokinetic force [98–107,152–154]	Charge separation by electric fields
Chemical: Fixed support	Functionalized fixed support [108–122,130,155–180]	EV capture by specific antibodies on fixed substrate
Chemical: Floating Beads	Magnetic beads [125–135,181–186]	EV capture by specific antibodies on beads for magnetic manipulation
	Polystyrene beads [80,99,124,137,139–141,187]	EV capture by specific antibodies on non-magnetic beads
	Magnetic nanoparticles [93,134,138,142–147,149,188]	EV capture or handling by specific antibodies on magnetic nanoparticles

4. Discussion

4.1. Physical and Chemical Microfluidic Approaches: Pros and Cons

When microfluidic devices are designed to improve the purification and/or isolation of specific analytes from complex starting samples, four main aspects must be considered to balance advantages and disadvantages: throughput of the process, quantity and quality of the output sample, automation capability, and complexity of the microfabrication. In the following section, we discuss the above-presented approaches for EV isolation in microfluidic devices, focusing on these four aspects. A schematic summary of the following discussion is also reported in Table 3.

Table 3. Summary of the comparison of different microfluidic techniques for EV isolation. Each of the four aspects is evaluated according to the following scale: (–) difficult/very bad, (-) non-trivial/mediocre, (+) easy/good, (++) very easy/excellent.

	Throughput	Output Sample (Quality and Quantity)	Possible Automation	Micro-Fabrication Simplicity
Physical: Passive	++	-	+	-
Physical: Active	-	+	+	-
Chemical: Fixed support	+	++	++	+
Chemical: Floating Beads	+	++	–	+

Physical approaches based on filtration, although based on principles similar to those of UF and SEC, allow for the processing of large volumes at high throughput. In fact, microfluidic protocols can be easily automated with the support of external pumps controlled by devoted software [189] to infuse biological samples at high rates throughout filters inside channels. However, the realization of these microfilters represents a critical aspect: as for their correspondent conventional methods, these devices are typically disposable, because they are prone to clogging, and the processing time for complete isolation is quite long compared to that of the other microfluidic techniques.

In contrast, physical approaches based on external stimuli, such as acoustic waves or electric fields, are not directly comparable to existing methods. These contact-free strategies prevent EV damage and preserve the EVs' functional properties [86,106]. In fact, the presented results show good vesicle integrity and a homogeneous size distribution, especially in terms of the acoustic approach [90]. On the other hand, captured EVs have a low purity, because other contaminants that have a similar size and/or density can be isolated together. Then, as in the previous case, the microfabrication requires both an appropriate environment (i.e., a clean room) and trained operators, since the electrodes must be integrated inside microfluidic devices, and high alternate electric fields must be applied and precisely controlled. The latter practice must be executed by equipped laboratories, which include costly facilities, such as photolithographic platforms, metal evaporators, electronic equipment, and related characterization instruments. However, compared with microfluidic filtering methods, cleaning protocols can be considered in the case of SAW or DEP devices, partially reducing the impact of the microfabrication. Most of the aforementioned strategies based on separation by size can be biased by the fact that other membrane-based components also have dimensions similar to those of EVs (e.g., lipoproteins).

Then, approaches rooted in chemical affinity have been integrated within microfluidic devices, with the aim of improving the anchor point by increasing the surface-to-volume ratio, isolation throughput, sample quality output, and process automation [190]. Compared to physical approaches, immunoaffinity provides a highly specific isolation that allows for the distinction between the EV subpopulation [191] and the high purity of the isolated sample. In the fixed support-based approach, molecules are coated on microchannel surfaces that typically present specific micro- and/or nanopatterns to improve the liquid mixing (microstructures) and/or acting as filters (nanostructures). In contrast, the use of micrometric beads as solid floating supports for EV capture allows for using more simple channel geometries, since the beads themselves act as traps that improve the surface-to-volume ratio. However, compared to other microfluidic approaches, the use of micrometric beads leads to difficulties in terms of the complete automation of microfluidic devices, since they are prone to sedimentation and can induce microchannel clogging [144].

Another disadvantage of this approach is the high costs of purchasing commercially available microbead kits devoted to EV capture; therefore, custom and home-made protocols are generally preferable [139]. A completely different perspective is provided using nanoparticles. They present a size comparable to or even smaller than a single EV, leading to an improvement in capture control and preventing damage to the EVs. Additionally, since several characterization methods are now based on optical and spectroscopic techniques (e.g.,: fluorescence microscopy, Raman spectroscopy), nanoparticles can be directly used to improve the signal-to-noise ratio [188]. A possible drawback of the use of NPs is that they require trained operators for their synthesis and, importantly, their stability is strongly influenced by the surrounding buffer, which may affect the EVs as well. Therefore, strong physical-chemical expertise is required to develop a working microfluidic device based on NPs' isolation.

In general, as summarized in Table 3, among the microfluidic approaches analyzed, we considered devices based on the chemical affinity of the fixed support to be the most promising, as they show good throughput and high-quality sample output, supported by simple microfabrication and automation. In this sense, the use of beads requires more precautions due to the possible sedimentation and clogging issues. In contrast, physical approaches suffer from complicated complex microfabrication requiring electrode integration, which, at the current state of the art, is not sufficiently balanced by appropriate throughput or sample quality.

4.2. *Microfluidic Isolation Techniques: Which Is the Most Popular*

To better appreciate how widespread these methods are, Figure 7 reports a statistical representation of all the published articles on EV isolation performed by microfluidic devices.

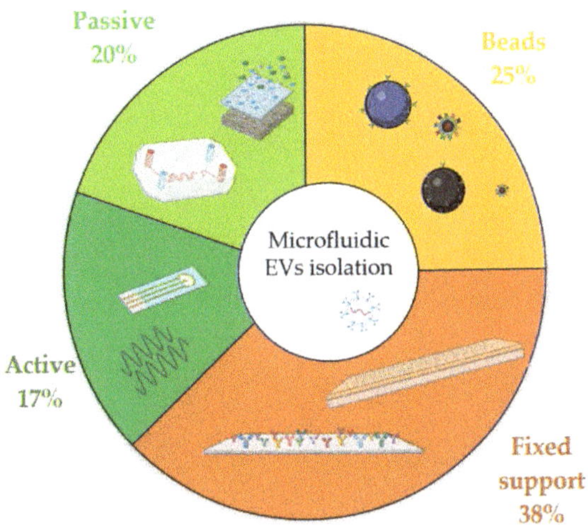

Figure 7. Frequency of the published works from 2000 to present (source: Web of Science database, excluding reviews, meeting abstracts, and retracted papers) related to each of the different isolation techniques that exploit microfluidic devices. Physical approaches are subdivided into active (dark green) and passive (green), whereas chemical immunocapture is separated by fixed supports (orange) and floating beads (yellow).

The pie chart in Figure 7 highlights that the predominance of microfluidic methods makes use of chemical immunocapture, which comprises two-thirds of the total output. This is probably due to the fact that physical approaches leverage concepts similar to conventional ones, such as DU, UF, or SEC, hence carrying similar drawbacks in their usage, and automation has not yet been implemented in the presented proof-of-concept device. Moreover, chemical affinity provides an easy implementation inside microfluidic devices, especially for flat channels that require a simple functionalization capable of binding to membrane proteins, such as tetraspanins. Additionally, these methods generally perform better than their physical counterparts, mainly in terms of throughput and recovery rate. Among chemical isolation devices, those exploiting fixed substrates are the most commonly used, probably due to the large amount of literature related to the functionalization of microfluidic devices [192], the automatization capability, and the simple microfabrication, despite the fact that the production of certain integrated nano-structures could lead to some difficulties. In contrast, there is no preferred physical approach for EV purification: passive devices are slightly more frequently used, despite the poor quality of final samples, probably because the users do not need specific training to control SAW and DEP, and these devices are more prone to miniaturization. However, these active approaches are relatively newer with respect to their passive counterparts, and therefore, further development is expected in the future to facilitate device production and handling.

4.3. Are Microfluidic Devices for EV Isolation Ready for Clinical Applications

Today, the clinicaltrials.gov database reports more than 350 trials (about 90 already completed) indexed by the keywords *exosomes* and *extracellular vesicles*; among them, only two, which have just started, involve microfluidic systems. The latter can be understood considering the still young character of the EV research field. However, to better analyze the current state of the art from a clinical perspective, we discuss the applicability of the presented microfluidic devices in real diagnostic conditions. In detail, the microfluidic devices devoted to purified and isolated EVs are validated using different starting samples (i.e.,: culture media, blood-derived fluids, urine, etc.), and, in some experiments, EVs are spiked in human fluids. Although we consider all these approaches to be fundamental for the validation of novel technologies, they represent a first proof of concept compared to real clinical assays, since EV isolation may depend on several factors, such as starting samples and EV biogenesis. Therefore, we divided the presented papers based on their validation methods, labeled as: (i) *proof of concept*, if they are limited to cell culture media or spiked EVs in body fluids, and (ii) *clinical sample*, when they directly employ body fluids for EV isolation (e.g., urine, plasma, serum, or whole blood). The result of this classification is reported in Table 4 and Figure 8, together with the isolation approach used.

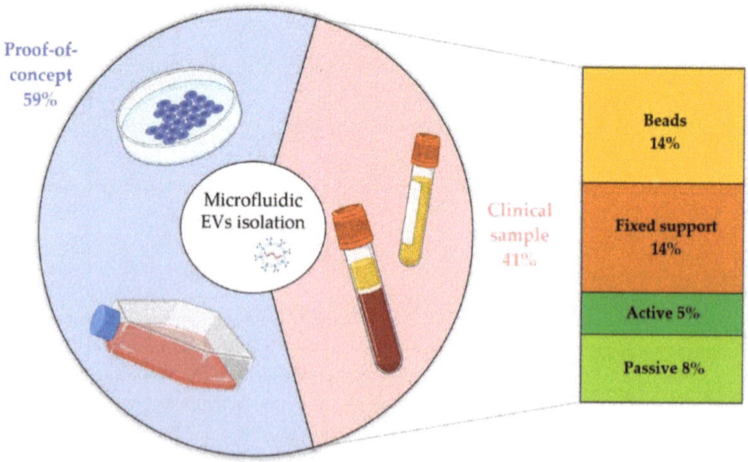

Figure 8. Frequency of published works from 2000 to present (source: Web of Science database, excluding reviews, meeting abstracts, and retracted papers) involving EV isolation from starting biofluids used as a *proof of concept* (cell culture media or EVs spiked in body fluids) or as *clinical samples* (body fluids directly employed) from healthy donors or patients for research studies or diagnostic purposes, further divided according to the isolation technique.

Table 4. Summary of articles published during the years 2000–2022 dealing with EV isolation from microfluidic devices processing samples for research purposes or clinical trials, according to techniques classified in Table 2.

	Proof of Concept		Clinical Sample	
	Starting Sample	**%**	**Starting Sample**	**%**
Physical: Passive	Plasma [52,62] Urine [44,60,95] Cell culture [41–43,50–52,54,55,61,68,71,72,77,88,150,151,193,194]	12	Plasma [39,195] Serum [43,46,71] Blood [45,53,67,69] Urine [38,39,44,46]	8
Physical: Active	Plasma [98,102,103,106,152] Serum [103] Saliva [88] Cell culture [86,89,91,98–100,104,105,107,153,154]	12	Plasma [87,91,94] Blood [90,101] Urine [91,92,95]	5
Chemical: Fixed support	Plasma [87,173] Serum [122,168] Cell culture [109,113,122,155–158,160–162,164–167,172,174–177,179,180,196–200]	24	Plasma [108,114,119,120,123,155,163,171,173,201] Serum [110,117,118,123,130,178,202,203] Blood [170] Urine [115,159,169]	14
Chemical: Floating beads	Plasma [130] Serum [130,183] Cell culture [99,137,139–141,143,147,182,186,188,193,204]	11	Plasma [80,124–127,129,132,135,138,144,185] Serum [131,134,145,147,148,181] Blood [133,142,146] Urine [128]	14
TOTAL		59		41

In Figure 8, it can be observed that 40% of microfluidic devices have been validated using body fluids taken from healthy donors or real patients, representing a good percentage of the total microfluidics-based studies. Among these works, most use immunoaffinity-based devices to capture vesicles, increasing the difference already noted in Figure 7 compared to the use of physical methods. Notably, passive physical methods are disposable and more prone to clogging than their non-microfluidic counterparts, whereas active interactions in real fluids must consider several parameters, such as the fact that viscous biofluids are not as easy to be manipulated as aqueous solution. It is worth noting that most of the works processing *proof-of-concept* samples, such as cell lines or spiked EVs, also include pre-purification steps (e.g., differential ultracentrifugation) before the injection of the biological fluids to favor better isolation of small vesicles after the elimination of heavier debris and cell fragments. In contrast, a device already tested with clinical samples without requiring pre-treatment can be considered ready to use for medical diagnostics, and this kind of validation must be the final goal of future studies.

4.4. Microfluidic Devices for EV Detection and Analysis

Extracellular vesicles collected and purified by the aforementioned approaches can be investigated by several detection techniques. The more conventional ones for EV size estimation require off-chip treatment and characterization by commercial instruments, mostly based on tracking such as nanoparticle tracking analysis (NTA) [205], dynamic light scattering (DLS) [206], or flow cytometry (FC) [207], which reveals information on EV morphology. Indeed, NTA is actually performed in a microfluidic chip for the estimation of size by scattered light based on Brownian particle fluctuations and similarly occurs for DLS; cytometric analysis requires hydrodynamic vesicle focusing that is otherwise achievable by the microfluidic channel.

Microfluidic devices have also been developed for EV detection, mainly based on two approaches: (i) self-standing, in which the device itself works as detector, and (ii) storage, in which the microchannels act as a storing chamber for EVs that are screened by an external microscope or probe. The first approach typically involves the integration of a devoted electrode, as occurs for analysis by field-effect transistors [185], electrochemical sensing [208], zeta potential estimation [102], or lateral flow immunoassay [172].

In contrast, in the second approach, the microfluidic device allows for several types of detection: (i) colorimetric for EV quantification [109], (ii) spectroscopy based on surface-enhanced Raman scattering (SERS) [94], (iii) optical properties by surface plasmon resonance (SPR) [200], (iv) resistive pulse sensing due to EV through nanopores [209], (v) detection by micro-nuclear magnetic resonance (NMR) [142], etc. Furthermore, some microfluidic platforms were engineered to also perform EV content analysis, such as on-chip quantitative PCR [181] or ELISA [44]. More details on detection methods based on microfluidic devices can be found in other specific reviews [20,210].

5. Conclusions and Perspective

This review looks at the most important techniques currently used to isolate EVs using microfluidic devices, analyzing their ability to be applied in clinics. In this perspective, despite the fact that microfluidic technologies have only recently been applied to the isolation of EVs, almost half (40%) of the presented devices use potentially clinical samples for their validation. Therefore, microfluidics seems to represent a promising strategy for medical investigation.

The comparison between physical and chemical microfluidic approaches for EV isolation emphasizes a slight preference for chemical methods over their physical counterparts, which becomes much more evident if one considers only the technologies validated with potential clinical samples. This might be the cause for two possible reasons: (i) the device performance itself (higher throughput, better capture efficiency, and simple microfabrication for mass production) or (ii) the general trend of the biomedical community to look for EVs that have specific phenotypes. Actually, we believe that the latter is dominant, even though there is probably a combination of both factors. Although the characterization of EVs was initially based on quantification and size classification, it is now clear that the specific EV subpopulations can provide useful information related to specific diseases [8]. Therefore, we expect that the gap between physical and chemical approaches will increase in the near future. A possible alternative to further improve the efficiency and purity in EV capture might be the implementation of both physical and chemical approaches combined together inside a single device, using passive EV separation from other massive particles, and then a specific immunocapture for distinguishing different subpopulations, providing also high throughput isolation.

In addition to EV isolation, microfluidic devices are also widely employed for EV detection and analysis. Whereas the first devices consisted of simple microchannels that aimed to store EVs for further analysis (as for the NTA), new approaches, which combine microfluidic systems, devoted optical systems, and antibody immunocapturing, point towards the analysis of single vesicles in order to achieve information about their heterogeneity within the same population [211].

In conclusion, we are confident that microfluidics, which is already employed in some gold-standard techniques for EVs analysis (i.e.,: NTA), can bring great support in terms of EV isolation by providing tools capable of performing repeatable and automated protocols, similar to other important biomarkers, such as CTCs and ctDNA. However, microfluidics experts exposed to the field for the first time must also consider biological and medical points of view. Therefore, to be really useful, new technologies must be developed in accordance with the requirements and expectations that, importantly, have progressively changed over the past five years, following the guidelines on the validation of the protocol and the outcomes established by the ISEV community [9].

Author Contributions: Writing—original draft preparation, A.M., V.M. and D.F.; writing—review and editing, A.M., V.M., P.B., M.P., G.M. and D.F.; funding acquisition, V.Z. and D.F. All authors have read and agreed to the published version of the manuscript.

Funding: This research was funded by University of Padova through a STARS grant—EXODROP and a BIRD grant 2021—BiodivSeq.

Institutional Review Board Statement: Not applicable.

Informed Consent Statement: Not applicable.

Data Availability Statement: Not applicable.

Acknowledgments: The authors are particularly grateful to Giorgio Delfitto for his valuable technical assistance.

Conflicts of Interest: The authors declare no conflict of interest.

References

1. Ginsburg, G.S.; Phillips, K.A. Precision Medicine: From Science to Value. *Health Aff.* **2018**, *37*, 694–701. [CrossRef] [PubMed]
2. Ellis, M.J.; Ding, L.; Shen, D.; Luo, J.; Suman, V.J.; Wallis, J.W.; Van Tine, B.A.; Hoog, J.; Goiffon, R.J.; Goldstein, T.C.; et al. Whole-Genome Analysis Informs Breast Cancer Response to Aromatase Inhibition. *Nature* **2012**, *486*, 353–360. [CrossRef] [PubMed]
3. Loughran, C.; Keeling, C.R. Seeding of Tumour Cells Following Breast Biopsy: A Literature Review. *Br. J. Radiol.* **2011**, *84*, 869–874. [CrossRef] [PubMed]
4. Crowley, E.; Di Nicolantonio, F.; Loupakis, F.; Bardelli, A. Liquid Biopsy: Monitoring Cancer-Genetics in the Blood. *Nat. Rev. Clin. Oncol.* **2013**, *10*, 472–484. [CrossRef] [PubMed]
5. Alix-Panabierèes, C.; Pantel, K. Circulating Tumor Cells: Liquid Biopsy of Cancer. *Clin. Chem.* **2013**, *59*, 110–118. [CrossRef]
6. Diaz, L.A.; Bardelli, A. Liquid Biopsies: Genotyping Circulating Tumor DNA. *J. Clin. Oncol.* **2014**, *32*, 579–586. [CrossRef]
7. El Andaloussi, S.; Mäger, I.; Breakefield, X.O.; Wood, M.J.A. Extracellular Vesicles: Biology and Emerging Therapeutic Opportunities. *Nat. Rev. Drug Discov.* **2013**, *12*, 347–357. [CrossRef]
8. Van Niel, G.; D'Angelo, G.; Raposo, G. Shedding Light on the Cell Biology of Extracellular Vesicles. *Nat. Rev. Mol. Cell Biol.* **2018**, *19*, 213–228. [CrossRef]
9. Théry, C.; Witwer, K.W.; Aikawa, E.; Alcaraz, M.J.; Anderson, J.D.; Andriantsitohaina, R.; Antoniou, A.; Arab, T.; Archer, F.; Atkin-Smith, G.K.; et al. Minimal Information for Studies of Extracellular Vesicles 2018 (MISEV2018): A Position Statement of the International Society for Extracellular Vesicles and Update of the MISEV2014 Guidelines. *J. Extracell. Vesicles* **2018**, *7*, 1535750. [CrossRef]
10. Raposo, G.; Stahl, P.D. Extracellular Vesicles: A New Communication Paradigm? *Nat. Rev. Mol. Cell Biol.* **2019**, *20*, 509–510. [CrossRef]
11. Jia, S.; Zocco, D.; Samuels, M.L.; Chou, M.F.; Chammas, R.; Skog, J.; Zarovni, N.; Momen-Heravi, F.; Kuo, W.P. Emerging Technologies in Extracellular Vesicle-Based Molecular Diagnostics. *Expert Rev. Mol. Diagn.* **2014**, *14*, 307–321. [CrossRef] [PubMed]
12. Jansen, F.; Nickenig, G.; Werner, N. Extracellular Vesicles in Cardiovascular Disease. *Circ. Res.* **2017**, *120*, 1649–1657. [CrossRef] [PubMed]
13. Turpin, D.; Truchetet, M.E.; Faustin, B.; Augusto, J.F.; Contin-Bordes, C.; Brisson, A.; Blanco, P.; Duffau, P. Role of Extracellular Vesicles in Autoimmune Diseases. *Autoimmun. Rev.* **2016**, *15*, 174–183. [CrossRef] [PubMed]
14. Thompson, A.G.; Gray, E.; Heman-Ackah, S.M.; Mäger, I.; Talbot, K.; El Andaloussi, S.; Wood, M.J.; Turner, M.R. Extracellular Vesicles in Neurodegenerative Disease-Pathogenesis to Biomarkers. *Nat. Rev. Neurol.* **2016**, *12*, 346–357. [CrossRef] [PubMed]
15. Vader, P.; Breakefield, X.O.; Wood, M.J.A. Extracellular Vesicles: Emerging Targets for Cancer Therapy. *Trends Mol. Med.* **2014**, *20*, 385–393. [CrossRef]
16. Weng, J.; Xiang, X.; Ding, L.; Wong, A.L.A.; Zeng, Q.; Sethi, G.; Wang, L.; Lee, S.C.; Goh, B.C. Extracellular Vesicles, the Cornerstone of next-Generation Cancer Diagnosis? *Semin. Cancer Biol.* **2021**, *74*, 105–120. [CrossRef]
17. Konoshenko, M.Y.; Lekchnov, E.A.; Vlassov, A.V.; Laktionov, P.P. Isolation of Extracellular Vesicles: General Methodologies and Latest Trends. *Biomed Res. Int.* **2018**, *2018*, 8545347. [CrossRef]
18. Whitesides, G.M. The Origins and the Future of Microfluidics. *Nature* **2006**, *442*, 368–373. [CrossRef]
19. Yang, Y.; Chen, Y.; Tang, H.; Zong, N.; Jiang, X. Microfluidics for Biomedical Analysis. *Small Methods* **2020**, *4*, 1–30. [CrossRef]
20. Serrano-Pertierra, E.; Oliveira-Rodríguez, M.; Matos, M.; Gutiérrez, G.; Moyano, A.; Salvador, M.; Rivas, M.; Blanco-López, M.C. Extracellular Vesicles: Current Analytical Techniques for Detection and Quantification. *Biomolecules* **2020**, *10*, 824. [CrossRef]
21. Livshts, M.A.; Khomyakova, E.; Evtushenko, E.G.; Lazarev, V.N.; Kulemin, N.A.; Semina, S.E.; Generozov, E.V.; Govorun, V.M. Isolation of Exosomes by Differential Centrifugation: Theoretical Analysis of a Commonly Used Protocol. *Sci. Rep.* **2015**, *5*, 17319. [CrossRef] [PubMed]

22. Torres Crigna, A.; Fricke, F.; Nitschke, K.; Worst, T.; Erb, U.; Karremann, M.; Buschmann, D.; Elvers-Hornung, S.; Tucher, C.; Schiller, M.; et al. Inter-Laboratory Comparison of Extracellular Vesicle Isolation Based on Ultracentrifugation. *Transfus. Med. Hemotherapy* **2021**, *48*, 48–59. [CrossRef] [PubMed]
23. Kain, S.R. *Methods and Protocols*; John Wiley & Sons: New York, NY, USA, 2005; Volume 47, ISBN 0471736821.
24. Grant, R.; Ansa-Addo, E.; Stratton, D.; Antwi-Baffour, S.; Jorfi, S.; Kholia, S.; Krige, L.; Lange, S.; Inal, J. A Filtration-Based Protocol to Isolate Human Plasma Membrane-Derived Vesicles and Exosomes from Blood Plasma. *J. Immunol. Methods* **2011**, *371*, 143–151. [CrossRef]
25. Nordin, J.Z.; Lee, Y.; Vader, P.; Mäger, I.; Johansson, H.J.; Heusermann, W.; Wiklander, O.P.B.; Hällbrink, M.; Seow, Y.; Bultema, J.J.; et al. Ultrafiltration with Size-Exclusion Liquid Chromatography for High Yield Isolation of Extracellular Vesicles Preserving Intact Biophysical and Functional Properties. *Nanomed. Nanotechnol. Biol. Med.* **2015**, *11*, 879–883. [CrossRef] [PubMed]
26. Böing, A.N.; van der Pol, E.; Grootemaat, A.E.; Coumans, F.A.W.; Sturk, A.; Nieuwland, R. Single-Step Isolation of Extracellular Vesicles by Size-Exclusion Chromatography. *J. Extracell. Vesicles* **2014**, *3*. [CrossRef]
27. Lobb, R.; Möller, A. Size Exclusion Chromatography: A Simple and Reliable Method for Exosome Purification. *Methods Mol. Biol.* **2017**, *1660*, 105–110. [CrossRef]
28. Mol, E.A.; Goumans, M.J.; Doevendans, P.A.; Sluijter, J.P.G.; Vader, P. Higher Functionality of Extracellular Vesicles Isolated Using Size-Exclusion Chromatography Compared to Ultracentrifugation. *Nanomed. Nanotechnol. Biol. Med.* **2017**, *13*, 2061–2065. [CrossRef]
29. Monguió-Tortajada, M.; Morón-Font, M.; Gámez-Valero, A.; Carreras-Planella, L.; Borràs, F.E.; Franquesa, M. Extracellular-Vesicle Isolation from Different Biological Fluids by Size-Exclusion Chromatography. *Curr. Protoc. Stem Cell Biol.* **2019**, *49*, e82. [CrossRef]
30. Kaddour, H.; Lyu, Y.; Shouman, N.; Mohan, M.; Okeoma, C.M. Development of Novel High-Resolution Size-Guided Turbidimetry-Enabled Particle Purification Liquid Chromatography (PPLC): Extracellular Vesicles and Membraneless Condensates in Focus. *Int. J. Mol. Sci.* **2020**, *21*, 5361. [CrossRef]
31. Alvarez, F.A.; Kaddour, H.; Lyu, Y.; Preece, C.; Cohen, J.; Baer, L.; Stopeck, A.T.; Thompson, P.; Okeoma, C.M. Blood Plasma Derived Extracellular Vesicles (BEVs): Particle Purification Liquid Chromatography (PPLC) and Proteomic Analysis Reveals BEVs as a Potential Minimally Invasive Tool for Predicting Response to Breast Cancer Treatment. *Breast Cancer Res. Treat.* **2022**, *196*, 423–437. [CrossRef]
32. Niu, Z.; Pang, R.T.K.; Liu, W.; Li, Q.; Cheng, R.; Yeung, W.S.B. Polymer-Based Precipitation Preserves Biological Activities of Extracellular Vesicles from an Endometrial Cell Line. *PLoS ONE* **2017**, *12*, e0186534. [CrossRef] [PubMed]
33. Zarovni, N.; Corrado, A.; Guazzi, P.; Zocco, D.; Lari, E.; Radano, G.; Muhhina, J.; Fondelli, C.; Gavrilova, J.; Chiesi, A. Integrated Isolation and Quantitative Analysis of Exosome Shuttled Proteins and Nucleic Acids Using Immunocapture Approaches. *Methods* **2015**, *87*, 46–58. [CrossRef] [PubMed]
34. Gemoll, T.; Rozanova, S.; Röder, C.; Hartwig, S.; Kalthoff, H.; Lehr, S.; Elsharawy, A.; Habermann, J.K. Protein Profiling of Serum Extracellular Vesicles Reveals Qualitative and Quantitative Differences after Differential Ultracentrifugation and Exoquicktm Isolation. *J. Clin. Med.* **2020**, *9*, 1429. [CrossRef] [PubMed]
35. Andreu, Z.; Yáñez-Mó, M. Tetraspanins in Extracellular Vesicle Formation and Function. *Front. Immunol.* **2014**, *5*, 1–12. [CrossRef]
36. Jakobsen, K.R.; Paulsen, B.S.; Bæk, R.; Varming, K.; Sorensen, B.S.; Jørgensen, M.M. Exosomal Proteins as Potential Diagnostic Markers in Advanced Non-Small Cell Lung Carcinoma. *J. Extracell. Vesicles* **2015**, *4*, 1–10. [CrossRef]
37. Sandfeld-Paulsen, B.; Jakobsen, K.R.; Bæk, R.; Folkersen, B.H.; Rasmussen, T.R.; Meldgaard, P.; Varming, K.; Jørgensen, M.M.; Sorensen, B.S. Exosomal Proteins as Diagnostic Biomarkers in Lung Cancer. *J. Thorac. Oncol.* **2016**, *11*, 1701–1710. [CrossRef]
38. Liang, L.G.; Kong, M.Q.; Zhou, S.; Sheng, Y.F.; Wang, P.; Yu, T.; Inci, F.; Kuo, W.P.; Li, L.J.; Demirci, U.; et al. An Integrated Double-Filtration Microfluidic Device for Isolation, Enrichment and Quantification of Urinary Extracellular Vesicles for Detection of Bladder Cancer. *Sci. Rep.* **2017**, *7*, 46224. [CrossRef]
39. Liu, F.; Vermesh, O.; Mani, V.; Ge, T.J.; Madsen, S.J.; Sabour, A.; Hsu, E.C.; Gowrishankar, G.; Kanada, M.; Jokerst, J.V.; et al. The Exosome Total Isolation Chip. *ACS Nano* **2017**, *11*, 10712–10723. [CrossRef]
40. Chen, Y.S.; Ma, Y.D.; Chen, C.; Shiesh, S.C.; Lee, G. Bin An Integrated Microfluidic System for On-Chip Enrichment and Quantification of Circulating Extracellular Vesicles from Whole Blood. *Lab Chip* **2019**, *19*, 3305–3315. [CrossRef]
41. Inci, F. Benchmarking a Microfluidic-Based Filtration for Isolating Biological Particles. *Langmuir* **2022**, *38*, 1897–1909. [CrossRef]
42. Riazanski, V.; Mauleon, G.; Lucas, K.; Walker, S.; Zimnicka, A.M.; McGrath, J.L.; Nelson, D.J. Real Time Imaging of Single Extracellular Vesicle PH Regulation in a Microfluidic Cross-Flow Filtration Platform. *Commun. Biol.* **2022**, *5*, 13. [CrossRef] [PubMed]
43. Casadei, L.; Choudhury, A.; Sarchet, P.; Mohana Sundaram, P.; Lopez, G.; Braggio, D.; Balakirsky, G.; Pollock, R.; Prakash, S. Cross-Flow Microfiltration for Isolation, Selective Capture and Release of Liposarcoma Extracellular Vesicles. *J. Extracell. Vesicles* **2021**, *10*, e12062. [CrossRef] [PubMed]
44. Woo, H.K.; Sunkara, V.; Park, J.; Kim, T.H.; Han, J.R.; Kim, C.J.; Choi, H.I.; Kim, Y.K.; Cho, Y.K. Exodisc for Rapid, Size-Selective, and Efficient Isolation and Analysis of Nanoscale Extracellular Vesicles from Biological Samples. *ACS Nano* **2017**, *11*, 1360–1370. [CrossRef] [PubMed]
45. Sunkara, V.; Kim, C.J.; Park, J.; Woo, H.K.; Kim, D.; Ha, H.K.; Kim, M.H.; Son, Y.; Kim, J.R.; Cho, Y.K. Fully Automated, Label-Free Isolation of Extracellular Vesicles from Whole Blood for Cancer Diagnosis and Monitoring. *Theranostics* **2019**, *9*, 1851–1863. [CrossRef] [PubMed]

46. Smith, J.T.; Wunsch, B.H.; Dogra, N.; Ahsen, M.E.; Lee, K.; Yadav, K.K.; Weil, R.; Pereira, M.A.; Patel, J.V.; Duch, E.A.; et al. Integrated Nanoscale Deterministic Lateral Displacement Arrays for Separation of Extracellular Vesicles from Clinically-Relevant Volumes of Biological Samples. *Lab Chip* **2018**, *18*, 3913–3925. [CrossRef]
47. Segre, G.; Silberg, A. Radial Particle Displacements in Poiseuille Flow of Suspensions. *Nature* **1961**, *189*, 209–210. [CrossRef]
48. Di Carlo, D. Inertial Microfluidics. *Lab Chip* **2009**, *9*, 3038–3046. [CrossRef]
49. Gou, Y.; Jia, Y.; Wang, P.; Sun, C. Progress of Inertial Microfluidics in Principle and Application. *Sensors* **2018**, *18*, 1762. [CrossRef]
50. Yeo, J.C.; Kenry; Zhao, Z.; Zhang, P.; Wang, Z.; Lim, C.T. Label-Free Extraction of Extracellular Vesicles Using Centrifugal Microfluidics. *Biomicrofluidics* **2018**, *12*, 024103. [CrossRef]
51. Kopp, M.R.G.; Linsenmeier, M.; Hettich, B.; Prantl, S.; Stavrakis, S.; Leroux, J.C.; Arosio, P. Microfluidic Shrinking Droplet Concentrator for Analyte Detection and Phase Separation of Protein Solutions. *Anal. Chem.* **2020**, *92*, 5803–5812. [CrossRef]
52. Han, B.H.; Kim, S.; Seo, G.; Heo, Y.; Chung, S.; Kang, J.Y. Isolation of Extracellular Vesicles from Small Volumes of Plasma Using a Microfluidic Aqueous Two-Phase System. *Lab Chip* **2020**, *20*, 3552–3559. [CrossRef] [PubMed]
53. Tay, H.M.; Leong, S.Y.; Xu, X.; Kong, F.; Upadya, M.; Dalan, R.; Tay, C.Y.; Dao, M.; Suresh, S.; Hou, H.W. Direct Isolation of Circulating Extracellular Vesicles from Blood for Vascular Risk Profiling in Type 2 Diabetes Mellitus. *Lab Chip* **2021**, *21*, 2511–2523. [CrossRef] [PubMed]
54. Teoh, B.Y.; Lim, Y.M.; Chong, W.Y.; Subramaniam, M.; Tan, Z.Z.; Misran, M.; Suk, V.R.E.; Lo, K.W.; Lee, P.F. Isolation of Exosome from the Culture Medium of Nasopharyngeal Cancer (NPC) C666-1 Cells Using Inertial Based Microfluidic Channel. *Biomed. Microdevices* **2022**, *24*, 1–10. [CrossRef] [PubMed]
55. Paganini, C.; Hettich, B.; Kopp, M.R.G.; Eördögh, A.; Capasso Palmiero, U.; Adamo, G.; Touzet, N.; Manno, M.; Bongiovanni, A.; Rivera-Fuentes, P.; et al. Rapid Characterization and Quantification of Extracellular Vesicles by Fluorescence-Based Microfluidic Diffusion Sizing. *Adv. Healthc. Mater.* **2022**, *11*, 2100021. [CrossRef] [PubMed]
56. Kuntaegowdanahalli, S.S.; Bhagat, A.A.S.; Kumar, G.; Papautsky, I. Inertial Microfluidics for Continuous Particle Separation in Spiral Microchannels. *Lab Chip* **2009**, *9*, 2973–2980. [CrossRef]
57. Tay, H.M.; Kharel, S.; Dalan, R.; Chen, Z.J.; Tan, K.K.; Boehm, B.O.; Loo, S.C.J.; Hou, H.W. Rapid Purification of Sub-Micrometer Particles for Enhanced Drug Release and Microvesicles Isolation. *NPG Asia Mater.* **2017**, *9*, e434. [CrossRef]
58. McGrath, J.; Jimenez, M.; Bridle, H. Deterministic Lateral Displacement for Particle Separation: A Review. *Lab Chip* **2014**, *14*, 4139–4158. [CrossRef]
59. Inglis, D.W.; Davis, J.A.; Austin, R.H.; Sturm, J.C. Critical Particle Size for Fractionation by Deterministic Lateral Displacement. *Lab Chip* **2006**, *6*, 655–658. [CrossRef]
60. Wunsch, B.H.; Smith, J.T.; Gifford, S.M.; Wang, C.; Brink, M.; Bruce, R.L.; Austin, R.H.; Stolovitzky, G.; Astier, Y. Nanoscale Lateral Displacement Arrays for the Separation of Exosomes and Colloids down to 20 Nm. *Nat. Nanotechnol. 2016 1111* **2016**, *11*, 936–940. [CrossRef]
61. Santana, S.M.; Antonyak, M.A.; Cerione, R.A.; Kirby, B.J. Microfluidic Isolation of Cancer-Cell-Derived Microvesicles from Hetergeneous Extracellular Shed Vesicle Populations. *Biomed. Microdevices* **2014**, *16*, 869–877. [CrossRef]
62. Laki, A.J.; Botzheim, L.; Iván, K.; Tamási, V.; Civera, P. Separation of Microvesicles from Serological Samples Using Deterministic Lateral Displacement Effect. *Bionanoscience* **2015**, *5*, 48–54. [CrossRef]
63. Derzsi, L.; Filippi, D.; Mistura, G.; Pierno, M.; Lulli, M.; Sbragaglia, M.; Bernaschi, M.; Garstecki, P. Fluidization and Wall Slip of Soft Glassy Materials by Controlled Surface Roughness. *Phys. Rev. E* **2017**, *95*, 1–6. [CrossRef] [PubMed]
64. Leshansky, A.M.; Bransky, A.; Korin, N.; Dinnar, U. Tunable Nonlinear Viscoelastic "Focusing" in a Microfluidic Device. *Phys. Rev. Lett.* **2007**, *98*, 1–4. [CrossRef] [PubMed]
65. Lim, E.J.; Ober, T.J.; Edd, J.F.; Desai, S.P.; Neal, D.; Bong, K.W.; Doyle, P.S.; McKinley, G.H.; Toner, M. Inertio-Elastic Focusing of Bioparticles in Microchannels at High Throughput. *Nat. Commun.* **2014**, *5*, 4120. [CrossRef] [PubMed]
66. Liu, C.; Xue, C.; Chen, X.; Shan, L.; Tian, Y.; Hu, G. Size-Based Separation of Particles and Cells Utilizing Viscoelastic Effects in Straight Microchannels. *Anal. Chem.* **2015**, *87*, 6041–6048. [CrossRef]
67. Varga, Z.; Fehér, B.; Kitka, D.; Wacha, A.; Bóta, A.; Berényi, S.; Pipich, V.; Fraikin, J.L. Size Measurement of Extracellular Vesicles and Synthetic Liposomes: The Impact of the Hydration Shell and the Protein Corona. *Colloids Surfaces B Biointerfaces* **2020**, *192*, 111053. [CrossRef]
68. Zhou, Y.; Ma, Z.; Tayebi, M.; Ai, Y. Submicron Particle Focusing and Exosome Sorting by Wavy Microchannel Structures within Viscoelastic Fluids. *Anal. Chem.* **2019**, *91*, 4577–4584. [CrossRef]
69. Nam, J.; Yoon, J.; Jee, H.; Jang, W.S.; Lim, C.S. High-Throughput Separation of Microvesicles from Whole Blood Components Using Viscoelastic Fluid. *Adv. Mater. Technol.* **2020**, *5*, 2000612. [CrossRef]
70. Liu, C.; Guo, J.; Tian, F.; Yang, N.; Yan, F.; Ding, Y.; Wei, J.; Hu, G.; Nie, G.; Sun, J. Field-Free Isolation of Exosomes from Extracellular Vesicles by Microfluidic Viscoelastic Flows. *ACS Nano* **2017**, *11*, 6968–6976. [CrossRef]
71. Liu, C.; Zhao, J.; Tian, F.; Chang, J.; Zhang, W.; Sun, J. I-DNA- A Nd Aptamer-Mediated Sorting and Analysis of Extracellular Vesicles. *J. Am. Chem. Soc.* **2019**, *141*, 3817–3821. [CrossRef]
72. Asghari, M.; Cao, X.; Mateescu, B.; Van Leeuwen, D.; Aslan, M.K.; Stavrakis, S.; Demello, A.J. Oscillatory Viscoelastic Microfluidics for Efficient Focusing and Separation of Nanoscale Species. *ACS Nano* **2020**, *14*, 422–433. [CrossRef] [PubMed]
73. Fraunhofer, W.; Winter, G. The Use of Asymmetrical Flow Field-Flow Fractionation in Pharmaceutics and Biopharmaceutics. *Eur. J. Pharm. Biopharm.* **2004**, *58*, 369–383. [CrossRef] [PubMed]

74. Wahlund, K.G.; Giddings, J.C. Properties of an Asymmetric Flow Field-Flow, Fractionation Channel Having One Permeable Wall. *Anal. Chem.* **1987**, *59*, 1332–1339. [CrossRef] [PubMed]
75. Zhang, H.; Freitas, D.; Kim, H.S.; Fabijanic, K.; Li, Z.; Chen, H.; Mark, M.T.; Molina, H.; Martin, A.B.; Bojmar, L.; et al. Identification of Distinct Nanoparticles and Subsets of Extracellular Vesicles by Asymmetric Flow Field-Flow Fractionation. *Nat. Cell Biol.* **2018**, *20*, 332–343. [CrossRef]
76. Yamada, M.; Nakashima, M.; Seki, M. Pinched Flow Fractionation: Continuous Size Separation of Particles Utilizing a Laminar Flow Profile in a Pinched Microchannel. *Anal. Chem.* **2004**, *76*, 5465–5471. [CrossRef]
77. Shin, S.; Han, D.; Park, M.C.; Mun, J.Y.; Choi, J.; Chun, H.; Kim, S.; Hong, J.W. Separation of Extracellular Nanovesicles and Apoptotic Bodies from Cancer Cell Culture Broth Using Tunable Microfluidic Systems. *Sci. Rep.* **2017**, *7*, 1–8. [CrossRef]
78. Bruus, H. Acoustofluidics 1: Governing Equations in Microfluidics. *Lab Chip* **2011**, *11*, 3742–3751. [CrossRef]
79. Bruus, H. Acoustofluidics 7: The Acoustic Radiation Force on Small Particles. *Lab Chip* **2012**, *12*, 1014–1021. [CrossRef]
80. Bai, Y.; Lu, Y.; Wang, K.; Cheng, Z.; Qu, Y.; Qiu, S.; Zhou, L.; Wu, Z.; Liu, H.; Zhao, J.; et al. Rapid Isolation and Multiplexed Detection of Exosome Tumor Markers Via Queued Beads Combined with Quantum Dots in a Microarray. *Nano-Micro Lett.* **2019**, *11*, 1–11. [CrossRef]
81. Sehgal, P.; Kirby, B.J. Separation of 300 and 100 Nm Particles in Fabry-Perot Acoustofluidic Resonators. *Anal. Chem.* **2017**, *89*, 12192–12200. [CrossRef]
82. Leibacher, I.; Reichert, P.; Dual, J. Microfluidic Droplet Handling by Bulk Acoustic Wave (BAW) Acoustophoresis. *Lab Chip* **2015**, *15*, 2896–2905. [CrossRef] [PubMed]
83. Gao, Y.; Wu, M.; Lin, Y.; Xu, J. Acoustic Microfluidic Separation Techniques and Bioapplications: A Review. *Micromachines* **2020**, *11*, 921. [CrossRef] [PubMed]
84. Ding, X.; Li, P.; Lin, S.C.S.; Stratton, Z.S.; Nama, N.; Guo, F.; Slotcavage, D.; Mao, X.; Shi, J.; Costanzo, F.; et al. Surface Acoustic Wave Microfluidics. *Lab Chip* **2013**, *13*, 3626–3649. [CrossRef]
85. Wu, M.; Ozcelik, A.; Rufo, J.; Wang, Z.; Fang, R.; Jun Huang, T. Acoustofluidic Separation of Cells and Particles. *Microsystems Nanoeng.* **2019**, *5*, 32. [CrossRef] [PubMed]
86. Lee, K.; Shao, H.; Weissleder, R.; Lee, H. Acoustic Purification of Extracellular Microvesicles. *ACS Nano* **2015**, *9*, 2321–2327. [CrossRef]
87. Gu, Y.; Chen, C.; Mao, Z.; Bachman, H.; Becker, R.; Rufo, J.; Wang, Z.; Zhang, P.; Mai, J.; Yang, S.; et al. Acoustofluidic Centrifuge for Nanoparticle Enrichment and Separation. *Sci. Adv.* **2021**, *7*. [CrossRef]
88. Wang, Z.; Li, F.; Rufo, J.; Chen, C.; Yang, S.; Li, L.; Zhang, J.; Cheng, J.; Kim, Y.; Wu, M.; et al. Acoustofluidic Salivary Exosome Isolation: A Liquid Biopsy Compatible Approach for Human Papillomavirus–Associated Oropharyngeal Cancer Detection. *J. Mol. Diagnostics* **2020**, *22*, 50–59. [CrossRef]
89. Habibi, R.; He, V.; Ghavamian, S.; De Marco, A.; Lee, T.H.; Aguilar, M.I.; Zhu, D.; Lim, R.; Neild, A. Exosome Trapping and Enrichment Using a Sound Wave Activated Nano-Sieve (SWANS). *Lab Chip* **2020**, *20*, 3633–3643. [CrossRef]
90. Wu, M.; Ouyang, Y.; Wang, Z.; Zhang, R.; Huang, P.H.; Chen, C.; Li, H.; Li, P.; Quinn, D.; Dao, M.; et al. Isolation of Exosomes from Whole Blood by Integrating Acoustics and Microfluidics. *Proc. Natl. Acad. Sci. USA* **2017**, *114*, 10584–10589. [CrossRef]
91. Ku, A.; Lim, H.C.; Evander, M.; Lilja, H.; Laurell, T.; Scheding, S.; Ceder, Y. Acoustic Enrichment of Extracellular Vesicles from Biological Fluids. *Anal. Chem.* **2018**, *90*, 8011–8019. [CrossRef]
92. Ku, A.; Fredsøe, J.; Sørensen, K.D.; Borre, M.; Evander, M.; Laurell, T.; Lilja, H.; Ceder, Y. High-Throughput and Automated Acoustic Trapping of Extracellular Vesicles to Identify MicroRNAs With Diagnostic Potential for Prostate Cancer. *Front. Oncol.* **2021**, *11*, 386. [CrossRef] [PubMed]
93. Ku, A.; Ravi, N.; Yang, M.; Evander, M.; Laurell, T.; Lilja, H.; Ceder, Y. A Urinary Extracellular Vesicle MicroRNA Biomarker Discovery Pipeline; from Automated Extracellular Vesicle Enrichment by Acoustic Trapping to MicroRNA Sequencing. *PLoS ONE* **2019**, *14*, e0217507. [CrossRef] [PubMed]
94. Hao, N.; Pei, Z.; Liu, P.; Bachman, H.; Downing Naquin, T.; Zhang, P.; Zhang, J.; Shen, L.; Yang, S.; Yang, K.; et al. Acoustofluidics-Assisted Fluorescence-SERS Bimodal Biosensors. *Small* **2020**, *16*, 2005179. [CrossRef]
95. Hao, N.; Liu, P.; Bachman, H.; Pei, Z.; Zhang, P.; Rufo, J.; Wang, Z.; Zhao, S.; Huang, T.J. Acoustofluidics-Assisted Engineering of Multifunctional Three-Dimensional Zinc Oxide Nanoarrays. *ACS Nano* **2020**, *14*, 6150–6163. [CrossRef] [PubMed]
96. Wang, C.; Wang, C.; Jin, D.; Yu, Y.; Yang, F.; Zhang, Y.; Yao, Q.; Zhang, G.J. AuNP-Amplified Surface Acoustic Wave Sensor for the Quantification of Exosomes. *ACS Sens.* **2020**, *5*, 362–369. [CrossRef] [PubMed]
97. Jubery, T.Z.; Srivastava, S.K.; Dutta, P. Dielectrophoretic Separation of Bioparticles in Microdevices: A Review. *Electrophoresis* **2014**, *35*, 691–713. [CrossRef]
98. Ibsen, S.D.; Wright, J.; Lewis, J.M.; Kim, S.; Ko, S.Y.; Ong, J.; Manouchehri, S.; Vyas, A.; Akers, J.; Chen, C.C.; et al. Rapid Isolation and Detection of Exosomes and Associated Biomarkers from Plasma. *ACS Nano* **2017**, *11*, 6641–6651. [CrossRef]
99. Zhao, W.; Zhang, L.; Ye, Y.; Li, Y.; Luan, X.; Liu, J.; Cheng, J.; Zhao, Y.; Li, M.; Huang, C. Microsphere Mediated Exosome Isolation and Ultra-Sensitive Detection on a Dielectrophoresis Integrated Microfluidic Device. *Analyst* **2021**, *146*, 5962–5972. [CrossRef]
100. Chen, Y.S.; Lai, C.P.K.; Chen, C.; Lee, G. Bin Isolation and Recovery of Extracellular Vesicles Using Optically-Induced Dielectrophoresis on an Integrated Microfluidic Platform. *Lab Chip* **2021**, *21*, 1475–1483. [CrossRef]
101. Davies, R.T.; Kim, J.; Jang, S.C.; Choi, E.J.; Gho, Y.S.; Park, J. Microfluidic Filtration System to Isolate Extracellular Vesicles from Blood. *Lab Chip* **2012**, *12*, 5202–5210. [CrossRef]

102. Akagi, T.; Kato, K.; Kobayashi, M.; Kosaka, N.; Ochiya, T.; Ichiki, T. On-Chip Immunoelectrophoresis of Extracellular Vesicles Released from Human Breast Cancer Cells. *PLoS ONE* **2015**, *10*, e0123603. [CrossRef] [PubMed]
103. Marczak, S.; Richards, K.; Ramshani, Z.; Smith, E.; Senapati, S.; Hill, R.; Go, D.B.; Chang, H.C. Simultaneous Isolation and Preconcentration of Exosomes by Ion Concentration Polarization. *Electrophoresis* **2018**, *39*, 2029–2038. [CrossRef] [PubMed]
104. Hadady, H.; Karamali, F.; Ejeian, F.; Haghjooy Javanmard, S.; Rafiee, L.; Nasr Esfahani, M.H. AC Electrokinetic Isolation and Detection of Extracellular Vesicles from Dental Pulp Stem Cells: Theoretical Simulation Incorporating Fluid Mechanics. *Electrophoresis* **2021**, *42*, 2018–2026. [CrossRef] [PubMed]
105. Cheung, L.S.; Sahloul, S.; Orozaliev, A.; Song, Y.A. Rapid Detection and Trapping of Extracellular Vesicles by Electrokinetic Concentration for Liquid Biopsy on Chip. *Micromachines* **2018**, *9*, 306. [CrossRef]
106. Cho, S.; Jo, W.; Heo, Y.; Kang, J.Y.; Kwak, R.; Park, J. Isolation of Extracellular Vesicle from Blood Plasma Using Electrophoretic Migration through Porous Membrane. *Sens. Actuators B Chem.* **2016**, *233*, 289–297. [CrossRef]
107. Tayebi, M.; Yang, D.; Collins, D.J.; Ai, Y. Deterministic Sorting of Submicrometer Particles and Extracellular Vesicles Using a Combined Electric and Acoustic Field. *Nano Lett.* **2021**, *21*, 6835–6842. [CrossRef]
108. Zhang, Y.; Tong, X.; Yang, L.; Yin, R.; Li, Y.; Zeng, D.; Wang, X.; Deng, K. A Herringbone Mixer Based Microfluidic Device HBEXO-Chip for Purifying Tumor-Derived Exosomes and Establishing MiRNA Signature in Pancreatic Cancer. *Sens. Actuators B Chem.* **2021**, *332*, 129511. [CrossRef]
109. Chen, Z.; Cheng, S.B.; Cao, P.; Qiu, Q.F.; Chen, Y.; Xie, M.; Xu, Y.; Huang, W.H. Detection of Exosomes by ZnO Nanowires Coated Three-Dimensional Scaffold Chip Device. *Biosens. Bioelectron.* **2018**, *122*, 211–216. [CrossRef]
110. Kanwar, S.S.; Dunlay, C.J.; Simeone, D.M.; Nagrath, S. Microfluidic Device (ExoChip) for on-Chip Isolation, Quantification and Characterization of Circulating Exosomes. *Lab Chip* **2014**, *14*, 1891–1900. [CrossRef]
111. Hisey, C.L.; Dorayappan, K.D.P.; Cohn, D.E.; Selvendiran, K.; Hansford, D.J. Microfluidic Affinity Separation Chip for Selective Capture and Release of Label-Free Ovarian Cancer Exosomes. *Lab Chip* **2018**, *18*, 3144–3153. [CrossRef]
112. Wang, Z.; Wu, H.J.; Fine, D.; Schmulen, J.; Hu, Y.; Godin, B.; Zhang, J.X.J.; Liu, X. Ciliated Micropillars for the Microfluidic-Based Isolation of Nanoscale Lipid Vesicles. *Lab Chip* **2013**, *13*, 2879–2882. [CrossRef] [PubMed]
113. Qi, R.; Zhu, G.; Wang, Y.; Wu, S.; Li, S.; Zhang, D.; Bu, Y.; Bhave, G.; Han, R.; Liu, X. Microfluidic Device for the Analysis of MDR Cancerous Cell-Derived Exosomes' Response to Nanotherapy. *Biomed. Microdevices* **2019**, *21*, 1–9. [CrossRef] [PubMed]
114. Kamyabi, N.; Abbasgholizadeh, R.; Maitra, A.; Ardekani, A.; Biswal, S.L.; Grande-Allen, K.J. Isolation and Mutational Assessment of Pancreatic Cancer Extracellular Vesicles Using a Microfluidic Platform. *Biomed. Microdevices* **2020**, *22*, 1–11. [CrossRef] [PubMed]
115. Sooriyaarachchi, D.; Maharubin, S.; Tan, G.Z. ZnO Nanowire-Anchored Microfluidic Device With Herringbone Structure Fabricated by Maskless Photolithography. *Biomed. Eng. Comput. Biol.* **2020**, *11*, 1179597220941431. [CrossRef]
116. Yang, L.; Tong, X.; Zhang, Y.; Li, Y.; Liu, J.; Yin, R.; Zeng, D.; Yuan, Y.; Deng, K. Tim4-Functionalized HBEV-Chip by Isolating Plasma-Derived Phosphatidylserine-Positive Small Extracellular Vesicles for Pan-Cancer Screening. *Adv. Mater. Technol.* **2022**, *7*, 2101115. [CrossRef]
117. Li, Q.; Wang, Y.; Xue, Y.; Qiao, L.; Yu, G.; Liu, Y.; Yu, S. Ultrasensitive Analysis of Exosomes Using a 3D Self-Assembled Nanostructured SiO2Microfluidic Chip. *ACS Appl. Mater. Interfaces* **2022**, *14*, 14693–14702. [CrossRef] [PubMed]
118. Chen, H.; Bian, F.; Guo, J.; Zhao, Y.; Chen, H.X.; Bian, F.K.; Guo, J.H.; Zhao, Y.J. Aptamer-Functionalized Barcodes in Herringbone Microfluidics for Multiple Detection of Exosomes. *Small Methods* **2022**, *6*, 2200236. [CrossRef]
119. Zhang, P.; He, M.; Zeng, Y. Ultrasensitive Microfluidic Analysis of Circulating Exosomes Using a Nanostructured Graphene Oxide/Polydopamine Coating. *Lab Chip* **2016**, *16*, 3033–3042. [CrossRef]
120. Kang, Y.-T.; Hadlock, T.; Lo, T.-W.; Purcell, E.; Mutukuri, A.; Fouladdel, S.; De Silva Raguera, M.; Fairbairn, H.; Murlidhar, V.; Durham, A.; et al. Dual-Isolation and Profiling of Circulating Tumor Cells and Cancer Exosomes from Blood Samples with Melanoma Using Immunoaffinity-Based Microfluidic Interfaces. *Adv. Sci.* **2020**, *7*, 2001581. [CrossRef]
121. Xiong, Y.; Kang, H.; Zhou, H.; Ma, L.; Xu, X. Recent Progress on Microfluidic Devices with Incorporated 1D Nanostructures for Enhanced Extracellular Vesicle (EV) Separation. *Bio-Design Manuf.* **2022**, *5*, 607–616. [CrossRef]
122. Chen, C.; Skog, J.; Hsu, C.H.; Lessard, R.T.; Balaj, L.; Wurdinger, T.; Carter, B.S.; Breakefield, X.O.; Toner, M.; Irimia, D. Microfluidic Isolation and Transcriptome Analysis of Serum Microvesicles. *Lab Chip* **2010**, *10*, 505–511. [CrossRef]
123. Zhang, P.; Zhou, X.; He, M.; Shang, Y.; Tetlow, A.L.; Godwin, A.K.; Zeng, Y. Ultrasensitive Detection of Circulating Exosomes with a 3D-Nanopatterned Microfluidic Chip. *Nat. Biomed. Eng.* **2019**, *3*, 438–451. [CrossRef]
124. Gwak, H.; Park, S.; Kim, J.; Lee, J.D.; Kim, I.S.; Kim, S.I.; Hyun, K.A.; Jung, H. Il Microfluidic Chip for Rapid and Selective Isolation of Tumor-Derived Extracellular Vesicles for Early Diagnosis and Metastatic Risk Evaluation of Breast Cancer. *Biosens. Bioelectron.* **2021**, *192*, 113495. [CrossRef]
125. He, M.; Crow, J.; Roth, M.; Zeng, Y.; Godwin, A.K. Integrated Immunoisolation and Protein Analysis of Circulating Exosomes Using Microfluidic Technology. *Lab Chip* **2014**, *14*, 3773–3780. [CrossRef]
126. Zhao, Z.; Yang, Y.; Zeng, Y.; He, M. A Microfluidic ExoSearch Chip for Multiplexed Exosome Detection towards Blood-Based Ovarian Cancer Diagnosis. *Lab Chip* **2016**, *16*, 489–496. [CrossRef]
127. Lu, Y.; Ye, L.; Jian, X.; Yang, D.; Zhang, H.; Tong, Z.; Wu, Z.; Shi, N.; Han, Y.; Mao, H. Integrated Microfluidic System for Isolating Exosome and Analyzing Protein Marker PD-L1. *Biosens. Bioelectron.* **2022**, *204*, 113879. [CrossRef]

128. Tian, Q.; He, C.; Liu, G.; Zhao, Y.; Hui, L.; Mu, Y.; Tang, R.; Luo, Y.; Zheng, S.; Wang, B. Nanoparticle Counting by Microscopic Digital Detection: Selective Quantitative Analysis of Exosomes via Surface-Anchored Nucleic Acid Amplification. *Anal. Chem.* **2018**, *90*, 6556–6562. [CrossRef]
129. Cheng, H.L.; Fu, C.Y.; Kuo, W.C.; Chen, Y.W.; Chen, Y.S.; Lee, Y.M.; Li, K.H.; Chen, C.; Ma, H.P.; Huang, P.C.; et al. Detecting MiRNA Biomarkers from Extracellular Vesicles for Cardiovascular Disease with a Microfluidic System. *Lab Chip* **2018**, *18*, 2917–2925. [CrossRef]
130. Reátegui, E.; Van Der Vos, K.E.; Lai, C.P.; Zeinali, M.; Atai, N.A.; Aldikacti, B.; Floyd, F.P.; Khankhel, A.; Thapar, V.; Hochberg, F.H.; et al. Engineered Nanointerfaces for Microfluidic Isolation and Molecular Profiling of Tumor-Specific Extracellular Vesicles. *Nat. Commun.* **2018**, *9*, 175. [CrossRef]
131. Xu, H.; Liao, C.; Zuo, P.; Liu, Z.; Ye, B.C. Magnetic-Based Microfluidic Device for On-Chip Isolation and Detection of Tumor-Derived Exosomes. *Anal. Chem.* **2018**, *90*, 13451–13458. [CrossRef]
132. Sharma, P.; Ludwig, S.; Muller, L.; Hong, C.S.; Kirkwood, J.M.; Ferrone, S.; Whiteside, T.L. Immunoaffinity-Based Isolation of Melanoma Cell-Derived Exosomes from Plasma of Patients with Melanoma. *J. Extracell. Vesicles* **2018**, *7*, 1435138. [CrossRef]
133. Chen, W.; Li, H.; Su, W.; Qin, J. Microfluidic Device for On-Chip Isolation and Detection of Circulating Exosomes in Blood of Breast Cancer Patients. *Biomicrofluidics* **2019**, *13*, 054113. [CrossRef]
134. Wang, Y.; Li, Q.; Shi, H.; Tang, K.; Qiao, L.; Yu, G.; Ding, C.; Yu, S. Microfluidic Raman Biochip Detection of Exosomes: A Promising Tool for Prostate Cancer Diagnosis. *Lab Chip* **2020**, *20*, 4632–4637. [CrossRef]
135. Sung, C.Y.; Huang, C.C.; Chen, Y.S.; Hsu, K.F.; Lee, G. Bin Isolation and Quantification of Extracellular Vesicle-Encapsulated MicroRNA on an Integrated Microfluidic Platform. *Lab Chip* **2021**, *21*, 4660–4671. [CrossRef]
136. Dueck, J. The Sedimentation Velocity of a Particle in a Wide Range of Reynolds Numbers in the Application to the Analysis of the Separation Curve. *Adv. Powder Technol.* **2013**, *24*, 150–153. [CrossRef]
137. Gwak, H.; Park, S.; Yu, H.; Hyun, K.A.; Jung, H. Il A Modular Microfluidic Platform for Serial Enrichment and Harvest of Pure Extracellular Vesicles. *Analyst* **2022**, *147*, 1117–1127. [CrossRef]
138. Kwon, S.; Oh, J.; Seok Lee, M.; Um, E.; Jeong, J.; Kang, J.H.; Kwon, S.; Oh, J.; Lee, M.S.; Kang, J.H.; et al. Enhanced Diamagnetic Repulsion of Blood Cells Enables Versatile Plasma Separation for Biomarker Analysis in Blood. *Small* **2021**, *17*, 2100797. [CrossRef]
139. Son, K.J.; Rahimian, A.; Shin, D.S.; Siltanen, C.; Patel, T.; Revzin, A. Microfluidic Compartments with Sensing Microbeads for Dynamic Monitoring of Cytokine and Exosome Release from Single Cells. *Analyst* **2016**, *141*, 679–688. [CrossRef]
140. Tayebi, M.; Zhou, Y.; Tripathi, P.; Chandramohanadas, R.; Ai, Y. Exosome Purification and Analysis Using a Facile Microfluidic Hydrodynamic Trapping Device. *Anal. Chem.* **2020**, *92*, 10733–10742. [CrossRef]
141. Dudani, J.S.; Gossett, D.R.; Tse, H.T.K.; Lamm, R.J.; Kulkarni, R.P.; Carlo, D. Di Rapid Inertial Solution Exchange for Enrichment and Flow Cytometric Detection of Microvesicles. *Biomicrofluidics* **2015**, *9*, 014112. [CrossRef]
142. Shao, H.; Chung, J.; Balaj, L.; Charest, A.; Bigner, D.D.; Carter, B.S.; Hochberg, F.H.; Breakefield, X.O.; Weissleder, R.; Lee, H. Protein Typing of Circulating Microvesicles Allows Real-Time Monitoring of Glioblastoma Therapy. *Nat. Med.* **2012**, *18*, 1835–1840. [CrossRef]
143. Hong, S.L.; Wan, Y.T.; Tang, M.; Pang, D.W.; Zhang, Z.L. Multifunctional Screening Platform for the Highly Efficient Discovery of Aptamers with High Affinity and Specificity. *Anal. Chem.* **2017**, *89*, 6535–6542. [CrossRef]
144. Ko, J.; Bhagwat, N.; Yee, S.S.; Ortiz, N.; Sahmoud, A.; Black, T.; Aiello, N.M.; McKenzie, L.; O'Hara, M.; Redlinger, C.; et al. Combining Machine Learning and Nanofluidic Technology to Diagnose Pancreatic Cancer Using Exosomes. *ACS Nano* **2017**, *11*, 11182–11193. [CrossRef]
145. Liu, Y.; Zhao, W.; Cheng, R.; Logun, M.; Zayas-Viera, M.D.M.; Karumbaiah, L.; Mao, L. Label-Free Ferrohydrodynamic Separation of Exosome-like Nanoparticles. *Lab Chip* **2020**, *20*, 3187–3201. [CrossRef]
146. Sancho-Albero, M.; Sebastián, V.; Sesé, J.; Pazo-Cid, R.; Mendoza, G.; Arruebo, M.; Martín-Duque, P.; Santamaría, J. Isolation of Exosomes from Whole Blood by a New Microfluidic Device: Proof of Concept Application in the Diagnosis and Monitoring of Pancreatic Cancer. *J. Nanobiotechnol.* **2020**, *18*, 150. [CrossRef]
147. Yu, Z.; Lin, S.; Xia, F.; Liu, Y.; Zhang, D.; Wang, F.; Wang, Y.; Li, Q.; Niu, J.; Cao, C.; et al. ExoSD Chips for High-Purity Immunomagnetic Separation and High-Sensitivity Detection of Gastric Cancer Cell-Derived Exosomes. *Biosens. Bioelectron.* **2021**, *194*, 113594. [CrossRef]
148. Yang, Y.; Kannisto, E.; Patnaik, S.K.; Reid, M.E.; Li, L.; Wu, Y. Ultrafast Detection of Exosomal RNAs via Cationic Lipoplex Nanoparticles in a Micromixer Biochip for Cancer Diagnosis. *ACS Appl. Nano Mater.* **2021**, *4*, 2806–2819. [CrossRef]
149. Vader, P.; Mol, E.A.; Pasterkamp, G.; Schiffelers, R.M. Extracellular Vesicles for Drug Delivery. *Adv. Drug Deliv. Rev.* **2016**, *106*, 148–156. [CrossRef]
150. Mason, H.G.; Bush, J.; Agrawal, N.; Hakami, R.M.; Veneziano, R. A Microfluidic Platform to Monitor Real-Time Effects of Extracellular Vesicle Exchange between Co-Cultured Cells across Selectively Permeable Barriers. *Int. J. Mol. Sci.* **2022**, *23*, 3534. [CrossRef]
151. Han, Z.; Peng, C.; Yi, J.; Zhang, D.; Xiang, X.; Peng, X.; Su, B.; Liu, B.; Shen, Y.; Qiao, L. Highly Efficient Exosome Purification from Human Plasma by Tangential Flow Filtration Based Microfluidic Chip. *Sensors Actuators B Chem.* **2021**, *333*, 129563. [CrossRef]
152. Gustafson, K.T.; Huynh, K.T.; Heineck, D.; Bueno, J.; Modestino, A.; Kim, S.; Gower, A.; Armstrong, R.; Schutt, C.E.; Ibsen, S.D. Automated Fluorescence Quantification of Extracellular Vesicles Collected from Blood Plasma Using Dielectrophoresis. *Lab Chip* **2021**, *21*, 1318–1332. [CrossRef]

153. Mogi, K.; Hayashida, K.; Yamamoto, T. Damage-Less Handling of Exosomes Using an Ion-Depletion Zone in a Microchannel. *Anal. Sci.* **2018**, *34*, 875–880. [CrossRef]
154. Vaidyanathan, R.; Naghibosadat, M.; Rauf, S.; Korbie, D.; Carrascosa, L.G.; Shiddiky, M.J.A.; Trau, M. Detecting Exosomes Specifically: A Multiplexed Device Based on Alternating Current Electrohydrodynamic Induced Nanoshearing. *Anal. Chem.* **2014**, *86*, 11125–11132. [CrossRef]
155. Kang, Y.-T.; Purcell, E.; Palacios-Rolston, C.; Lo, T.-W.; Ramnath, N.; Jolly, S.; Nagrath, S.; Kang, Y.; Purcell, E.; Palacios-Rolston, C.; et al. Isolation and Profiling of Circulating Tumor-Associated Exosomes Using Extracellular Vesicular Lipid–Protein Binding Affinity Based Microfluidic Device. *Small* **2019**, *15*, 1903600. [CrossRef]
156. Zhou, Z.; Chen, Y.; Qian, X. Target-Specific Exosome Isolation through Aptamer-Based Microfluidics. *Biosensors* **2022**, *12*, 257. [CrossRef]
157. Kang, Y.T.; Purcell, E.; Hadlock, T.; Lo, T.W.; Mutukuri, A.; Jolly, S.; Nagrath, S. Multiplex Isolation and Profiling of Extracellular Vesicles Using a Microfluidic DICE Device. *Analyst* **2019**, *144*, 5785–5793. [CrossRef]
158. Zhu, L.; Wang, K.; Cui, J.; Liu, H.; Bu, X.; Ma, H.; Wang, W.; Gong, H.; Lausted, C.; Hood, L.; et al. Label-Free Quantitative Detection of Tumor-Derived Exosomes through Surface Plasmon Resonance Imaging. *Anal. Chem.* **2014**, *86*, 8857–8864. [CrossRef]
159. Yasui, T.; Yanagida, T.; Ito, S.; Konakade, Y.; Takeshita, D.; Naganawa, T.; Nagashima, K.; Shimada, T.; Kaji, N.; Nakamura, Y.; et al. Unveiling Massive Numbers of Cancer-Related Urinary-MicroRNA Candidates via Nanowires. *Sci. Adv.* **2017**, *3*. [CrossRef]
160. Wang, J.; Li, W.; Zhang, L.; Ban, L.; Chen, P.; Du, W.; Feng, X.; Liu, B.F. Chemically Edited Exosomes with Dual Ligand Purified by Microfluidic Device for Active Targeted Drug Delivery to Tumor Cells. *ACS Appl. Mater. Interfaces* **2017**, *9*, 27441–27452. [CrossRef]
161. Wang, Y.; Yuan, W.; Kimber, M.; Lu, M.; Dong, L. Rapid Differentiation of Host and Parasitic Exosome Vesicles Using Microfluidic Photonic Crystal Biosensor. *ACS Sensors* **2018**, *3*, 1616–1621. [CrossRef]
162. Algarni, A.; Greenman, J.; Madden, L.A. Procoagulant Tumor Microvesicles Attach to Endothelial Cells on Biochips under Microfluidic Flow. *Biomicrofluidics* **2019**, *13*, 064124. [CrossRef]
163. Zhang, P.; Zhou, X.; Zeng, Y. Multiplexed Immunophenotyping of Circulating Exosomes on Nano-Engineered ExoProfile Chip towards Early Diagnosis of Cancer. *Chem. Sci.* **2019**, *10*, 5495–5504. [CrossRef]
164. Cavallaro, S.; Horak, J.; HÅÅg, P.; Gupta, D.; Stiller, C.; Sahu, S.S.; Görgens, A.; Gatty, H.K.; Viktorsson, K.; El Andaloussi, S.; et al. Label-Free Surface Protein Profiling of Extracellular Vesicles by an Electrokinetic Sensor. *ACS Sens.* **2019**, *4*, 1399–1408. [CrossRef]
165. Lv, X.; Geng, Z.; Su, Y.; Fan, Z.; Wang, S.; Fang, W.; Chen, H. Label-Free Exosome Detection Based on a Low-Cost Plasmonic Biosensor Array Integrated with Microfluidics. *Langmuir* **2019**, *35*, 9816–9824. [CrossRef]
166. Wijerathne, H.; Witek, M.A.; Jackson, J.M.; Brown, V.; Hupert, M.L.; Herrera, K.; Kramer, C.; Davidow, A.E.; Li, Y.; Baird, A.E.; et al. Affinity Enrichment of Extracellular Vesicles from Plasma Reveals MRNA Changes Associated with Acute Ischemic Stroke. *Commun. Biol.* **2020**, *3*, 613. [CrossRef]
167. Zhou, S.; Hu, T.; Han, G.; Wu, Y.; Hua, X.; Su, J.; Jin, W.; Mou, Y.; Mou, X.; Li, Q.; et al. Accurate Cancer Diagnosis and Stage Monitoring Enabled by Comprehensive Profiling of Different Types of Exosomal Biomarkers: Surface Proteins and MiRNAs. *Small* **2020**, *16*, 2004492. [CrossRef]
168. Han, S.; Xu, Y.; Sun, J.; Liu, Y.; Zhao, Y.; Tao, W.; Chai, R. Isolation and Analysis of Extracellular Vesicles in a Morpho Butterfly Wing-Integrated Microvortex Biochip. *Biosens. Bioelectron.* **2020**, *154*, 112073. [CrossRef]
169. Yang, Q.; Cheng, L.; Hu, L.; Lou, D.; Zhang, T.; Li, J.; Zhu, Q.; Liu, F. An Integrative Microfluidic Device for Isolation and Ultrasensitive Detection of Lung Cancer-Specific Exosomes from Patient Urine. *Biosens. Bioelectron.* **2020**, *163*, 112290. [CrossRef]
170. Zhou, S.; Hu, T.; Zhang, F.; Tang, D.; Li, D.; Cao, J.; Wei, W.; Wu, Y.; Liu, S. Integrated Microfluidic Device for Accurate Extracellular Vesicle Quantification and Protein Markers Analysis Directly from Human Whole Blood. *Anal. Chem.* **2020**, *92*, 1574–1581. [CrossRef]
171. Kashefi-Kheyrabadi, L.; Kim, J.; Chakravarty, S.; Park, S.; Gwak, H.; Kim, S.I.; Mohammadniaei, M.; Lee, M.H.; Hyun, K.A.; Jung, H. Il Detachable Microfluidic Device Implemented with Electrochemical Aptasensor (DeMEA) for Sequential Analysis of Cancerous Exosomes. *Biosens. Bioelectron.* **2020**, *169*, 112622. [CrossRef]
172. Yu, Q.; Zhao, Q.; Wang, S.; Zhao, S.; Zhang, S.; Yin, Y.; Dong, Y. Development of a Lateral Flow Aptamer Assay Strip for Facile Identification of Theranostic Exosomes Isolated from Human Lung Carcinoma Cells. *Anal. Biochem.* **2020**, *594*, 113591. [CrossRef]
173. Sun, N.; Lee, Y.T.; Zhang, R.Y.; Kao, R.; Teng, P.C.; Yang, Y.; Yang, P.; Wang, J.J.; Smalley, M.; Chen, P.J.; et al. Purification of HCC-Specific Extracellular Vesicles on Nanosubstrates for Early HCC Detection by Digital Scoring. *Nat. Commun.* **2020**, *11*, 4489. [CrossRef]
174. Nikoloff, J.M.; Saucedo-Espinosa, M.A.; Kling, A.; Dittrich, P.S. Identifying Extracellular Vesicle Populations from Single Cells. *Proc. Natl. Acad. Sci. USA* **2021**, *118*, e2106630118. [CrossRef]
175. Chiodi, E.; Daaboul, G.G.; Marn, A.M.; Ünlü, M.S. Multiplexed Affinity Measurements of Extracellular Vesicles Binding Kinetics. *Sensors* **2021**, *21*, 2634. [CrossRef]
176. Suwatthanarak, T.; Thiodorus, I.A.; Tanaka, M.; Shimada, T.; Takeshita, D.; Yasui, T.; Baba, Y.; Okochi, M. Microfluidic-Based Capture and Release of Cancer-Derived Exosomes via Peptide–Nanowire Hybrid Interface. *Lab Chip* **2021**, *21*, 597–607. [CrossRef] [PubMed]

177. Radnaa, E.; Richardson, L.S.; Sheller-Miller, S.; Baljinnyam, T.; De Castro Silva, M.; Kumar Kammala, A.; Urrabaz-Garza, R.; Kechichian, T.; Kim, S.; Han, A.; et al. Extracellular Vesicle Mediated Feto-Maternal HMGB1 Signaling Induces Preterm Birth. *Lab Chip* **2021**, *21*, 1956–1973. [CrossRef] [PubMed]
178. Lo, T.W.; Figueroa-Romero, C.; Hur, J.; Pacut, C.; Stoll, E.; Spring, C.; Lewis, R.; Nair, A.; Goutman, S.A.; Sakowski, S.A.; et al. Extracellular Vesicles in Serum and Central Nervous System Tissues Contain MicroRNA Signatures in Sporadic Amyotrophic Lateral Sclerosis. *Front. Mol. Neurosci.* **2021**, *14*, 246. [CrossRef] [PubMed]
179. Wang, F.; Gui, Y.; Liu, W.; Li, C.; Yang, Y. Precise Molecular Profiling of Circulating Exosomes Using a Metal–Organic Framework-Based Sensing Interface and an Enzyme-Based Electrochemical Logic Platform. *Anal. Chem.* **2022**, *94*, 875–883. [CrossRef] [PubMed]
180. Paisrisarn, P.; Yasui, T.; Zhu, Z.; Klamchuen, A.; Kasamechonchung, P.; Wutikhun, T.; Yordsri, V.; Baba, Y. Tailoring ZnO Nanowire Crystallinity and Morphology for Label-Free Capturing of Extracellular Vesicles. *Nanoscale* **2022**, *14*, 4484–4494. [CrossRef] [PubMed]
181. Shao, H.; Chung, J.; Lee, K.; Balaj, L.; Min, C.; Carter, B.S.; Hochberg, F.H.; Breakefield, X.O.; Lee, H.; Weissleder, R. Chip-Based Analysis of Exosomal MRNA Mediating Drug Resistance in Glioblastoma. *Nat. Commun.* **2015**, *6*, 6999. [CrossRef]
182. Zhao, Z.; McGill, J.; Gamero-Kubota, P.; He, M. Microfluidic On-Demand Engineering of Exosomes towards Cancer Immunotherapy. *Lab Chip* **2019**, *19*, 1877–1886. [CrossRef] [PubMed]
183. Niu, F.; Chen, X.; Niu, X.; Cai, Y.; Zhang, Q.; Chen, T.; Yang, H. Integrated Immunomagnetic Bead-Based Microfluidic Chip for Exosomes Isolation. *Micromachines* **2020**, *11*, 503. [CrossRef]
184. Song, Z.; Mao, J.; Barrero, R.A.; Wang, P.; Zhang, F.; Wang, T. Development of a CD63 Aptamer for Efficient Cancer Immunochemistry and Immunoaffinity-Based Exosome Isolation. *Molecules* **2020**, *25*, 5585. [CrossRef]
185. Huang, C.C.; Kuo, Y.H.; Chen, Y.S.; Huang, P.C.; Lee, G. Bin A Miniaturized, DNA-FET Biosensor-Based Microfluidic System for Quantification of Two Breast Cancer Biomarkers. *Microfluid. Nanofluidics* **2021**, *25*, 1–12. [CrossRef]
186. Zhao, Y.; Fang, X.; Bai, M.; Zhang, J.; Yu, H.; Chen, F.; Zhao, Y. A Microfluidic Surface-Enhanced Raman Scattering (SERS) Sensor for MicroRNA in Extracellular Vesicles with Nucleic Acid-Tyramine Cascade Amplification. *Chin. Chem. Lett.* **2022**, *33*, 2101–2104. [CrossRef]
187. Choi, Y.; Park, U.; Koo, H.J.; Park, J.S.; Lee, D.H.; Kim, K.; Choi, J. Exosome-Mediated Diagnosis of Pancreatic Cancer Using Lectin-Conjugated Nanoparticles Bound to Selective Glycans. *Biosens. Bioelectron.* **2021**, *177*, 112980. [CrossRef]
188. Piffoux, M.; Silva, A.K.A.; Lugagne, J.B.; Hersen, P.; Wilhelm, C.; Gazeau, F. Extracellular Vesicle Production Loaded with Nanoparticles and Drugs in a Trade-off between Loading, Yield and Purity: Towards a Personalized Drug Delivery System. *Adv. Biosyst.* **2017**, *1*, 1700044. [CrossRef] [PubMed]
189. Ferraro, D.; Serra, M.; Filippi, D.; Zago, L.; Guglielmin, E.; Pierno, M.; Descroix, S.; Viovy, J.L.; Mistura, G. Controlling the Distance of Highly Confined Droplets in a Capillary by Interfacial Tension for Merging On-Demand. *Lab Chip* **2019**, *19*, 136–146. [CrossRef] [PubMed]
190. Salva, M.L.; Rocca, M.; Niemeyer, C.M.; Delamarche, E. Methods for Immobilizing Receptors in Microfluidic Devices: A Review. *Micro Nano Eng.* **2021**, *11*, 100085. [CrossRef]
191. Kang, Y.T.; Kim, Y.J.; Bu, J.; Cho, Y.H.; Han, S.W.; Moon, B.I. High-Purity Capture and Release of Circulating Exosomes Using an Exosome-Specific Dual-Patterned Immunofiltration (ExoDIF) Device. *Nanoscale* **2017**, *9*, 13495–13505. [CrossRef]
192. Nielsen, J.B.; Hanson, R.L.; Almughamsi, H.M.; Pang, C.; Fish, T.R.; Woolley, A.T. Microfluidics: Innovations in Materials and Their Fabrication and Functionalization. *Anal. Chem.* **2020**, *92*, 150–168. [CrossRef] [PubMed]
193. Kim, J.; Sahloul, S.; Orozaliev, A.; Do, V.Q.; Pham, V.S.; Martins, D.; Wei, X.; Levicky, R.; Song, Y.A. Microfluidic Electrokinetic Preconcentration Chips: Enhancing the Detection of Nucleic Acids and Exosomes. *IEEE Nanotechnol. Mag.* **2020**, *14*, 18–34. [CrossRef]
194. Shiri, F.; Feng, H.; Petersen, K.E.; Sant, H.; Bardi, G.T.; Schroeder, L.A.; Merchant, M.L.; Gale, B.K.; Hood, J.L. Separation of U87 Glioblastoma Cell-Derived Small and Medium Extracellular Vesicles Using Elasto-Inertial Flow Focusing (a Spiral Channel). *Sci. Rep.* **2022**, *12*, 6146. [CrossRef] [PubMed]
195. Leong, S.Y.; Ong, H.B.; Tay, H.M.; Kong, F.; Upadya, M.; Gong, L.; Dao, M.; Dalan, R.; Hou, H.W. Microfluidic Size Exclusion Chromatography (MSEC) for Extracellular Vesicles and Plasma Protein Separation. *Small* **2022**, *18*, 2104470. [CrossRef] [PubMed]
196. Raju, D.; Bathini, S.; Badilescu, S.; Ouellette, R.J.; Ghosh, A.; Packirisamy, M. LSPR Detection of Extracellular Vesicles Using a Silver-PDMS Nano-Composite Platform Suitable for Sensor Networks. *Enterp. Inf. Syst.* **2018**, *14*, 532–541. [CrossRef]
197. Wu, T.; Yang, Y.; Cao, Y.; Huang, Y.; Xu, L.P.; Zhang, X.; Wang, S. Enhanced Lateral Flow Assay with Double Conjugates for the Detection of Exosomes. *Sci. China Chem.* **2018**, *61*, 1423–1429. [CrossRef]
198. Yokota, S.; Kuramochi, H.; Okubo, K.; Iwaya, A.; Tsuchiya, S.; Ichiki, T. Extracellular Vesicles Nanoarray Technology: Immobilization of Individual Extracellular Vesicles on Nanopatterned Polyethylene Glycol-Lipid Conjugate Brushes. *PLoS ONE* **2019**, *14*, e0224091. [CrossRef]
199. Zhang, P.; Crow, J.; Lella, D.; Zhou, X.; Samuel, G.; Godwin, A.K.; Zeng, Y. Ultrasensitive Quantification of Tumor MRNAs in Extracellular Vesicles with an Integrated Microfluidic Digital Analysis Chip. *Lab Chip* **2018**, *18*, 3790–3801. [CrossRef]
200. Zhou, Q.; Rahimian, A.; Son, K.; Shin, D.S.; Patel, T.; Revzin, A. Development of an Aptasensor for Electrochemical Detection of Exosomes. *Methods* **2016**, *97*, 88–93. [CrossRef]

201. Dong, J.; Zhang, R.Y.; Sun, N.; Smalley, M.; Wu, Z.; Zhou, A.; Chou, S.J.; Jan, Y.J.; Yang, P.; Bao, L.; et al. Bio-Inspired NanoVilli Chips for Enhanced Capture of Tumor-Derived Extracellular Vesicles: Toward Non-Invasive Detection of Gene Alterations in Non-Small Cell Lung Cancer. *ACS Appl. Mater. Interfaces* **2019**, *11*, 13973–13983. [CrossRef]
202. Chen, W.; Cao, R.; Su, W.; Zhang, X.; Xu, Y.; Wang, P.; Gan, Z.; Xie, Y.; Li, H.; Qin, J. Simple and Fast Isolation of Circulating Exosomes with a Chitosan Modified Shuttle Flow Microchip for Breast Cancer Diagnosis. *Lab Chip* **2021**, *21*, 1759–1770. [CrossRef] [PubMed]
203. Kang, Y.T.; Hadlock, T.; Jolly, S.; Nagrath, S. Extracellular Vesicles on Demand (EVOD) Chip for Screening and Quantification of Cancer-Associated Extracellular Vesicles. *Biosens. Bioelectron.* **2020**, *168*, 112535. [CrossRef] [PubMed]
204. Chutvirasakul, B.; Nuchtavorn, N.; Suntornsuk, L.; Zeng, Y. Exosome Aggregation Mediated Stop-Flow Paper-Based Portable Device for Rapid Exosome Quantification. *Electrophoresis* **2020**, *41*, 311–318. [CrossRef] [PubMed]
205. Gardiner, C.; Ferreira, Y.J.; Dragovic, R.A.; Redman, C.W.G.; Sargent, I.L. Extracellular Vesicle Sizing and Enumeration by Nanoparticle Tracking Analysis. *J. Extracell. Vesicles* **2013**, *2*. [CrossRef] [PubMed]
206. Palmieri, V.; Lucchetti, D.; Gatto, I.; Maiorana, A.; Marcantoni, M.; Maulucci, G.; Papi, M.; Pola, R.; De Spirito, M.; Sgambato, A. Dynamic Light Scattering for the Characterization and Counting of Extracellular Vesicles: A Powerful Noninvasive Tool. *J. Nanoparticle Res.* **2014**, *16*. [CrossRef]
207. Welsh, J.A.; Holloway, J.A.; Wilkinson, J.S.; Englyst, N.A. Extracellular Vesicle Flow Cytometry Analysis and Standardization. *Front. Cell Dev. Biol.* **2017**, *5*, 78. [CrossRef]
208. Ortega, F.G.; Piguillem, S.V.; Messina, G.A.; Tortella, G.R.; Rubilar, O.; Jiménez Castillo, M.I.; Lorente, J.A.; Serrano, M.J.; Raba, J.; Fernández Baldo, M.A. EGFR Detection in Extracellular Vesicles of Breast Cancer Patients through Immunosensor Based on Silica-Chitosan Nanoplatform. *Talanta* **2019**, *194*, 243–252. [CrossRef]
209. Cimorelli, M.; Nieuwland, R.; Varga, Z.; van der Pol, E. Standardized Procedure to Measure the Size Distribution of Extracellular Vesicles Together with Other Particles in Biofluids with Microfluidic Resistive Pulse Sensing. *PLoS ONE* **2021**, *16*, e0249603. [CrossRef]
210. Wang, S.; Khan, A.; Huang, R.; Ye, S.; Di, K.; Xiong, T.; Li, Z. Recent Advances in Single Extracellular Vesicle Detection Methods. *Biosens. Bioelectron.* **2020**, *154*, 112056. [CrossRef]
211. Bordanaba-Florit, G.; Royo, F.; Kruglik, S.G.; Falcón-Pérez, J.M. Using Single-Vesicle Technologies to Unravel the Heterogeneity of Extracellular Vesicles. *Nat. Protoc.* **2021**, *16*, 3163–3185. [CrossRef]

biosensors

Article

Red Blood Cell Sedimentation Index Using Shear Stress of Blood Flow in Microfluidic Channel

Yang Jun Kang

Department of Mechanical Engineering, Chosun University, 309 Pilmun-daero, Dong-gu, Gwangju 61452, Korea; yjkang2011@chosun.ac.kr; Tel.: +82-62-230-7052; Fax: +82-62-230-7055

Abstract: Red blood cell sedimentation has been used as a promising indicator of hematological diseases and disorders. However, to address several issues (i.e., syringe installation direction, blood on-off flow control, image-based quantification, and hemodilution) raised by the previous methods, it is necessary to devise a new method for the effective quantification of red blood cell sedimentation under a constant blood flow. In this study, the shear stress of a blood flow is estimated by analyzing an interface in a co-flowing channel to quantify the red blood cell sedimentation in blood syringes filled with blood (hematocrit = 50%). A red blood cell sedimentation index is newly suggested by analyzing the temporal variations in the shear stress. According to the experimental investigation, the sedimentation index tends to decrease at a higher flow rate. A higher level of hematocrit has a negative influence on the sedimentation index. As a performance demonstration of the present method, the red blood cell sedimentation processes of various test bloods were quantitatively compared in terms of the shear stress, image intensity, and sedimentation velocity. It was found that the proposed index provided a more than 10-fold increase in sensitivity over the previous method (i.e., image intensity). Additionally, it provided more consistent results than another conventional sedimentation method (sedimentation velocity). In conclusion, the present index can be effectively adopted to monitor the red blood cell sedimentation in a 10-min blood delivery.

Keywords: red blood cell sedimentation index; shear stress; microfluidic coflowing channel; erythrocyte sedimentation rate; blood syringe; microscopic image intensity

Citation: Kang, Y.J. Red Blood Cell Sedimentation Index Using Shear Stress of Blood Flow in Microfluidic Channel. *Biosensors* **2022**, *12*, 547. https://doi.org/10.3390/bios12070547

Received: 7 July 2022
Accepted: 19 July 2022
Published: 21 July 2022

1. Introduction

Red blood cells (RBCs) are thought to be the most common components of blood and are considered as extremely deformable. RBCs have substantial impacts on micro blood flows. Alterations of RBCs can be monitored physically by quantifying certain mechanical properties (RBC aggregation, RBC deformability, viscoelasticity, etc.) under a capillary blood flow [1–3]. Among the mechanical properties of blood, RBC aggregation occurs at a sufficiently low shear flow or stasis. In contrast, RBCs disaggregate at higher shear flow rates. RBC aggregation is highly dependent on aspects such as the membrane viscoelasticity, surface charge, and plasma protein [4,5]. As RBC aggregation provides quantitative information on the individual or interaction effects of plasma and RBCs [6], it has been regarded as a promising indicator for detecting infections, cardiovascular diseases, metabolic disorders, and hematological diseases [7–9]. The erythrocyte sedimentation rate (ESR), as calculated using the Westergren technique, is considered as a simple and gold standard for determining RBC aggregation. Without an external actuator, RBCs are sedimented owing to the gravitational force in a tube. According to this measurement method, the ESR is obtained as a sedimentation velocity (mm/h) from a visual detection of the sedimentation front with an elapse of 1 h. During the sedimentation of the RBCs, they aggregate and form continuous networks [10]. The networks of the colloidal gels then collapse [11].

Several techniques have been applied to quantify RBC aggregation or the ESR, including those based on electrical impedance [5,12–15], optical light intensity [7,16,17], microscopic image intensity [18,19], interface detection in the tube [11,20,21], a holographic laser tweezer [6], and the shear stress in microfluidic device [22]. Additionally, an external mechanism (e.g., a pinch valve [16], vibration motor [7], driving syringe [20], or vacuum pump [22]) can be adopted to periodically run or stop the blood flow in a microfluidic channel. Ultrasonic transducers have been suggested as an option for accelerating the RBC sedimentation in reservoirs [23]. Previous methods have measured the RBC aggregation or ESR in dextran-induced blood [24] or clinical disease blood [7,19,20,25]. More recently, our group suggested simple methods for quantifying the ESR using a microfluidic platform. By controlling a driving syringe filled with blood in a periodic on-off fashion, the ESR and RBC aggregation were obtained by analyzing the microscopic image intensity [26]. Notably, when blood is supplied into a microfluidic device from a driving syringe that has been installed horizontally, the hematocrit (Hct) of the RBCs tends to decrease over time. After a certain amount of time, the diluent is separated from the blood in the syringe. The blood and diluent are supplied sequentially into the microfluidic device. The contributions of the Hct can then be quantified by measuring the viscosity, as well as the junction pressure [27]. However, conventional and/or modified ESR techniques adopt a vertical installation of a driving syringe or tube [19]. During the RBC sedimentation in the syringe, the RBCs become concentrated at the bottom position, and diluted at the sedimentation front [28]. When blood is suppled into the microfluidic channel from the blood syringe, the Hct tends to increase over time. Additionally, when quantifying the ESR using the microscopic image intensity, it is necessary to periodically turn the syringe pump on and off. Lastly, as RBC sedimentation increases at lower levels of Hct, diluted blood (i.e., Hct = 25%) is generally prepared to quantify the ESR within a short time [27,29]. In view of the several issues raised by the previous studies (i.e., syringe installation direction, on-off blood flow, image intensity, blood dilution), it is necessary to devise a new and simple method for quantifying RBC sedimentation under a constant micro blood flow.

In this study, by referring to the threshold shear stress for aggregating RBCs under a blood flow [22,30], the shear stress in microfluidic channel is suggested as an index for quantifying the RBC sedimentation in the blood syringe. The shear stress represents fluidic resistance of blood flow under continuous blood flow. It is determined by blood viscosity as well as blood flow rate. The shear stress could be considered as better effective than image intensity of blood flow. Thus, the shear stress is newly suggested to quantify red blood cell sedimentation under continuous blood flow. In this case, the Hct is set to a normal range (i.e., Hct = 50%). A driving syringe filled with blood is installed vertically. The RBC sedimentation in the syringe contributes to increasing the Hct of the blood supplied into the microfluidic channel over time. The shear stress of the blood flow is estimated by analyzing an interface in a coflowing channel partially filled with blood and a reference fluid. Based on the temporal variations of shear stress, a new index is then suggested for RBC sedimentation. The index can be used to estimate the RBC sedimentation in a blood syringe effectively.

In contrast to the previous studies, the present method does not require stopping or restarting the blood flow (i.e., it works with a continuous blood flow). To provide settings similar to those of the conventional ESR technique, a driving syringe is installed vertically (i.e., vertical installation). Instead of the microscopic image intensity, the shear stress is used to obtain the RBC sedimentation index, especially under continuous blood flow. The present study does not require hemodilution to increase the sensitivity. In particular, the proposed index provides a more than 10-fold increase in sensitivity over the previous method (i.e., image intensity).

2. Materials and Methods

2.1. Microfluidic-Based Experimental Setup

As shown in Figure 1A, microfluidic device reported in a previous study [27] was reused to quantify the ESR in a driving syringe. The microfluidic device was designed with two inlets for two fluids (i.e., blood and reference fluid), two guiding channels (blood channel and reference fluid channel), a coflowing channel (width = 1000 μm, length = 3500 μm) filled with the blood and reference fluid, and an outlet. The lower panel of Figure 1A shows 3D CAD model of a microfluidic device. The depth of all channels was fixed at 50 μm. A polydimethylsiloxane (Sylgard 184, Dow Corning, Midland, MI, USA) device was fabricated using a soft lithography technique. The device was placed on an inverted optical microscope (IX53, Olympus, Tokyo, Japan) equipped with a 4× objective lens (numerical aperture = 0.1). One Tygon tubing (Cole-Parmer, Vernon Hills, IL, USA, ID = 0.01 inch, OD = 0.03 inch, and length = 300 mm) was inserted into each inlet. Another Tygon tubing (length = 200 mm) was fitted to the outlet. To repel air in the device and tubing, 1× phosphate-buffered saline (PBS) was injected through the tubing connected to the outlet. A reference syringe was filled by adding a glycerin solution (30%) to a disposable syringe (~1 mL). Simultaneously, a blood syringe was filled by adding blood into a disposable syringe. Thereafter, as depicted in Figure 1B, both syringes were installed into a syringe pump (neMESYS, Cetoni GmbH, Korbußen, Germany) aligned along the gravitational direction. The flow rates of each individual syringe (Q_r: flow rate of reference fluid, Q_b: flow rate of blood) were set to constant values. As a preliminary study, to understand the contribution of the RBC sedimentation in the blood syringe to the blood flow in the microfluidic channel, the RBC sedimentation was monitored by quantifying the sedimentation front in the blood syringe. The blood (Hct = 50%) was prepared by adding normal RBCs to a dextran solution (10 mg/mL). The flow rate of the blood syringe was set to 0.5 mL/h. Figure 1C shows snapshots of the RBC sedimentation in the blood syringe over 90 min. As blood was supplied to the microfluidic device from the blood syringe, the blood volume (or height) in the blood syringe decreased linearly over time. The RBC sedimentation contributed to the gradual increase in the diluent volume (or height) over time.

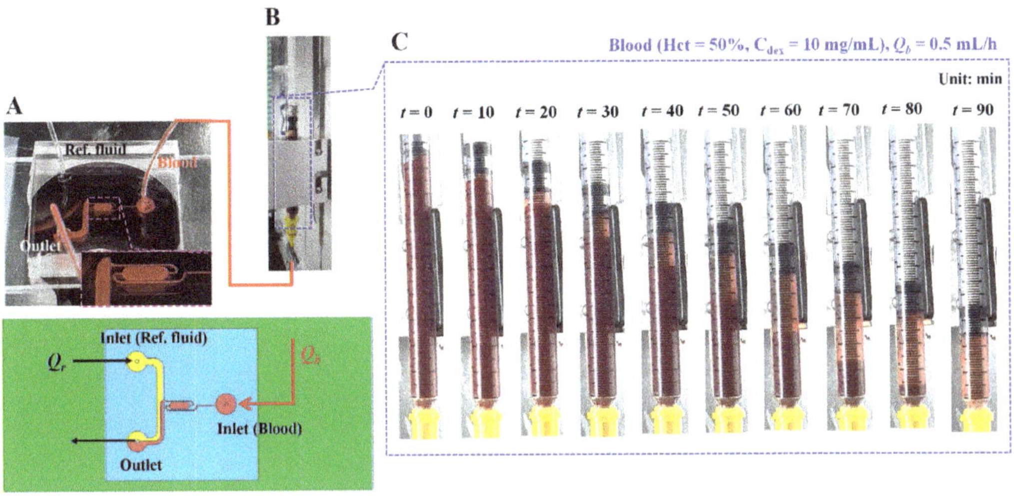

Figure 1. Microfluidic platform for quantifying red blood cell (RBC) sedimentation in terms of previous suggested factors. (**A**) Schematics of a microfluidic device (i.e., prototype, and 3D model). Microfluidic device with two inlets, outlet, two guiding channels (reference fluid channel and blood channel), and coflowing channel. Lower-side panel shows blood channel filled with blood. (**B**) Syringe pump for delivering blood as well as reference fluid. Q_r and Q_b denote flow rates of reference fluid and blood sample, respectively. (**C**) Snapshots for showing RBC sedimentation in the blood syringe for 90 min. In this case, blood (Hct = 50%) was prepared by mixing normal RBCs into dextran solution (10 mg/mL). Flow rate of syringe pump set to Q_b = 0.5 mL/h.

2.2. RBC Sedimentation Index Using Temporal Variation of Shear Stress

Instead of the image intensity of the blood flow in the microfluidic channel, the shear stress was suggested as an index for quantifying the RBC sedimentation in the blood syringe. As shown in Figure 2A, the co-flowing channel was modeled with discrete fluidic circuit elements, such as the fluidic resistance and flow rate. A single co-flowing channel was partially filled with the reference fluid and blood. The width of each fluid was given as $(1 - \beta) \times w$ and $\beta \times w$. According to previous studies [31–34], the virtual wall concept was used to simplify complex problems. In particular, the co-flowing channel was assumed to be two independent channels (i.e., a reference fluid channel and blood channel) connected in parallel. A correction factor (C_P), which was expressed as interface (β), was then suggested to compensate for the mathematical modeling error resulting from the difference between the real physical model and simple mathematical model [35,36]. The frictional losses of the fluids were represented by two fluidic resistances (R_r and R_b), where the subscripts r and b represented the reference fluid and the blood, respectively. The flow rates of the two fluids were denoted by Q_r and Q_b, respectively. A symbol ($\bigtriangledown$) represented the zero value of the gauge pressure (i.e., ground, p = 0). At the distance of L from the ground, the pressure of each fluid was designated as P_r or P_b, respectively. In this case, both pressures had the same values in the straight and rectangular channels (i.e., $P_r \approx P_b$). The pressure of each fluid was then derived as [35],

$$P_r = \frac{12\,\mu_r \times L \times Q_r}{(1-\beta) \times w \times h^3} \tag{1}$$

In addition,

$$P_b = \frac{12\,\mu_b \times L \times Q_b}{C_P \times \beta \times w \times h^3} \tag{2}$$

In Equations (1) and (2), μ_r and μ_b denoted the viscosities of the reference fluid and blood, respectively. L represented channel length of coflowing channel. Based on the same pressure condition, the blood viscosity formula could be derived as,

$$\mu_b = \mu_r \times \left(\frac{\beta}{1-\beta}\right) \times \left(\frac{Q_r}{Q_b}\right) \times C_P(\beta) \tag{3}$$

According to the force balance between the pressure-induced force and viscous shear force along reference fluid stream [35], the relationship between shear stress and pressure difference was given as,

$$\tau_r \times (2(1-\beta\,)\,w\,L) = P_r \times ((1-\beta\,)\,w\,h) \tag{4}$$

Similarly, the relationship between shear stress and pressure difference along blood stream was expressed as,

$$\tau_b \times (2\,\beta\,w\,L) = P_b \times (\beta\,w\,h) \tag{5}$$

By substituting Equations (4) and (5) into Equations (1) and (2), the shear stress of each fluid stream was then derived as,

$$\tau_r = \frac{6\,\mu_r \times Q_r}{(1-\beta) \times w \times h^2} \tag{6}$$

In addition,

$$\tau_b = \frac{6\,\mu_b \times Q_b}{C_P \times \beta \times w \times h^2} \tag{7}$$

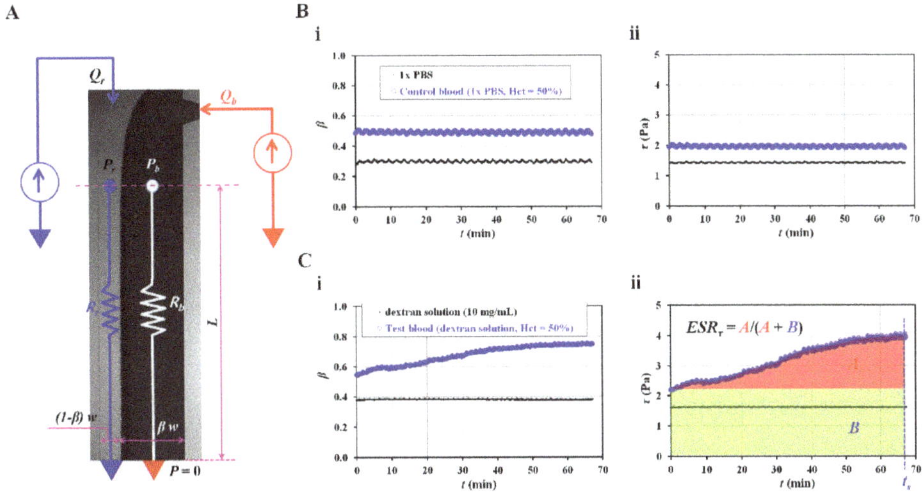

Figure 2. RBC sedimentation index in terms of shear stress of blood flow. (**A**) Discrete fluidic circuit model for estimating shear stress of blood flow in coflowing channel. (**B**) Variations of shear stress for control blood with no RBC sedimentation. In this case, control blood (hematocrit = 50%) was prepared by adding normal RBCs into 1× phosphate-buffered solution (PBS). (**i**) Temporal variations of interface (β) with respect to 1× PBS as well as control blood. (**ii**) Temporal variations of shear stress with respect to 1× PBS as well as control blood. (**C**) Variations of shear stress for sedimentation-enhanced test blood. Test blood (hematocrit = 50%) was prepared by adding normal RBCs into dextran solution (10 mg/mL). (**i**) Temporal variations of interface (β) with respect to dextran solution and test blood. (**ii**) Temporal variations of shear stress with respect to dextran solution and test blood. From the results, RBC sedimentation index was suggested as $ESR_\tau = A/(A+B)$. Based on the temporal variations of τ, A and B were calculated as $A = \int_0^{t_s} (\tau(t) - \tau(t=0))dt$ and $B = \int_0^{t_s} \tau(t=0)dt$.

By substituting Equation (3) into Equation (7), both fluid streams satisfied with the same shear stress condition (i.e., $\tau_r = \tau_b = \tau$). According to Equations (3) and (6), the blood viscosity and shear stress of blood stream could be quantified by monitoring the interface (β) at a specific flow rate of the two fluids. For a rectangular microfluidic channel (i.e., width = 1000 μm, depth = 50 μm), the correction factor was given as $C_P = -9.014\ \beta^4 + 21.273\ \beta^3 - 18.403\ \beta^2 + 7.051\ \beta - 0.168$ ($R^2 = 0.99$) [27]. According to Equations (1) and (6), pressure of reference fluid (P_r) was expressed as $P_r = \tau_r \times L/h$. Namely, pressure was proportional to shear stress. As L/h was fixed in the identical microfluidic channel, the accuracy of shear stress formula was the same as the accuracy of pressure formula. Based on the previous work [27], normalized difference between analytical formula and numerical simulation was less than 6%. Thus, the Equation (6) could be used to monitor change in shear stress in the coflowing channel with enough accuracy.

To validate the contribution of the RBC sedimentation to the shear stress of the blood flow in the microfluidic channel, control blood (Hct = 50%) was prepared by adding normal RBCs to 1× PBS. As the diluent (1× PBS) did not include plasma proteins and the Hct was set to a higher level of 50%, the variations in the RBC sedimentation could be negligible within a short duration of 2 h [19]. The flow rate of both fluids was set to 0.5 mL/h. As shown in Figure 2(Bi), temporal variations of the interface (β) were identified with respect to the pure diluent (1× PBS) and control blood. As expected, neither fluid contributed to the varying β over time. Based on Equation (6), the shear stresses of both fluids were calculated over time. As shown in Figure 2(Bii), the temporal variations in shear stress were obtained with respect to the 1× PBS and control blood. The shear stresses of both fluids remained unchanged over time (i.e., shear stress of 1× PBS = 1.42 ± 0.01 Pa, shear stress of control blood = 1.95 ± 0.03 Pa). Then, to enhance the RBC aggregation in the control blood, the 1× PBS was replaced with a dextran solution (10 mg/mL). Test blood (Hct = 50%) was prepared by adding normal RBCs to the dextran solution. As shown in Figure 2(Ci), temporal variations in β were identified with respect to the pure dextran solution and test blood. As the pure dextran solution did not include RBCs, it did not contribute to the change in β over time. However, the test blood exhibited variations of β over time. Namely, the pure dextran solution influenced the RBC aggregation in the test blood. Figure 2(Cii) shows the temporal variations in τ with respect to the pure dextran solution and test blood. In this case, at an initial time (t = 0), the corresponding shear stresses of each fluid were obtained as τ = 1.61 ± 0.01 Pa (pure dextran solution) and τ = 2.18 ± 0.01 Pa (test blood), respectively. The shear stress of the test fluid increased with time. Referring to an RBC aggregation index reported in a previous study [7,37], the RBC sedimentation index was newly suggested as $ESR_\tau = A/(A + B)$. Based on the temporal variations in τ, A and B were calculated as $A = \int_0^{t_s} (\tau(t) - \tau(t=0))dt$ and $B = \int_0^{t_s} \tau(t=0)dt$. For the control blood (Figure 2(Bii)), ESR_τ was calculated as zero, because A was zero. However, the RBC sedimentation index of the test blood was estimated as ESR_τ = 0.30 ± 0.02 (n = 3).

From the preliminary results, the RBC sedimentation index (ESR_τ) showed substantial differences between the control and test blood.

2.3. Quantification of Flow Rate of Reference Fluid with Micro Particle Image Velocimetry

To visualize the velocity fields of the reference fluid, RBCs (20 µL) were added to the reference fluid (1 mL). A high-speed camera (FASTCAM MINI, Photron, Tokyo, Japan) was used to capture microscopic blood flow images at intervals of 1 s. The frame rate of the high-speed camera was set to 2000 frames/s. Two sequential images were continuously captured at intervals of 2 s. As shown in Figure A1A, a region of interest (ROI) of 300 × 510 pixels was selected in the straight channel. The velocity fields of the reference fluid were obtained by setting the integration size to 32 × 16 pixels (i.e., a 50% overlap between two sequential images) [38]. The average velocity ($<U>$) was then obtained by averaging the velocity fields distributed over the ROI. The flow rate of the reference fluid was calculated using $Q_{PIV} = A_c \times <U>$. In this case, A_c denoted the cross-sectional area of the rectangular channel (A_c = width × depth).

2.4. Blood Preparation for Stimulating RBC Sedimentation in the Driving Syringe

An RBC bag (~320 mL) filled with concentrated RBCs was provided by the Gwangju–Chonnam Blood Bank (Gwangju, South Korea). It was stored in a refrigerator at 4°. According to the specific washing procedures [23], normal RBCs were then collected from the concentrated RBCs. To enhance the aggregation within the normal RBCs, normal RBCs were added to the dextran solution. Nine dextran solutions (C_{dex} = 5, 7.5, 10, 15, 20, 30, 40, 60, and 80 mg/mL) were prepared by dissolving dextran powder (*Leuconostoc* spp., MW = 450–650 kDa; Sigma-Aldrich, St. Louis, MO, USA) in 1× PBS. Unless otherwise stated, the Hct was set at 50%.

3. Results and Discussion

3.1. Red Blood Cell (RBC) Sedimentation in Driving Blood Syringe and Its Contribution to Blood Flow

Three representative factors (diluent volume/blood volume, Hct, and image intensity) were used to quantify the effects of the RBC sedimentation in the blood syringe. First, based on the snapshots (Figure 1C), the diluent height (h) and blood height (H) were obtained by inspecting the sedimentation front in the blood syringe over time. Figure 3A shows the temporal variations in h, H, and h/H. h gradually increased for up to 80 min. The right panel shows the diluent height (h) and blood height (H). H gradually decreased over time. The RBC sedimentation caused an increase in h/H ranging from 0 to 1. The RBC sedimentation caused a decrease in the Hct of the blood in the blood syringe. As the total blood volume was fixed in the blood syringe, it was expected that the Hct of the blood supplied from the blood syringe would increase over time. Second, to measure the Hct of the blood in the microfluidic channel, blood was collected from the outlet at intervals of 10 min. Thereafter, the variations in Hct were identified using a micro hemocytometer (Microhematocrit, VS-12000, Vison Scientific Co., Daejeon, South Korea). Figure 3(Bi) shows an image of the capillary tubes captured after operation of the micro hemocytometer from t = 10 min to t = 90 min. As shown in Figure 3(Bii), the Hct values were obtained at intervals of 10 min. As expected, the Hct increased gradually from t = 10 min to t = 80 min. The Hct values at specific times were measured as Hct = 43.7 ± 1.5% (t = 10 min), 51.5 ± 2% (t = 20 min), 64.4 ± 1.6% (t = 40 min), 70.5 ± 2.8% (t = 60 min), 70.6 ± 2.8% (t = 80 min), and 27.4 ± 0.8% (t = 90 min). This indicated that the RBC sedimentation contributed to a continuous increase in the Hct of the blood in the microfluidic channel. Finally, the image intensity of the blood flow (I_b) was obtained by analyzing the image intensity of the blood flows within the blood channel (Region-Of-Interest, 2 × 1 mm^2). Figure 3C shows the temporal variations in I_b over time. I_b remained unchanged over time. From these results, it was considered as impossible to monitor the RBC sedimentation in terms of the image intensity of the blood flow, especially under a continuous blood flow. Thus, it was inferred that the RBC aggregation [14] should be measured after stopping the blood flow [7,16].

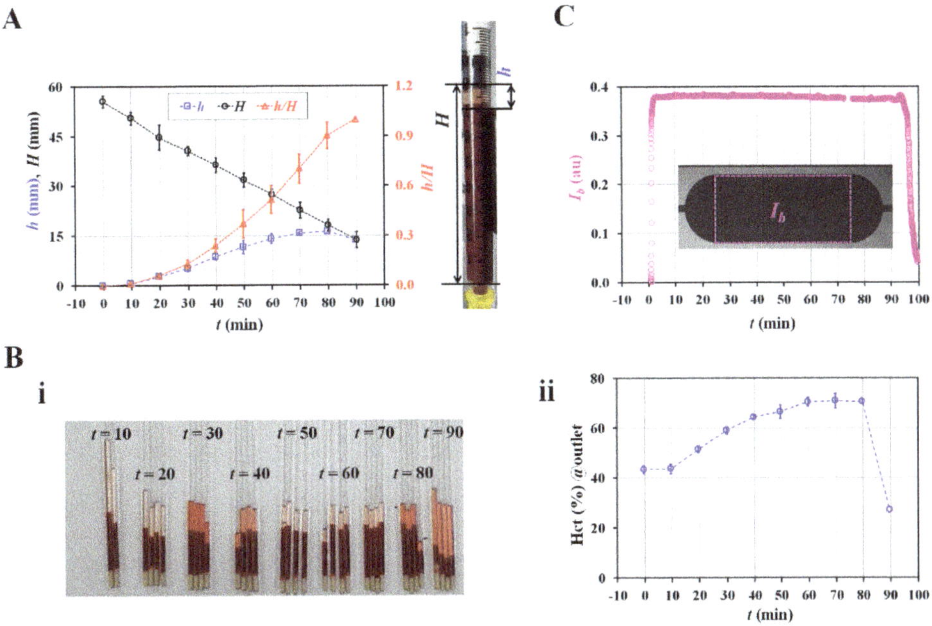

Figure 3. Quantification of RBC sedimentation in terms of three representative factors (i.e., sedimentation distance, hematocrit, and image intensity). (**A**) Temporal variations of RBC sedimentation distance in blood syringe. Right-side panel shows diluent height (h) as well as blood height (H). (**B**). Measurement of hematocrit with hemocytometer. (**i**) Capillary tubes captured after operation of micro hemocytometer. (**ii**) Temporal variations of hematocrit of blood collected at outlet. (**C**) Temporal variation of image intensity (I_b) in blood channel.

3.2. Contributions of Flow Rate and Hematocrit to RBC Sedimentation Index

As shown in Figure 1C, it was expected that the RBC sedimentation in the blood syringe could vary according to the flow rate and Hct. Thus, it was necessary to evaluate the contributions of these two representative factors to the RBC sedimentation index.

According to Equation (6), to obtain a consistent value of the shear stress, the flow rate of the reference fluid should remain consistent over a sufficient period. The fluctuations in the flow rate controlled by the syringe pump were validated using microparticle image velocimetry. The flow rate of the syringe pump varied from Q_{sp} = 0.25 mL/h to Q_{sp} = 1.5 mL/h. Figure A1B shows the temporal variations in Q_{PIV} with respect to Q_{sp}. As a result, it can be seen that the flow rate of the reference fluid remained constant over time. To quantify the fluctuations of the flow rate, the coefficient of variance (COV) was calculated as COV = standard deviation/mean. Figure A1C shows the variations of Q_{PIV} and the COV with respect to setting the flow rate of the syringe pump (Q_{sp}). A linear regression analysis gives a higher value of the regression coefficient (i.e., R^2 = 0.9997). By increasing the flow rate of the syringe pump from 0.25 mL/h to 1.5 mL/h, the COV decreased substantially, from 7.7% to 1.5%. At a flow rate of Q_{sp} = 0.5 mL/h, the flow rate of the reference fluid was measured as Q_{PIV} = 0.56 ± 0.03 mL/h. The COV was then calculated as 4.5%. These results indicate that syringe pump can consistently maintain the flow rate of the reference fluid for a long period of 70 min.

Next, the contribution of the flow rate the RBC sedimentation index (ESR_τ) was validated by varying the flow rates from 0.25 mL/h to 1.5 mL/h. The flow rates of both fluids were set to the same value (i.e., $Q_r = Q_b$). The test blood (Hct = 50%) was prepared by adding normal RBCs to a dextran solution (10 mg/mL). Figure 4(Ai) shows the temporal variations in the shear stress (τ) with respect to the blood flow rate (Q_b). At t = 0, the shear stress (τ) increases substantially at higher flow rates. A higher flow rate causes a decrease in the delivery time. Based on the delivery time of Q_b =1.5 mL/h, the integration time for calculating A and B was limited to 30 min (i.e., t_s = 30 min). Figure 4(Aii) shows the variations in the RBC sedimentation index (ESR_τ) with Q_b. The ESR_τ is represented as the mean ± standard deviation (n = 4). The sedimentation index does not exhibit a substantial difference between Q_b = 0.25 and Q_b = 0.5 mL/h. In addition, it includes large fluctuations at Q_b = 0.25 mL/h. Above Q_b = 0.5 mL/h, the sedimentation index tends to decrease at higher flow rates. Namely, when the syringe pump is set to a higher flow rate, the RBC sedimentation decreases. Therefore, the RBC sedimentation index decreases. The results indicate that the flow rate of the syringe pump has a strong influence on the RBC sedimentation index. Thus, the flow rates of both fluids were fixed as $Q_r = Q_b$ = 0.5 mL/h for the following experiments.

According to previous studies [13,19,20,27], a higher Hct level could retard RBC sedimentation in the tube or syringe. To evaluate the contribution of the Hct to the RBC sedimentation index, the sedimentation index was measured with respect to Hct = 30%, 40%, and 50%. Figure 4(Bi) shows the temporal variations in the shear stress with respect to the Hct. Using the temporal variations in τ, the RBC sedimentation index was obtained with respect to Hct. The integration time was fixed at t_s = 30 min. As shown in Figure 4(Bii), the ESR_τ gradually decreases with respect to Hct. The sedimentation index is expressed as the mean ± standard deviation (n = 3). As expected, a higher Hct level negatively affects RBC sedimentation. Compared with previous studies [7,20,24], the present results exhibited sufficiently reliable trends with respect to Hct. From the results, and for obtaining a consistent RBC sedimentation index, the Hct of test fluid was fixed at 50% for the following experiments.

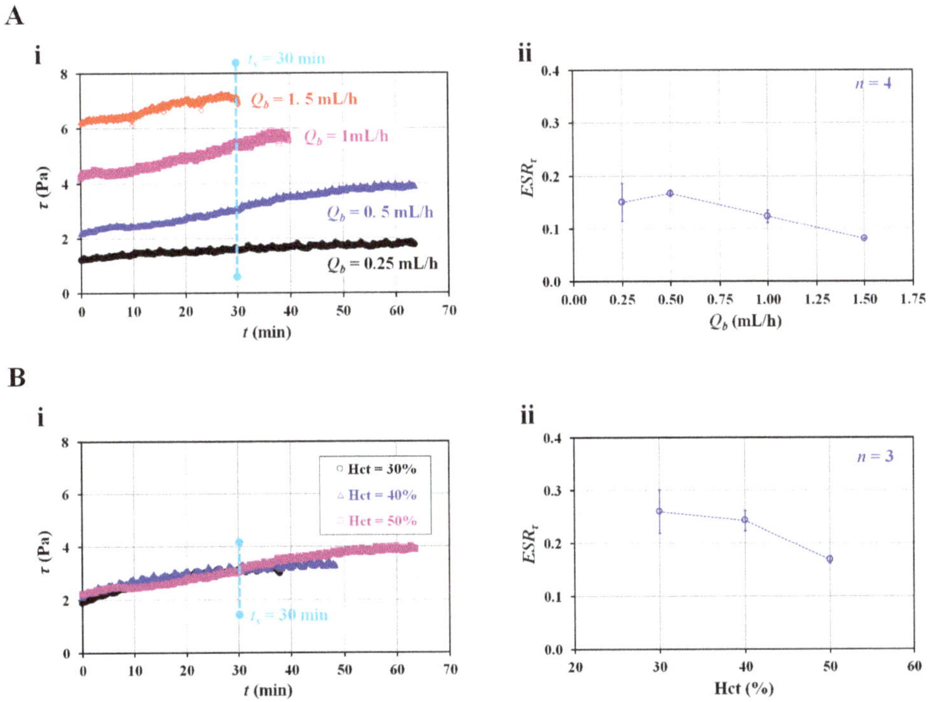

Figure 4. Contributions of flow rate and hematocrit to RBC sedimentation index (ESR_τ). (**A**) Effect of blood flow rate on RBC sedimentation index. (**i**) Temporal variation of shear stress with respect to blood flow rate (Q_b). (**ii**) Variation of ESR_τ with respect to Q_b. (**B**) Effect of hematocrit on RBC sedimentation index. (**i**) Temporal variation of shear stress with respect to Hct = 30%, 40%, and 50%. (**ii**) Variation of ESR_τ with respect to Hct.

3.3. *Quantitative Validation of Suggested RBC Sedimentation Index*

As the last demonstration, it was necessary to validate the suggested RBC sedimentation index by comparing it with previous methods (i.e., sedimentation height per 1 h and image intensity). A quantitative comparison between the present method and previous methods was conducted using RBC sedimentation-enhanced blood. Based on previous studies [39–41], the dextran solution contributed to increasing the RBC aggregation or sedimentation.

As shown in Figure 5(Ai), the temporal variations in τ were identified by increasing the concentrations of the dextran solution. The initial shear stress τ (t = 0) increases substantially at higher dextran solution concentrations. For two test blood samples (C_{dex} = 10 and 20 mg/mL), the shear stress tends to increase significantly over time. However, above C_{dex} = 40 mg/mL, the test blood exhibits fluctuations in the shear stress over time, rather than a substantial increase over time. To determine the effect of the pure diluent, the viscosity of the test blood and initial shear stress were obtained with respect to C_{dex}.

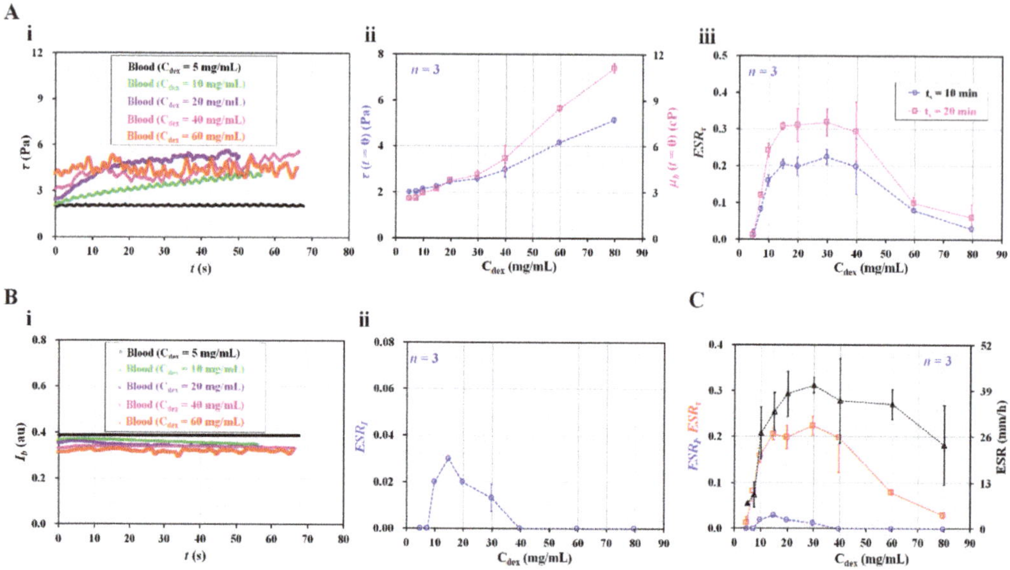

Figure 5. Quantitative comparison between present method and previous methods. (**A**) Contribution of dextran solution to RBC sedimentation index. (**i**) Temporal variations of shear stress with respect to C_{dex}. (**ii**) Variations of initial shear stress (τ (t = 0)) as well as blood viscosity (μ_b (t = 0)) with respect to C_{dex}. (**iii**) Variations of RBC sedimentation index with respect to C_{dex} and t_s. (**B**) Evaluation of RBC sedimentation using image intensity of blood flow (I_b). (**i**) Temporal variations of I_b with respect to C_{dex}. (**ii**) Variations of ESR_I with respect to C_{dex}. (**C**) Quantitative comparison of RBC sedimentation index in terms of shear stress, image intensity, and sedimentation velocity.

As shown in Figure 5(Aii), the variations in τ (t = 0) and μ_b (t = 0) are represented with respect to C_{dex}. At t =0, the blood viscosity and shear stress increase at higher dextran solution concentrations. From the results, it can be seen that the dextran solution contributes significantly to increasing the blood viscosity, as well as the shear stress. According to Equation (7), the shear stress is proportional to the blood viscosity. Thus, these results can be considered as sufficiently reliable. Based on the temporal variations in τ, the RBC sedimentation index (ESR_τ) was obtained with respect to C_{dex}. Furthermore, the test blood was saturated for 20 min (C_{dex} = 20 mg/mL). In this case, the integration time was selected as t_s = 10 or 20 min. Figure 5(Aiii) shows the variations in the RBC sedimentation index with respect to C_{dex} and t_s. The sedimentation index increases up to a C_{dex} of 15 mg/mL. It remains constant from C_{dex} = 15 mg/mL to C_{dex} = 40 mg/mL. The sedimentation index gradually decreases above C_{dex} = 40 mg/mL. Based on the peak value of the sedimentation index, the longer integration time (t_s = 20 min) causes a 50% increase in sensitivity relative to the shorter integration time (t_s = 10 min).

In previous studies [7,20,23,37], the RBC aggregation or sedimentation was quantified by analyzing the image intensity of the blood flow over time. Following such previous methods, as shown in Figure 3C, the image intensity of the blood flow (I_b) was obtained under a continuous blood flow (Q_b = 0.5 mL/h). Figure 5(Bi) shows the temporal variations in I_b with respect to C_{dex}. The image intensity is higher at lower concentrations of dextran solution. The time variation of I_b is not distinct with respect to C_{dex}. Based on the definition of the RBC aggregation index [37], the ESR index (ESI_I) was calculated using the temporal variations in I_b. Figure 5(Bii) shows the variations in ESR_I with respect to C_{dex}. The ESR_I exhibits a peak value of 0.03 at C_{dex} = 10 mg/mL. The ESR_I has a value of zero for C_{dex} = 40 mg/mL. As the variation in I_b is much smaller over time, the ESR_I provides extremely low sensitivity within C_{dex} = 40 mg/mL. For this reason, the image intensity is ineffective for quantifying RBC sedimentation, especially under a continuous blood flow. In addition, a conventional ESR was quantified by monitoring the RBC sedimentation in disposable syringe. After injecting 1 mL of blood into the disposable syringe (~1 mL), the sedimentation front was monitored for 30 min. The conventional ESR was then obtained by dividing the sedimentation height at 0.5 h (i.e., sedimentation velocity, mm/h). Figure 5C shows a quantitative comparison of the RBC sedimentation index in terms of the shear stress (ESR_τ), image intensity (ESR_I), and sedimentation velocity (mm/h). Among the two data of the ESR_τ as shown in Figure 5(Aiii), the ESRτ calculated at a shorter integrating time (t_s = 10 min) was redrawn with respect to C_{dex}. The present sedimentation index exhibits very similar trends to those of the conventional ESR technique (i.e., sedimentation velocity). However, the ESR_τ shows more consistent trends than the sedimentation velocity. The sedimentation velocity exhibits a large scattering. Furthermore, for the test blood (C_{dex} = 20 mg/mL), the sensitivity of ESR_τ is more than 10-fold higher than that of ESR_I.

From the experimental investigation, it can be concluded that the RBC sedimentation index proposed in this study can be effectively used to quantify RBC sedimentation in a driving syringe. Furthermore, the sedimentation index yields consistent results when compared with the conventional ESR (sedimentation velocity). Thus, while supplying blood (Hct = 50%) into the microfluidic channel continuously for 10 min, it is possible to quantify RBC sedimentation effectively in terms of the shear stress, rather than the image intensity. One limitation of the present study is that the present index was not applied to test clinical blood. Furthermore, as the present method was demonstrated in a well-equipped laboratory, it will be necessary to update the present method for in-situ diagnoses.

4. Conclusions

In this study, to quantify the RBC sedimentation in blood syringes filled with blood (Hct = 50%), a new RBC sedimentation index was suggested, based on the shear stress of the blood flow in the microfluidic channel. Under a constant blood flow, the shear stress was estimated by analyzing an interface in a coflowing channel. According to an experimental investigation of the flow rate and Hct, the sedimentation index tended to decrease at higher flow rates. A higher Hct level had a negative effect on the RBC sedimentation index. As a performance demonstration, the RBC sedimentation values of various test blood samples were quantitatively compared in terms of the shear stress, image intensity, and sedimentation velocity. As a result, it was found that the proposed sedimentation index provides more than 10-fold higher sensitivity than the image intensity. It has the ability to provide more consistent results than the conventional ESR technique (sedimentation velocity). In the near future, the present sedimentation index will be applied to test clinical blood, and the method will be updated for in in-situ diagnosis.

Funding: This work was supported by a research fund from the Chosun University in 2022.

Institutional Review Board Statement: This study was conducted in accordance with the Declaration of Helsinki and approved by the ethical committee of Chosun University under the reference code (2-1041055-AB-N-01-2021-80).

Informed Consent Statement: Not applicable.

Data Availability Statement: Not applicable.

Conflicts of Interest: The author declares no conflict of interest.

Appendix A

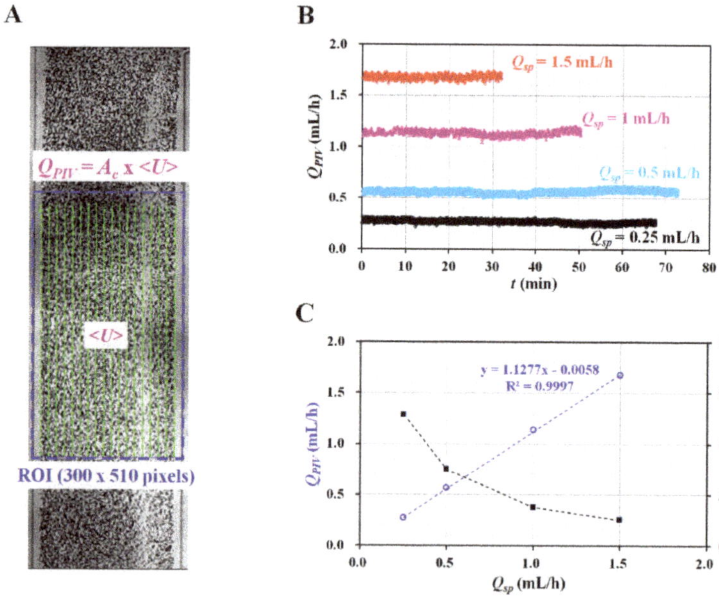

Figure A1. Validation of flow rate of reference fluid controlled by syringe pump. (**A**) Quantification of flow rate of reference fluid with microparticle image velocimetry. (**B**) Temporal variations of Q_{PIV} with respect to flow rate of syringe pump (Q_{sp}) = 0.25, 0.5, 1, and 1.5 mL/h. (**C**) Variations of Q_{PIV} and coefficient of variance (COV) with respect to Q_{sp}.

References

1. Baskurt, O.K.; Meiselman, H.J. Blood rheology and hemodynamics. *Semin. Thromb. Hemost.* **2003**, *29*, 435–450. [PubMed]
2. Piety, N.Z.; Stutz, J.; Yilmaz, N.; Xia, H.; Yoshida, T.; Shevkoplyas, S.S. Microfuidic capillary networks are more sensitive than ektacytometry to the decline of red blood cell deformability induced by storage. *Sci. Rep.* **2021**, *11*, 604. [CrossRef] [PubMed]
3. Trejo-Soto, C.; Lázaro, G.R.; Pagonabarraga, I.; Hernández-Machado, A. Microfluidics approach to the mechanical properties of red blood cell membrane and their effect on blood rheology. *Membranes* **2022**, *12*, 217. [CrossRef] [PubMed]
4. Nader, E.; Skinner, S.; Romana, M.; Fort, R.; Lemonne, N.; Guillot, N.; Gauthier, A.; Antoine-Jonville, S.; Renoux, C.; Hardy-Dessources, M.-D.; et al. Blood rheology: Key parameters, impact on blood flow, role in sickle cell disease and effects of exercise. *Front. Physiol.* **2019**, *10*, 1329. [CrossRef]
5. Sabuncu, A.C.; Muldur, S.; Cetin, B.; Usta, O.B.; Aubry, N. β-dispersion of blood during sedimentation. *Sci. Rep.* **2021**, *11*, 2642. [CrossRef]
6. Lee, K.; Shirshin, E.; Rovnyagina, N.; Yaya, F.; Boujja, Z.; Priezzhev, A.; Wagner, C. Dextran adsorption onto red blood cells revisited: Single cell quantification by laser tweezers combined with microfluidics. *Biomed. Opt. Express* **2018**, *6*, 324551. [CrossRef]
7. Namgung, B.; Lee, T.; Tan, J.K.S.; Poh, D.K.H.; Park, S.; Chng, K.Z.; Agrawal, R.; Park, S.-Y.; Leo, H.L.; Kim, S. Vibration motor-integrated low-cost, miniaturized system for rapid quantification of red blood cell aggregation. *Lab Chip* **2020**, *20*, 3930–3937. [CrossRef]
8. Baskur, O.K.; Meiselman, H.J. Erythrocyte aggregation: Basic aspects and clinical importance. *Clin. Hemorheol. Microcirc.* **2013**, *53*, 23–37. [CrossRef] [PubMed]
9. Lapić, I.; Padoan, A.; Bozzato, D.; Plebani, M. Erythrocyte sedimentation rate and C-reactive protein in acute inflammation. *Am. J. Clin. Pathol.* **2020**, *153*, 14–29. [CrossRef]
10. Darras, A.; Breunig, H.G.; John, T.; Zhao, R.; Koch, J.; Kummerow, C.; König, K.; Wagner, C.; Kaestner, L. Imaging erythrocyte sedimentation in whole blood. *Front. Physiol.* **2022**, *12*, 729191. [CrossRef]
11. Dasanna, A.K.; Darras, A.; John, T.; Gompper, G.; Kaestner, L.; Wagner, C.; Fedosov, D.A. Erythrocyte sedimentation: Effect of aggregation energy on gel structure during collapse. *Phys. Rev. E* **2022**, *105*, 024610. [CrossRef] [PubMed]
12. Wen, J.; Wan, N.; Bao, H.; Li, J. Quantitative measurement and evaluation of red blood cell aggregation in normal blood based on a modified hanai equation. *Sensors* **2019**, *19*, 1095. [CrossRef] [PubMed]

13. Zhbanov, A.; Yang, S. Effects of aggregation on blood sedimentation and conductivity. *PLoS ONE* **2015**, *10*, e0129337. [CrossRef] [PubMed]
14. Baskurt, O.K.; Uyuklu, M.; Meiselman, H.J. Time Course of Electrical Impedance During Red Blood Cell Aggregation in a Glass Tube: Comparison With Light Transmittance. *IEEE Trans. Biomed. Eng.* **2010**, *57*, 969–978. [CrossRef]
15. Zhbanov, A.; Yang, S. Electrochemical impedance spectroscopy of blood for sensitive detection of blood hematocrit, sedimentation and dielectric properties. *Anal. Methods* **2017**, *9*, 3302–3313. [CrossRef]
16. Isiksacan, Z.; Erel, O.; Elbuken, C. A portable microfluidic system for rapid measurement of the erythrocyte sedimentation rate. *Lab Chip* **2016**, *16*, 4682–4690. [CrossRef]
17. Shin, S.; Yang, Y.; Suh, J.-S. Measurement of erythrocyte aggregation in a microchip stirring system by light transmission. *Clin. Hemorheolo. Microcir.* **2009**, *41*, 197–207. [CrossRef]
18. Kang, Y.J. Microfluidic-based measurement of RBC aggregation and the ESR using a driving syringe system. *Anal. Methods* **2018**, *10*, 1805–1816. [CrossRef]
19. Kang, Y.J.; Ha, Y.-R.; Lee, S.-J. Microfluidic-based measurement of erythrocyte sedimentation rate for biophysical assessment of blood in an in vivo malaria-infected mouse. *Biomicrofluidics* **2014**, *8*, 044114. [CrossRef]
20. Hong, H.; Song, J.M.; Yeom, E. Micro-vibrational erythrocyte sedimentation rate (ESR) for sensitive measurement of erythrocyte aggregation. *J. Vis.* **2021**, *24*, 749–760. [CrossRef]
21. Darras, A.; Peikert, K.; Rabe, A.; Yaya, F.; Simionato, G.; John, T.; Dasanna, A.K.; Buvalyy, S.; Geisel, J.; Hermann, A.; et al. Acanthocyte Sedimentation Rate as a Diagnostic Biomarker for Neuroacanthocytosis Syndromes: Experimental Evidence and Physical Justification. *Cells* **2021**, *10*, 788. [CrossRef] [PubMed]
22. Shin, S.; Nam, J.-H.; Hou, J.-X.; Suh, J.-S. A transient microfluidic approach to the investigation of erythrocyte aggregation: The threshold shear-stress for erythrocyte disaggregation. *Clin. Hemorheol. Microcirc.* **2009**, *42*, 117–125. [CrossRef] [PubMed]
23. Kim, G.; Jeong, S.; Kang, Y.J. Ultrasound standing wave-based cell-to-liquid separation for measuring viscosity and aggregation of blood sample. *Sensors* **2020**, *20*, 2284. [CrossRef]
24. Kang, Y.J. Simultaneous measurement method of erythrocyte sedimentation rate and erythrocyte deformability in resource-limited settings. *Physiol. Meas.* **2020**, *41*, 025009. [CrossRef]
25. Yeom, E.; Lee, S.-J. Microfluidic-based speckle analysis for sensitive measurement of erythrocyte aggregation: A comparison of four methods for detection of elevated erythrocyte aggregation in diabetic rat blood. *Biomicrofluidics* **2015**, *9*, 024110. [CrossRef] [PubMed]
26. Kang, Y.J. Microfluidic-based measurement method of red blood cell aggregation under hematocrit variations. *Sensors* **2017**, *17*, 2037. [CrossRef] [PubMed]
27. Kang, Y.J. Contributions of red blood cell sedimentation in a driving syringe to blood flow in capillary channels. *Micromachines* **2022**, *13*, 909. [CrossRef]
28. Rallabandi, B.; Nune, J.K.; Perazzo, A.; Gershtei, S.; Ston, H.A. Representative subsampling of sedimenting blood. *Proc. R. Soc. A* **2019**, *475*, 20190223. [CrossRef]
29. Kang, Y.J. Sequential quantification of blood and diluent using red cell sedimentation-based separation and pressure-induced work in a microfluidic channel. *Anal. Methods* **2022**, *14*, 1194–1207. [CrossRef]
30. Lim, H.-J.; Lee, Y.-J.; Nam, J.-H.; Chung, S.; Shin, S. Temperature-dependent threshold shear stress of red blood cell aggregation. *J. Biomech.* **2010**, *43*, 546–550. [CrossRef]
31. Hintermüller, M.A.; Offenzeller, C.; Jakoby, B. A microfluidic viscometer with capacitive readout using screen-printed electrodes. *IEEE Sens. J.* **2021**, *21*, 2565–2572. [CrossRef]
32. Solomon, D.E.; Vanapalli, S.A. Multiplexed microfluidic viscometer for high-throughput complex fluid rheology. *Microfluid. Nanofluid.* **2014**, *16*, 677–690. [CrossRef]
33. Mehri, R.; Mavriplis, C.; Fenech, M. Red blood cell aggregates and their effect on non-Newtonian blood viscosity at low hematocrit in a two-fluid low shear rate microfluidic system. *PLoS ONE* **2018**, *13*, e0199911. [CrossRef] [PubMed]
34. Guillot, P.; Moulin, T.; Kotitz, R.; Guirardel, M.; Dodge, A.; Joanicot, M.; Colin, A.; Bruneau, C.-H.; Colin, T. Towards a continuous microfluidic rheometer. *Microfluid. Nanofluid.* **2008**, *5*, 619–630. [CrossRef]
35. Kang, Y.J. Quantitative monitoring of dynamic blood flows using coflowing laminar streams in a sensorless approach. *App. Sci.* **2021**, *11*, 7260. [CrossRef]
36. Kang, Y.J. Continuous and simultaneous measurement of the biophysical properties of blood in a microfluidic environment. *Analyst* **2016**, *141*, 6583–6597. [CrossRef]
37. Kang, Y.J. Assessment of blood biophysical properties using pressure sensing with micropump and microfluidic comparator. *Micromachines* **2022**, *13*, 483. [CrossRef]
38. Thielicke, W.; Stamhuis, E.J. PIVlab—Towards user-friendly, affordable and accurate digital particle image velocimetry in MATLAB. *J. Open Res. Softw.* **2014**, *2*, e30. [CrossRef]
39. Neu, B.; Wenby, R.; Meiselman, H.J. Effects of dextran molecular weight on red blood cell aggregation. *Biophys. J.* **2008**, *95*, 3059–3065. [CrossRef] [PubMed]

40. Neu, B.; Meiselman, H.J. Sedimentation and electrophoretic mobility behavior of human red blood cells in various dextran solutions. *Langmuir* **2001**, *17*, 7973–7975. [CrossRef]
41. Neu, B.; Meiselman, H.J. Depletion-mediated red blood cell aggregation in polymer solutions. *Biophys. J.* **2002**, *83*, 2482–2490. [CrossRef]

biosensors

MDPI

Article

An Asymmetric Microfluidic/Chitosan Device for Sustained Drug Release in Guided Bone Regeneration Applications

Xin Shi [1], Beibei Ma [2], Hongyu Chen [1], Wei Tan [1], Shiqing Ma [3,*] and Guorui Zhu [1,*]

1 School of Chemical Engineering and Technology, Tianjin University, Tianjin 300350, China
2 School and Hospital of Stomatology, Tianjin Medical University, Tianjin 300070, China
3 Department of Stomatology, The Second Hospital of Tianjin Medical University, Tianjin 300211, China
* Correspondence: mashiqing@tmu.edu.cn (S.M.); zhuguorui@tju.edu.cn (G.Z.)

Abstract: One of the major challenges of guided bone regeneration (GBR) is infections caused by pathogen colonization at wound sites. In this paper, an asymmetric microfluidic/chitosan device was developed to release drugs to inhibit infections and to ensure that guided bone regeneration can be realized. The microfluidic technique was introduced into the GBR membrane for the first time, which demonstrated more controllable drug release, more flexible clinical use and had a lower cost compared with surface treatments and embedded nanoparticles. Based on the theory of diffusion and Fick's first law, the contact area and concentration gradient were adjusted to realize sustained drug release. The standard deviation of minocycline release over 5 days was only 12.7%, which was lower than the joint effect of porous chitosan discs and nanospheres. The in vitro experiments against *E. coli* and Streptococcus mutans showed the excellent antibacterial performance of the device (>95%). The in vitro experiments for fibroblasts at the microfluidic side and osteoblasts at the chitosan side showed the satisfactory biocompatibility and the ability of the device to enhance bone regeneration. Therefore, this microfluidic/chitosan device is a promising therapeutic approach to prevent infection and guide bone regeneration.

Keywords: microfluidic; drug release; guided bone regeneration

Citation: Shi, X.; Ma, B.; Chen, H.; Tan, W.; Ma, S.; Zhu, G. An Asymmetric Microfluidic/Chitosan Device for Sustained Drug Release in Guided Bone Regeneration Applications. *Biosensors* **2022**, *12*, 847. https://doi.org/10.3390/bios12100847

Received: 28 August 2022
Accepted: 6 October 2022
Published: 9 October 2022

1. Introduction

The guided bone regeneration (GBR) technique has been commonly performed to repair bone defects caused by pathologic lesions or to augment alveolar bones for dental implant treatment [1]. The core of this technique involves using a barrier membrane to cover the bone defect area, which prevents the migration of epithelial cells and the surrounding fibroblasts from interfering with bone formation. Current GBR membranes fulfill the following five design criteria: (1) biocompatibility; (2) proper stiffness for space maintenance; (3) prevent epithelial cell migration; (4) tissue integration and (5) clinical manageability [2–5].

In addition to these basic characteristics, GBR membranes are supposed to have some additional functional characteristics. Collagen membranes have advantages such as weak immunogenicity and weak cytotoxicity compared with polytetrafluoroethylene (PTFE) membranes and titanium mesh [6,7]. Bio-Gide® is the leading collagen membrane for oral bone regeneration. The smooth side of this membrane prevents soft tissue from growing into the defect, meanwhile the rough side serves as a framework for bone cells and blood vessels [8,9]. However, collagen membranes usually involve a complex manufacturing process and it is hard to deal with wound infections when using these membranes.

In order to promote antibacterial properties, various GBR membranes were developed [10–12]. A common method involves combining some inorganic or organic ingredients with substrates. Chitosan has been recognized as an antimicrobial barrier membrane in GBR and guided tissue regeneration (GTR) [13,14]. Choi et al. showed that chitosan effectively inhibited Actinobacillus actinomycetemcomitans, the representative oral pathogen [15]. Shao et al. [16] fabricated a chitosan/poly(ethylene oxide) membrane with silver nanoparticles to inhibit Porphyromonas gingivalis and Fusobacterium nucleatum. In addition, graphene oxide (GO) can reduce bacteria viability by degrading their cell membranes [17,18]. Zhang et al. [19] reported a multifunctional bilayer GBR membrane using CS, GO, and a type of CaSi (xonotlite) nanowires as a building block. Although this method has advantages in its design and manufacture, the antibacterial properties of the membranes are limited by their intrinsic properties and these are difficult to adjust.

Loading drugs into the membranes directly or via nanoparticles can efficiently inhibit bacterial proliferation [20–22]. Chang et al. [23] developed gelatin/hyaluronic acid-based hydrogel membranes loaded with hinokitiol. The membranes were immersed in hinokitiol solution for loading and showed antibacterial properties after 48 h. Lian et al. [24] fabricated a bi-layered GBR membrane, whose loose layer was incorporated with dexamethasone-loaded mesoporous silica nanoparticles (DEX@MSNs), while the dense layer was loaded with the broad-spectrum antibiotic doxycycline hyclate (DCH). In vitro drug release tests indicated that both DEX and DCH followed a favorable sustained release profile. In our previous work [25], we prepared a novel asymmetric microfluidic/chitosan GBR membrane that contained minocycline-loaded nanoparticles. The bacteriostatic rates of this membrane against Porphyromonas gingivalis and Fusobacterium nucleatum were 95.3% and 92.1%, respectively.

Although many drug-loaded membranes have shown their potential, there are still many challenges in manufacturing an ideal antibacterial GBR membrane, such as (1) more precise control of the antibacterial drug release; (2) customized and flexible use in clinical treatments; and (3) low manufacturing cost. For most conventional drug-loaded membranes, the drug is easily burst-released when it is loaded on the surface, while it is difficult to be released when it is loaded inside. In addition, the type and dosage of the drugs are determined along with the manufacture process. Dentists cannot customize the membrane according to the characteristics of the patients. The manufacturing cost should also be considered. Some reported antibacterial GBR membranes need to be manufactured following complex processes, which are only suitable for laboratory scale tests and it is difficult for them to be mass produced.

In recent years, the microfluidic technique has received increasing attention with regard to drug release due to its precise control of fluids and suspended particles [26–28]. Meng et al. [29] proposed a micro-high intensity focused ultrasound (MHIFU) generated by a microfluidic device, which is able to control the drug release from temperature-sensitive liposomes (TSL). The flow cytometry results show that the drug delivery under MHIFU sonication leads to a significant increase in apoptosis. Jiang et al. [30] presented a low-cost, passive and flexible microfluidic device that uses wound pH as a trigger for localized drug release. The device can dispense 50 μL onto a 160 mm^2 dermal coverage within 4 h, showing the precision and controllability of the microfluidic technique. Trani et al. [31] described a subcutaneously implantable remote-controlled nanofluidic device that is capable of sustained drug release with adjustable dosing and timing, demonstrating the advantage of the microfluidic technique in treatment precisely tailored to individual needs.

In this study, a novel asymmetric microfluidic/chitosan device was developed for GBR applications. As shown in Figure 1, the microfluidic side prevented fibroblasts from invading bone defects using dense polylactic acid (PLA) substrates (PLA has been approved by European authorities and the Food and Drug Administration for the treatment of periodontal disease and is widely used [32–34]), and sustained release of antibacterial drugs from the embedded reservoir at the gingival region was reported. The reservoir can be preloaded with drug-loaded nanoparticles and further filled with customized solutions in clinical use. The chitosan side could facilitate osteoblast adhesion and proliferation at the bone defect region. Minocycline was selected to explore the in vitro drug release profiles under different drug-loaded strategies. The antibacterial performance of the microfluidic/chitosan device was tested against Escherichia coli and Streptococcus mutans. The fibrogenic and osteogenic properties of the microfluidic/chitosan device were further examined using

L929 fibroblasts and MC3T3-E1 osteoblasts, respectively. The microfluidic/chitosan device can be manufactured by simple methods using common and commercial materials, which leads to their wider application. The developed microfluidic/chitosan device with both osteogenic and antibacterial properties may be a promising candidate for the GBR technique.

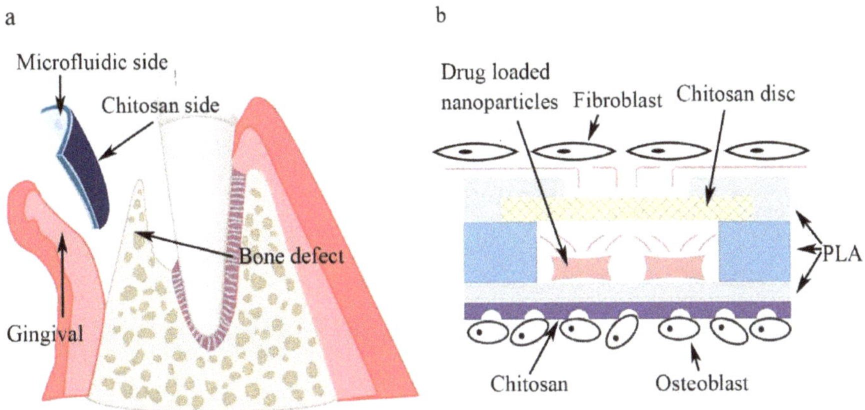

Figure 1. Conceptual illustration of the microfluidic/chitosan device for guided bone regeneration. (**a**) The microfluidic/chitosan device is placed between the gingival region and the bone defect region. (**b**) The working principle of the microfluidic/chitosan device. It can inhibit bacteria proliferation on the microfluidic side and promote bone growth on the chitosan side.

2. Materials and Methods

2.1. Device Design and Manufacture

The microfluidic/chitosan device shown in Figure 1a consisted of three PLA layers, porous chitosan discs, a sealing film and a chitosan layer. The top and the bottom PLA layers were cut from the PLA film (Hongjian Bio-medical Products Co. Ltd., Guangzhou, China) using a laser (GMA6040, Yueming Laser CO., Dongguan, China). The cutting speed was 250 mm/s. The middle PLA layer was 800 μm and constructed using a fused deposition modeling (FDM) 3D printer (JGAURORA A6, JGMAKER CO., Shenzhen, China). The structure of the middle PLA layer was designed with AutoCAD 2018 and then transferred to the 3D printer. The printing temperature was 210 °C and the printing velocity was 80 mm/s. The XY printing accuracy was 0.05 mm and the minimum Z plane resolution was 0.1 mm. The cavity within the three PLA layers was the drug reservoir, whose volume was 265.38 mm^3. The porous chitosan discs were placed between the top PLA layer and the middle PLA layer. For chitosan layer synthesis, 2% chitosan solution was prepared using 2 g of chitosan powder and 100 mL of 2% ice acetic acid and left overnight at 4 °C to remove the bubbles. Then, the chitosan solution was slowly poured into a polytetrafluoroethylene mold (60 mm in diameter, 15 mm in depth) to obtain a homogeneous film. The films were then immersed into liquid nitrogen for 10 s. Subsequently, these chitosan films were lyophilized at −80 °C for 24 h to produce a porous structure. The inlets were used to inject the drug solution. After injection, the inlets were blocked by a sealing film. The structure of the microfluidic/chitosan device is shown in Figure 2. All components are bonded by tissue adhesive. The total thickness of the device was about 2 mm (50 μm for the top and bottom PLA layer, 800 μm for the middle PLA layer and about 1 mm for the chitosan layer). The area of the device can be adjusted according to the bone defect area and the internal structure can be scaled proportionally. To facilitate testing, the radius of the device in the in vitro experiments was 10.2 mm. When placed into the oral position, the radius of the device can be reduced to 2–5 mm.

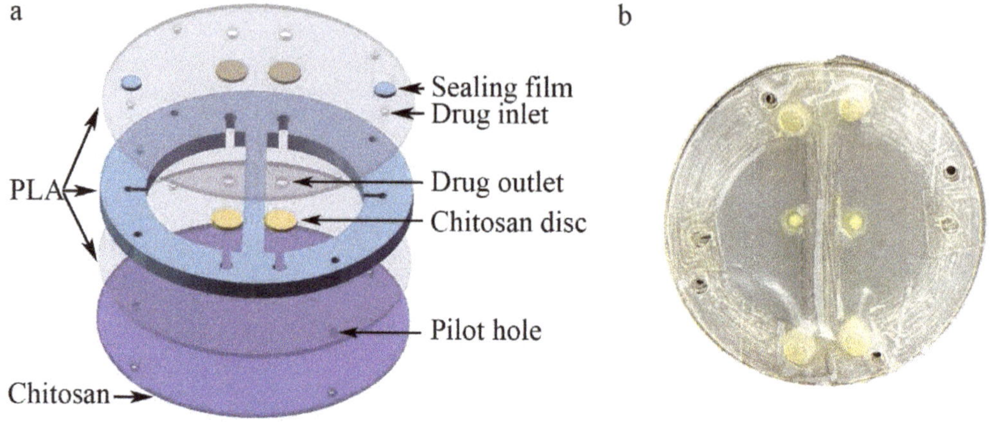

Figure 2. Structure of the microfluidic/chitosan device. (**a**) Three-dimensional exploded view of the device, from top to bottom: top PLA layer, drug reservoir, bottom PLA layer, and chitosan layer. (**b**) Photograph of the microfluidic/chitosan device.

2.2. Drug Release Characterizations

To characterize the drug release capability, the microfluidic/chitosan devices loaded with minocycline solution and nanoparticles (PERIOCLINE, Sunstar INC., Osaka, Japan) were placed in beakers with 10 mL PBS (pH 7.4) and gently shaken at a constant temperature of 37 °C. Subsequently, the supernatant was collected at various time points and inspected by a liquid chromatograph mass spectrometer (Waters E2695). The minocycline concentration was calculated based on the standard curve and used to calculate cumulative release [35]. All calculations were repeated more than three times and averaged.

2.3. In-Vitro Assessment of Bacterial Infections

The proliferation of the Escherichia coli and Streptococcus mutans were evaluated to measure the antibacterial performance by measuring the OD value [36]. The bacteria were seeded in 96-well plates (1×10^7 colony forming units (CFU)/mL) using the leachate of the microfluidic/chitosan device). Then, 10 μL of bacterial suspension was used to obtain the OD value at particular timepoints. The morphologies of the top PLA layer and the bacteria were observed by scanning electron microscopy (SEM; Nova NanoSEM 430, FEI, Hillsboro, OR, USA).

2.4. In-Vitro Assessment on Fibroblast Cell Proliferation

The proliferation of the fibroblast L929 on the microfluidic side was evaluated using the normal CCK-8 assay [37]. The fibroblasts in different conditions were seeded in 96-well plates (2000 cells/well). Then, 10 μL of CCK-8 reagent (Solarbio, Beijing, China) was added to the parallel wells every 24 h, followed by incubation for 4 h at 37 °C. The OD value was calculated at 450 nm absorbance. Live/dead cell double staining (Solarbio, Beijing, China) was used to distinguish dead cells (red staining) from live ones (green staining) to evaluate the viability of fibroblasts and the biocompatibility of the microfluidic/chitosan device. The fibroblasts were seeded on the device in 24-well plates (40,000 cells/well). Then, the cells were washed with PBS three times, and then stained with AO/EB (acridine orange/ethidium bromide) solution (0.5 mL/well) at various timepoints. The media were refreshed every 24 h. The fibroblasts were observed using laser confocal microscopy (CMLS) (Fv-1000, Olympus, Tokyo, Japan). The morphology of the fibroblasts on the top PLA layer surface was observed by SEM. All tests were repeated at least 3 times.

2.5. In-Vitro Assessment of Osteoblast Cell Proliferation

The proliferation and the viability of the osteoblasts MC3T3-E1 on the chitosan side were also evaluated by using the CCK-8 assay. The method was the same as the method mentioned in Section 2.4. The morphology of the osteoblast on the chitosan device surface was observed by SEM.

3. Results and Discussion

3.1. Mechanism of Drug Delivery at the Microfluidic Side

The antimicrobial effect of the microfluidic/chitosan device is based on the imbedded drug. When the microfluidic/chitosan device was immersed into culture medium in vitro (or pasted to gum tissue in vivo), minocycline was released due to molecular diffusion. The release amount Q is determined as follows:

$$Q = JAt \tag{1}$$

where J is the diffusion flux, A is the contact area and t is the time. The diffusion flux, J, can be calculated as Fick's first law, which is as follows:

$$J = -D\frac{dC}{dx} \tag{2}$$

where D is the diffusion coefficient, C is the concentration of the solute and x is the position. Thus, $\frac{dC}{dx}$ is the concentration gradient. It shows that the solute will move from a region of high concentration to a region of low concentration. The diffusion coefficient, D, can be estimated by the Wilke–Chang formula, which is as follows:

$$D = 7.4 \times 10^{-8}\left[(\varphi\mu_{\mathrm{B}})^{1/2}\frac{T}{\eta_{\mathrm{B}}V_{\mathrm{A}}^{0.6}}\right] \tag{3}$$

where φ is the parameter of association of the solvent, μ_{B} is the molecular mass of the solvent B, V_{A} is the molar volume of solute A at its boiling point under normal conditions, η_{B} is the substance viscosity and T is the temperature. According to the formulas above, for a certain drug solution at a certain period and temperature, the amount of drug released depends on the contact area and drug concentration gradient. In this microfluidic/chitosan device, the contact area is controlled by the porosities of chitosan discs. The drug concentration gradient can be adjusted flexibly by drug dosage and the proportion of solution and nanoparticles.

3.2. Minocycline Release Performance

In order to control the release velocity, chitosan films with various porosities were synthesized and cut into discs. Chitosan solution was slowly poured into a polytetrafluoroethylene mold at room temperature to obtain an even liquid film. Then, the liquid film was pre-heated at 45 °C for 2 h (Figure 3a)/1 h (Figure 3b)/0.5 h (Figure 3c) for dry phase separation, before immersion into the sodium hydroxide solution for wet phase separation. Subsequently, the obtained chitosan film was washed repeatedly with the distilled water until the pH value was neutral. Based on different operation parameters, three types of chitosan films were synthesized and punched into discs, whose diameters were 1.5 mm. As shown in Figure 3, the size and total area of the pores on disc C's surface were smaller than that on A and B. In addition, the internal structure of the disc became more complicated as the pore size decreased. In summary, the disc C can minimize the contact area between the embedded reservoir and external PBS.

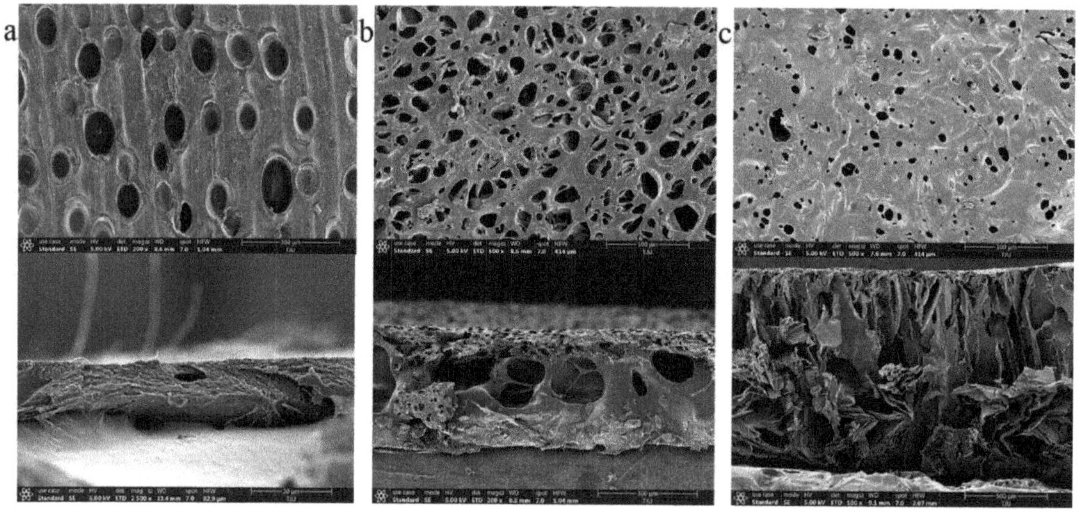

Figure 3. Images of chitosan discs using a scanning electron microscope (SEM). (**a**) SEM image of chitosan disc A, left for 2 h for dry phase separation. (**b**) SEM image of chitosan disc B, left for 1 h for dry phase separation. (**c**) SEM image of chitosan disc C, left for 0.5 h for dry phase separation.

The three types of chitosan discs were then placed into microfluidic/chitosan devices. After injected with the same volume of minocycline solution (0.25 mg/mL), the devices were submerged into 10 mL PBS (pH 7.4). The cumulative release curves of different devices were calculated and are shown in Figure 4. For the control with no chitosan discs, the contact area was the whole drug outlet. The mass transfer efficiency reached the theoretical maximum of this device. The total release was defined as 100% at 72 h. About 90% of minocycline was released over 12 h. For the device with disc A, the release velocity was similar to the control. The large holes induced an insufficient decrease in the contact area. Thus, minocycline still burst-released within 12 h. For the device with disc B, the 12 h release and 72 h release of minocycline were decreased to 73% and 88%, respectively. The decrease in the 72 h release may be caused by a higher dead volume in the internal caves, which were connected to the drug reservoir. About 12% minocycline remained in the device and was not easily released. Thus, the relative release rate, the ratio of 12 h release to 72 h release, is a more accurate parameter to measure sustained-release performance. The relative release rates of devices with disc A and disc B were calculated as 89.9% and 83.0%, respectively. It indicated that disc B had an indistinctive effect on reducing the mass transfer rate. For the device with disc C, the 12 h release and the 72 h release rates were 54.9% and 84.6%, respectively. The 72 h release showed that the dead volume of disc C was similar to that of disc B. The 12 h release showed that the disc had the capability to control the mass transfer rate by adjusting the contact area with limited drug retention. Thus, the microfluidic/chitosan device with disc C was selected for further tests. The original data of this experiment from the liquid chromatograph mass spectrometer are presented in Table S1 in the Supplementary Materials.

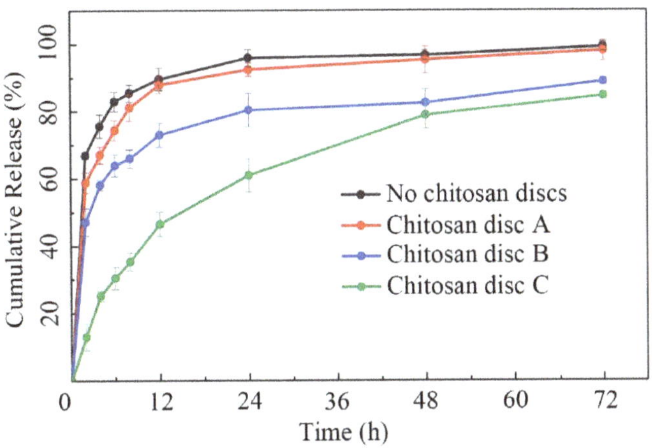

Figure 4. Cumulative release plot of minocycline in 72 h. The minocycline burst-released within 12 h in the microfluidic/chitosan device with no discs/disc A/disc B. The minocycline demonstrated sustained release over 48 h in the microfluidic/chitosan device with disc C. Error bars represent the standard deviation of the mean.

Besides contact area, the mass transfer was also affected by the minocycline concentration gradient. To simulate the metabolism of gum tissue, we refreshed 50% of the PBS (5 mL) every 24 h. Minocycline residue in the dish before and after refresh was acquired and standardized. As shown in Figure 5, when the minocycline solution (0.25 mg/mL) was directly added to the dish without the microfluidic/chitosan device (control A, black belt), the amount of minocycline residue was halved with each PBS refresh. On the other hand, the amount of minocycline residue quickly increases due to the gradient shock caused by the first PBS refresh when the minocycline solution (0.25 mg/mL) was added into the device (Group A, blue belt). In the period of 24 h to 48 h, the minocycline release rate decreased steeply due to insufficient minocycline reserves. After 4 refreshes, the 120 h minocycline residue was only 8.9%. The standard deviation of minocycline residue at different time points was calculated to quantify the release fluctuation. For the solution in the dish and solution in the microfluidic/chitosan device, the SD values were 38.1% and 32.8%, respectively. The high SD value indicated that minocycline did not demonstrate sustained release throughout the whole test period. To overcome the burst-release in the initial 48 h, nanoparticles loaded with minocycline were selected (Periocline, Sunstar INC). The release character of the nanoparticles (loaded with 80 μg minocycline) in the dish was referred to as Control B, the red belt. Although it decreased the release rate in the early stage, the dramatic decline in minocycline residue over 120 h led to a large SD of 29.2%. Then, we injected a mixture of PBS and nanoparticles (loaded with 80 μg minocycline) into the microfluidic/chitosan device. Confined by the nanoparticle and chitosan disc, it was difficult for minocycline molecules to diffuse to the external environment in the initial 48 h. The residue did not reach its peak until 96 h (Group B, purple belt). The SD decrease to 14.9% reflected a more stable release process. However, in order to avoid surgical site infection (SSI), the local minocycline concentration should increase to the same value as the bacteriostatic concentration as soon as possible. This means that the characteristics of Group B do not match the clinical requirements. Finally, a mixture of minocycline solution (0.25 mg/mL) and nanoparticles (loaded with 80 μg minocycline) was injected into the microfluidic/chitosan device (Group C, green belt). In the initial 48 h, the minocycline concentration rapidly increased to the same value as the bacteriostatic concentration. The release in the initial period was dominated by the minocycline solution, compared to Group B. From 48 h to 120 h, the minocycline residue increased after each PBS refresh, which is different from the rapid decrease in Group A. It can be deduced that the release in this period was dominated by minocycline-loaded nanoparticles. At the same time, the SD was only 12.7%, which indicated a stable drug release with less flush. In conclusion, the joint loaded method of minocycline solution and nanoparticles in the microfluidic/chitosan device showed its advantages of timeliness and stationarity. The original data of this experiment from the liquid chromatograph mass spectrometer are presented in Table S2 in the Supplementary Information.

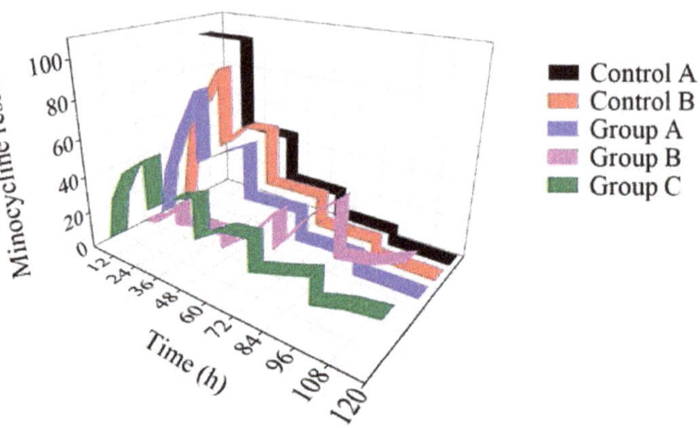

Figure 5. Minocycline residue over 120 h. For the solution (black belt) in the dish, the minocycline residue decreased rapidly. For the nanoparticles (red belt) in the dish and the solution in the microfluidic/chitosan device (blue belt), the amount of minocycline residue in the initial period was much higher than that in the terminal period. For the nanoparticles in the microfluidic/chitosan device (purple belt), the amount of minocycline residue in the initial period was much lower than that in the terminal period. When a mixture of the solution and the nanoparticles was loaded into the microfluidic/chitosan device (green belt), the minocycline residue was stable throughout the whole period.

3.3. Antibacterial Performance of the Microfluidic/Chitosan Device

The antibacterial efficiency of the microfluidic/chitosan device for GBR treatment was evaluated by investigating the optical density (OD) values of the tested bacteria. *Escherichia coli*, the most common strain, and *Streptococcus mutans*, which exists only in the oral environment, were selected for the bacteriostatic tests. The concentration of minocycline solution in the microfluidic/chitosan device was 0.25 mg/mL. The mass of the minocycline loaded by the nanoparticles was 80 μg.

As shown in Figure 6a,b, after the cultivation of these bacteria with the PLA substrate (blue stripe) for 24 h, the OD values were similar to those of the blank controls (red stripe). As expected from reports [38,39], the PLA substrate had limited influence on antibacterial activity in our experiments. When minocycline was introduced, the proliferation rate of *E. coli* was inhibited by 98.1% ($(1 - \frac{\text{proliferation rate of device}}{\text{proliferation rate of control}}) \times 100\%$). The SEM images of surfaces of the PLA substrate and microfluidic/chitosan device are shown in Figure 6c and Figure 6d, respectively. For *S. mutans*, the 24 h inhibition rate reached 96.0% in the presence of the microfluidic/chitosan device. The SEM images of surfaces of the PLA substrate and microfluidic/chitosan device are shown in Figure 6e and Figure 6f, respectively. These results indicated that the bacteriostasis of the microfluidic/chitosan device can be mainly attributed to minocycline. The antibacterial mechanisms of minocycline include inhibiting protein synthesis by binding at the decoding center of the small subunit [40,41].

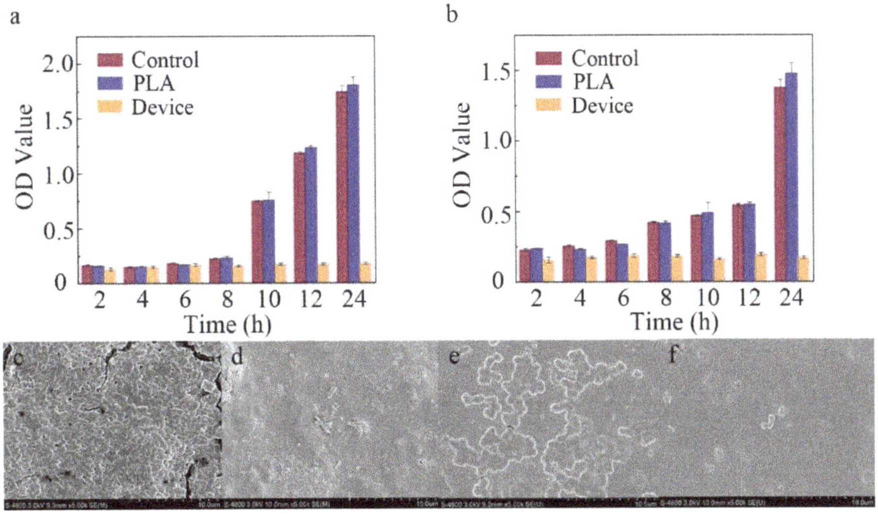

Figure 6. Characterization of antibacterial activity. (**a**) The OD value of *E. coli* after 24 h. (**b**) The OD value of *S. mutans* after 24 h. (**c**) SEM image of PLA substrate surface with *E. coli*. (**d**) SEM image of microfluidic/chitosan device surface with *E. coli*. (**e**) SEM image of PLA substrate surface with *S. mutans*. (**f**) SEM image of microfluidic/chitosan device surface with *S. mutans*.

3.4. Fibrogenic Performance of the Microfluidic/Chitosan Device

Figure 7a shows the CCK8 assay results of the fibroblasts seeded on the PLA surface of the microfluidic/chitosan device after 1–5 days. With the same initial seeding density, the OD values of the control and test groups were on a similar level on day 5. The blue line represents the minocycline condition that was same as the condition used in the bacterial test. Under this loading condition, bacterial proliferation was inhibited while the fibroblasts grew normally, which demonstrated the antibacterial and non-cytotoxic feature of the device. The red and green curves represent another two minocycline conditions with lower (0.125 mg/mL for solution and 80 μg for nanoparticles) and higher (0.75 mg/mL for solution and 160 μg for nanoparticles) minocycline loads, respectively. These similar proliferation curves showed a wide range for minocycline load and flexibility in the use of our device. As shown in the SEM results, the dense PLA layer could facilitate cell attachment (Figure 7b). The results of live/dead cell staining using the AO/EB kit are shown in Figure 7c–f. Both cells in the control and minocycline condition (0.25 mg/mL for solution and 80 μg for nanoparticles) spread normally over 5 days and demonstrated spindle morphology.

3.5. Osteogenic Performance of the Microfluidic/Chitosan Device

Figure 8a shows the CCK8 assay results of the osteoblasts seeded on the chitosan surface of the microfluidic/chitosan device after 1–5 days. The cells rapidly proliferated and presented an increasing trend, suggesting that the chitosan layer possessed good biocompatibility (red curve). The blue curve and green curve represent the minocycline conditions (0.25 mg/mL for solution and 80 μg for nanoparticles), which suggests that this loading condition was also acceptable for osteoblasts. SEM micrographs of the chitosan surface without osteoblasts are shown in Figure 8b. Based on our previous study, the loose and porous morphology can promote osteoblast adhesion. SEM micrographs of the chitosan surface with osteoblasts (red circles) are shown in Figure 8c. After 3 days of seeding, the osteoblasts showed a round shape and were anchored to the chitosan surface by discrete filopodia, as shown in Figure 8d, suggesting that the device had satisfactory cytocompatibility.

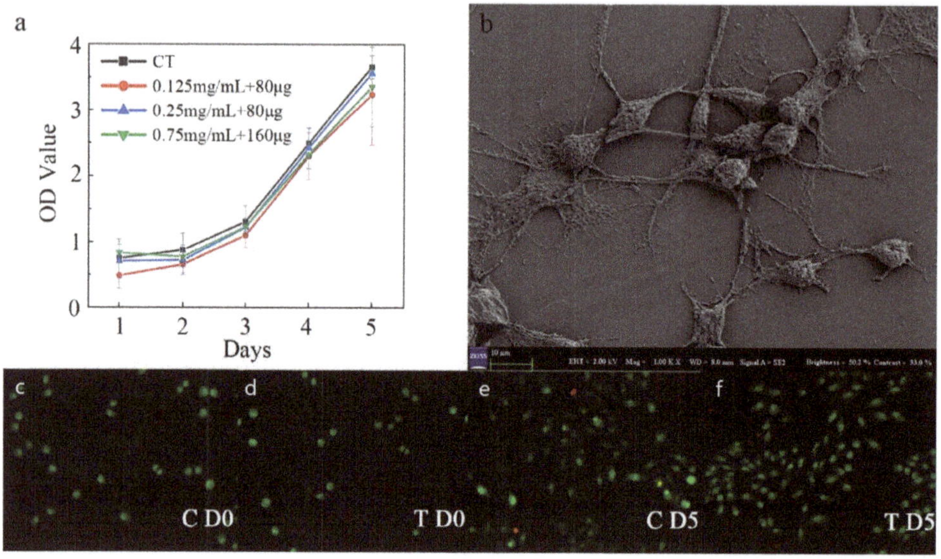

Figure 7. Characterization of fibrogenic performance. (**a**) The OD value of the control and minocycline conditions across 5 days. (**b**) SEM image of PLA substrate surface with fibroblast. (**c**) Live/dead staining for control on day 0. (**d**) Live/dead staining for minocycline condition on day 0. (**e**) Live/dead staining for control on day 5. (**f**) Live/dead staining for minocycline condition on day 5.

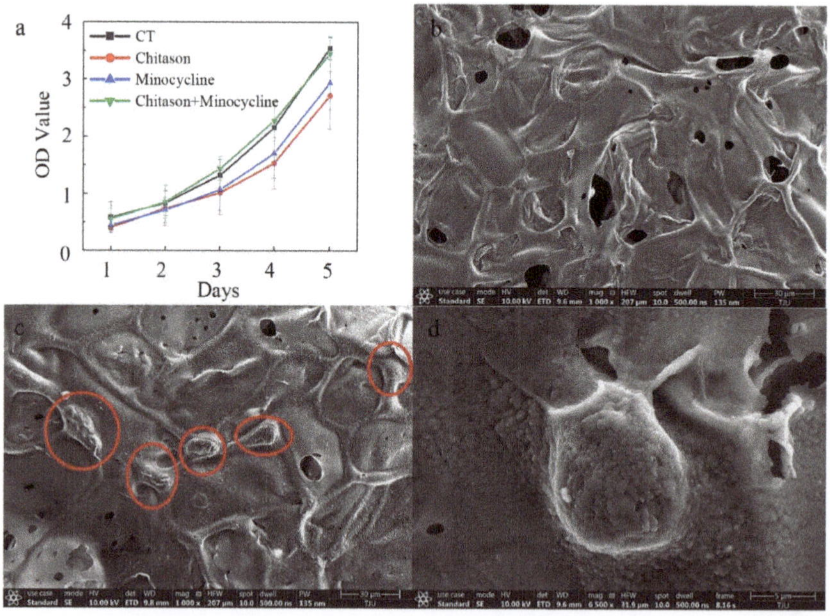

Figure 8. Characterization of osteogenic performance. (**a**) The OD value of the control and minocycline conditions across 5 days. (**b**) SEM image of chitosan substrate surface without osteoblasts. (**c**) SEM image of chitosan substrate surface with osteoblasts (30 μm scale). (**d**) SEM image of chitosan substrate surface with osteoblasts (5 μm scale).

4. Conclusions

The novel asymmetric antibacterial microfluidic/chitosan GBR device can be manufactured by PLA layer preparation, chitosan layer synthesis and further assembly. Porous chitosan discs were used to adjust the contact area between the embedded reservoir and external environment. The 12 h release and the 72 h release of the test group decreased to 54.9% and 84.6%, compared to the control group. When injecting a mixture of minocycline solution and minocycline-loaded nanoparticles into the reservoir, the standard deviation of minocycline release was only 12.7% after 5 days, which showed demonstrated sustained release. The 24 h inhibition rate of *E. coli* and *S. mutan* reached 98.1% and 96.0%, respectively. Both fibroblasts at the microfluidic side and osteoblasts at the chitosan side were able to proliferate normally and showed excellent fibrogenic and osteogenic performance. In conclusion, this microfluidic/chitosan device can efficiently prevent infection and the introduction of microfluidics into the GBR technique show potential regarding clinical requirements.

Supplementary Materials: The following supporting information can be downloaded at: https://www.mdpi.com/article/10.3390/bios12100847/s1, Table S1: The minocycline concentration from liquid chromatography-mass spectrometry; Table S2: The minocycline concentration from liquid chromatography-mass spectrometry.

Author Contributions: X.S. initiated and developed the idea. H.C. performed the drug release experiments. B.M. performed the biological experiments. X.S. and W.T. prepared the manuscript. G.Z. and S.M. supervised the research. All authors have read and agreed to the published version of the manuscript.

Funding: This research was funded by the Science and Technology Plan Project of Tianjin (21JCZDJC00420), Tianjin Municipal Science and Technology Bureau.

Institutional Review Board Statement: Not applicable.

Informed Consent Statement: Not applicable.

Data Availability Statement: Not applicable.

Conflicts of Interest: There are no conflicts to declare.

References

1. Lee, S.W.; Kim, S.G. Membranes for the Guided Bone Regeneration. *Maxillofac. Plast. Reconstr. Surg.* **2014**, *36*, 239–246. [CrossRef] [PubMed]
2. Rakhmatia, Y.D.; Ayukawa, Y.; Furuhashi, A.; Koyano, K. Current barrier membranes: Titanium mesh and other membranes for guided bone regeneration in dental applications. *J. Prosthodont. Res.* **2013**, *57*, 3–14. [CrossRef] [PubMed]
3. Hitti, R.A.; Kerns, D.G. Guided bone regeneration in the oral cavity: A review. *Open Pathol. J.* **2011**, *5*, 33–45. [CrossRef]
4. Retzepi, M.; Donos, N. Guided bone regeneration: Biological principle and therapeutic applications. *Clin. Oral Implant. Res.* **2010**, *21*, 567–576. [CrossRef]
5. Scantlebury, T.V. 1982–1992: A decade of technology development for guided tissue regeneration. *J. Periodontol.* **1993**, *64*, 1129–1137. [CrossRef] [PubMed]
6. Jung, R.E.; Fenner, N.; Hämmerle, C.H.; Zitzmann, N.U. Long-term outcome of implants placed with guided bone regeneration (GBR) using resorbable and non-resorbable membranes after 12–14 years. *Clin. Oral Implant. Res.* **2013**, *24*, 1065–1073. [CrossRef] [PubMed]
7. Piattelli, A.; Scarano, A.; Paolantonio, M. Bone formation inside the material interstices of e-PTFE membranes: A light microscopical and histochemical study in man. *Biomaterials* **1996**, *17*, 1725–1731. [CrossRef]
8. Schlegel, A.; Möhler, H.; Busch, F.; Mehl, A. Preclinical and clinical studies of a collagen membrane (Bio-Gide®). *Biomaterials* **1997**, *18*, 535–538. [CrossRef]
9. Schwarz, F.; Rothamel, D.; Herten, M.; Wüstefeld, M.; Sager, M.; Ferrari, D.; Becker, J. Immunohistochemical characterization of guided bone regeneration at a dehiscence-type defect using different barrier membranes: An experimental study in dogs. *Clin. Oral Implant. Res.* **2008**, *19*, 402–415. [CrossRef] [PubMed]
10. Tokuda, S.; Obata, A.; Kasuga, T. Preparation of poly(lactic acid)/siloxane/calcium carbonate composite membranes with antibacterial activity. *Acta Biomater.* **2009**, *5*, 1163–1168. [CrossRef] [PubMed]

11. He, Y.; Jin, Y.; Wang, X.; Yao, S.; Li, Y.; Wu, Q.; Ma, G.; Cui, F.; Liu, H. An antimicrobial peptide-loaded gelatin/chitosan nanofibrous membrane fabricated by sequential layer-by-layer electrospinning and electrospraying techniques. *Nanomaterials* **2018**, *8*, 327. [CrossRef] [PubMed]
12. Bottino, M.C.; Thomas, V.; Schmidt, G.; Vohra, Y.K.; Chu, T.-M.G.; Kowolik, M.J.; Janowski, G.M. Recent advances in the development of GTR/GBR membranes for periodontal regeneration—A materials perspective. *Dent. Mater.* **2012**, *28*, 703–721. [CrossRef]
13. Husain, S.; Al-Samadani, K.H.; Najeeb, S.; Zafar, M.S.; Khurshid, Z.; Zohaib, S.; Qasim, S. Chitosan biomaterials for current and potential dental applications. *Materials* **2017**, *10*, 602. [CrossRef]
14. Xu, C.; Lei, C.; Meng, L.; Wang, C.; Song, Y. Chitosan as a barrier membrane material in periodontal tissue regeneration. *J. Biomed. Mater. Res. B Appl. Biomater.* **2012**, *100*, 1435–1443. [CrossRef]
15. Choi, B.-K.; Kim, K.-Y.; Yoo, Y.-J.; Oh, S.-J.; Choi, J.-H.; Kim, C.-Y. In vitro antimicrobial activity of a chitooligosaccharide mixture against Actinobacillus actinomycetemcomitans and Streptococcus mutans. *Int. J. Antimicrob. Agents* **2001**, *18*, 553–557. [CrossRef]
16. Shao, J.; Yu, N.; Kolwijck, E.; Wang, B.; Tan, K.W.; Jansen, J.A.; Walboomers, X.F.; Yang, F. Biological evaluation of silver nanoparticles incorporated into chitosan-based membranes. *Nanomedicine* **2017**, *12*, 2771–2785. [CrossRef]
17. Tu, Y.; Lv, M.; Xiu, P.; Huynh, T.; Zhang, M.; Castelli, M.; Liu, Z.; Huang, Q.; Fan, C.; Fang, H.; et al. Destructive extraction of phospholipids from Escherichia coli membranes by graphene nanosheets. *Nat. Nanotechnol.* **2013**, *8*, 594–601. [CrossRef]
18. Zou, X.; Zhang, L.; Wang, Z.; Luo, Y. Mechanisms of the antimicrobial activities of graphene materials. *J. Am. Chem. Soc.* **2016**, *138*, 2064–2077. [CrossRef] [PubMed]
19. Zhang, K.-R.; Gao, H.-L.; Pan, X.-F.; Zhou, P.; Xing, X.; Xu, R.; Pan, Z.; Wang, S.; Zhu, Y.; Hu, B.; et al. Multifunctional Bilayer Nanocomposite Guided Bone Regeneration Membrane. *Matter* **2019**, *1*, 770–781. [CrossRef]
20. Xue, J.; He, M.; Liu, H.; Niu, Y.; Crawford, A.; Coates, P.D.; Chen, D.; Shi, R.; Zhang, L. Drug loaded homogeneous electrospun PCL/gelatin hybrid nanofiber structures for anti-infective tissue regeneration membranes. *Biomaterials* **2014**, *35*, 9395–9405. [CrossRef]
21. Xue, J.; He, M.; Niu, Y.; Liu, H.; Crawford, A.; Coates, P.; Chen, D.; Shi, R.; Zhang, L. Preparation and in vivo efficient anti-infection property of GTR/GBR implant made by metronidazole loaded electrospun polycaprolactone nanofiber membrane. *J. Int. J. Pharm.* **2014**, *475*, 566–577. [CrossRef] [PubMed]
22. Murali, V.P.; Guerra, F.D.; Ghadri, N.; Christian, J.M.; Stein, S.H.; Jennings, J.A.; Smith, R.A.; Bumgardner, J.D. Simvastatin loaded chitosan guided bone regeneration membranes stimulate bone healing. *J. Periodontal Res.* **2021**, *56*, 877–884. [CrossRef] [PubMed]
23. Chang, K.C.; Lin, D.J.; Wu, Y.R.; Chang, C.W.; Chen, C.H.; Ko, C.L.; Chen, W.C. Characterization of genipin-crosslinked gelatin/hyaluronic acid-based hydrogel membranes and loaded with hinokitiol: In vitro evaluation of antibacterial activity and biocompatibility. *Mater. Sci. Eng. C Mater. Biol. Appl.* **2019**, *105*, 110074. [CrossRef] [PubMed]
24. Lian, M.; Sun, B.; Qiao, Z.; Zhao, K.; Zhou, X.; Zhang, Q.; Zou, D.; He, C.; Zhang, X. Bi-layered electrospun nanofibrous membrane with osteogenic and antibacterial properties for guided bone regeneration. *Colloids Surf. B Biointerfaces* **2019**, *176*, 219–229. [CrossRef]
25. Ma, S.; Adayi, A.; Liu, Z.; Li, M.; Wu, M.; Xiao, L.; Sun, Y.; Cai, Q.; Yang, X.; Zhang, X.; et al. Asymmetric Collagen/chitosan Membrane Containing Minocycline-loaded Chitosan Nanoparticles for Guided Bone Regeneration. *Sci. Rep.* **2016**, *6*, 31822. [CrossRef]
26. Jiang, N.; Montelongo, Y.; Butt, H.; Yetisen, A.K.J.S. Microfluidic contact lenses. *Small* **2018**, *14*, 1704363. [CrossRef]
27. Iyer, M.A.; Eddington, D. Storing and releasing rhodamine as a model hydrophobic compound in polydimethylsiloxane microfluidic devices. *Lab Chip* **2019**, *19*, 574–579. [CrossRef]
28. Shakeri, A.; Khan, S.; Didar, T.F. Conventional and emerging strategies for the fabrication and functionalization of PDMS-based microfluidic devices. *Lab Chip* **2021**, *21*, 3053–3075. [CrossRef]
29. Meng, L.; Deng, Z.; Niu, L.; Li, F.; Yan, F.; Wu, J.; Cai, F.; Zheng, H. A Disposable Microfluidic Device for Controlled Drug Release from Thermal-Sensitive Liposomes by High Intensity Focused Ultrasound. *Theranostics* **2015**, *5*, 1203–1213. [CrossRef]
30. Jiang, H.; Ochoa, M.; Waimin, J.F.; Rahimi, R.; Ziaie, B. A pH-regulated drug delivery dermal patch for targeting infected regions in chronic wounds. *Lab Chip* **2019**, *19*, 2265–2274. [CrossRef]
31. Di Trani, N.; Silvestri, A.; Bruno, G.; Geninatti, T.; Chua, C.Y.X.; Gilbert, A.; Rizzo, G.; Filgueira, C.S.; Demarchi, D.; Grattoni, A. Remotely controlled nanofluidic implantable platform for tunable drug delivery. *Lab Chip* **2019**, *19*, 2192–2204. [CrossRef] [PubMed]
32. Nieminen, T.; Kallela, I.; Keränen, J.; Hiidenheimo, I.; Kainulainen, H.; Wuolijoki, E.; Rantala, I. In Vivo and in vitro degradation of a novel bioactive guided tissue regeneration membrane. *Int. J. Oral Maxillofac. Surg.* **2006**, *35*, 727–732. [CrossRef] [PubMed]
33. Caffesse, R.G.; Mota, L.F.; Quiñones, C.R.; Morrison, E.C. Clinical comparison of resorbable and non-resorbable barriers for guided periodontal tissue regeneration. *J. Clin. Periodontol.* **1997**, *24*, 747–752. [CrossRef]
34. Gottlow, J.; Laurell, L.; Lundgren, D.; Mathisen, T.; Nyman, S.; Rylander, H.; Bogentoft, C.; Dentistry, R. Periodontal tissue response to a new bioresorbable guided tissue regeneration device: A longitudinal study in monkeys. *Int. J. Periodontics Restor. Dent.* **1994**, *14*, 437–449.
35. Ghica, M.; Albu, M.; Leca, M.; Popa, L.; Moisescu, S.T. Design and optimization of some collagen-minocycline based hydrogels potentially applicable for the treatment of cutaneous wound infections. *Die Pharm. Int. J. Pharm. Sci.* **2011**, *66*, 853–861.

36. Zhang, Y.; Liu, X.; Wang, Y.; Jiang, P.; Quek, S. Antibacterial activity and mechanism of cinnamon essential oil against Escherichia coli and Staphylococcus aureus. *Food Control* **2016**, *59*, 282–289. [CrossRef]
37. Yu, Z.; Yu, M.; Zhang, Z.; Hong, G.; Xiong, Q. Bovine serum albumin nanoparticles as controlled release carrier for local drug delivery to the inner ear. *Nanoscale Res. Lett.* **2014**, *9*, 1–7. [CrossRef]
38. Fu, S.; Zhang, P. Chitosan–gelatin enhanced antibacterial and biological properties of PLA and PGA braided threads for juvenile pseudomyopia treatment. *Text. Res. J.* **2021**, *91*, 2053–2062. [CrossRef]
39. Ahmed, J.; Mulla, M.Z.; Arfat, Y.A. Thermo-mechanical, structural characterization and antibacterial performance of solvent casted polylactide/cinnamon oil composite films. *Food Control* **2016**, *69*, 196–204. [CrossRef]
40. Elewa, H.F.; Hilali, H.; Hess, D.C.; Machado, L.S.; Fagan, S.C.; Therapy, D. Minocycline for short-term neuroprotection. *J. Hum. Pharmacol. Drug Ther.* **2006**, *26*, 515–521. [CrossRef]
41. Olson, M.W.; Ruzin, A.; Feyfant, E.; Rush, T.S., III; O'Connell, J.; Bradford, P.A. Functional, biophysical, and structural bases for antibacterial activity of tigecycline. *Antimicrob. Agents Chemother.* **2006**, *50*, 2156–2166. [CrossRef] [PubMed]

biosensors

MDPI

Article

Generation of Dynamic Concentration Profile Using A Microfluidic Device Integrating Pneumatic Microvalves

Chang Chen [1], Panpan Li [2], Tianruo Guo [3], Siyuan Chen [2], Dong Xu [2] and Huaying Chen [1,*]

1 School of Mechanical Engineering and Automation, Harbin Institute of Technology, Shenzhen 518055, China
2 School of Science, Harbin Institute of Technology, Shenzhen 518055, China
3 Graduate School of Biomedical Engineering, The University of New South Wales, Sydney, NSW 2052, Australia
* Correspondence: chenhuaying@hit.edu.cn; Tel.: +86-755-8615-3249

Abstract: Generating and maintaining the concentration dilutions of diffusible molecules in microchannels is critical for high-throughput chemical and biological analysis. Conventional serial network microfluidic technologies can generate high orders of arbitrary concentrations by a predefined microchannel network. However, a previous design requires a large occupancy area and is unable to dynamically generate different profiles in the same chip, limiting its applications. This study developed a microfluidic device enabling dynamic variations of both the concentration in the same channel and the concentration distribution in multiple channels by adjusting the flow resistance using programmable pneumatic microvalves. The key component (the pneumatic microvalve) allowed dynamic adjustment of the concentration profile but occupied a tiny space. Additionally, a Matlab program was developed to calculate the flow rates and flow resistance of various sections of the device, which provided theoretical guidance for dimension design. *In silico* investigations were conducted to evaluate the microvalve deformation with widths from 100 to 300 μm and membrane thicknesses of 20 and 30 μm under the activation pressures between 0 and 2000 mbar. The flow resistance of the deformed valve was studied both numerically and experimentally and an empirical model for valve flow resistance with the form of $R_h = ae^{bP}$ was proposed. Afterward, the fluid flow in the valve region was characterized using Micro PIV to further demonstrate the adjustment mechanism of the flow resistance. Then, the herringbone structures were employed for fast mixing to allow both quick variation of concentration and minor space usage of the channel network. Finally, an empirical formula-supported computational program was developed to provide the activation pressures required for the specific concentration profile. Both linear ($C_k = -0.2k + 1$) and nonlinear ($C_k = \left(\frac{1}{\sqrt{10}}\right)^k$) concentration distribution in four channels were varied using the same device by adjusting microvalves. The device demonstrated the capability to control the concentration profile dynamically in a small space, offering superior application potentials in analytical chemistry, drug screening, and cell biology research.

Keywords: concentration profile; dynamic; programmable; pneumatic microvalves; microfluidic; fluid resistance

Citation: Chen, C.; Li, P.; Guo, T.; Chen, S.; Xu, D.; Chen, H. Generation of Dynamic Concentration Profile Using A Microfluidic Device Integrating Pneumatic Microvalves. *Biosensors* **2022**, *12*, 868. https://doi.org/10.3390/bios12100868

Received: 30 August 2022
Accepted: 9 October 2022
Published: 13 October 2022

1. Introduction

Microfluidic technologies have been widely explored to create microenvironments with a concentration profile due to their prodigious advantages in automation, integration, and control of samples with extremely tiny volumes [1–6]. Existing microfluidic devices generate the concentration profile based on either passive self-diffusion [7–10], or continuous-flow mixing networks [11–13] before the observation region. The passive self-diffusion devices usually utilize diffusion across two parallelly laminar flows with chemical concentration discrepancy to generate a local concentration gradient [14–16]. To stabilize subjects, for example, cells in the middle observation chamber, the low-height channel [17,18], the membrane [19] and gels [20,21] were employed to form the obstacles.

The unique advantage of high integration makes it possible for microfluidics to contain an array of chambers for high-through gradient analysis using limited samples [22–25]. However, the self-diffusion-based approach can only maintain a simple gradually decreasing concentration between a chemical source and a sink channel.

For more complex concentration profiles, the device requires intricate channel design such as proportional, pyramidal, and serial networks [26]. A proportional network controls the volumetric proportions of two streams by adjusting the flow resistance ratios of the channels before mixing [27,28]. However, the mixing ratio in a proportional network approach is limited by the channel design, making it hard to achieve dynamic adjustment. In contrast, the other two networks are more flexible. The pyramidal network repeats the splitting, mixing, and recombination of fluid streams to generate the concentration gradients. Since the concentrations are developed before the permutation combination, the pyramidal multistage-dilution device can produce more diverse concentration profiles [29]. For example, the Christmas tree network was designed to generate fixed gradient profiles by integrating multiple T-shaped mixing grades [11,30]. This approach has been extended to generate parabola [31], inverse proportion, cube root function [32], exponent [33], and linear [34] concentration variation across the width of the downstream observation channel. The concentration profiles were able to be dynamically tuned by adjusting the inlet flow rates [29,35]. Nevertheless, such networks are limited to a dilution range with few orders of magnitude, which confines the applications for dose–response experiments or high-throughput drug screening and optimization.

Comparably, the serial network, comprising a series of stepwise dilutions to mimic conventional manual serial dilutions, can generate logarithmic concentrations [36–38]. However, multiple serial dilution steps usually result in long flow channels, which require a large occupancy area. Additionally, the output concentrations profile highly relies on the network design, the majority of applications that use such networks failed to dynamically generate different profiles in the same chip. Since high order and dynamic adjustment are the two major requirements of the chemical environment in comparative experiments [39], there is a strong unmet need for modulating the arbitrary large range concentration profile dynamically in a diversity of experimental conditions. Thus, it is important to design a device empowering the serial network with dynamic regulation capability.

Since the pioneering report in 2000 [40], pneumatic microvalves have been extensively explored due to the overwhelming advantages of high integration [41], precision [42], and compatibility with PDMS-based devices [43]. They have been widely applied for multiple purposes including on/off switching [44] or even hydrodynamic trap [43,45]. Their key advantage is real-time adjustability. Moreover, manipulating microvalves can change the local flow resistance. Consequently, integrating the serial network device and pneumatic microvalves can be a promising approach for dynamically adjusting concentration profiles by precisely regulating the mixing ratio.

This paper proposed a prototype microfluidic device to dynamically change the chemical concentration in each of the four branch channels using programmed microvalves. The influence of activation pressure on the valve deformation and the resulting flow resistance was studied both experimentally and numerically. The bypass manometry measurement technique and the empirical model for the flow resistance provided a deep understanding of the irregular cross-section channel study in microfluidics. Then, a Matlab program was developed to predict the required activation pressures for generating varied custom dilution profiles. The herringbone microstructures were utilized to decrease the channel length and the response time for dynamic concentration variation. Finally, both linear and nonlinear concentration profiles were achieved in four branch channels and swapped by adjusting the pressure combination on four valves. This paper demonstrated the dynamic adjustment of the arbitrary concentration profile by active components in one microfluidic platform. The advantage of the device in the programmable alteration of the concentration will possess high application potentials in both chemical and biological analysis including *in vitro* cell culture and cytotoxicity test, drug screening as well as immunoassay.

2. Materials and Methods

2.1. Design and Fabrication

A microfluidic system (Figure 1A) was developed to dynamically regulate the chemical concentration in multiple branch channels by varying the pressure applied to the microvalves. The microfluidic system included a pressure controller, two syringe pumps, and a microfluidic chip (Figure S1). The chip was composed of a gas layer, a membrane layer, and a fluid layer from top to bottom (Figure 1B). There were four isolated gas channels (300 μm width and 50 μm depth) for individual control of the microvalves. The 20 μm thick membrane was deformed to modulate the cross-section (thus flow resistance) of the channel in the fluid layer after pressurization by the gas channels. The fluid layer had one chemical inlet, four inlets for diluting liquid (e.g., DI water), and four identical branch channels for concentrations generation. Specifically, there were five components: the inlet region (0.2 mm wide and 3 mm long), the mixing region with herringbone microstructures (0.2 mm wide and 20 mm long), the flow-shunting channel (0.2 mm wide and 20 mm long), the microvalve region (50 μm width and 300 μm length, with different flow resistance channels) and the observation region (1 mm width and 10 mm length).

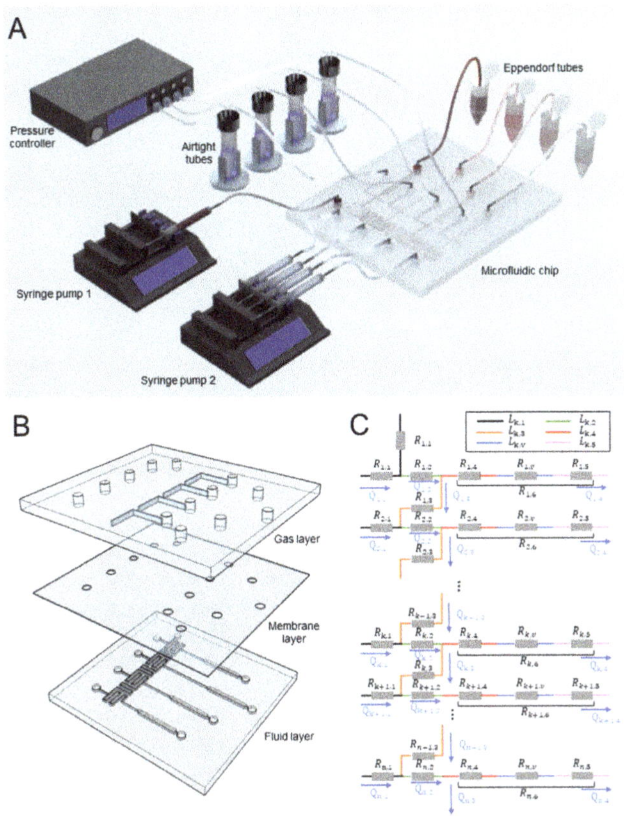

Figure 1. The microfluidic system for dynamic generation of concentration profiles in four parallel channels. (**A**) The schematic of the experimental platform consisting of a pressure controller, two syringe pumps, and a microfluidic. (**B**) The explosion view of the device with the gas, membrane, and fluid layers. (**C**) An equivalent circuit of the chip with n branch channels.

The SU-8 molds were manufactured by standard lithography as described in the literature [41,45–48]. Briefly, after 30-min dehydration at 150 °C, two wafers were spin-coated with SU-8 2025 (MicroChem, Newton, MA, USA) at 1670 rpm to form 50 μm thick films. Following 3-min and 6-min baking at 65 and 95 °C, the wafers were exposed to UV light (URE-2000, Chinese Academy of Sciences, Beijing, China) through the photomask with the pattern of either gas channels or flow channels with a dose of 170 mJ/cm^2. After the post-exposure baking at 65 and 95 °C for 1 and 6 min, the wafer with the flow-channel layer was spin-coated with SU-8 2025 at 1080 rpm to form a 75 μm thick film. After 3 min of 65 °C pre-bake and 9 min of 95 °C soft bake, the second photoresist layer was exposed to UV light with the dose of 210 mJ/cm^2 through the mask with herringbone mixing structures and baked at 65 and 95 °C for 2 and 7 min, respectively. Finally, all wafers with photoresist films were developed in the SU-8 developer (MicroChem, Round Rock, TX, USA) to produce the molds of the gas and fluid layers.

Sylgard® 184 elastomer base and curing agent (Dow Corning® Corporation, Midland, TX, USA) were mixed with the weight ratio of 10:1 and completely degassed before being cast onto the SU-8 molds and baked at 70 °C for 2 h. Then, the Sylgard® mixture with the curing ratio of 25:1 (base: curing agent) was spin-coated on the silicon wafer at 3750 rpm to form the 20 μm thick membrane before curing [45]. Afterward, the Polydimethylsiloxane (PDMS) replicas of the gas and fluid layers were peeled off the wafer. The inlet and outlet ports were manufactured using a puncher with an inner diameter of 0.75 mm. Three layers were then irreversibly bonded together after 30-second oxidation using oxygen plasma (PDC-002, Harrick Plasma, Ithaca, NY, USA). Finally, the inlet and outlet ports were connected to the tubing through 21-gauge, flat-end syringe needles. Before each experiment, the gas channel was filled with DI water to avoid activation pressure loss. To ascertain precise dimensions, the cross-sections of both membrane and microchannels were imaged using an inverted microscope (IX83 Olympus, Tokyo, Japan) equipped with a CCD camera (C11440-36U, Hamamatsu, Hamamatsu, Japan).

2.2. Concentration Prediction

A MATLAB-based customer program (Program 1) was developed to calculate the flow resistance in valve regions required for various concentration combinations in n (the branch number, $n \geq 2$) parallel branch channels. The flow resistance was employed to guide the design of valve size. Figure 1C shows an equivalent circuit [26] of the chip. $L_{k,1}$ to $L_{k,3}$ and $R_{k,1}$ to $R_{k,3}$ represent the length as well as the flow resistance of the channel in front of the mixing region, the serpentine mixing channels before the microvalve and the channel between the kth microvalve and the $(k + 1)$th inlet, respectively. The flow resistance $R_{k,6}$ of the branch channel consisted of the flow resistance of the region between bifurcation and microvalve ($R_{k,4}$), the microvalve ($R_{k,v}$) and the observation regions ($R_{k,5}$).

To control the chip size, the dimensions of other components were fixed and listed in Table S1. Each observation region had an identical dimension (1 mm width and 10 mm length) and their flow resistance was referred to as $R_{k,5}$. The $L_{k,4}$ was determined as the minimum length required for the chemical concentration profiles, including $\sqrt{10}$ times decrease and linear decline with the slope of -0.2, in the four channels (from top to bottom). To generate the corresponding concentration profiles, the flow rates directed to the next mixing region ($Q_{k,3}$) were controlled by the required $R_{k,v}$, which were obtained by adjusting pneumatic valves. All detailed parameters in this design were listed in Table S1 in the Supporting Information (SI).

The flow resistance of the valve regions was calculated based on the desired concentration ratio. The concentration ratio between the kth observation channel and the chemical inlet was represented by C_k. As the flow rates in all dilution inlets were set to be a constant (Q_{in}), the concentration after mixing (C_k) depends on the split flow ($Q_{k,3}$), which was calculated by Equation (1):

$$Q_{k,3} = \frac{Q_{in} \cdot C_{k+1}}{C_k - C_{k+1}} \tag{1}$$

The flow rate ($Q_{k,2}$) in the serpentine mixing region before the microvalve was the sum of a corresponding inlet flow (Q_{in}) and the flow ($Q_{k-1,3}$) from the former shunt

(Equation (S1)). The flow rate in the observation region ($Q_{k,4}$) was the difference between $Q_{k,2}$ and $Q_{k,3}$ (the flow to the next level) (Equation (S2)). The flow resistance in all non-valve regions was calculated by Equation (S3).

The flow resistance in the branch channels ($R_{k,6}$) can be given by Equation (2):

$$R_{k,6} = \frac{R_{k,3} \cdot Q_{k,3} + R_{k+1,2} \cdot Q_{k+1,2} + R_{k+1,6} \cdot R_{k+1,6}}{Q_{k,6}} \quad (2)$$

Finally, the flow resistance in the microvalve region ($R_{k,v}$) can be calculated by Equation (3):

$$R_{k,v} = R_{k,6} - R_{k,4} - R_{k,5} \quad (3)$$

The $R_{k,v}$ can provide guidance to microvalve dimension design and pressure manipulation both before and after chip manufacturing.

2.3. Simulation of The Valve Deformation and Flow Resistance

The deflection of the PDMS membrane was simulated to study its influence on the flow resistance across the valve using COMSOL Multiphysics 5.5 (COMSOL Inc.,Stockholm, Sweden). Membrane deflection was firstly simulated using a three-dimensional (3D) solid mechanics model (no fluid flow) consisting of a PDMS membrane with thicknesses (t) of 20 and 30 μm on a constant fluid channel (50 μm in both width and height) (see Figure S2A). The material details of the model can be found in Table S2. Briefly, the four vertical faces of the membrane were set to be fixed boundaries. A uniform pressure of 400 to 2000 mbar was applied on the top surface of the membrane. As reported in our previous paper [45], the contact boundaries between the vertical walls and the membrane were elastic support.

Based on the deflection results above, a second simulation was conducted to predict the flow resistance of the channel under the bending membrane. In the 3D model, the cross-section of semi-closed microchannels was built from the PDMS membrane deflection simulation (see Figure S2B), when the valve widths (w) varied from 100 to 300 μm. The 300 μm long channels (50 μm in both height and width) were set in front and behind the semi-closed microchannels. After applying a developed flow with a maximum velocity of 1.2 μm/s at the inlet of the whole channel, the flow rates and pressure drops in different semi-closed microvalve regions were obtained from the simulation. The relevant flow resistance (R_h) was calculated by Equation (4):

$$R_h = k\frac{\Delta P}{Q} \quad (4)$$

where ΔP is the pressure drop between the inlet and outlet of the microvalve region and Q is the flow rate in the microchannel. k (see Equation (S6)) is the correction parameter to correct the error between simulation and experimental results.

2.4. Flow Resistance Measurement

A microfluidic chip with one straight flow channel and three pneumatic valves was employed to study the relationship between activation pressure and flow resistance. As shown in Figure S3, the bottom layer of the chip contained a straight fluid channel (50 μm width and 50 μm height) with two branches connecting a differential pressure sensor (DPS, Honeywell, Morris Plains, NJ, USA). The top layer consisted of three gas channels with the width of either 100, 200 or 300 μm and the height of 50 μm. The membrane between the gas layer and the flow layer was 20 μm thick. After device fabrication, the tubing connecting two ports of a pressure sensor was filled up with insulation oil (Yinglida, Shenzhen, China) to insulate the inner components and transfer the hydraulic pressure in microchannels [46]. The sensor was mounted on a sensor evaluation kit (Honeywell, Morris Plains, NJ, USA) with a microcontroller board (Arduino Uno Rev3, Somerville, MA, USA) to dynamically read the hydraulic pressure of the channel.

The DI water was injected into the flow channel at 1 μL/min (Q) using a syringe pump (neMESYS 290N, CETONI, Korbussen, Germany). One of the three pneumatic valves was activated by the pressure ranging from 0 to 2000 mbar with an increment of 400 mbar using the pressure pump (Elveflow OB1 MK3+, Paris, France). The resulting pressure drop was recorded by the DPS when the membrane valve deflected at various pressure. The pressure drops (ΔP) were finally used for flow resistance calculation. The flow resistance of the valve activated at different pressure could be simply calculated by Equation (5):

$$R_h = \frac{\Delta P}{Q} \tag{5}$$

2.5. Characterization of the Flow in Microvalves

The flow field under the valve region was characterized by the Micro-Particle Imaging Velocimetry system (Micro-PIV). Briefly, the 200 nm-diameter fluorescent particles (Huizhi Technology, Shanghai, China) were diluted in DI water with 0.1% Triton X-100. Under a flow rate of 1.3 μL/min, the pneumatic valve was controlled by activation pressures ranging from 0 to 2000 mbar with an increment of 400 mbar. A microscope (IX73, Olympus, Tokyo, Japan) integrating the double-pulse Nd-YAG laser (Vlite-135, Beamtech, San Francisco, CA, USA) was used for imaging. The excitation interval between the two lasers was set to 20 ms and the nanoparticles were recorded by a CCD camera (630091 PowerView 4MP-HS camera, TSI, Shoreview, MN, USA). The non-activated valves were imaged on the midplane, whereas the act-activated valves were imaged on the plane under the deformed membrane. Finally, the local fluid velocity was calculated by analyzing the displacement of the fluorescent particles.

2.6. Mixing Characterization

Standard herringbone microstructures were employed to enhance the mixing of fluids from different inlets. The mixing capability was visualized by the intensity of the fluorescein solution. The DI water and Fluorescein sodium (Energy Chemical Technology Co., Ltd. Shanghai, China) diluted in DI water (0.05%) were injected into the chemical inlet at 2.16 μL/min using syringe pumps. Both the mixing and the observation regions were imaged using a CCD camera with an exposure time of 10 ms. Finally, the fluorescence intensities in the mixing region on the micro photos were analyzed using ImageJ [49].

2.7. Generation of Dynamic Concentration and Data Analysis

Dynamically varied concentration was generated in four parallel channels by changing the pressure activating four pneumatic microvalves. Briefly, the fluorescein solution (0.05%) was injected into the chemical inlet at 0.46 μL/min or 4 μL/min for the nonlinear or linear profiles. The DI water was injected at 1 μL/min into the rest four inlets. The predefined pressures (see Table S3) acquired using the codes developed in Section 3.1 were applied to four microvalves to generate either linear decline or nonlinear increase concentration profiles. The fluorescence images in mixing regions were obtained via the CCD camera when the fluid flow became stable after the adjustment of the valve deflection. Finally, ImageJ was used to assess pixel intensity in the four mixing regions. Unless stated otherwise, the error bars in all figures represent the standard deviation (SD) of three replicates (N = 3).

3. Results and Discussion

3.1. Flow Resistance Calculation for Dynamic Concentration Profile

A Matlab program (Program 1) was developed based on the equivalent circuit principle to calculate flow resistance in valve regions for distinct concentrations in the parallel branch channels. Since a four-branch chip was applied to the proof-of-concept experiments, the dimensions of four channels between bifurcation and microvalve ($L_{k,4}$, k = 1 to 4) were determined by the program to achieve the minimum flow resistance required for the concentration profiles (Figure 2A). Basically, $L_{k,4}$ were determined as the minimum length required for both linear ($C_k = -0.2k + 1$) and nonlinear ($C_k = \left(\frac{1}{\sqrt{10}}\right)^k$) chemical concentration profiles (Figure 2B). The corresponding flow resistance of these four valves is shown in Figure 2C and Table S4.

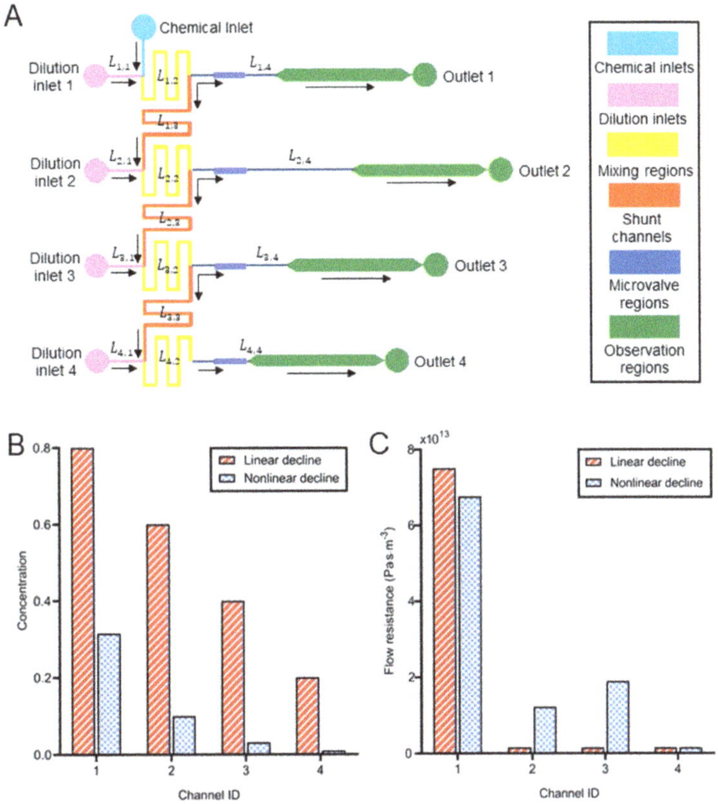

Figure 2. The calculation of the flow resistance for the dynamic concentration generation. (**A**) The schematic of the microchannel network in the flow layer. (**B**) The designed concentration in four parallel channels if the concentration at the chemical inlet was 1 (n = 4). The concentration in each channel was normalized to that in the first channel. (**C**) The flow resistance in the valve regions to acquire the concentration ratios in Figure (**B**).

The flow resistance in the first channel was significantly higher than the others. For example, to acquire a linearly declining concentration, the flow resistance in the first channel was 75.1×10^{12} Pa·s/m^3, which was 48.14 to 48.77 times that in the rest three channels. Otherwise, most liquid might be directed to the first observation channel owing to the flow resistance distribution of the whole chip. Similarly, for the nonlinear concentration decline, the flow resistance in the first channels was 3.58 to 43.68 times that in the other channels. Moreover, the flow resistance in the second and third channels was 7.87 and 12.19 times bigger than in the last channel.

A similar calculation method and chip design had been applied for both linear, logarithmic or even six orders of magnitude concentration profiles generation in different devices [12,50]. Nevertheless, each device can only provide a constant profile. Active components such as pneumatic valves were introduced for dynamic control of 'step-down', 'step-up' or 'gradient flip' but the profiles were still constant [51]. The principle was the combination of two switchable pyramidal networks thus the changes were not arbitrary. To the best of our knowledge, this paper is the first demonstration of dynamic adjustment of the arbitrary concentration profile by active components in one serial network microfluidic platform.

3.2. Simulation of The Valve Deformation and Flow Resistance

The key components of the concentration profile generator are the pneumatic microvalves, since they are employed to dynamically and independently tune the local flow resistance of each branch channel. Therefore, it is critical to study the deformation of the valve and the resulting flow resistance. The deflected membrane at 800 mbar is shown in Figure 3A. The membrane deflection increased as the augment of the activation pressure when t and w were 20 μm and 300 μm, respectively (Figure 3B). Although the membrane profile was parabolic, the maximum deflection had a linear relationship with the applied pressures as reported in our previous study [45]. The membrane deformed maximally for 5.8 μm (29.1 μm) and covered 9.2% (45.9%) of the channel cross-section when the activation pressure was 400 mbar (2000 mbar). Importantly, higher pressure formed the sharper corner between the vertical wall and the membrane (Figure 3A,B), which might greatly impact the flow resistance.

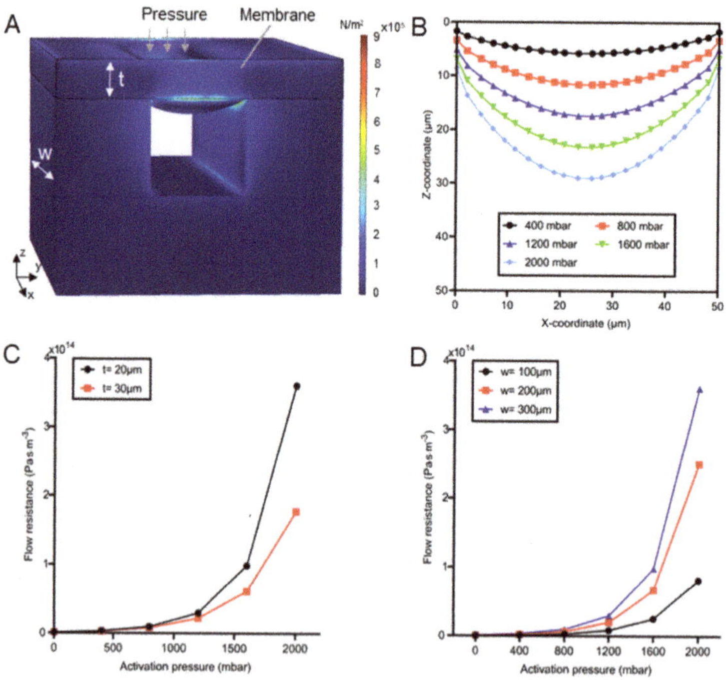

Figure 3. Study of the deformation and resulting flow resistance of pneumatic microvalves. (**A**) A 3D model of a microvalve. The width and the thickness of the membrane are represented by w and t, respectively. Color legend indicates the surface stress. (**B**) The cross-sectional profile of the membrane deflected at various pressure when w = 300 μm and t = 20 μm. (**C**) The flow resistance as a function of activation pressure for a valve with different membrane thicknesses and fixed width (300 μm). (**D**) The flow resistance as a function of activation pressure at different valve widths and a given membrane thickness (20 μm).

The flow resistance was inversely related to the membrane thickness and proportional to the activation pressure (see Figure 3C). The thicker membrane resulted in smaller flow resistance at the same pressure. For instance, the resistance in the valve with a 20 μm thick membrane was 2.04 times that with a 30 μm-thick membrane when the pressure was 2000 mbar. That is because different deflections of the membrane with distinct thickness resulted in the flow resistance difference [45]. Whereas for the given membrane, the larger pressure might induce greater membrane deformation and significant raise in the flow resistance [52]. Figure 3C indicates that when the pressure increased from 400 to 2000 mbar, the resistance of the 20 μm-thick membrane dramatically rose by 109.4 times. This might be related to the difference in the cross-sectional shape of the channel. The channel cross-section with sharp edges (such as the triangle) normally has a higher flow resistance than that with a smooth boundary (such as the circle). Since the higher activation pressure resulted in a sharper corner between the membrane and channel walls, the flow resistance increased rapidly when larger pressure was applied.

Meanwhile, the flow resistance was also proportional to valve width (Figure 3D). Regardless of the valve widths, large pressure caused significant raise in flow resistance. When the pressure increased from 400 to 2000 mbar, the flow resistance increased from 1.06×10^{12} (3.27×10^{12}) to 8.1×10^{13} (3.61×10^{14}) $Pa{\cdot}s/m^3$ when the width was 100 μm (300 μm). When the width tripled from 100 to 300 μm, the resistance of the membrane under 400 mbar was raised by 2.08 times, whereas it dramatically increased 3.45 times when the pressure was 2000 mbar. Nevertheless, the flow resistance was not raised linearly with valve width. The phenomenon could be explained by the Equation (6) of flow resistance (R_h) for a rectangular channel in microfluidics [52]:

$$R_h = \frac{12\mu L}{(1 - 0.63h/w) \cdot wh^3} \tag{6}$$

where μ is the dynamic viscosity of the fluid, L, w and h are the length, width and height of the rectangular microchannel, respectively.

The valve width (w) in this paper was defined as the width of the gas channel, corresponding to the length of the liquid channel, which is L in Equation (6). Similarly, w in this equation corresponds to the channel width in the valve region (w_v, see Table S1). Since the flow resistance is inversely proportional to the fourth power of height, the effect of height reduction due to membrane deflection was more pronounced than the growth of valve width. Simulation results indicated the deflection of a 20 μm-thick membrane under a 2000 mbar pressure initially rose with the increase in the microvalve width (Figure S4A) and then plateaued at approximately 150 μm (Figure S4B). As a result, the flow resistance may increase following the activation pressure with a rising slope when the valve width was smaller than 150 μm and be proportional to the valve width when it is bigger than 150 μm.

The influence of pressure and membrane dimension (width and thickness) on the flow resistance was important for device design. It can be used to estimate the valve size required to generate predefined concentration ratios (see Equation (S3)). Previous studies calculated the flow resistance of the channel with the regular cross-section including circle, ellipse, and triangle [52]. However, the flow resistance of the pneumatic valve cannot be solved analytically, since the deflected membrane created a parabolic cross-section for fluid flow. To facilitate the control of flow resistance, valves enabling wider resistance ranges were preferred. Therefore, the valve with a 20 μm-thick membrane was chosen for the concentration profile generation in the following experiments.

3.3. Flow Resistance Measurement

The flow resistance in the microvalve region increased nonlinearly as a function of the activation pressure value. A custom system including a pressure sensor and a microfluidic chip (Figure S3) was developed to measure the flow resistance of the microvalve activated at various pressures. Figure 4A shows the deformations (top view) of the membranes in the 200 μm-wide microvalves activated at either 400 or 1200 mbar. As indicated by the shadow, the deflection under 1200 mbar was much larger than that under 400 mbar.

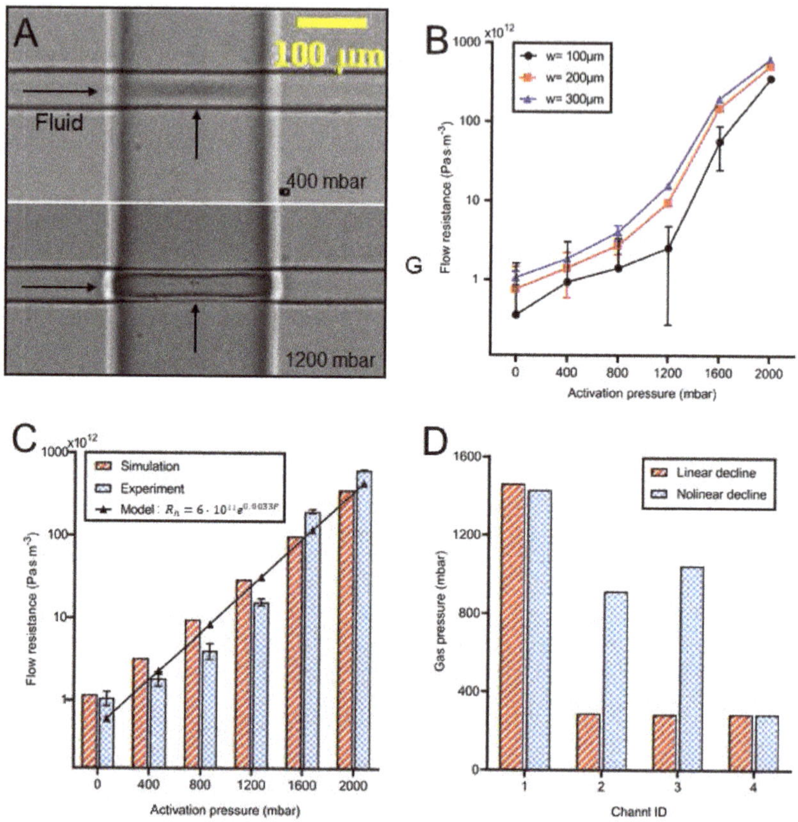

Figure 4. Measuring flow resistance in microvalves at various pressures. (**A**) The images of the deflecting membrane (20 μm thick) in a 200 μm long valve. (**B**) The flow resistance in the valve regions at various pressures. (**C**) The flow resistance (of a 300 μm valve) acquired either from experiment or simulation. The black line was a model describing the relationship between activation pressure and resistance (goodness: 0.86). (**D**) Activation pressures required to generate linear and nonlinear declining concentration profiles in four parallel channels (see Section 3.1).

Similar to the findings in Section 3.2 (Figure 3D), the flow resistance was proportional to both the activation pressure and the valve width (Figure 4B). Under the same pressure, the wider valve induced higher resistance. For the 100 μm valve, the flow resistance increased 386 times from 9.3×10^{11} to 3.6×10^{14} $\text{Pa}{\cdot}\text{s/m}^3$, as the activation pressure rose from 400 to 2000 mbar. However, for the 300 μm valve, the flow resistance enhanced 344.4 times from 1.83×10^{12} to 6.32×10^{14} $\text{Pa}{\cdot}\text{s/m}^3$, when the activation pressure varied in the same range. This suggested shorter valves were preferable if a significantly higher adjustable range of flow resistance was of interest. However, the standard deviation for shorter valves (100 μm) was much larger than that for long valves, indicating that higher accuracy of flow resistance was obtained for the long valve. Therefore, the 300 μm valves were used in the following experiments to generate a desirable dynamic concentration profile.

The flow resistance in a 300 μm valve acquired by either simulation or experiment was further compared in Figure 4C. Resistance from both methods had a difference of less than 57% for low activation pressure, whereas large errors (102% and 75%) occurred when 1600 and 2000 mbar were applied. This error may be related to the expansion of the gas channel, which has not been considered in the simulation. Since the walls of the gas channel were also formed by elastic PDMS, they may have a deformation under the pressure. Similar to the big wall collapse (8.6 μm for 2000 mbar) on the liquid channel (Figure 3A), there was a horizontal collapse on the gas channel, which was already observable under 1200 mbar pressure (Figure 4A). The pneumatic expansion of the gas channel under high pressure (more than 1200 mbar) prolonged the valve width, which significantly improved the experimental value of the flow resistance. An empirical model (R^2 = 0.86) was fitted to describe the relationship (Equation (7)) between the flow resistance (R_h) and the activation pressure (P):

$$R_h = ae^{bP} \tag{7}$$

where e is the Euler number, a (6×10^{11}) and b (0.0033) are constants.

To the best of our knowledge, the flow resistance performance in the deforming microvalve with a non-regular cross-section has rarely been studied previously. Measurement of the flow resistance is essential for accurate generation of the concentration, owing to the error induced by the simplification in the simulation models and difficulty in the analytical calculation of the flow resistance in channels with irregular cross-section. The bypass manometry technique [53] was improved with a pressure sensor to accurately measure the pressure drop in this study. The measurement technique and the empirical model of the flow resistance and activation pressure may contribute to the microfluidic field.

Based on the model above, the required gas pressure for the desired concentration profile could be estimated at any predefined flow resistance. Take a four-branch channel chip as an example, the gas pressure on each microvalve for linear ($C_k = -0.2k + 1$) and non-linear ($C_k = \left(\frac{1}{\sqrt{10}}\right)^k$) concentration profiles were calculated by Program 2 (see Figure 4D). Thus, the device integrated with active pneumatic microvalves can be altered to acquire the varied concentration profiles only in one chip which can achieve a spatial concentration dilution and minimize the manufactural error.

3.4. Characterization of The Flow in Microvalves

Micro-PIV was employed to measure the velocity under the 300 μm-width valve on the device mentioned in Section 3.3 and analyze the influence of the deformed membrane on flow resistance. When no pressure was applied to the valve, the velocity was uniform along the whole channel (see Figure 5A top). Nevertheless, there was a high-speed region under the membrane as the activation pressure on the valve increased from 400 to 1600 mbar (see Figure 5A), which was caused by the parabolic profile of the cross-sectional area (see Figure 3A). In addition, high-velocity regions expanded at higher activation pressure values, indicating a flatter membrane surface under higher pressure.

The velocity along the channel center (velocity distribution along the dotted line in Figure 5A) is shown in Figure 5B. Even though there was a low-speed layer on channel sides caused by the wall effect, flow velocity was uniform along the channel center line when no pressure was applied (see Figure 5B). However, under the activation pressure, there was a peak of speed in the valve region. That is because the flow rate was constant thus the local speed in the valve region, with a smaller cross-section, may be increased. In addition, the midplane (the velocity plane of no-pressure condition) had the highest speed and thus the no-pressured channel showed a higher speed in the no-valve region than that under an activation pressure. As the activation pressure increased from 400 to 1600 mbar, the peak velocity value rose from 10.96 to 23.36 mm/s.

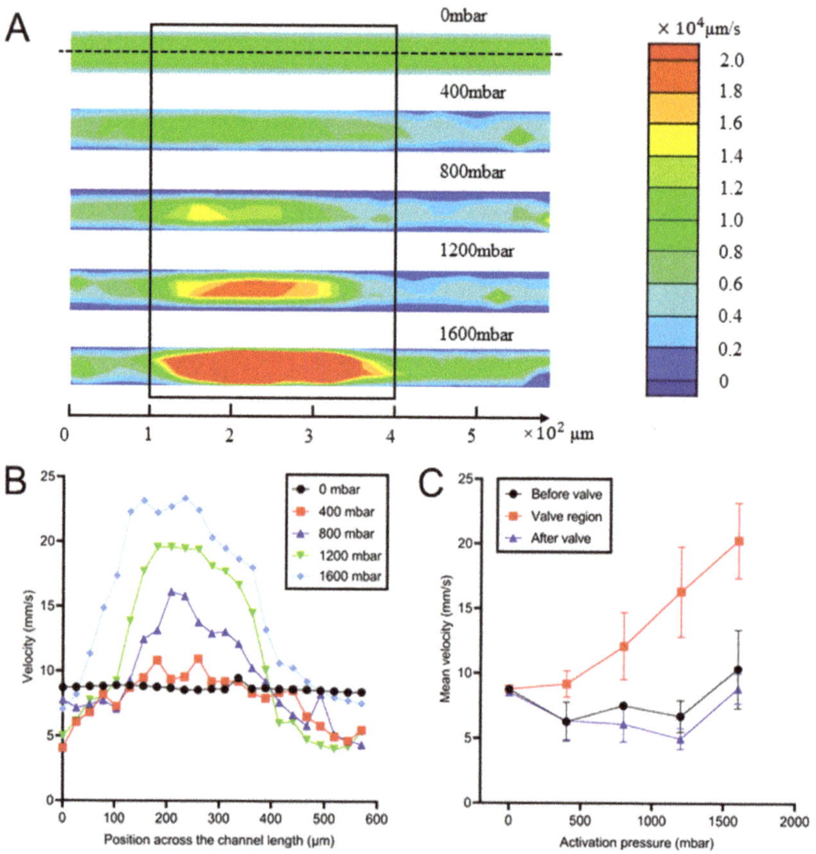

Figure 5. Characterization of flow in the valve zone. (**A**) The flow velocity contour in the valve region when the activation pressure was increased from 0 mbar to 1600 mbar. The black box represents the position of the valve. (**B**) Velocity distribution along the channel center under distinct activation pressures. (**C**) The mean velocity before, in and after the valve area at different activation pressures.

The comparison of the flow velocity between the valve region and other regions of the same channel is shown in Figure 5C. The velocity in the valve region was proportional to the activation pressure whereas there was an inverse relationship between the two factors for the channels outside of the valve when activation pressure was smaller than 1600 mbar. The mean velocity in the valve dramatically rose from 9.2 to 22.0 mm/s (Figure 5C) as the activation pressure increased from 400 to 1600 mbar. By contrast, the velocities outside of the valve were even smaller than the midplane velocity when the valves were not activated. As the 1600 mbar pressure was applied, the membrane was deformed to about 23 μm (see Figure 3B), which is around the midplane of the channel. As a consequence, the velocities outside the valve in this plane were similar to that at the midplane in no-pressure conditions.

The experimental studies of flow fields when activation pressure was applied further explained the mechanism of flow resistance control. On the one hand, the deformed pneumatic valve may enhance the local speed as the flow rate was constant. On the other hand, the deformed valve was demonstrated to be less influential for the flow velocity profile out of the valve region. As a consequence, the activation pressure can only adjust the cross-section in the valve region and affect the local speed, without changing the flow rate in other regions. Since the flow variation outside the valve region was independent of activation pressure, it is indicated the addition of the valves may not affect fluid flow in this concentration profile generation device.

3.5. Mixing Characterization

To increase the throughput of mixed liquid, a higher inlet flow rate was required. However, at a high flow rate, the fluorescein was hardly mixed in the observation region, since the pure diffusion-based mixing required much longer channels. Therefore, herringbone structures were used to enhance mixing at a high inlet flow rate. In Figure 6A, the clear dark region in the vicinity of the fluorescence suggested uneven mixing between the dye and water at 15 mm away from the water-fluorescence junction when herringbone structures were absent. In contrast, the dye in the channel with herringbone structures was mixed only 5 mm away from the junction (Figure 6B), suggesting the efficiency of herringbone structures in mixing both liquids. The mixing characterization of the herringbone structures was analyzed quantitatively by measuring fluorescence intensity in the microchannel. In particular, the mixing efficiency (insets in Figure 6C,D) was calculated by dividing the mean intensity of the region from position 30 to 60 μm by 0.5. Figure 6C,D showed the fluorescence intensity on the lines (across the channel) 0, 5, 10 and 15 mm away from the merging point of the DI water and fluorescein in devices either with or without herringbone structures. In devices without herringbones (Figure 6C) there was only a 28.9% mixing after flowing for 15 mm. However, in devices with herringbone structures, the mixing efficiency reached 51.6% after 5 mm flowing and achieved 99.0% at the position 15 mm away from the mixing junction (see Figure 6D). The results indicated that the herringbone structures well enhanced the mixing in microchannel.

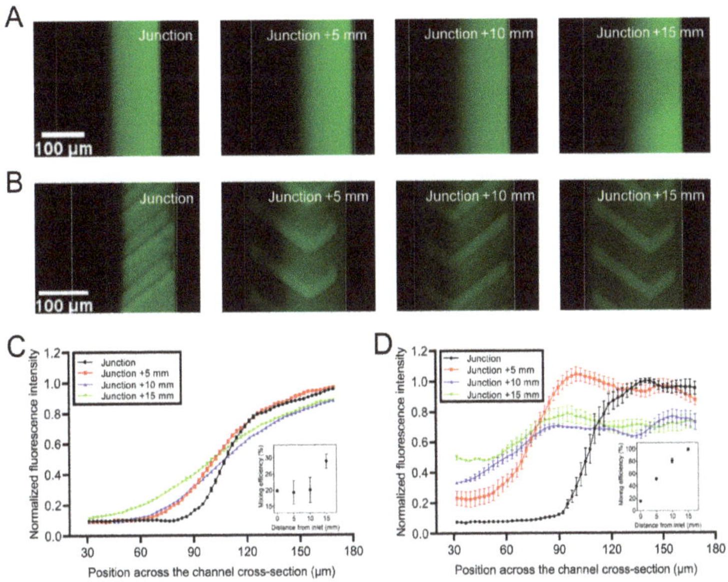

Figure 6. The mixing efficiency study. The images of the channels at different locations after dye and water were injected into devices (**A**) without and (**B**) with herringbone microstructures. The normalized fluorescence intensity based on the highest value of the fluorescent across the channels for the mixing characterization after flowing various distances in devices (**C**) without and (**D**) with herringbone microstructures. (Insets) Mixing efficiency after flowing varied distances in devices (**C**) without and (**D**) with herringbone microstructures. Error bars represent the standard error of the mean of five replicates.

The performance of liquid mixing depends on both diffusion and convection. Most microfluidic devices generating concentration profiles employed pure diffusion across laminar flows. Therefore, the devices required long channels [54] and a low flow rate [36] to obtain steady concentration profiles. Plenty of technologies such as magnetic [55] or acoustic [56] have been exploited to enhance mixing efficiency [57,58]. Compared with complex active components, passive mixing can simplify both the device structure and the operation process. For instance, cylinder microstructures array [59] has been demonstrated to effectively enhance the mixing, utilizing just a 3 cm-length channel. Compared with the method of controlling fluid transformation sequences by cylinder microstructures, mixing with herringbone-inspired microstructures was able to overcome the diffusion limit in co-laminar microfluidic devices [60]. Previous evidence suggested that the scaling regime transited from a purely laminar regime to an entrance region turbulent regime with increasing Reynolds numbers [60]. Therefore, the herringbone microstructures overcome the limitation of pure diffusion and reduce the channel length for mixing [61]. As proved in Figure 6B, the herringbone microstructures induced the effective mixing in a 15 mm-length channel, which not only saved space but increased response speed for dynamic concentration variation.

3.6. Generation of Dynamic Concentration

The device's capabilities in generating dynamically varying concentrations were validated by creating both linear ($C_k = -0.2k + 1$) and nonlinear decline ($C_k = \left(\frac{1}{\sqrt{10}}\right)^k$) concentration profiles. Figure 7A and B showed the fluorescent images of four observation zones 2 min after the activation of four valves (see Table S3 for the pressure combinations). The uniform fluorescence intensity demonstrated thorough and quick mixing since the images were taken only 2 min after valve activation. When the valve states combinations were swapped, the normalized concentration in four channels quickly varied between linear and nonlinear decline as shown in Figure 7C,D. Moreover, there was less than 26.6% difference in the normalized concentration acquired between the experiment and the desired value for both the nonlinear and the linear cases, except no signal was detected in Channel 4 for nonlinear profile, which may be because of the excitation limitation of the dye. The errors might be related to manufacturing accuracy or flow resistance calculation, as some structures such as herringbone parts were ignored in the calculation.

However, these results suggested the ability of the proposed microfluidic networks in generating multiple concentration profiles in different channels. Limited by the serial dilution principle, it is difficult to generate the peak [11] or step-up [51] profiles such as those generated in the pyramidal network. Even though it is possible to acquire such profiles by switching some downstream dilution inlets to chemical inlets, the step-down profile demonstrated in this paper is enough for arbitrary concentration-dependent experiments. Previous studies generated temporal or spatial concentration profiles based on the serial dilution network [12]. However, long channels [37] or channels with low depth [50] were applied to provide a high-flow resistance network, which increased either the occupancy area or fabrication complexity. The pneumatic valves in this paper were able to generate a large range of flow resistance (variation in 594 times) within a small space (Section 3.3), which highly decreased the occupancy area. On the other hand, previous devices lack flexibility in the concentration variation, which is critical for comparative studies such as cell response experiments. Multiple components have been introduced in microfluidics for the fluid flow variation [62–65]. For concentration profile adjustment purposes, the active components include electronic ion pump [66], finger actuation button [67] and pinch [68] or pneumatic valve [69]. Among them, the pneumatic valve is ideal for dynamic control in microfluidics due to the excellence of accuracy, response time and integration [40,45,48]. In this paper, the addition of pneumatic enables the dynamic adjustment of flow resistance. As the result, the device was able to generate programmable arbitrary concentration profiles.

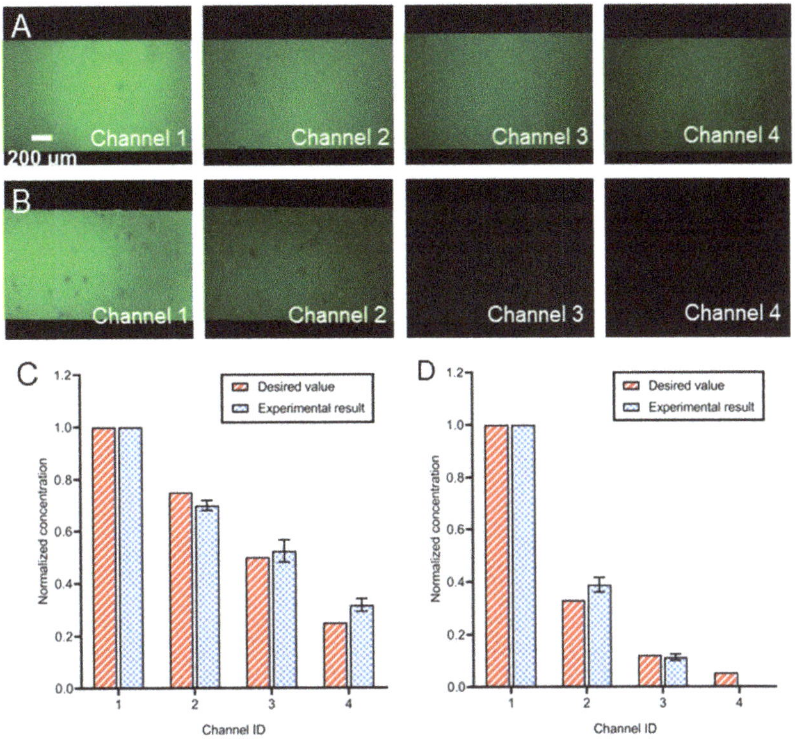

Figure 7. Generation of dynamic concentration using the microfluidic device. (**A**) Fluorescence images of the observation region with (**A**) linear and (**B**) nonlinear decline of concentration. The (**C**) linear or (**D**) nonlinear decline profile of the normalized concentration in four parallel channels. At least three different microchips were prepared and examined for each experimental condition.

4. Conclusions

This paper reports a microfluidic platform that can dynamically generate the concentration profile in four channels by adjusting the pneumatic microvalves. A program was developed to provide the theoretical guidance for the design in dimension determination. Simulation results showed the deformation profile with varied activation pressures and valve dimensions. Afterwards, flow resistance under different conditions was studied by both numerical and experimental methods. An exponential rising tendency of flow resistance increasing by activation pressure was found. Moreover, local flow characterization in the valve region further demonstrated the flow resistance mechanism. Supported by the empirical formula from the experimental flow resistance study above, a computational program for predicting the activation pressure adjustment was developed to guide the concentration profile generator manipulation. To enhance the mixing efficiency, the herringbone structures were introduced and proved an efficient mixing at a 15 mm distance. Finally, a microfluidic device was demonstrated by generating both linear ($C_k = -0.2k + 1$) and nonlinear ($C_k = \left(\frac{1}{\sqrt{10}}\right)^k$) concentration profiles in the same chip with less than 26.6% error. In comparison to previous devices, the pneumatic valves are simple, reliable and easy to integrate and operate. The platform in this paper will be of great application potential in analytical chemistry, drug screening and cell biology research requiring concentration profile control.

Supplementary Materials: The following supporting information can be downloaded at: https://www.mdpi.com/article/10.3390/bios12100868/s1, Figure S1: the photo of the experimental platform; Figure S2: the 3D models in the numerical study; Figure S3: the schematic of the microfluidic chip for flow resistance measurement; Figure S4: simulation results of the deflection of membranes with distinct valve widths; Table S1: details for four-branch design; Table S2: the parameters in simulation models; Table S3: the gas pressures required for different concentration profiles; Table S4: the flow resistance of microvalves for different concentration profiles; Program_1_Mixerdesign.m; Program_2_Pressurecalculator.m.

Author Contributions: Conceptualization, C.C. and H.C.; methodology, C.C., P.L., S.C. and D.X.; software, C.C.; validation, C.C. and D.X.; formal analysis, C.C., P.L. and H.C.; investigation, C.C., P.L. and H.C.; resources, H.C.; data curation, C.C., P.L. and S.C.; writing—original draft preparation, C.C., P.L., T.G. and H.C.; writing—review and editing, C.C., T.G. and H.C.; visualization, C.C.; supervision, H.C.; project administration, H.C.; funding acquisition, H.C. All authors have read and agreed to the published version of the manuscript.

Funding: This research was funded by the National Natural Science Foundation of China (Grant No. 32171400), Natural Science Foundation of Guangdong Province (Grant No. 2020A1515010706), and the Shenzhen Science and Technology Innovation Committee (Grant No. GXWD20201230155427003-20200823111304001 and JCYJ20180306172109024).

Institutional Review Board Statement: Not applicable.

Informed Consent Statement: Not applicable.

Data Availability Statement: Not applicable.

Acknowledgments: This study was conducted at the Center for Microflows and Nanoflows, Harbin Institute of Technology, Shenzhen.

Conflicts of Interest: The authors declare no conflict of interest.

References

1. Lucchetta, E.; Lee, J.; Fu, L.; Patel, N.H.; Ismagilov, R.F. Dynamics of Drosophila embryonic patterning network perturbed in space and time using microfluidics. *Nature* **2005**, *434*, 1134–1138. [CrossRef]
2. Wang, X.; Christov, I.C. Theory of the flow-induced deformation of shallow compliant microchannels with thick walls. *Proc. Math. Phys. Eng. Sci.* **2019**, *475*, 20190513. [CrossRef] [PubMed]
3. Chen, H.; Yu, Z.; Bai, S.; Lu, H.; Xu, D.; Chen, C.; Liu, D.; Zhu, Y. Microfluidic models of physiological or pathological flow shear stress for cell biology, disease modeling and drug development. *TrAC Trends Anal. Chem.* **2019**, *117*, 186–199. [CrossRef]
4. Ayuso, J.M.; Virumbrales-Muñoz, M.; Lang, J.M.; Beebe, D.J. A role for microfluidic systems in precision medicine. *Nat. Commun.* **2022**, *13*, 3086. [CrossRef] [PubMed]
5. Hua, D.; Xiong, R.; Braeckmans, K.; Scheid, B.; Huang, C.; Sauvage, F.; De, S.C. Concentration Gradients in Material Sciences: Methods to Design and Biomedical Applications. *Adv. Funct. Mater.* **2021**, *31*, 2009005. [CrossRef]
6. Berthier, E.; Beebe, D.J. Gradient generation platforms: New directions for an established microfluidic technology. *Lab Chip* **2014**, *14*, 3241–3247. [CrossRef]
7. Boneschansker, L.; Yan, J.; Wong, E.; Briscoe, D.M.; Irimia, D. Microfluidic platform for the quantitative analysis of leukocyte migration signatures. *Nat. Commun.* **2014**, *5*, 4787. [CrossRef] [PubMed]
8. Meier, B.; Zielinski, A.; Weber, C.; Arcizet, D.; Youssef, S.; Franosch, T.; Rädler, J.O.; Heinrich, D. Chemotactic cell trapping in controlled alternating gradient fields. *Proc. Natl. Acad. Sci. USA* **2011**, *108*, 11417–11422. [CrossRef]
9. Berendsen, J.T.W.; Kruit, S.A.; Atak, N.; Willink, E.; Segerink, L.I. Flow-Free Microfluidic Device for Quantifying Chemotaxis in Spermatozoa. *Anal. Chem.* **2020**, *92*, 3302–3306. [CrossRef] [PubMed]
10. Xiao, Z.; Nsamela, A.; Garlan, B.; Simmchen, J. A Platform for Stop-Flow Gradient Generation to Investigate Chemotaxis. *Angew. Chem. Int. Ed. Engl.* **2022**, *61*, e202117768. [CrossRef] [PubMed]
11. Jeon, N.L.; Dertinger, S.K.; Chiu, D.T.; Choi, I.S.; Stroock, A.D.; Whitesides, G.M. Generation of Solution and Surface Gradients Using Microfluidic Systems. *Langmuir* **2000**, *16*, 8311–8316. [CrossRef]
12. Hattori, K.; Sugiura, S.; Kanamori, T. Generation of arbitrary monotonic concentration profiles by a serial dilution microfluidic network composed of microchannels with a high fluidic-resistance ratio. *Lab Chip* **2009**, *9*, 1763–1772. [CrossRef]
13. Jiang, X.; Xu, Q.; Dertinger, S.K.; Stroock, A.D.; Fu, T.M.; Whitesides, G.M. A general method for patterning gradients of biomolecules on surfaces using microfluidic networks. *Anal. Chem.* **2005**, *77*, 2338–2347. [CrossRef]
14. Mao, H.; Cremer, P.S.; Manson, M.D. A sensitive, versatile microfluidic assay for bacterial chemotaxis. *Proc. Natl. Acad. Sci. USA* **2003**, *100*, 5449–5454. [CrossRef]

15. Holden, M.A.; Kumar, S.; Castellana, E.T.; Beskok, A.; Cremer, P.S. Generating fixed concentration arrays in a microfluidic device. *Sens. Actuators B Chem.* **2003**, *92*, 199–207. [CrossRef]
16. Walker, G. Cell infection within a microfluidic device using virus gradients. *Sens. Actuators B Chem.* **2004**, *98*, 347–355. [CrossRef]
17. Keenan, T.M.; Frevert, C.W.; Wu, A.; Wong, V.; Folch, A. A new method for studying gradient-induced neutrophil desensitization based on an open microfluidic chamber. *Lab Chip* **2010**, *10*, 116–122. [CrossRef]
18. Shamloo, A.; Ma, N.; Poo, M.M.; Sohn, L.L.; Heilshorn, S.C. Endothelial cell polarization and chemotaxis in a microfluidic device. *Lab Chip* **2008**, *8*, 1292–1299. [CrossRef]
19. Diao, J.; Young, L.; Kim, S.; Fogarty, E.A.; Heilman, S.M.; Zhou, P.; Shuler, M.L.; Wu, M.; DeLisa, M.P. A three-channel microfluidic device for generating static linear gradients and its application to the quantitative analysis of bacterial chemotaxis. *Lab Chip* **2006**, *6*, 381–388. [CrossRef]
20. Xu, H.; Ferreira, M.M.; Heilshorn, S.C. Small-molecule axon-polarization studies enabled by a shear-free microfluidic gradient generator. *Lab Chip* **2014**, *14*, 2047–2056. [CrossRef]
21. Kim, M.; Kim, T. Diffusion-based and long-range concentration gradients of multiple chemicals for bacterial chemotaxis assays. *Anal. Chem.* **2010**, *82*, 9401–9409. [CrossRef] [PubMed]
22. Bhattacharjee, N.; Folch, A. Large-scale microfluidic gradient arrays reveal axon guidance behaviors in hippocampal neurons. *Microsyst. Nanoeng.* **2017**, *3*, 17003. [CrossRef] [PubMed]
23. Chandrasekaran, A.; Ellett, F.; Jorgensen, J.; Irimia, D. Temporal gradients limit the accumulation of neutrophils towards sources of chemoattractant. *Microsyst. Nanoeng.* **2017**, *3*, 16067. [CrossRef]
24. Lin, S.J.; Chao, P.H.; Cheng, H.W.; Wang, J.K.; Wang, Y.L.; Han, Y.Y.; Huang, N.T. An antibiotic concentration gradient microfluidic device integrating surface-enhanced Raman spectroscopy for multiplex antimicrobial susceptibility testing. *Lab Chip* **2022**, *22*, 1805–1814. [CrossRef]
25. Zeng, W.; Chen, P.; Li, S.; Sha, Q.; Li, P.; Zeng, X.; Feng, X.; Du, W.; Liu, B.F. Hand-powered vacuum-driven microfluidic gradient generator for high-throughput antimicrobial susceptibility testing. *Biosens. Bioelectron.* **2022**, *205*, 114100. [CrossRef]
26. Oh, K.W.; Lee, K.; Ahn, B.; Furlani, E.P. Design of pressure-driven microfluidic networks using electric circuit analogy. *Lab Chip* **2012**, *12*, 515–545. [CrossRef]
27. Yamada, M.; Hirano, T.; Yasuda, M.; Seki, M. A microfluidic flow distributor generating stepwise concentrations for high-throughput biochemical processing. *Lab Chip* **2006**, *6*, 179–184. [CrossRef]
28. Walker, G.M.; Monteiro-Riviere, N.; Rouse, J.; O'Neill, A.T. A linear dilution microfluidic device for cytotoxicity assays. *Lab Chip* **2007**, *7*, 226–232. [CrossRef]
29. Paduthol, G.; Korma, T.S.; Agrawal, A.; Paul, D. Dynamic generation of power function gradient profiles in a universal microfluidic gradient generator by controlling the inlet flow rates. *Lab Chip* **2022**, *22*, 592–604. [CrossRef]
30. Liu, X.; Jia, Y.; Han, Z.; Hou, Q.; Zhang, W.; Zheng, W.; Jiang, X. Integrating a Concentration Gradient Generator and a Single-Cell Trapper Array for High-Throughput Screening the Bioeffects of Nanomaterials. *Angew. Chem. Int. Ed. Engl.* **2021**, *60*, 12319–12322. [CrossRef]
31. Dertinger, S.K.W.; Chiu, D.T.; Jeon, N.L.; Whitesides, G.M. Generation of Gradients Having Complex Shapes Using Microfluidic Networks. *Anal. Chem.* **2001**, *73*, 1240–1246. [CrossRef]
32. Irimia, D.; Geba, D.A.; Toner, M. Universal Microfluidic Gradient Generator. *Anal. Chem.* **2006**, *78*, 3472–3477. [CrossRef] [PubMed]
33. Campbell, K.; Groisman, A. Generation of complex concentration profiles in microchannels in a logarithmically small number of steps. *Lab Chip* **2007**, *7*, 264–272. [CrossRef]
34. Amarie, D.; Glazier, J.A.; Jacobson, S.C. Compact Microfluidic Structures for Generating Spatial and Temporal Gradients. *Anal. Chem.* **2007**, *79*, 9471. [CrossRef] [PubMed]
35. Lin, F.; Saadi, W.; Rhee, S.W.; Wang, S.J.; Mittal, S.; Jeon, N.L. Generation of dynamic temporal and spatial concentration gradients using microfluidic devices. *Lab Chip* **2004**, *4*, 164–167. [CrossRef] [PubMed]
36. Lee, K.; Kim, C.; Ahn, B.; Panchapakesan, R.; Full, A.R.; Nordee, L.; Kang, J.Y.; Oh, K.W. Generalized serial dilution module for monotonic and arbitrary microfluidic gradient generators. *Lab Chip* **2009**, *9*, 709–717. [CrossRef]
37. Hosokawa, M.; Hayashi, T.; Mori, T.; Yoshino, T.; Nakasono, S.; Matsunaga, T. Microfluidic device with chemical gradient for single-cell cytotoxicity assays. *Anal. Chem.* **2011**, *83*, 3648–3654. [CrossRef]
38. Zheng, G.; Lu, L.; Yang, Y.; Wei, J.; Han, B.; Zhang, Q.; Wang, Y. Development of Microfluidic Dilution Network-Based System for Lab-on-a-Chip Microalgal Bioassays. *Anal. Chem.* **2018**, *90*, 13280–13289. [CrossRef]
39. Skoge, M.; Yue, H.; Erickstad, M.; Bae, A.; Levine, H.; Groisman, A.; Loomis, W.F.; Rappel, W.J. Cellular memory in eukaryotic chemotaxis. *Proc. Natl. Acad. Sci. USA* **2014**, *111*, 14448–14453. [CrossRef]
40. Unger, M.A.; Chou, H.P.; Thorsen, T.; Scherer, A.; Quake, S.R. Monolithic microfabricated valves and pumps by multilayer soft lithography. *Science* **2000**, *288*, 113–116. [CrossRef]
41. Chen, H.; Cornwell, J.; Zhang, H.; Lim, T.; Resurreccion, R.; Port, T.; Rosengarten, G.; Nordon, R.E. Cardiac-like flow generator for long-term imaging of endothelial cell responses to circulatory pulsatile flow at microscale. *Lab Chip* **2013**, *13*, 2999–3007. [CrossRef] [PubMed]
42. Zeng, S.; Li, B.; Qin, J.; Lin, B. Microvalve-actuated precise control of individual droplets in microfluidic devices. *Lab Chip* **2009**, *9*, 1340–1343. [CrossRef] [PubMed]

43. Jeong, H.H.; Jin, S.H.; Lee, B.J.; Kim, T.; Lee, C.S. Microfluidic static droplet array for analyzing microbial communication on a population gradient. *Lab Chip* **2015**, *15*, 889–899. [CrossRef] [PubMed]
44. Kim, J.; Kang, M.; Jensen, E.C.; Mathies, R.A. Lifting Gate Polydimethylsiloxane Microvalves and Pumps for Microfluidic Control. *Anal. Chem.* **2012**, *84*, 2067–2071. [CrossRef]
45. Chen, C.; Xu, D.; Bai, S.; Yu, Z.; Zhu, Y.; Xing, X.; Chen, H. Dynamic screening and printing of single cells using a microfluidic chip with dual microvalves. *Lab Chip* **2020**, *20*, 1227–1237. [CrossRef]
46. Chen, Z.; Zhu, Y.; Xu, D.; Alam, M.M.; Shui, L.; Chen, H. Cell elasticity measurement using a microfluidic device with real-time pressure feedback. *Lab Chip* **2020**, *20*, 2343–2353. [CrossRef]
47. Chen, H.; Sun, J.; Wolvetang, E.; Cooper-White, J. High-throughput, deterministic single cell trapping and long-term clonal cell culture in microfluidic devices. *Lab Chip* **2014**, *15*, 1072–1083. [CrossRef]
48. Chen, C.; Zhu, Y.; Ho, J.W.; Chen, H. The method to dynamically screen and print single cells using microfluidics with pneumatic microvalves. *MethodsX* **2021**, *8*, 101190. [CrossRef]
49. Abràmoff, M.D.; Magalhães, P.J.; Ram, S.J. Image processing with ImageJ. *Biophotonics Int.* **2004**, *11*, 36–42.
50. Sugiura, S.; Hattori, K.; Kanamori, T. Microfluidic Serial Dilution Cell-Based Assay for Analyzing Drug Dose Response over a Wide Concentration Range. *Anal. Chem.* **2010**, *82*, 8278–8282. [CrossRef]
51. Irimia, D.; Liu, S.Y.; Tharp, W.G.; Samadani, A.; Toner, M.; Poznansky, M.C. Microfluidic system for measuring neutrophil migratory responses to fast switches of chemical gradients. *Lab Chip* **2006**, *6*, 191–198. [CrossRef] [PubMed]
52. Bruus, H. *Theoretical Microfluidics*; Oxford University Press: New York, NY, USA, 2008.
53. Suteria, N.S.; Nekouei, M.; Vanapalli, S.A. Microfluidic bypass manometry: Highly parallelized measurement of flow resistance of complex channel geometries and trapped droplets. *Lab Chip* **2018**, *18*, 343–355. [CrossRef]
54. Kim, C.; Lee, K.; Kim, J.H.; Shin, K.S.; Lee, K.-J.; Kim, T.S.; Kang, J.Y. A serial dilution microfluidic device using a ladder network generating logarithmic or linear concentrations. *Lab Chip* **2008**, *8*, 473–479. [CrossRef] [PubMed]
55. Chen, H.; Chen, C.; Bai, S.; Gao, Y.; Metcalfe, G.; Cheng, W.; Zhu, Y. Multiplexed detection of cancer biomarkers using a microfluidic platform integrating single bead trapping and acoustic mixing techniques. *Nanoscale* **2018**, *10*, 20196–20206. [CrossRef] [PubMed]
56. Ahmed, D.; Chan, C.Y.; Lin, S.-C.S.; Muddana, H.S.; Nama, N.; Benkovic, S.J.; Jun Huang, T. Tunable, pulsatile chemical gradient generation via acoustically driven oscillating bubbles. *Lab Chip* **2013**, *13*, 328–331. [CrossRef]
57. Liu, B.; Ma, Z.; Yang, J.; Gao, G.; Liu, H. A Concentration Gradients Tunable Generator with Adjustable Position of the Acoustically Oscillating Bubbles. *Micromachines* **2020**, *11*, 827. [CrossRef] [PubMed]
58. Liang, J.; Chen, K.; Xia, Y.; Gui, J.; Wu, Z.; Cui, H.; Wu, Z.; Liu, W.; Zhao, X.; Guo, S. A localized surface acoustic wave applied spatiotemporally controllable chemical gradient generator. *Biomicrofluidics* **2020**, *14*, 024106. [CrossRef]
59. Amini, H.; Sollier, E.; Masaeli, M.; Xie, Y.; Ganapathysubramanian, B.; Stone, H.A.; di Carlo, D. Engineering fluid flow using sequenced microstructures. *Nat. Commun.* **2013**, *4*, 1826. [CrossRef]
60. Marschewski, J.; Jung, S.; Ruch, P.; Prasad, N.; Mazzotti, S.; Michel, B.; Poulikakos, D. Mixing with herringbone-inspired microstructures: Overcoming the diffusion limit in co-laminar microfluidic devices. *Lab Chip* **2015**, *15*, 1923–1933. [CrossRef]
61. Jiang, X.; Ng, J.M.K.; Stroock, A.; Dertinger, S.K.W.; Whitesides, G.M.J. A miniaturized, parallel, serially diluted immunoassay for analyzing multiple antigens. *J. Am. Chem. Soc.* **2003**, *125*, 5294–5295. [CrossRef]
62. Chen, C.; Meng, H.; Guo, T.; Deshpande, S.; Chen, H. Development of Paper Microfluidics with 3D-Printed PDMS Barriers for Flow Control. *ACS Appl. Mater. Interfaces* **2022**, *14*, 40286–40296. [CrossRef]
63. Chen, C.; Zhao, L.; Zhang, H.; Shen, X.; Zhu, Y.; Chen, H. Novel Wax Valves To Improve Distance-Based Analyte Detection in Paper Microfluidics. *Anal. Chem.* **2019**, *91*, 5169–5175. [CrossRef] [PubMed]
64. Meng, H.; Chen, C.; Zhu, Y.; Li, Z.; Ye, F.; Ho, J.W.; Chen, H. Automatic flow delay through passive wax valves for paper-based analytical devices. *Lab Chip* **2021**, *21*, 4166–4176. [CrossRef]
65. Liu, B.; Ran, B.; Chen, C.; Shi, L.; Liu, Y.; Chen, H.; Zhu, Y. A low-cost and high-performance 3D micromixer over a wide working range and its application for high-sensitivity biomarker detection. *React. Chem. Eng.* **2022**. [CrossRef]
66. Isaksson, J.; Kjäll, P.; Nilsson, D.; Robinson, N.; Berggren, M.; Richter-Dahlfors, A. Electronic control of Ca^{2+} signalling in neuronal cells using an organic electronic ion pump. *Nat. Mater.* **2007**, *6*, 673–679. [CrossRef] [PubMed]
67. Park, J.; Roh, H.; Park, J.-K. Finger-Actuated Microfluidic Concentration Gradient Generator Compatible with a Microplate. *Micromachines* **2019**, *10*, 174. [CrossRef]
68. Moorjani, S.; Nielson, R.; Chang, X.A.; Shear, J.B. Dynamic remodeling of subcellular chemical gradients using a multi-directional flow device. *Lab Chip* **2010**, *10*, 2139–2146. [CrossRef] [PubMed]
69. Lin, B.; Holmes, W.R.; Wang, C.J.; Ueno, T.; Harwell, A.; Edelstein-Keshet, L.; Inoue, T.; Levchenko, A. Synthetic spatially graded Rac activation drives cell polarization and movement. *Proc. Natl. Acad. Sci. USA* **2012**, *109*, E3668–E3677. [CrossRef] [PubMed]

biosensors

MDPI

Article

Effective Enrichment of Plasmonic Hotspots for SERS by Spinning Droplets on a Slippery Concave Dome Array

Jialin Wu [1], Jianpeng Cai [1,2], Yuan Fan [1,2], Ying Zhang [1,2], Hui Fang [1,*] and Sheng Yan [3,*]

[1] Nanophotonics Research Center, Institute of Microscale Optoelectronics, Shenzhen University, Shenzhen 518060, China; 2070496010@email.szu.edu.cn (J.W.); 1900453026@email.szu.edu.cn (J.C.); 2176285320@email.szu.edu.cn (Y.F.); 1950453026@email.szu.edu.cn (Y.Z.)
[2] College of Physics and Optoelectronics Engineering, Shenzhen University, Shenzhen 518060, China
[3] Institute for Advanced Study, Shenzhen University, Shenzhen 518060, China
* Correspondence: fhui79@szu.edu.cn (H.F.); shengyan@szu.edu.cn (S.Y.)

Abstract: Surface-enhanced Raman scattering (SERS) detection requires dense hotspots and a uniform distribution of analytes to obtain a stable signal with good repeatability. However, due to the coffee-ring effect on the hydrophilic substrate, and the difficulty of droplet manipulation on the superhydrophobic substrate, few substrates can ensure that the analytes are evenly distributed. In this work, we develop a method that can efficiently enrich plasmonic hotspots for SERS measurement on the superhydrophobic concave dome array (SCDA). The SCDA is formed by spraying hydrophobic silica nanoparticles onto a polydimethylsiloxane (PDMS) slab with a concave dome array that can physically confine the droplets and overcome the coffee-ring effect. During droplet evaporation, the SCDA is driven by a horizontal spinner, and the droplets spin on the SCDA, enabling the plasmonic nanoparticles to become closely packed to form the SERS hotspots. The limit of detection (LOD) of the dynamic-enriched SERS hotspots for crystal violet and methylene blue can reach up to 10^{-11} M. Moreover, the LOD for melamine in milk can reach 5×10^{-7} M, which is lower than the safety threshold defined by the Food and Drug Administration (FDA). Based on this SERS platform, an effective, low-cost, and simple method for SERS detection in analytical chemistry and food safety is highly expected.

Keywords: droplet manipulation; SERS; coffee-ring effect; biosensing

Citation: Wu, J.; Cai, J.; Fan, Y.; Zhang, Y.; Fang, H.; Yan, S. Effective Enrichment of Plasmonic Hotspots for SERS by Spinning Droplets on a Slippery Concave Dome Array. *Biosensors* **2022**, *12*, 270. https://doi.org/10.3390/bios12050270

Received: 30 March 2022
Accepted: 22 April 2022
Published: 24 April 2022

1. Introduction

When a droplet of suspension situated on a hydrophilic surface with a low contact angle evaporates, its suspended matter will predominately deposit at the outer boundary; this process is called the coffee-ring effect [1]. During evaporation, the droplet features not only an air–water–solid interface, forming a spherical cap, but also an internal Marangoni flow due to small temperature gradients, thus bringing the suspended matter (which is usually composed of small particles) to the outer ring [2]. A natural consequence is that the suspended particles will be highly concentrated along the edge of the original droplet, but loosely distributed at the center. The coffee-ring effect occurs in a wide variety of particles, ranging from large colloids [3,4] to nanoparticles [5]. In many applications based on sample preparation through droplet evaporation, the uneven particle distribution will affect the quality of the measurement. For example, in surface-enhanced Raman scattering (SERS) detection, strong uniformity of the Raman signal occurs [6].

One of the effective approaches to suppress the coffee-ring effect is to replace the hydrophilic surface with a superhydrophobic surface on which a droplet with a high contact angle will form [7]. With the advancement of fabrication technologies, such as photolithography [8], e-beam lithography [9], and stereolithography [10], some nature-inspired superhydrophobic micro/nanostructures have been developed for SERS detection [11,12]. However, these technologies rely on bulky and expensive facilities, which somewhat limits their accessibility. As an alternative approach, electrochemical deposition can deposit metal nanostructures onto a specific substrate to create superhydrophobic surface [13]. In addition, the SERS substrate with plasmonic nanostructures created using the electrochemical deposition method has high stability and enhances SERS signal [14]. The order of the plasmonic nanostructures is highly dependent of the precise control of the chemical reaction.

Several smart methods used to prepare the superhydrophobic surfaces are reported that can enrich the mixture of the plasmonic nanoparticles and analytes. For example, a nanotexture surface infiltrated with a perfluorinated liquid to form the slippery liquid-infused porous surface (SLIPS) can concentrate the analytes and plasmonic particles for attomole-level detection [15]. Any disturbance will cause the droplets to slip away from the surface due to the superhydrophobicity. Therefore, a superhydrophobic and magnetically functionalized surface was developed by spraying silica nanoparticles onto the surface of a magnetorheological elastomer, where several droplets can steadily stand in the magnetically-induced deformation area [16,17]. Unfortunately, the functionalized surface is unable to process abundant droplets simultaneously due to the interference of the magnets. Therefore, it is still challenging to develop a facile and efficient way to achieve multi-droplet manipulation to prepare the scalable SERS hotspots for multiplexing detection.

Here, we propose a superhydrophobic concave dome array (SCDA) for the dynamic enrichment of plasmonic nanoparticles. Each droplet with plasmonic nanoparticles can stay in each concave dome during the spontaneous evaporative process to generate the SERS hotspots. Unlike the plain superhydrophobic surface that allows droplets to slip from the substrate, the concave dome can physically confine the droplets. Therefore, multiple SERS hotspots can be prepared simultaneously on the same substrate. Similar to the commercial multi-well Petri dish, this SERS array is able to realize multi-concentration and multi-target detection.

2. Materials and Methodology

2.1. Preparation of SCDA

Figure 1a,b describes the preparation process of SCDA. First, the polydimethylsiloxane (PDMS) concave dome array was replicated from the 3D-printed convex dome array, as shown in Figure 1a. Specifically, a mold with an overall size of 50 mm $\times$ 50 mm and a 6 $\times$ 6 convex dome array on the surface was printed with a 3D printer. The convex dome was designed as a semi-sphere with a radius of 3.5 mm and a distance between domes of 8 mm. The printed mold will be treated with the release agent. The PDMS gel and curing agent were mixed evenly with the mass ratio of 10:1. After the bubbles in the mixed gel were removed under vacuum, the mixed gel was poured onto the 3D-printed convex dome array. To avoid the deformation of 3D-printed mold, the mixture gel was cured at 50 °C for 10 h. After the PDMS was fully cured, a concave dome array was formed after peeling off.

Next, a superhydrophobic surface was prepared by spraying silica nanoparticle suspension, as shown in Figure 1b. The silica nanoparticle suspension was prepared by mixing 0.8 g hydrophobic nanosilica, 0.45 g PDMS mixture, and 40 mL cyclohexane, and was then sealed to avoid cyclohexane evaporation. To avoid the agglomeration of the nanosilica, the silica nanoparticle suspension was put into the ultrasonic oscillator for 30 min (frequency is 40 kHz). Significantly, the temperature should be kept below 30 °C to avoid partial solidification of PDMS in the solution. Subsequently, the suspension was evenly mixed by magnetic stirring for 30 min (at a speed of 700 revolutions per minute (RPM)). After complete mixing, 10 mL mixed suspension was transferred into an airbrush cup, and then the suspension was evenly sprayed onto the surface of the concave PDMS substrate. During spraying, the airbrush was connected to the air compressor under a constant pressure of 4 bar, and the distance between the nozzle (diameter of 0.3 mm) and the concave dome array was kept at 15 cm. The working principle of suspension is that cyclohexane enables the concave dome array to swell, and hydrophobic silica nanoparticles adhere to the swollen surface; PDMS allows the hydrophobic silica nanoparticles to firmly adhere to the surface of concave dome array. After spraying, the concave dome array was horizontally placed on a hotplate at 150 °C for 40 min to enhance the evaporation of cyclohexane and cure the PDMS. The excess silica nanoparticles on the SCDA surface were rinsed with distilled (DI) water to obtain a clean substrate.

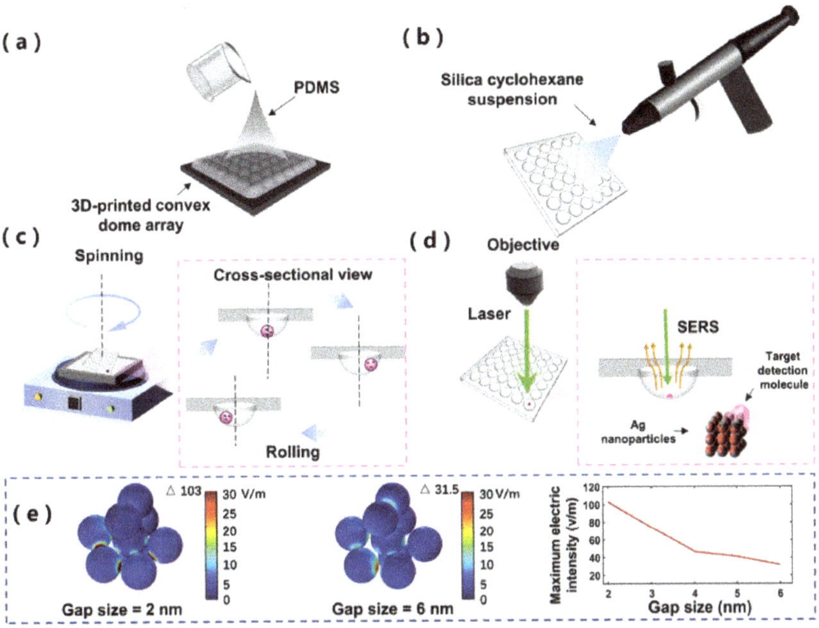

Figure 1. (**a**) Schematic diagram of concave dome array preparation; (**b**) schematic diagram of superhydrophobic surface coating; (**c**) schematic diagram of working mechanism of dynamic enrichment on SCDA; (**d**) schematic diagram of surface-enhanced Raman scattering (SERS) detection on SCDA; (**e**) numerical simulation of electric field intensity of Ag nanoparticles with different gap sizes in three-dimensional space.

2.2. Working Mechanism

Figure 1c shows the working principle of enriching plasmonic nanoparticles and analytes by rotating the SCDA. Because of the properties of the high static water contact angle and concave dome structure of the SCDA, droplets can easily roll in the concave dome, but will not leave the substrate due to the concave structure (Videos S1 and S2). When a droplet is dropped into one of the wells of the SCDA, the plasmonic nanoparticles in the droplet can be fully and evenly mixed with the analytes through periodic rotation (Video S3). During the evaporation process, nanoparticles follow the movement of the droplets, and are not easy to deposit at the contact line between the droplet and the substrate so as to effectively suppress the coffee-ring effect. [18]. Therefore, the nanoparticles can be concentrated into a small area to enhance the Raman signal. As the droplet volume gradually decreases during the evaporative process (Figure S1), the microdroplets begin to adhere to SCDA, and cannot rotate freely and dynamically. At this time, three-dimensional SERS hotspots will be formed inside the microdroplet. At the same time, the analytes are actively captured in small gaps (hotspots) to further improve the Raman signal, as shown in Figure 1d [19,20].

2.3. Materials

PDMS gel and curing agent were purchased from Dow Corning (Midland, USA). Hydrophobic silica nanoparticles with an average size of 250 nm were purchased from

Macklin (Shanghai, China), and used to make superhydrophobic coating suspension. The Ag colloid was from Beijing Biotyscience company (Beijing, China), with a diameter of 60 nm and concentration of 0.1 mg/mL. Crystal violet (CV), methylene blue (MB), and melamine were all purchased from Macklin (Shanghai, China) to test the SERS detection on the SCDA.

2.4. Characterization and Experimental Setup

The images of enriched nanoparticles, after the evaporation of droplets, on different substrates and different enrichment modes were analyzed using a high-resolution field emission scanning electron microscope (SEM) (ZEISS SUPRA® 55, Carl Zeiss, Oberkochen, Germany). CA values were obtained with an optical contact angle measuring instrument (Theta, Biolin scientific, Gothenburg, Sweden). An optical magnifier (HAG0950, SHOCREX, Shenzhen, China) was used to record the evaporative process of the droplets and observe the laser point. A horizontal rotator (NSP-300, NuoMi, Suzhou, China) provided the periodic rotation for droplet spinning on the SCDA.

The Raman signal detection system included a Raman spectrometer (iHR550, Horiba, Kyoto, Japan), two long-pass filters (RL-532nm, Shanghai-optics, Nanjing, China), a laser (532 nm, Ventus, Konstanz, Germany), a lens with a focal length of 2 mm, and an objective lens of 20× magnification and numerical aperture (NA) of 0.4. The diameter of detection spot was about 1.6 μm. The exposure time was 5 s, and the laser power was about 12 mW.

As shown in Figure S2, the side view is similar to the diagram in Figure 1d. The SCDA was placed under the objective lens of 20× magnification. The top view illustrates the collection optical path of the Raman signal. A filter was placed in the collection optical path to filter Rayleigh light, and then a 10× objective lens (NA is 0.3) was used to focus the Raman signal to the optical fiber port to collect the Raman signal directly.

2.5. Numerical Simulation for SERS

To verify that evaporation of droplets on SCDA can form effective SERS hotspots, the wave optical module of COMSOL was used for numerical simulation of electric field distribution of Ag particle aggregation, with the particle gap either at 2 nm or 6 nm. Figure 1e shows the electric field intensity of such configurations, with randomly distributed Ag particles in three-dimensional space, where nine Ag particles all with a diameter of 60 nm are illuminated by the 532 nm optical plane wave. In the simulation model, a sphere domain with a diameter of 400 nm was set as the water, and an outer layer of 800 nm thickness was set as the perfect match layer (PML).

3. Results and Discussion

3.1. Investigation of SCDA

To verify the superhydrophobicity of SCDA, we measured the droplet contact angle on the plane substrate coated with the silica suspension, and compared with contact angle on the glass surface and PDMS surface. Since our optical contact angle measuring instrument cannot measure the contact angle in the concave dome, the data for the contact angle on the superhydrophobic concave dome are not available. Figure 2a shows an optical image of 10 μL pure water on these three types of surfaces. The superhydrophobic surface shows the average contact angle of 152.8°, the PDMS surface 100.7°, and the glass surface 57° (all from multiple measurements).

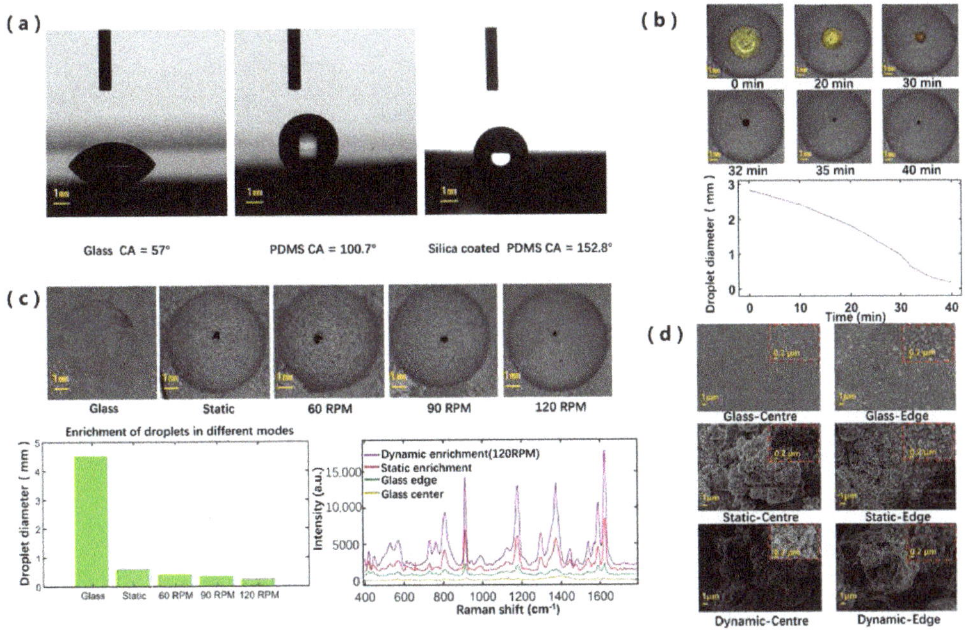

Figure 2. (**a**) Optical images of the 10 μL pure water droplet on different substrates showing different contact angles; (**b**) optical images and time-dependent curve showing the volume change of the droplet, containing 13 μL droplet during rotary evaporation at a spinning speed of 120 RPM; (**c**) optical images, related histogram, and SERS signals for the evaporation-ended droplet on glass and on SCDA spun at different speeds; (**d**) SEM images of droplets vaporized on different substrates and at different locations.

To further study the enrichment capability during droplet dynamic evaporation on the SCDA, we recorded the optical images displaying the diameter change of the droplets (13 μL), as shown in Figure 2b. The SCDA was spun at a speed of 120 RPM. To speed up the evaporation, we illuminated the SCDA with an incandescent lamp. Over time, the volume of the droplet gradually decreased; finally, the droplet was concentrated into a small area of about 0.17 mm in diameter.

3.2. Comparison of the Deposition Patterns and SERS Performance Based on Different Enrichment Modes

To study the deposition property and SERS performance of droplets under different spinning speeds applied to the SCDA, we consistently started with the droplet composed of 10 μL CV solution (10^{-8} M) and 3 μL Ag colloid. As shown in Figure 2c, the smallest enrichment area can be reached when the spinning speed is set as high as 120 RPM. This phenomenon may be attributed to the fact that the higher spinning speed reduces the contact time between the moving droplet and the substrate, thus effectively reducing the droplet adhesion with the substrate surface [21]. Previous studies found that there is a vortex zone above the contact line for stationary evaporating droplets, where most of the nanoparticles are deposited [18]. For moving droplets, there is a vortex zone that accounts for most of the droplets. The interior nanoparticles rotate with the vortex zone, so the nanoparticles are more concentrated and enriched on the substrate during the evaporative process. In the experiment, when the spinning speed exceeded 120 RPM, the area of the final deposition of the droplet was not further reduced. The reason for this may be that when the speed exceeds 120 RPM, the droplets do not follow the movement of the SCDA, and are unable to rotate synchronously with the SCDA. Therefore, the higher speed cannot further concentrate the plasmonic nanoparticles into a smaller area. The spinning speed of 120 RPM was therefore used here. Furthermore, as expected, the coffee-ring effect was evident when the droplets evaporated on the glass sheet.

In the next step, the SERS measurements were performed on the prepared samples due to the dynamic enrichment processes of the droplets. In Figure 2c, the SERS signals corresponding to the droplets undergoing different enrichment processes are presented. The SERS signal for the dynamic enrichment on the SCDA was the strongest, followed by, several times lower, the signal for the static enrichment on the SCDA; those signals measured from the glass plate (one from center and the other from the edge) were both much lower. The CV molecules usually show the typical Raman peaks at 807 cm^{-1} and 915 cm^{-1} (both from the symmetric stretch of the dimethylamino bond C-N-C), 1182 cm^{-1} (from the stretching vibration of C-N), 1373 cm^{-1} (from the stretching vibration of C=C), and 1624 cm^{-1} (from the stretching vibration of C=C) [22]. Here, we quantitatively studied the Raman peak at 915 cm^{-1}, and straightforward calculation revealed the following: the SERS signal for the SCDA dynamic enrichment was 2.2 times that for the SCDA static enrichment, 7.7 times that for the edge of the glass sheet, and 39.7 times that for the center of the glass sheet.

We further continued on the estimation of the Raman enhancement factor for the SCDA dynamic enrichment case, based on the following formula [23]:

$$\mathrm{EF} = \frac{I_{SERS}/C_{SERS}}{I_{Raman}/C_{Raman}} \quad (1)$$

where I_{SERS} and I_{Raman} represent the Raman intensity, respectively, for the SERS measurement and the conventional Raman measurement (without Ag nanoparticles), and C_{SERS} and C_{Raman}, respectively, represent the corresponding solution concentrations of analyte by plugging in the values $C_{SERS} = 10^{-8}$ M and $C_{Raman} = 10^{-3}$ M (refer to Figure S3); we subsequently obtained EF = 3.55×10^5.

As the final step, we took the SEM images for all the above samples, as shown in Figure 2d. Each of the large images were taken at the 5000 magnification, while each inlet small images were taken at 30,000 magnification. For the glass sheet, there were obviously more Ag nanoparticles at the edge compared to the center; for the static enrichment case, the Ag nanoparticles were still denser at the edge, but the contrast was much weaker. Finally, for the dynamic enrichment case, the Ag nanoparticles at both the edge and the center appear closely packed, indicating the resulting even distribution of the Ag nanoparticles.

3.3. SERS Performance Study of the SCDA

To further study the limit of detection (LOD) and the signal uniformity of the dynamic enrichment mode of SCDA, we used CV and MB, respectively, as the probe molecules to prepare multiple detection hotspots with different concentrations of these two molecules on the same SCDA. Figure 3a,e shows the SERS signals for different molar concentrations, ranging from 10^{-7} M to 10^{-11} M, while Figure 3c,g shows the SERS signals for 20 randomly chosen positions when the molar concentration is at 10^{-8} M.

For the CV molecules, several Raman characteristic peaks can still be discerned even for concentrations as low as 10^{-11} M (Figure 3a). In Figure 3b, we further plotted the data of the Raman peak intensity versus the molecular concentration for the Raman peaks at 915 cm^{-1} and 1624 cm^{-1}, respectively. Apparently, the intensity points can be separated into two linearly fitting regions, and at the higher concentration region, the slope is much larger. The resulting linear fits for the 915 cm^{-1} peak are as follows: for the higher concentration region, $Y_{915cm^{-1}} = 11823 \cdot \lg C_{CV} + 108157$, and for the lower concentration region, $Y_{915cm^{-1}} = 1078 \cdot \lg C_{CV} + 12138$, with correlation coefficients (R^2) of 0.987 and 0.953, respectively. These results imply that SCDA is suitable for quantitative analysis at high concentrations, and qualitative analysis at low concentrations.

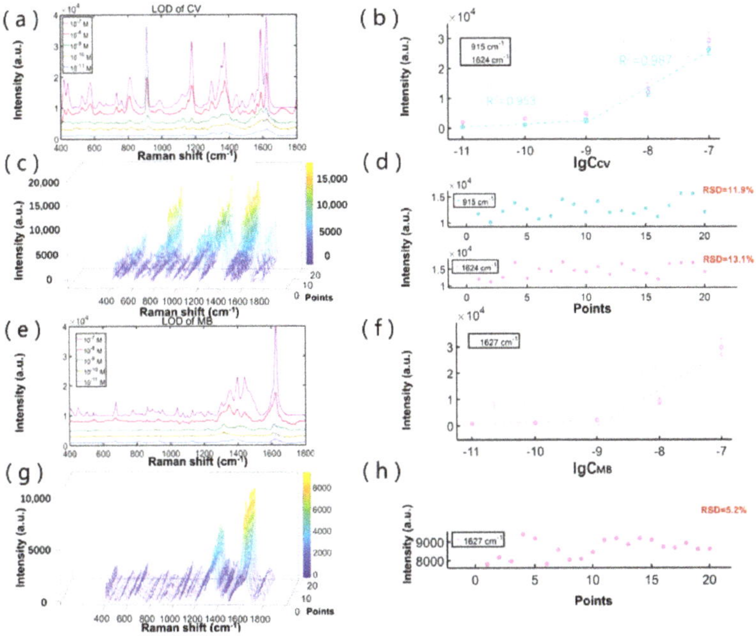

Figure 3. (**a**) SERS signals from dynamically enriching 13 μL droplet (10 μL CV solution and 3 μL Ag colloid) with different concentrations; (**b**) the data for SERS intensity peaks at 915 cm^{-1} and 1624 cm^{-1} versus the CV concentration; (**c**) SERS signals of 10^{-8} M CV/AgNPs from 20 random positions based on SCDA dynamic enrichment mode; (**d**) the data for CV SERS intensity at 915 cm^{-1} and 1624 cm^{-1} from 20 random positions; (**e**) SERS signals from dynamically enriching 13 μL droplet (10 μL MB solution and 3 μL Ag colloid) with different concentrations; (**f**) the data for SERS intensity peaks at 1627 cm^{-1} versus the MB concentration; (**g**) SERS signals of 10^{-8} M MB/AgNPs from 20 random positions based on SCDA dynamic enrichment mode; (**h**) the data for MB SERS intensity at 1627 cm^{-1} from 20 random positions.

The 20 random points of the SERS signals show similar spectral structures and signal strengths, demonstrating the signal uniformity of our method (Figure 3c). The detailed examination of the intensity points for the Raman peak at 915 cm^{-1} and at 1624 cm^{-1} are plotted in Figure 3d. The statistical calculation conducted on these intensity points revealed that the relative standard deviations (RSD) are 11.9% and 13.1%, respectively, showing high signal uniformity. The slight difference in Raman signals may be attributed to the slight vibration of the experimental table, and the pulse type of the laser.

We further performed the exact same measurements on the MB molecules. In Figure 3e, the SERS signals show the MB characteristic peaks at 1397 cm^{-1} (from the asymmetric C–N stretching) and 1627 cm^{-1} (from the C–C ring stretching) [24]. As shown in Figure 3f, the linear fits for the 1627 cm^{-1} appear as $Y_{1627cm^{-1}} = 13{,}735 \cdot lgC_{MB} + 123{,}467$ (R^2 equals to 0.927) for the higher concentration region, and $Y_{1627cm^{-1}} = 597.45 \cdot lgC_{MB} + 7324$ (R^2 equals to 0.833) for the lower concentration region. As presented in Figure 3g–h, the RSD of the SERS intensity at 1627 cm^{-1} for the 20 measuring points was calculated to be 5.2%, which was attributed to the relatively uniform distribution of the CV molecules and Ag colloids. The experimental results show that the dynamic enrichment mode can not only concentrate the analytes and plasmonic nanoparticles into a smaller area to obtain dense hotspots for better SERS measurement, but it can also obtain a uniform signal with good repeatability.

As shown in Table 1, compared with other substrates, SCDA has the ability to manipulate abundant droplets and make multiple hotspots at the same time. Within the same time period, where other methods can only prepare one droplet, our method can prepare numerous droplets. Therefore, when a large number of hotspots is demanded, SCDA is more applicable. Moreover, the LOD of our method can be improved by using an objective lens with a higher numerical aperture for collecting the intensified Raman signals.

Table 1. Comparing performance of SERS substrates with other reported methods.

Method	Analyte	Time	LOD	Droplet Manipulation	Number of Droplets Prepared Simultaneously	Ref.
Light-trapping SERS substrate	R6G	Not available	10^{-13} M	Not applicable	1	[12]
Taro-leaf@Ag	R6G	120 min (4 μL)	10^{-8} M	Not applicable	3	[25]
Slippery liquid-infused porous surface	R6G	5 min (50 μL)	10^{-18} M	Not applicable	1	[15]
Continuous-rolling-assisted evaporation on a superhydrophobic surface	CV	9 min (50 μL)	10^{-15} M	Feasible	1	[18]
Superhydrophobic magnetically functionalized PDMS	R6G	180 min (20 μL)	10^{-17} M	Feasible	9	[16,17]
Superhydrophobic concave dome array	CV	40 min (13 μL)	10^{-11} M	Feasible	36 (can be scalable)	This work

3.4. SERS Detection of Melamine

Melamine is a triazine heterocyclic organic compound that is widely used in the production of melamine resin, flame retardants, fertilizers, and other products. Due to its high nitrogen content (66% mass nitrogen), melamine is sometimes illegally added to the dairy products to increase their apparent protein content [26]. To prohibit such a problem, the Food and Drug Administration (FDA) has set a safe threshold of 1 part per million (ppm) (8×10^{-6} M) for melamine intake in infant formula, and 2.5 ppm for food and dairy products [27].

To verify the feasibility of SCDA application in the field of food safety, we performed the SERS study by mixing melamine into milk as the sample. In a controlled experiment, the Raman spectrum of solid melamine was first measured by ourselves (Figure 4a) to identify the characteristic peaks of 685 cm^{-1} (associated with ring breathing mode II of the in-plane triazine ring). We then measured the SERS signals of melamine solutions at the various concentrations, as shown in Figure 4b,c. The results show that the LOD is about 5×10^{-7} M. From Figure 4c, the intensity points can also be separated into two linearly fitting regions, with the higher concentration region as $Y_{685cm}{}^{-1} = 10281 \cdot lgC_{melamine} + 65484$, and the lower concentration region as $Y_{685cm}{}^{-1} = 1546 \cdot lgC_{melamine} + 12009$. The results imply that the SCDA can be used for quantitative analysis when the concentration of melamine is higher than 10^{-6} M.

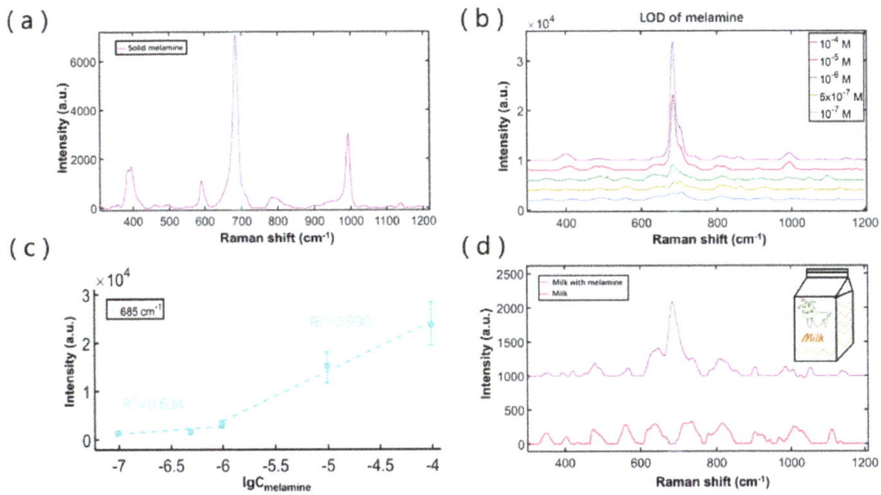

Figure 4. (**a**) Raman spectrum of solid melamine measured on the glass sheet; (**b**) SERS signals from dynamically enriching 13 μL droplet (10 μL melamine solution and 3 μL Ag colloid) with different concentrations; (**c**) data for SERS intensity peaks at 685 cm^{-1} versus melamine concentration; (**d**) Raman spectrum of pure milk and SERS signal of contaminated milk with melamine.

We then prepared the milk sample by mixing melamine at a concentration of 8×10^{-6} M, which is just within the safety threshold. As shown in Figure 4d, the Raman spectrum of the pure milk (the red curve) does not display the 685 cm^{-1} peak. In contrast, the Raman spectrum of the mixture (the violet curve) displays a pronounced peak at 685 cm^{-1}, indicating a strong capability for practical melamine detection.

4. Conclusions

In this work, we developed a superhydrophobic concave dome array (SCDA) for the dynamic enrichment of plasmonic nanoparticles that can achieve the uniform SERS signal measurement. Due to low surface adhesion, the SCDA can effectively suppress the coffee-ring effect by dynamically enriching the droplets. A droplet with the initial volume of 13 μL can be condensed into a small region of about 0.17 mm in diameter after the complete evaporation of the solvent. The LOD of CV and MB molecules can reach up to 10^{-11} M. As a proof-of-concept application, these dynamic-enriched SERS hotspots can detect the melamine in milk at concentrations lower than the FDA standard. We expect our method have more potential applications in analytical chemistry and biomedicine.

Supplementary Materials: The following supporting information can be downloaded at: https://www.mdpi.com/article/10.3390/bios12050270/s1, Figure S1: On the same SCDA, different kinds and concentrations of analytes are made into multiple hot-spots for SERS detection using the enrichment mode of dynamic evaporation.; Figure S2: Images of experimental setup.; Figure S3: Raman spectrum of a 10 μL CV droplet (10^{-3} M). Video S1: The SCDA shows the excellent superhydrophobic property.; Video S2: The concave dome can physically confine the droplets.; Video S3: The analytes rotate periodically on the SCDA.

Author Contributions: Conceptualization: S.Y. and H.F.; methodology, S.Y. and H.F.; software, J.W. and Y.Z.; validation, J.W., J.C. and Y.F.; investigation and resources, S.Y., H.F. and J.W.; writing—original draft preparation, J.W.; writing—review and editing, H.F., S.Y. and J.W.; visualization, J.W.; supervision, H.F. and S.Y.; funding acquisition, H.F. and S.Y. All authors have read and agreed to the published version of the manuscript.

Funding: Sheng Yan thanks the financial support from Guangdong Basic and Applied Basic Research Foundation (2021A1515110277) and the Shenzhen Natural Science Fund (the Stable Support

Plan Program: 20200811205344001). Hui Fang thanks the financial support from the National Natural Science Foundation of China (12074268), the Nature Science Foundation of Guangdong Province (2018B030311048), and the Shenzhen Natural Science Fund (the Stable Support Plan Program: 20200814164819001).

Institutional Review Board Statement: Not applicable.

Data Availability Statement: Not applicable.

Conflicts of Interest: The authors declare no conflict of interest.

References

1. Deegan, R.D.; Bakajin, O.; Dupont, T.F.; Huber, G.; Nagel, S.R.; Witten, T.A. Capillary flow as the cause of ring stains from dried liquid drops. *Nature* **1997**, *389*, 827–829. [CrossRef]
2. Yunker, P.J.; Still, T.; Lohr, M.A.; Yodh, A.G. Suppression of the coffee-ring effect by shape-dependent capillary interactions. *Nature* **2011**, *476*, 308–311. [CrossRef] [PubMed]
3. Deegan, R.D.; Bakajin, O.; Dupont, T.F.; Huber, G.; Nagel, S.R.; Witten, T.A. Contact line deposits in an evaporating drop. *Phys. Rev. E* **2000**, *62*, 756. [CrossRef]
4. Deegan, R.D. Pattern formation in drying drops. *Phys. Rev. E* **2000**, *61*, 475. [CrossRef] [PubMed]
5. Bigioni, T.P.; Lin, X.-M.; Nguyen, T.; Corwin, E.I.; Witten, T.A.; Jaeger, H.M. Kinetically driven self assembly of highly ordered nanoparticle monolayers. *Nat. Mater.* **2006**, *5*, 265–270. [CrossRef] [PubMed]
6. Pan, X.; Dong, J.; Li, Y.; Sun, X.; Yuan, C.; Qian, W. The strategy of two-scale interface enrichment for constructing ultrasensitive SERS substrates based on the coffee ring effect of AgNP@β-CD. *RSC Adv.* **2016**, *6*, 29586–29591. [CrossRef]
7. Hertaeg, M.J.; Rees-Zimmerman, C.; Tabor, R.F.; Routh, A.F.; Garnier, G. Predicting coffee ring formation upon drying in droplets of particle suspensions. *J. Colloid Interface Sci.* **2021**, *591*, 52–57. [CrossRef]
8. Chen, C.; Hirdes, D.; Folch, A. Gray-scale photolithography using microfluidic photomasks. *Proc. Natl. Acad. Sci. USA* **2003**, *100*, 1499–1504. [CrossRef]
9. Groves, T.R.; Pickard, D.; Rafferty, B.; Crosland, N.; Adam, D.; Schubert, G. Maskless electron beam lithography: Prospects, progress, and challenges. *Microelectron. Eng.* **2002**, *61*, 285–293. [CrossRef]
10. Melchels, F.P.W.; Feijen, J.; Grijpma, D.W. A review on stereolithography and its applications in biomedical engineering. *Biomaterials* **2010**, *31*, 6121–6130. [CrossRef]
11. Zhu, H.; Huang, Y.; Lou, X.; Xia, F. Bioinspired superwetting surfaces for biosensing. *VIEW* **2021**, *2*, 20200053. [CrossRef]
12. Jin, X.; Zhu, Q.; Feng, L.; Li, X.; Zhu, H.; Miao, H.; Zeng, Z.; Wang, Y.; Li, Y.; Wang, L.; et al. Light-Trapping SERS Substrate with Regular Bioinspired Arrays for Detecting Trace Dyes. *ACS Appl. Mater. Interfaces* **2021**, *13*, 11535–11542. [CrossRef] [PubMed]
13. Li, H.; Yang, Q.; Hou, J.; Li, Y.; Li, M.; Song, Y. Bioinspired Micropatterned Superhydrophilic Au-Areoles for Surface-Enhanced Raman Scattering (SERS) Trace Detection. *Adv. Funct. Mater.* **2018**, *28*, 1800448. [CrossRef]
14. Kozhina, E.P.; Bedin, S.; Nechaeva, N.; Podoynitsyn, S.; Tarakanov, V.; Andreev, S.; Grigoriev, Y.; Naumov, A. Ag-Nanowire Bundles with Gap Hot Spots Synthesized in Track-Etched Membranes as Effective SERS-Substrates. *Appl. Sci.* **2021**, *11*, 1375. [CrossRef]
15. Yang, S.; Dai, X.; Stogin, B.B.; Wong, T.S. Ultrasensitive surface-enhanced Raman scattering detection in common fluids. *Proc. Natl. Acad. Sci. USA* **2016**, *113*, 268–273. [CrossRef]
16. Chen, G.; Dai, Z.; Ji, B.; Li, S.; Chen, X.; Gao, Y.; Wen, W.; Zhou, B. Dynamic enrichment of plasmonic hot-spots and analytes on superhydrophobic and magnetically functionalized platform for surface-enhanced Raman scattering. *Sens. Actuators B Chem.* **2020**, *319*, 128297. [CrossRef]
17. Chen, G.; Dai, Z.; Li, S.; Huang, Y.; Xu, Y.; She, J.; Zhou, B. Magnetically Responsive Film Decorated with Microcilia for Robust and Controllable Manipulation of Droplets. *ACS Appl. Mater. Interfaces* **2021**, *13*, 1754–1765. [CrossRef]
18. Sun, Y.; Chen, X.; Zheng, Y.; Song, Y.; Zhang, H.-R.; Zhang, S. Surface-Enhanced Raman Scattering Trace-Detection Platform Based on Continuous-Rolling-Assisted Evaporation on Superhydrophobic Surfaces. *Acs Appl. Nano Mater.* **2020**, *3*, 4767–4776. [CrossRef]
19. Liu, H.; Yang, Z.; Meng, L.; Sun, Y.; Wang, J.; Yang, L.; Liu, J.; Tian, Z. Three-Dimensional and Time-Ordered Surface-Enhanced Raman Scattering Hotspot Matrix. *J. Am. Chem. Soc.* **2014**, *136*, 5332–5341. [CrossRef]
20. Ge, M.; Li, P.; Zhou, G.; Chen, S.; Han, W.; Qin, F.; Nie, Y.; Wang, Y.; Qin, M.; Huang, G.; et al. General Surface-Enhanced Raman Spectroscopy Method for Actively Capturing Target Molecules in Small Gaps. *J. Am. Chem. Soc.* **2021**, *143*, 7769–7776. [CrossRef]
21. Bormashenko, E.; Bormashenko, Y.; Oleg, G. On the Nature of the Friction between Nonstick Droplets and Solid Substrates. *Langmuir* **2010**, *26*, 12479–12482. [CrossRef] [PubMed]
22. Zhang, Z.; Yu, J.; Ma, L.; Sun, Y.; Wang, P.; Wang, T.; Peng, S. Preparation of the plasmonic Ag/AgBr/ZnO film substrate for reusable SERS detection: Implication to the Z-scheme photocatalytic mechanism. *Spectrochim. Acta Part A-Mol. Biomol. Spectrosc.* **2020**, *224*, 117381. [CrossRef] [PubMed]
23. Le Ru, E.C.; Blackie, E.; Meyer, M.; Etchegoin, P.G. Surface enhanced Raman scattering enhancement factors: A comprehensive study. *J. Phys. Chem. C* **2007**, *111*, 13794–13803. [CrossRef]

24. Xiao, G.-N.; Man, S.-Q. Surface-enhanced Raman scattering of methylene blue adsorbed on cap-shaped Silver nanoparticles. *Chem. Phys. Lett.* **2007**, *447*, 305–309. [CrossRef]
25. Huang, J.A.; Zhang, Y.-L.; Zhao, Y.; Zhang, X.-L.; Sun, M.-L.; Zhang, W. Superhydrophobic SERS chip based on a Ag coated natural taro-leaf. *Nanoscale* **2016**, *8*, 11487–11493. [CrossRef]
26. Mauer, L.J.; Chernyshova, A.L.; Hiatt, A.; Deering, A.; Davis, R. Melamine Detection in Infant Formula Powder Using Near- and Mid-Infrared Spectroscopy. *J. Agric. Food Chem.* **2009**, *57*, 3974–3980. [CrossRef]
27. Kim, A.; Barcelo, S.J.; Williams, R.S.; Li, Z. Melamine sensing in milk products by using surface enhanced Raman scattering. *Anal. Chem.* **2012**, *84*, 9303–9309. [CrossRef]

biosensors

MDPI

Review

Fluorescent Biosensors for the Detection of Viruses Using Graphene and Two-Dimensional Carbon Nanomaterials

Ahmed M. Salama [1], Ghulam Yasin [2], Mohammed Zourob [3,*] and Jun Lu [1,*]

[1] State Key Laboratory of Chemical Resource Engineering, Beijing University of Chemical Technology, Beisanhuan East Road 15, Beijing 100029, China; 2019420035@mail.buct.edu.cn
[2] Institute for Advanced Study, Shenzhen University, Shenzhen 518060, China; yasin@mail.buct.edu.cn
[3] Department of Chemistry, Alfaisal University, Riyadh 11533, Saudi Arabia
* Correspondence: mzourob@alfaisal.edu (M.Z.); lujun@mail.buct.edu.cn (J.L.)

Abstract: Two-dimensional carbon nanomaterials have been commonly employed in the field of biosensors to improve their sensitivity/limits of detection and shorten the analysis time. These nanomaterials act as efficient transducers because of their unique characteristics, such as high surface area and optical, electrical, and magnetic properties, which in turn have been exploited to create simple, quick, and low-cost biosensing platforms. In this review, graphene and two-dimensional carbon material-based fluorescent biosensors are covered between 2010 and 2021, for the detection of different human viruses. This review specifically focuses on the new developments in graphene and two-dimensional carbon nanomaterials for fluorescent biosensing based on the Förster resonance energy transfer (FRET) mechanism. The high-efficiency quenching capability of graphene via the FRET mechanism enhances the fluorescent-based biosensors. The review provides a comprehensive reference for the different types of carbon nanomaterials employed for the detection of viruses such as Rotavirus, Ebola virus, Influenza virus H_3N_2, HIV, Hepatitis C virus (HCV), and Hepatitis B virus (HBV). This review covers the various multiplexing detection technologies as a new direction in the development of biosensing platforms for virus detection. At the end of the review, the different challenges in the use of fluorescent biosensors, as well as some insights into how to overcome them, are highlighted.

Keywords: FRET sensing; 2D carbon material-based sensors; multiplexing virus detection; limit of detection; recognition element

Citation: Salama, A.M.; Yasin, G.; Zourob, M.; Lu, J. Fluorescent Biosensors for the Detection of Viruses Using Graphene and Two-Dimensional Carbon Nanomaterials. *Biosensors* **2022**, *12*, 460. https://doi.org/10.3390/bios12070460

Received: 6 May 2022
Accepted: 14 June 2022
Published: 27 June 2022

1. Introduction

There is competition in the development of graphene-based biosensors in the market [1]. A large number of research articles have been published, exceeding 3600 articles since 2010, focusing on fluorescent biosensors. Biosensors have enormous applications in various areas, including (a) diagnostic applications [2], bioprocess monitoring [3], and improving the quality of new pharmaceutics [4]; (b) environmental testing [5]; and (c) food quality [6].

Biosensors can be defined as miniaturized devices incorporating recognition elements for identifying and/or quantifying either a synthetic or biochemical analyte. The recognition element can be an antibody, whole cell, aptamer, peptides, or DNA [7–9]. Biosensors can provide real-time outputs with high sensitivity, high specificity, and a low limit of detection [10,11]. Optical biosensors are a major group of transducers with unique features in comparison with other biosensors [12,13]. The optical detection mechanism relies upon the interaction between a recognition element and the optical transducer [14]. Optical biosensing can be characterized into two general categories: label-based and label-free sensors [15,16].

In the label-free sensors, the detected signal is the result of the direct interaction of the analyte with the immobilized recognition element on the transducer's surface [17,18].

On the other hand, the label-based sensors require the employment of labels, such as colorimetric, fluorescent, or luminescent, to produce the optical signal [19–21]. There are various types of optical biosensors reported in the literature [22]. This review will cover the fluorescent-based biosensors used for the detection or identification of viruses.

To the best of the authors' knowledge, no review has been published so far covering the use of carbon nanomaterials for virus detection using fluorescence. As a consequence, this recent literature review reports the virus detection using fluorescent biosensors based on carbon nanomaterials [22]. Moreover, this review will focus on the recent improvements in graphene-based biosensors for human virus detection using multiplexed detection based on a FRET–graphene oxide biosensor [23].

2. Graphene Oxide Fluorescent Biosensor for Human Virus Detection

In 2004, Novoselov and Geim synthesized and characterized a single sheet of graphene with covalently linked atoms arranged in one or multiple atomic layer planes, which form bulk layered materials via weak van der Waals gaps [24–26]. Since then, the application of graphene has grown exponentially in several biomedical fields owing to its well-defined properties and high specific surface area [27]. Therefore, graphene research has progressed to the development of more adaptable and customizable 2D alternatives with a higher composition, structure, and diversity in functionality [28]. Graphene-derived materials, including graphene oxide (GO) and reduced graphene oxide (rGO), are extensively used in various biosensing devices and assays. In-vivo and in-vitro, the target biomolecule is fluorescently labeled, which is identified using fluorescence microscopy or fluorescence spectrometry [12].

Recently, graphene oxide (GO) has gained a great deal of interest for virus detection due to its large surface area, 2D carbon structure, electrical conductivity, thermal robustness, flexibility, optical transparency, and mechanical characteristics, with high chemical stability [29–34]. Aptamers have demonstrated that graphene possesses unique attachment, molecular recognition, and biocompatibility via linking them with graphene to enhance the sensitivity and selectivity of the manufactured biosensors. Antigen–antibody and aptamer–target interactions, essentially were previously widely employed to create biosensors for the detection of analytes. The aptamer–target system-based aptasensors have several superior features over antigen–antibody system-based biosensors, including higher affinity, reduced costs, simpler fabrication, higher sensitivity, and wide range of analyte-sensing applications [35].

2.1. Fluorescence Resonance Energy Transfer Mechanism (FRET)

FRET is an effective approach to the quantitative determination of biomolecules with high specificity and sensitivity [36]. A fluorophore probe is absorbed on the surface of a quencher (graphene) to constitute a FRET pair, as shown in Figure 1. Graphene-like 2D nanomaterial is a powerful fluorescence quencher, that increased the application of fluorescent sensors to achieve a sensitive detection platform, for various targets via an assembly or conjugation mechanism. The binding force between the fluorophore and target biomolecules decreases the fluorophore–graphene interaction in the presence of the target biomolecules, resulting in the fluorophore's release from the graphene surface to restore the dye's fluorescence [37].

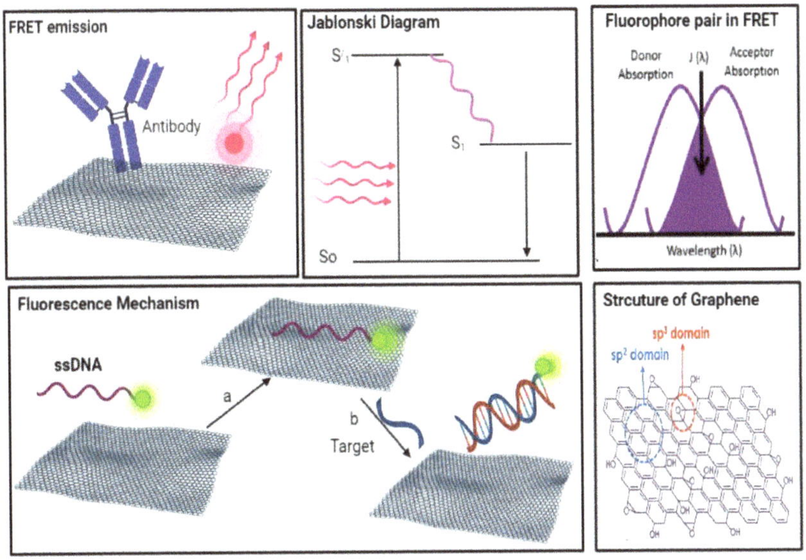

Figure 1. Illustration of the FRET emission mechanism on graphene oxide related to Jablonski diagram.

GO has combined electron–hole pairs positioned inside sp^2 carbon clusters, incorporated within the sp^3 matrix, and exhibits UV to near-infrared (NIR) light absorption [38–40]. The large dislocations in the p-electrons of GO display an exceptional ability to quench the fluorescence emitted by fluorescent dyes or quantum dots. Therefore, GO can be considered a good option as a quencher in FRET-based biosensors, as it offers low background noise and a high signal-to-noise ratio [41].

2.2. Characteristics of Graphene Material and Biomolecule Interaction

There are several functional groups—for instance, carboxyl, epoxide, and hydroxyl groups—at the surface and edges of GO that can be used for the immobilization of recognition elements via covalent bonding, electrostatic interaction, and hydrogen interactions.

Due to the honeycomb structure of the graphene oxide lattice, inserting suitable functional groups enhances the optical properties. The chemically active, soluble, hybrid graphene surfaces result from the functionalization of pure graphene via covalent and noncovalent attachment [42,43].

Hydrogen bonding, van der Waals interactions, and π–π interactions are forms of noncovalent functionalization, whereas covalent bonding using 1-ethyl-3-(3-dimethylaminop ropyl)carbodiimide hydrochloride/N-hydroxy sulfosuccinimide (EDC/NHS) chemistry and others has been utilized to attach a variety of recognition receptors onto the graphene biosensor surface (Figure 2) [44]. Fluorescent-labeled single-stranded DNA (ssDNA) was adsorbed onto the GO molecule surface due to π–π stacking [12,45–47]. Through van der Waals interaction, partially reduced GO(p-rGO) or rGO interacts with numerous biomolecules, such as proteins and DNA [46]. The functionalization of GO by inserting different functional groups with high oxygen content helped in developing a functional pathogen biosensor.

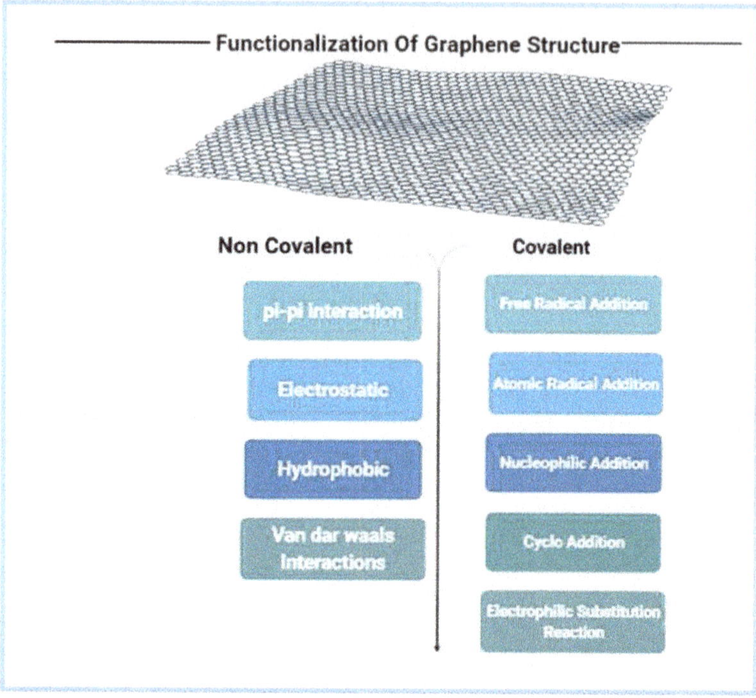

Figure 2. Different categories of functionalization of graphene oxide.

GO sheets have higher and more diverse adsorption capacities, which may increase the interaction with the pathogen and virus, which in turn enhances the sensing performance [48]. rGO lacks hydroxyl groups, so it has lower solubility in cell culture media, limiting its application to the evaluation of cellular biomaterials on substrates and nanoparticles [49].

The surface chemistry of graphene has a great role in its cytotoxicity. By adhesion or bonding with cell receptors, bare graphene or GO can restrict the flow of nutrients, generate stress, and trigger apoptotic pathways, resulting in significant toxicity even at low concentrations. The use of biocompatible polymers or chemicals to functionalize graphene-based materials will dramatically minimize this cytotoxicity [50]. As a result, the nanomaterial's surface properties, including shape, aggregation state, coatings, or functionalization, that may be present could play a part in their harmful impacts. The surface area/mass ratio and retention time of particles impact their toxicity; particles with a higher surface area and retention time interact more with cell membranes, allowing for increased absorption and transit into cells. As a result, the surface termination, size, and aggregation state of nanomaterials must be considered in any investigation, as they may have a major impact on the outcome of the biocompatibility tests.

2.3. *Virus Detection*

Currently, the COVID-19 pandemic represents a widespread health risk. Reducing COVID-19 transmission and infection in pre-symptomatic and asymptomatic persons needs ultrasensitive and early infection detection [51,52]. In order to return to normal routines in business, industry, schools, and universities, multiple strategies for reducing the risk that is associated with COVID-19 are required, including rapid, specific, easy, low-cost, and effective virus testing and monitoring [53,54]

In comparison with conventional virus detection techniques, the unique characteristics of graphene biosensors have been effectively utilized for the diagnosis of diseases [55–57]. Graphene oxide (GO) has played a vital role in the detection of different types of viruses, such as Rotavirus, Ebola virus, Influenza virus H3N2 hemagglutinin gene, AIDS (HIV), Hepatitis B virus (HBV), and Hepatitis C virus (HCV). The various assays and diagnostic tests that use the fluorescent biosensing mechanism are reviewed below [28,58]. Moreover, g-C3N4 was used to detect HBV via a simple, flexible, sensitive, and low-cost fluorescent biosensing device due to its high specific surface area and strong affinity [59].

2.3.1. Rotavirus

Rotavirus is a gastrointestinal virus that causes infections and symptoms in newborns, children, and adults all over the world, making it a major cause of diarrhea [60]. Jung and coworkers developed a FRET-based GO biosensor for rapid Rotavirus detection. The AuNP–DNA–antibody formed a covalent bond with GO to prepare a sandwich structure, as shown in Figure 3 [61–65]. Virus binding onto the immobilized antibody took place via a specific antigen–antibody interaction. As a result of this, the fluorescence was quenched due to the interaction of the GO and AuNPs. The cross-reactivity of the biosensor was tested using Poliovirus and Variola virus. No emission of fluorescence was observed. However, a 15-fold increase in the fluorescence signal was observed with the addition of the Rotavirus. This biosensor achieved a linear dynamic range (10^3 to 10^5 pfu.mL^{-1}). The unique fluorescence quenching property of GO sheets proves that GO can be applied in molecular diagnostic biosensors [64,65].

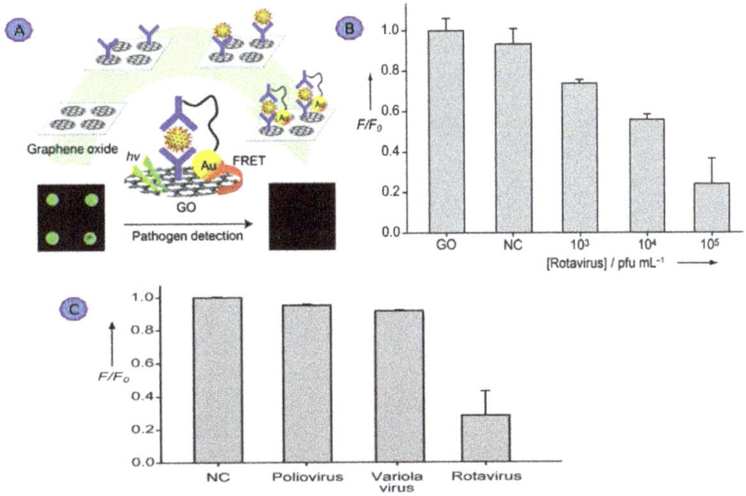

Figure 3. (**A**) Illustration of the GO-based immuno-biosensor; (**B**) dynamic range of the GO-based sensor; (**C**) cross-reactivity of the GO-based immunosensor was tested with Poliovirus and Variola virus. Reprinted with permission from Ref. [65]. Copyright 2010, Wiley-VCH.

2.3.2. Ebolavirus

The Ebola virus epidemic caused around 28,000 cases and 11,323 recorded deaths between 2014 and 2016. Ebola virus disease, widely known as a fatal disease, which can cause severe infection [66]. Wen and coworkers targeted the Ebola virus protein VP40, which interacts with the GO sheets, to develop a sensitive and selective fluorescent detection assay [51], using rolling circle amplification (RCA) (Figure 4). This fluorescent biosensor exhibits 1.4 pM LOD with a linear dynamic range from 30 fM to 3 nM. GO is negatively charged, and it directly interacts with the positively charged virus through π–π stacking. The GO sharp edges, was applied to lyse the virus envelope to start the amplification without the need for the addition of lysing agents or instrumentation [67]. The GO-based fluorescent biosensor offers an integrated lysing and sensing tool for the rapid detection of viruses [51].

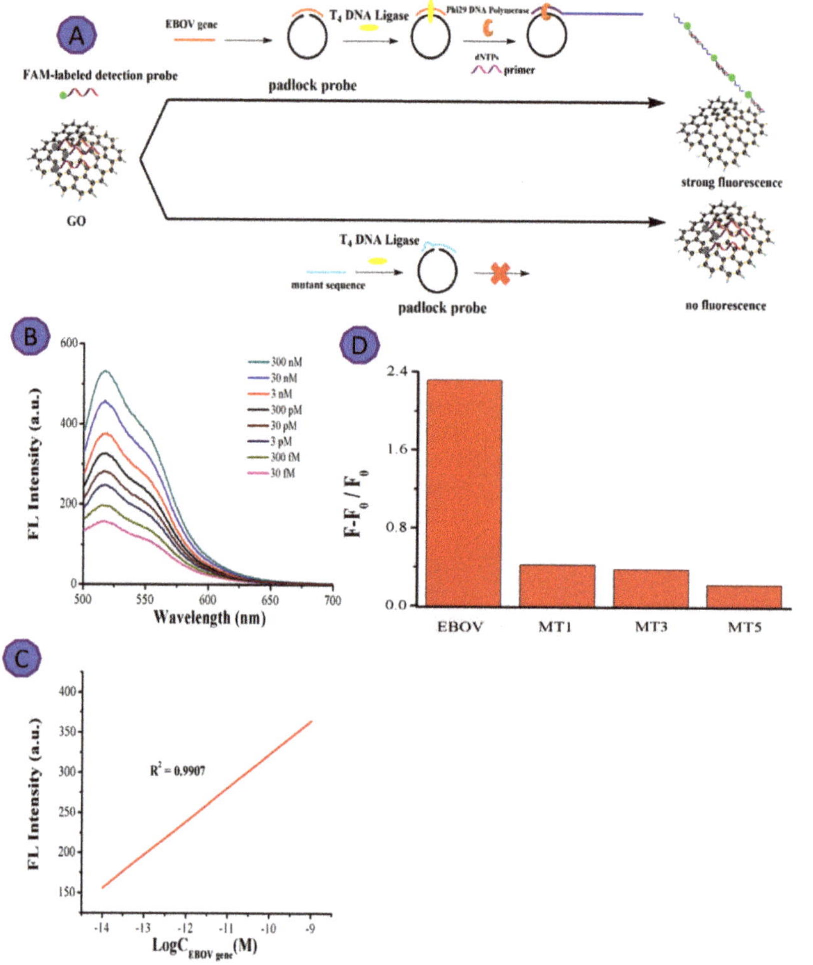

Figure 4. (**A**) The graphical representation shows the GO-assisted amplified biosensor for EBOV detection; (**B**) fluorescence increase as a function of virus increase; (**C**) the relationship between fluorescent intensity and the concentration of EBOV gene; (**D**) the relationship between the $(F - F_0)/F_0$ of GO and EBOV gene, with single-base mismatched sequence (MT1), three-base mismatched sequence (MT3), and five-base mismatched sequence (MT5). Reprinted with permission from Ref. [67]. Copyright 2022, Elsevier B.V.

2.3.3. Influenza

Researchers are increasingly concerned about Influenza as a health threat. Influenza is a member of Orthomyxoviridae, consisting of a single-stranded RNA genome with four types (A, B, C, and D). Influenza type A consists of several surface antigens—for instance, hemagglutinin (H) and neuraminidase (N) [68]. Jeong et al. created a simple fluorometric platform using graphene oxide (GO) to detect Influenza. This platform relies on employing fluorescent DNA that is directly absorbed onto the GO via π–π stacking and hydrogen bonding (Figure 5), which results in the quenching of the FAM fluorescent molecule. The emission of the fluorescence of the FAM was increased upon addition of the Influenza RNA virus in a range from 37 to 9400 pg. This platform was able to detect as little as 3.8 pg Influenza RNA [38].

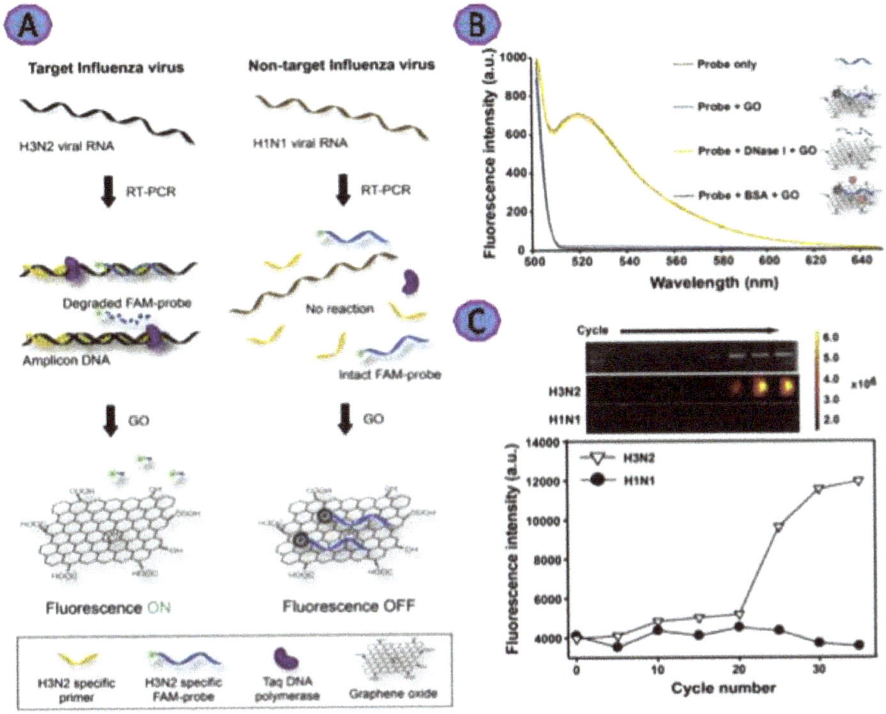

Figure 5. (**A**) Representation of fluorescence detection of Influenza virus RNA via GO and Taq polymerase's 5′ to 3′ exonuclease activity during RT-PCR; (**B**) representation of the relationship between the fluorescence emission spectra of the FAM-DNA probe and the DNase I after GO incubation (λex = 485 nm); (**C**) with each PCR cycle, the fluorescence of the FAM-DNA probe increased. RT-PCR was performed on H3N2 (target) and H1N1 (non-target) viral RNAs using H3N2 hemagglutinin gene-specific primers and the FAM-DNA probe. Reprinted with permission from Ref. [38]. Copyright 2022, Elsevier B.V.

2.3.4. HIV

A large number of patients over the world have died or became infected due to HIV/AIDS. As a result, a number of researchers have developed ultrasensitive diagnostic techniques for the early detection of HIV during the infection cycle [69,70]. Qaddare and Salimi developed a FRET-based biosensor to detect HIV, where carbon dots (CDs) were used as fluorescent NPs quenched with AuNPs, as shown in Figure 6A. This biosensor detected as little as 15 fM of the target oligonucleotides and had a dynamic range between 50.0 fM and 1.0 nM (Figure 6) [71].

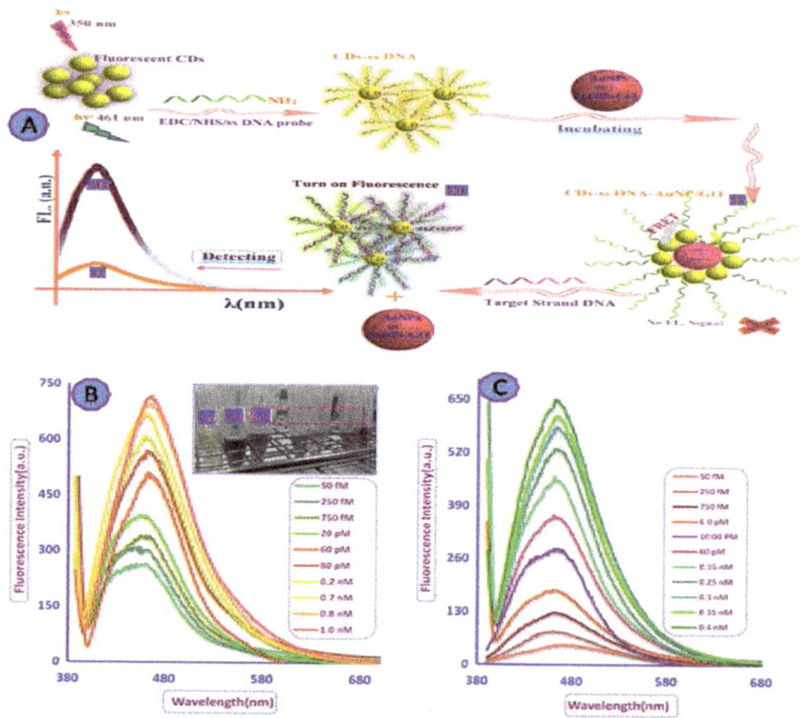

Figure 6. GO-fluorescent biosensor for HIV viral detection. (**A**) The scheme of the FRET-based detection system; (**B**) fluorescence recovery spectra of the capture probe ssDNA-CD/AuNP system in the presence of various amounts of target DNA at λ 350 nm excitation and 0.22 nM AuNPs as a quencher; (**C**) the fluorescence recovery spectra of the ssDNA-CD-AuNP/GO system in the presence of different concentrations of target DNA at excitation wavelength of 350 nm and 5 μg/mL of GO/AuNPs as quencher. Reprinted with permission from Ref. [71]. Copyright 2017, Elsevier.

HIV antibody detection was used as a sensitive and specific assay. Wu et al. designed a highly efficient FRET biosensor for the detection of anti-HIV-1 gp120 antibody via the interaction of GO with the peptide-functionalized UCNPs (Figure 7). The upconversion fluorescence intensity was increased linearly as the antibody concentration increased. The reported dynamic range was 5 to 150 nM of antibody concentration and D.L. was 2 nM [72].

Zhang et al. reported a new assay to detect HIV by labeling two partially complementary DNA probes (hairpin probe 1 (H1) and hairpin probe 2 (H2)), linked at one end with silver nanoclusters, as shown in Figure 8. They attached hairpin probes with silver nanoclusters (AgNCs) to GO. The developed system was exposed to the products of the hybridization chain reaction (HCR). The limit of detection for the reported assay was 1.18 nM [73,74].

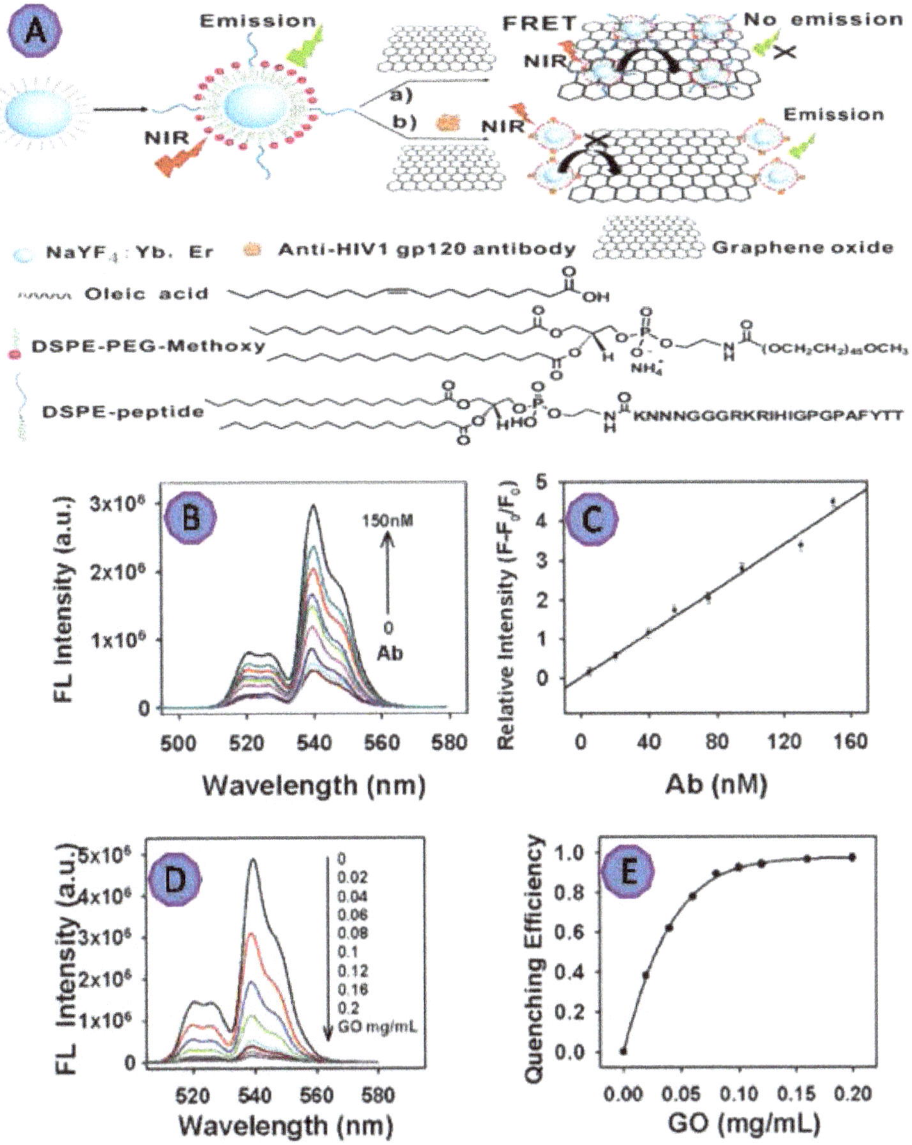

Figure 7. (**A**) Scheme of the upconversion FRET-based biosensor for detecting anti-HIV-1 gp120 antibody; (**B**) the upconversion fluorescence emission of the biosensor; (**C**) the calibration curve of the UCNPs showed a linear response with the various antibody concentrations in the range of 5–150 nM; (**D**) the fluorescence spectra of the peptide-functionalized UCNPs after incubation with varying concentrations of GO; (**E**) fluorescence quenching efficiency against GO concentration. Reprinted with permission from Ref. [72]. Copyright 2022, ROYAL SOCIETY OF CHEMISTRY.

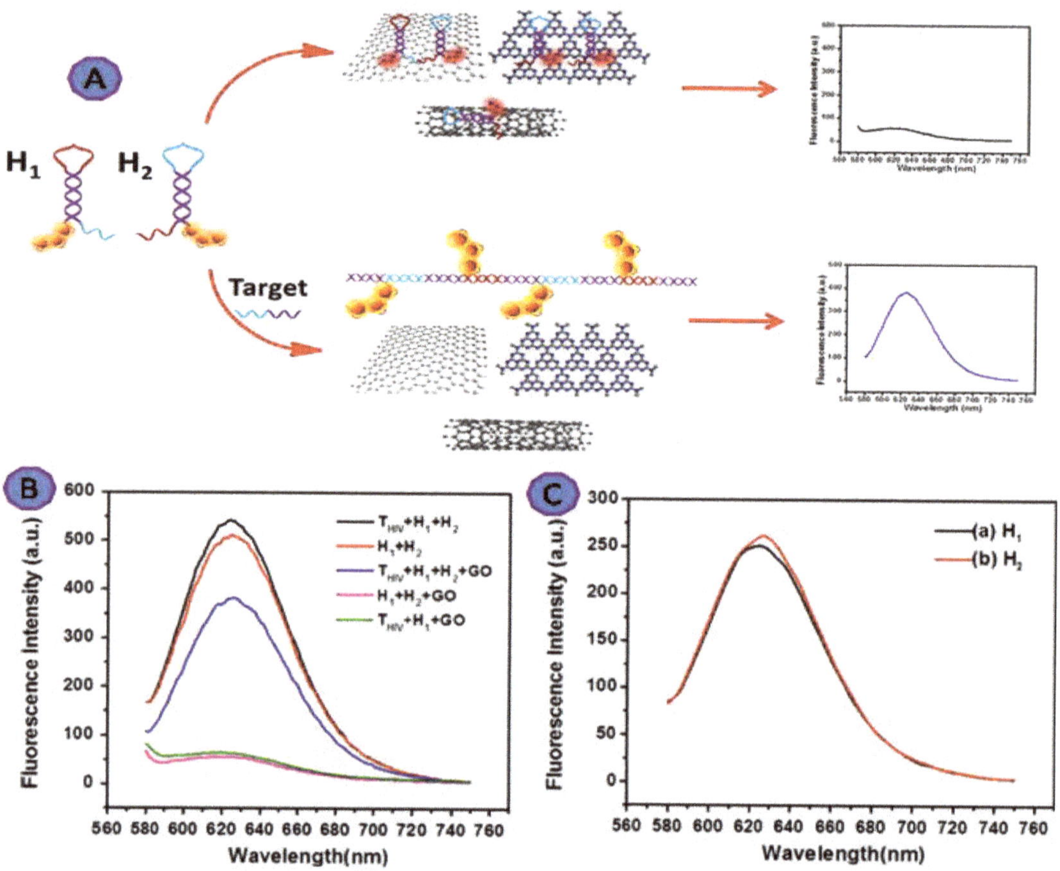

Figure 8. (**A**) Scheme showing graphene oxide (GO)-based HCR for DNA detection and the fluorescence response of the system under different conditions; (**B**) THIV + H1 + H2; H1 + H2; THIV + H1 + H2 + GO; H1 + H2 + GO; THIV + H1 + GO; (**C**) H1; H2. [H1] = 100 nM, [H2] = 100 nM, [THIV] = 100 nM. Reprinted with permission from Ref. [73]. Copyright 2017, Elsevier.

Moreover, Zhang et al. developed a fluorescence biosensing technique to detect HIV-1 protease based on GO, using covalently bonded, fluorescently labeled HIV-1 protease target peptide molecules with GO. The fluorescence was increased upon the addition of HIV-1 protease as a result of the peptide cleavage and the release of the fluorescent peptide fragment away from the GO, as shown in Figure 9. This fluorescence-based detection assay was successfully detected as little as 18 ng/mL of HIV-1 protease [75].

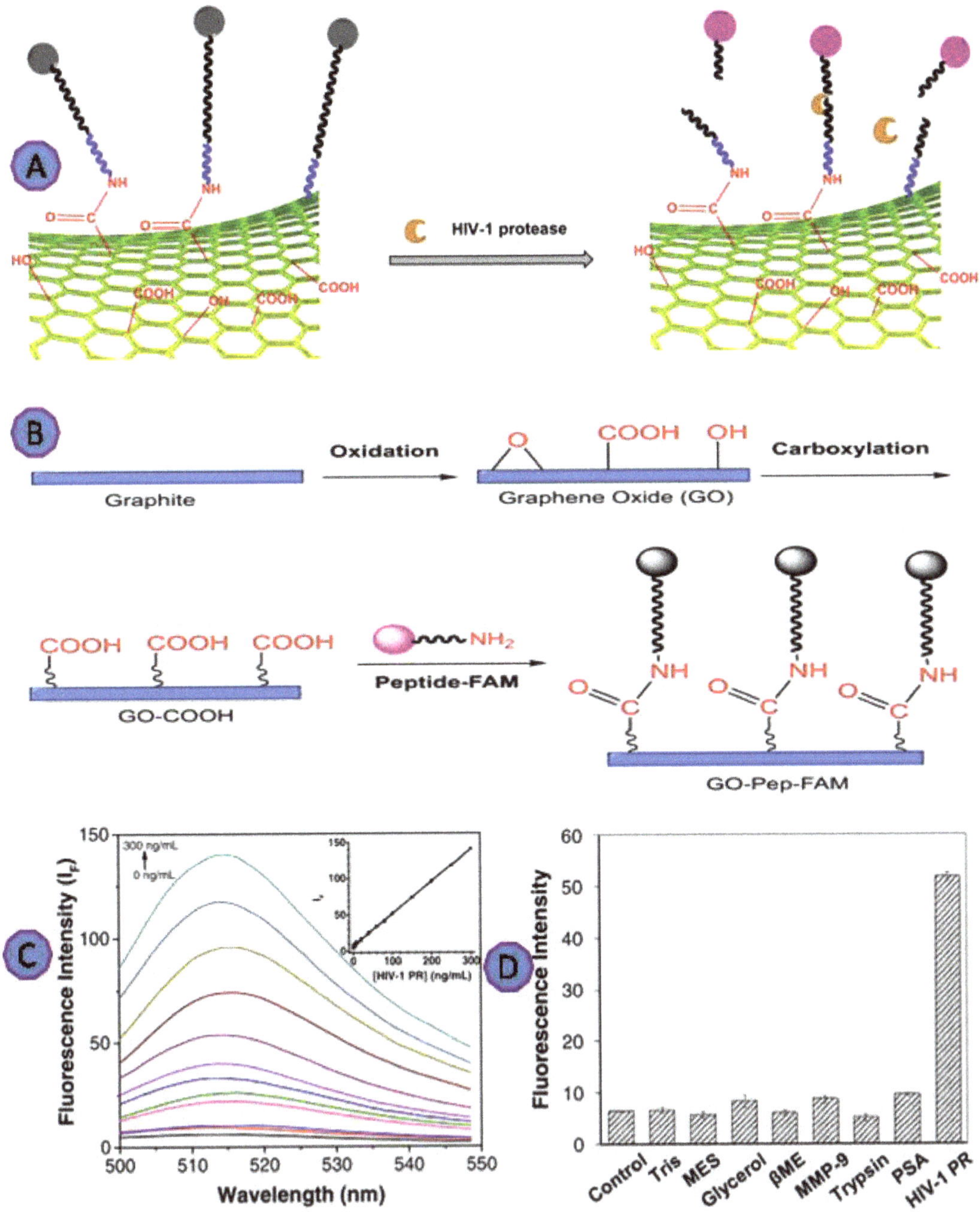

Figure 9. (**A**) Scheme depicting the principles of detecting HIV-1 protease; (**B**) scheme depicting the production of an HIV-1 protease sensor using graphite powder; (**C**) fluorescence spectra of the GO-Pep-FAM sensor in the presence of HIV-1 protease at various doses; (**D**) plot of fluorescence intensity versus various species, demonstrating the selectivity of the GO-Pep-FAM sensor. Reprinted with permission from Ref. [75]. Copyright 2018, Springer Nature.

2.3.5. Hepatitis B Virus

Hepatitis is a liver inflammatory disease. It is often caused by a viral infection known as viral hepatitis that leads to 686,000 deaths per year [74]. HBV is a Hepadnaviridae virus that has its viral double-stranded DNA enclosed. Scientists have developed various biosensors with high sensitivity to be used in the early diagnosis of HBV [76].

Carbon nitrides, g-C3N4, a new family of carbon functional materials, have recently attracted the attention of researchers in the field of biosensing. This is a metal-free carbonaceous substance with a high N:C ratio and a structure that ranges from polymeric to graphitic, with heptazine or triazine rings serving as structural motifs [77]. The properties of g-C3N4, such as large specific area, optical properties, low-cost, biocompatibility, and the stability of the photoluminescence, make it ideal for the fluorescence sensing. Therefore, Xiao et al. developed an innovative fluorescent biosensor to detect the HBV gene based on the fluorescence quenching of g-C3N4 nanosheets (Figure 10). This unique design shows a great deal of potential for monitoring disease markers, especially HBV genes, because it offers a low-cost and quick response, with a 1.0 nM LOD [59].

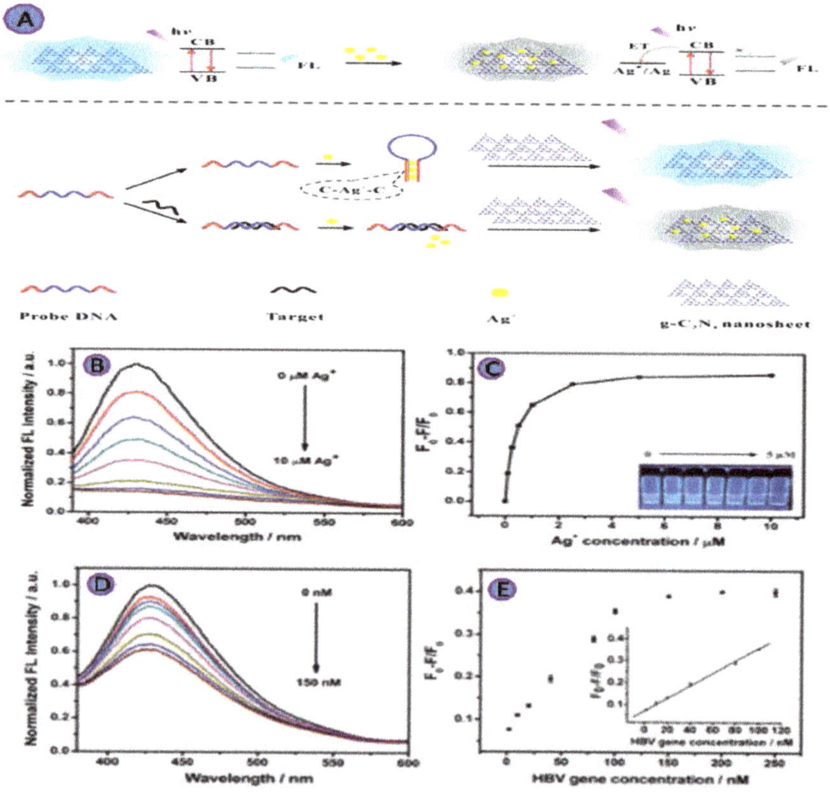

Figure 10. (**A**) A diagram of the method for detecting pathogenic DNA based on metal ion binding-induced fluorescence quenching of graphitic carbon nitride nanosheets; (**B**) fluorescence spectra of g-C3N4 nanosheets (1.0 gmL^{-1}) in the presence of various silver ion concentrations; (**C**) the fluorescence intensity of g-C3N4 nanosheets was measured in the presence of various numbers of silver ions; (**D**) fluorescence spectra of g-C3N4 nanosheets in the presence of different HBV gene concentrations; (**E**) the relative fluorescence intensity of g-C3N4 nanosheets in response to various HBV gene concentrations. Reprinted with permission from Ref. [60]. Copyright 2017, ROYAL SOCIETY OF CHEMISTRY.

2.3.6. Hepatitis C Virus (HCV)

The Hepatitis C virus (HCV) is the most widespread cause of chronic liver disease, affecting 2–3% of the world population. In the past, early detection of HCV was difficult, especially for those living in underdeveloped nations. Jialong et al. employed reduced graphene oxide nanosheets (rGONS) with the hybridization chain reaction (HCR) amplification technique, to develop an ultrasensitive method for the detection of HCV RNA. The reported LOD was as low as 10 fM, which is substantially lower than the commonly used fluorescence approach based on GO [78] (Figure 11).

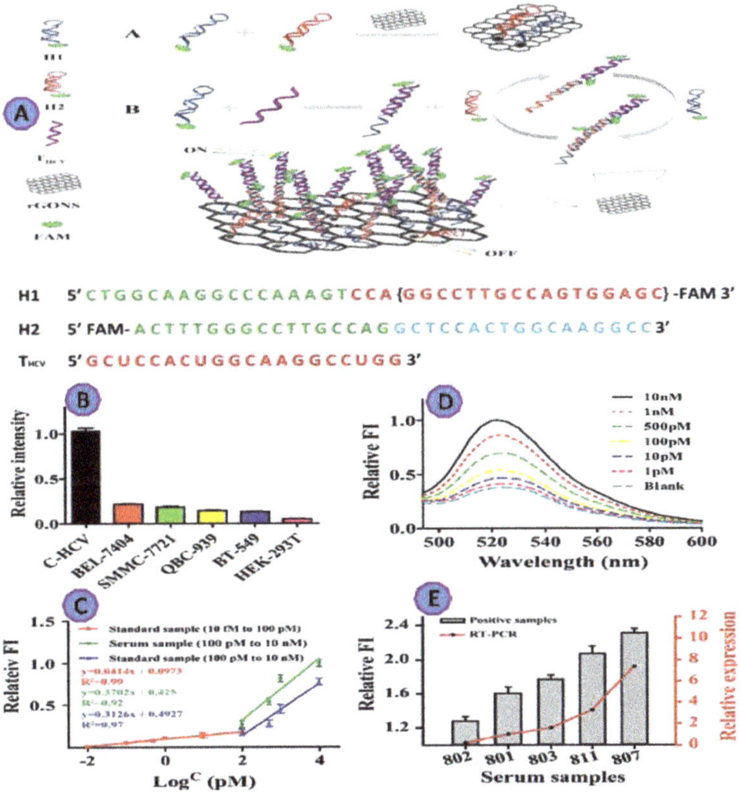

Figure 11. (**A**) Graphic depicting the proposed rGONS-HCR system for HCV detection; (**B**) THCV detection in various cell lysates; (**C**) fluorescence spectra of the rGONS-HCR in the presence of THCV at various concentrations in normal human serum; (**D**) the relative fluorescence intensity (FI) vs. THCV concentration; (**E**) THCV detection in clinical serum samples. Reprinted with permission from Ref. [79]. Copyright 2019, ROYAL SOCIETY OF CHEMISTRY.

2.4. Multiplexing Viruses Detection

One of the most important requirements for a biosensing platform is the simultaneous detection of multiple viruses infection (multiplexing) [80–82]. Different diseases can manifest with similar symptoms. Multiplex detection methodologies have become increasingly useful to detect a particular virus or viruses that cause an infection and their severity (quantitative) [51,82,83].

Chen et al. introduced a scalable, fluorescent biosensor for multiplex virus detection using graphene and 2D carbon materials. For the multiplexed analysis of virus genes, Chen et al. used an N, S co-doped GO platform for the detection of HBV as well as the detection of

HIV. In this study, two complementary sequences were conjugated with various fluorescent dyes to the HPV and HIV virus genes, as shown in Figure 12A. This platform showed an improvement in the detection limit to as little as 2.4 nM (Figure 12) [79].

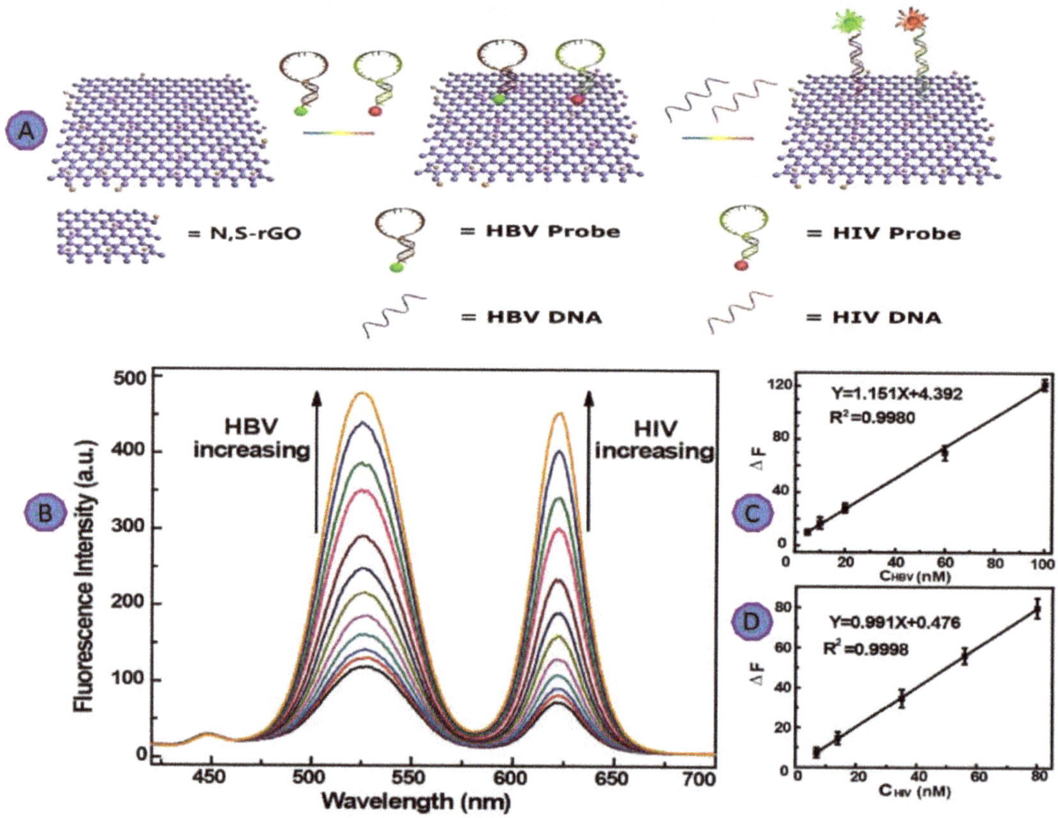

Figure 12. (**A**) Schematic illustration of N, S-rGO for multiplexing detection of HBV and HIV; (**B**) fluorescence spectra of N, S-rGO; (**C**) linear relationship for the detection of HBV DNA; (**D**) linear relationship for the detection of HIV DNA. Reprinted with permission from Ref. [77]. Copyright 2016, American Chemical Society.

Another fluorescent multiplexing viruses detection method has been designed, through a hybrid system consisting of GO and nucleic acid-stabilized silver nanoclusters (AgNCs), for the detection of the HIV and HBV genes, as shown in Figure 13. This design can be used for the multiplexing detection due to the tunable fluorescence properties of AgNCs and the combination of AgNCs/DNA with GO. The conjugation system of ssDNA sequences and the DNA-stabilized AgNCs has demonstrated that adsorption to GO induces AgNC fluorescence quenching. The reported LODs for the genes of HIV and HBV were 1 nM and 0.5 nM, respectively [84].

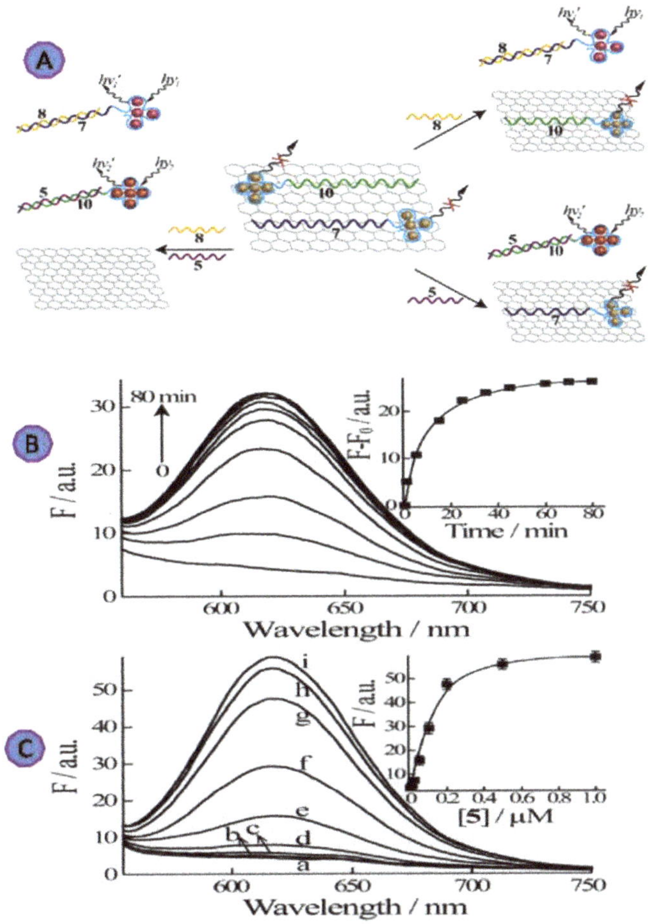

Figure 13. (**A**) Schematic diagram of the multiplexed detection platform for the HBV gene (5) and the HIV gene using near-infrared and red-emitting AgNC probes (8); (**B**) time-dependent fluorescence spectra of the red-emitting AgNCs upon challenging the (6)-AgNCs/GO with the HBV target gene (5), 100 nM. Inset: Time-dependent fluorescence changes at λ = 616 nm upon analyzing (5) by the (6)-AgNCs/GO. (**C**) Fluorescence spectra of the AgNCs upon analyzing different concentrations of the HBV gene (5) by the (6)-AgNCs/GO system. (a) 0, (b) 2 nM, (c)10 nM, (d) 20 nM, (e) 50 nM, (f) 100 nM, (g) 200 nM, (h) 500 nM, and (i) 1000 nM. Fluorescence spectra were recorded after a fixed time-interval of 80 min. Inset: Calibration curve corresponding to the luminescence of the related AgNCs (at λ = 616 nm) in the presence of different concentrations of (5). Reprinted with permission from Ref. [84]. Copyright 2013, American Chemical Society.

3. Literature on 2D Fluorescent Biosensors for the detection of Viruses

Two-dimensional materials have been used extensively by various researchers to detect different viruses. Table 1 summarizes some of the developed 2D fluorescent biosensors for the detection of various viruses using various recognition elements, their mechanisms of operation, and their limits of detection.

Table 1. List of the 2D fluorescent biosensors available for virus detection, types of employed recognition element, dynamic range, and limits of detection (LOD).

2D Carbon Structure	Virus Sensing	Recognition Element	Type of Interaction	Dynamic Range	Limit of Detection	REF
GO	Rotavirus	Antibodies	Carbodiimide-assisted amidation reaction	10^3 to 10^5 pfu mL^{-1}	10^5 pfu mL^{-1}	[64,65]
GO	Ebola virus gene Influenza	dsDNA	π–π stacking interaction	30 fM–3 nM	1.4 Pm	[67]
GO	Influenza virus H3N2 hemagglutinin gene	RNA	π–π stacking interaction and hydrogen bonding	37–9400 pg	3.8 pg	[38]
GO	HIV-1 gene	dsDNA	carbodiimide-assisted amidation reaction	50.0 fM–1.0 nM	15 fM	[71]
GO	HIV	Antibodies	π–π interaction	5–150 nM	2 nM	[72]
GO	HIV	dsDNA	π–π interaction	-	1.18 nM	[73,74]
GO	HIV	Enzyme	π–π stacking and/or electrostatic interaction between	5–300 ng/mL	109 pM	[75]
rGO	HCV	dsDNA	π–π interaction		10 fM	[78]
g-C3N4	HBV gene	DNA	π–π stacking	2–100 nM	1.0 nM	[59]
rGO	HIV	Aptamer	π–π interaction	-	3.0 nM	[79]
rGO	HBV	Aptamer	π–π interaction	-	2.4 nM	[79]
GO	HBV	ssDNA	π–π stacking	-	0.5 nM	[84]
GO	HIV	ssDNA	π–π stacking		1 nm	[84]

3.1. Overcoming the Drawbacks in Carbon Nanomaterials

The performance of the fluorescent biosensors can be influenced by different characteristics, such as sensitivity, the limit of detection (LOD), repeatability, as well as reproducibility. The fluorescence transduction mechanism depends on the morphological parameters, and coordination surface chemistry can be used to examine the reliability and capability of sensors [49].

The drawbacks of stability and nonspecific absorption of GO-based DNA have attracted more attention due to the decreasing electron transfer efficiency [35]. Some works have shown that these problems can be overcome by using the heteroatom doping technique. Heteroatom doping is a characteristic technique was used to influence the electronic properties within the GO surface. The possibility of developing a new electronic structure on the sp^2-hybridized carbon has gained a large amount of interest in terms of co-doping multiplex heteroatoms to enhance the activity of bare GO. For example, N, S co-doped GO with significant sulfur loading has been shown to efficiently break carbon material's inertness, activate the sp^2-hybridized carbon lattice, and permit electron transport from electron donor functional groups to GO sheets [82].

3.2. Detection Limit and Analysis Time

In short, the literature suggests that the detection limit is strongly determined by the sensitivity of the transducer, the mass transport kinetics and how fast the binding of the analyte with the recognition element [85]. Firstly, the sensitivity of the transducer can be determined by detecting the smallest concentration of the analyte that binds to the sensor's surface, taking into consideration the differentiation of the signal-to-noise ratio that occurs due to the nonspecific binding. The mass transport kinetics and reaction of the analyte with the recognition element can influence the analysis time [86].

Improving the detection limit and shortening the analysis time can be achieved by selecting highly efficient recognition elements that bind tightly and rapidly to the analyte(s) in question. Furthermore, the recognition element should have high specificity to the analyte in question, to reduce the cross-reactivity, and the sensor surface should be blocked well, to reduce the non-specific binding and to reduce false positive signals. Another means for shortening the analysis time is to enhance the transfer of the analyte to the immobilized recognition element onto the sensor surface. This includes integrating the sensor surface with various electrokinetic techniques, ultrasound standing waves, optical trapping and concentration, and the usage of pre-concentration techniques prior to analysis.

3.3. Non-Specific Binding

Generally, non-specific binding of the analyte, can cause some limitations in the biosensing platforms. Reliable biosensor devices must have the ability to distinguish between the target analyte of interest and the other species present in the sample. The majority of the competing species can interact non-specifically with the sensing platform, which can cause a false positive signal. To overcome this challenge, some approaches have developed 2D material-based sensors that show more specific properties for the species in question. Graphene is one of these 2D materials that introduces specific binding, and the resulting fluorescence comes only from the binding species, and not from the interfering species [87]. Graphene is functionalized via either non-covalent or covalent binding. Most of the covalent methods present very complex and difficult conditions to control.

External parameters such as temperature, pH, salt content, etc. that might affect the interaction of the biomolecules with the surface of the graphene, as well as their unique characteristics, represent a challenge for improving the efficiency. In addition, the heterogeneity of the graphene has great significance for the performance of fluorescent biosensors based on different factors, such as size, thickness, and the number of functional groups on the surface [31].

One mean of improving the sensor's specificity is to develop high-affinity and specific biosensing platforms via non-covalent attachment of the recognition elements to the surfaces of 2D carbon nanomaterials [88].

3.4. Cytotoxicity

The carbon element is intrinsically compatible with living systems, therefore, carbon-based nanomaterials are considered "safe". Although graphene-based sensors have received a great deal of attention in health monitoring and biomedical applications. Numerous studies have investigated the graphene-based nanomaterials' biocompatibility, water-solubility, toxicity, and potential environmental risks. As a result, the increased use of graphene in a variety of applications and industries will require different toxicological profiles in-vitro and in-vivo [64,89]. However, a variety of carbon nanomaterials, such as carbon nanotubes, nanodiamonds, and graphene oxide, have been examined in living cell lines and found to have little cytotoxicity [41]. The morphology, size, shape, purity, and surface chemistry have a powerful impact on the cytotoxicity, while the functionalization of graphene can adjust its interactions with the living cells [90]. Due to the lack of in-vivo applications, the cytotoxicity properties of graphene are still unknown. Several factors, including the lateral scale, oxygen content, and dose, need to be carefully researched [20,43,64].

A cytotoxicity evaluation of graphene using rat pheochromocytoma cells (PC12) observed 40% cell death using the MTT assay [91]. The cytotoxicity of GO on human lung carcinoma (A549) cells was observed with no cellular uptake, which resulted in decreased viability at higher concentrations. Moreover, the cytotoxicity of graphene depends on the quality of its dispersion [89]. Zhang et al. noticed an increase in the cytotoxicity of graphene when applied above 10 μg/mL. The study observed no pathological alterations with the use of distributed single-layered GO sheets of 10–800 nm in lateral size in Kunming mice, at a dose of either 1 mg/kg or 10 mg/kg [92]. It was noticed that the presence of 10% FBS in the cell media decreased the cytotoxicity of GO because of the direct physical damage of the plasma membrane caused by GO [93,94].

4. Future Outlook and Challenges

Great efforts have been devoted to designing new graphene-based FRET sensing systems for clinical testing, demonstrating inexpensive biosensing platforms with the eco-friendly, sustainable synthesis of graphene to solve the existing stability issues and efficiently enhance the electron transfer rate in enzymatic biosensors [95,96].

Graphene-like 2D nanomaterials such as graphene, GO, and rGO, with their functional reactive groups such as –OH, –COOH, –CO, and –C–O–C, are crucial for the immobilization of recognition elements. The strong quenching efficiency of GO and rGO in comparison with graphene offers exciting opportunities for the development of fluorescent biosensors with significant improved sensitivity and selectivity, and low limits of detection. The amphiphilic character of the GO facilitates the adsorption of diverse biomolecules on its surface due to its strong binding affinity with the biomolecules via pi–pi stacking. Several challenges still hinder the practical application of graphene-based smart materials—for example, the lack of techniques that can produce high-quality graphene materials at low-cost and in a more controllable and scalable fashion. It is necessary to investigate the toxicity of these materials—both in-vitro and in-vivo—via the functionalization of graphene-based smart materials. Because graphene assemblies are lightweight, flexible, and elastic, it is of great interest to develop graphene-based smart materials that can sensitively respond to multiple stimuli or are operational under harsh conditions (for example, high temperatures, high strength, and strongly corrosive media [97]). One of the most critical scientific challenges is to produce ultrasensitive fluorescent biosensors for rapid and specific detection at low-cost. Another challenge facing the graphene-based biosensors is the reproducibility and reliability of the materials production and processes [90].

To meet the ever-increasing demands of the fluorescent biosensors in nanomedicine applications, fluorescent biosensors must be able to distinguish the signals of the target analyte from the signal of the interferences. By applying different efficient recognition elements, they are expected to facilitate highly specific multiplexing viruses detection. Many of the current biorecognition elements are very reliable, while some of them have poor chemical stability, have a limited shelf-life, suffer from cross-reactivity, and are expensive, necessitating further advances in the synthesis and selection of emerging recognition elements to overcome the shortcomings. Multiplexing detection technologies are considered one of the main areas of future development.

Enhancing the signal-to-noise ratio, which is normally calculated by the receptor–nanomaterial–transducer interface, is another significant factor to consider. More research into the mechanisms of interactions between the DNA probes or modified DNA probes and the graphene-based transducer is required to provide more reliable and accurate measurements. The full potential of currently researched biorecognition elements is limited due to the lack of reproducibility and reusability of the biosensing platforms performance characteristics in the literature.

Despite the massive increase in the use of 2D carbon materials in point-of-care diagnostics, future studies considering their toxicology and biocompatibility are still needed.

To simplify healthcare and medicine in the future, a massive increase in the use of 2D carbon materials in point-of-care diagnostics studies will be necessary, such as touchscreen devices for epidermal electronic devices [98]. Huang et al. developed a cost-effective, fast, accurate, and specific microfluidic chip for point-of-care testing that allows for the simultaneous identification of various clinical pneumonia-related bacteria in 1.45 μL reactions, without cross-contamination, in 45 min [99]. Pang et al. designed a novel, rapid, and sensitive self-priming PDMS/paper hybrid (SPH) microfluidic chip for the detection of specific multiplex foodborne pathogens via mixed-dye loading. This device can detect eight pathogens simultaneously by increasing the number of reaction chambers in the SPH chip, demonstrating its significant potential for on-site food safety screening and wider application [100].

Point-of-care (POC) devices enable multiplex, portable, rapid, and economically affordable instrumentation, with respect to traditional culture or PCR-based assays. Such instrumentation can be further miniaturized into point-of-care (POC) devices. All these features can be effectively utilized in the detection of pathogenic microorganisms [29,101]. Using a smartphone translation tool, point-of-care (POC) devices look to be a promising step for physicians in lab analysis. Researchers are looking at new smartphone-connected POCT devices to improve data sensitivity, with additional advantages such as privacy and theft protection, medical recommendations, and time- and cost-intensive laboratory processing without sample pretreatment. They should be portable and user-friendly while still delivering appropriate analytical results and clinical relevance [102]. The sensitivity, low detection limit, and selectivity of the point-of-care (POC) devices must be improved by utilizing a sensing prototype based on a smart optoelectronic nanosystems, using precise diagnostics without interferents or loss to choose a real sample source. These procedures are used to validate and scale up the sensor for clinical use [103]. Due to the present COVID-19 pandemic, a reliable, rapid, cost-effective, and simple testing technique is urgently required. One improvement could be integration with lab-on-chip to clean up the sample and pre-concentrate the analyte of interest. The application of graphene in LOC devices has greatly increased their sensing capabilities [104].

As a result, graphene-based LOC devices can provide a robust microenvironment for advanced detection techniques at a low-cost [105]. Xiang et al. developed a microscale fluorescence-based colorimetric sensor employing a graphene nanoprobe for multiplexed DNA analysis [106]. A highly flexible microfluidic tactile sensor based on a GO nanosuspension was reported by Kenry et al. [107]. A microfluidic chip integrated with a novel GO-based Forster resonance energy transfer biosensor was fabricated by Cao et al. to detect cancer cells in-situ [96].

Graphene has great mechanical strength and also has the feature of flexibility. Due to its remarkable mechanical strength, graphene is a useful material in the production of wearable LOC devices. Graphene also has a large specific surface area, which helps surfaces to load the necessary ligand in the LOC devices for single-molecule detection. Furthermore, graphene has tunable optical characteristics, which are highly useful for optical readouts in fluorescent LOCs to detect a wide variety of viruses, ranging from Influenza virus to the present SARS-CoV-2 virus [105]. For the detection of H1N1, Joshi et al. fabricated a stable, reproducible, and green device based on rGO flakes [108]. For the detection of avian Influenza, Roberts et al. developed a functionalized graphene-based field-effect transistor (GFET) with an LOD of 10 fM [109]. For the detection of Ebola virus, Jin et al. obtained an LOD of 2.4 pg/mL based on the response of rGO with GFET as a function of the Dirac voltage [110]. For the detection of SARS-CoV-2, Simone et al. [111] theoretically studied the possibility of using a graphene-based optical sensor via employing Raman scattering, generating results within 1 min and having an LOD of $1.68 \times 10 -22$ μg mL^{-1}. Torrente-Rodríguez et al. [53] designed a highly selective, unique, ultrasensitive, multiplexed, and rapid graphene-based immunosensor and telemedicine system for the detection of SARS-CoV-2 in blood and saliva samples.

5. Conclusions

The excellent properties of 2D carbon nanomaterials, such as high sensitivity and specificity, ease of operation, and fast detection, make them suitable to detect viruses, e.g., Influenza, Hepatitis B, HIV, and Rotavirus, using fluorescent biosensing techniques. This review focused on the advantages of 2D nanomaterial-based biosensors. Despite the developments and advancements in laboratory-based virus detection, practical implementations face challenges in terms of in-field applications. Additional excellent 2D carbon fluorescent biosensors will be proposed in the future, and new solutions will arise from collaborative efforts in the field of nanomedicine. As a result, further research into the feasibility and broad practical applicability of fluorescent biosensors for multiplexing detection is needed to augment the detection efficiency. The challenges for the researchers in this domain are (a) the need to improve efficiency through production and biofunctionalization, (b) the improvement of the systems' robustness, (c) the application of biosensors in implantable systems, and (d) advancements in large-scale manufacturing strategies with small-sized devices.

In summary, the development of fluorescent biosensors based on graphene nanomaterials represents a major turning point in the medical field, which could improve the quality of life of many patients.

Author Contributions: A.M.S., G.Y., J.L. conceptualization, methodology, writing first draft. A.M.S., G.Y., M.Z., J.L. proof read, edit and finalizing the review. All authors have read and agreed to the published version of the manuscript.

Funding: This research received no external funding.

Acknowledgments: We would like to thank all the co-workers cited in the references for their invaluable contributions to the work described here.

Conflicts of Interest: The authors declare no conflict of interest.

References

1. Ahmed, S.R.; Mogus, J.; Chand, R.; Nagy, E.; Neethirajan, S. Optoelectronic fowl adenovirus detection based on local electric field enhancement on graphene quantum dots and gold nanobundle hybrid. *Biosens. Bioelectron.* **2018**, *103*, 45–53. [CrossRef] [PubMed]
2. Bianying, F.; Linjie, G.; Lihua, W.; Fan, L.; Jianxin, L.; Jimin, G.; Chunhai, F.; Qing, H. A graphene oxide-based fluorescent biosensor for the analysis of peptide-receptor interactions and imaging in somatostatin receptor subtype 2 overexpressed tumor cells. *Anal. Chem.* **2013**, *85*, 7732–7737. [CrossRef] [PubMed]
3. Constantinou, A.; Polizzi, K.M. Opportunities for bioprocess monitoring using FRET biosensors. *Biochem. Soc. Trans.* **2013**, *41*, 1146–1151. [CrossRef] [PubMed]
4. Dong, Y.; Zhang, T.; Lin, X.; Feng, J.; Luo, F.; Gao, H.; Wu, Y.; Deng, R.; He, Q. Graphene/aptamer probes for small molecule detection: From in vitro test to in situ imaging. *Microchim. Acta* **2020**, *187*, 179. [CrossRef]
5. Ravikumar, A.; Panneerselvam, P.; Radhakrishnan, K. Fluorometric determination of lead(II) and mercury(II) based on their interaction with a complex formed between graphene oxide and a DNAzyme. *Microchim. Acta* **2018**, *185*, 2. [CrossRef]
6. Chen, J.; Nugen, S.R. Detection of protease and engineered phage-infected bacteria using peptide-graphene oxide nanosensors. *Anal. Bioanal. Chem.* **2019**, *411*, 2487–2492. [CrossRef]
7. Nguyen, E.P.; de Carvalho Castro Silva, C.; Merkoçi, A. Recent advancement in biomedical applications on the surface of two-dimensional materials: From biosensing to tissue engineering. *Nanoscale* **2020**, *12*, 19043–19067. [CrossRef]
8. Sohail Haroone, M.; Li, L.; Ahmad, A.; Huang, Y.; Ma, R.; Zhang, P.; Hu, Y.; Muhammad Kaleem, Q.; Lu, J. Luminous composite ultrathin films of CdTe quantum dots/silk fibroin co-assembled with layered doubled hydroxide: Enhanced photoluminescence and biosensor application. *J. Mater.* **2018**, *4*, 165–171. [CrossRef]
9. Ryoo, S.R.; Yim, Y.; Kim, Y.K.; Park, I.S.; Na, H.K.; Lee, J.; Jang, H.; Won, C.; Hong, S.; Kim, S.Y.; et al. High-throughput chemical screening to discover new modulators of microRNA expression in living cells by using graphene-based biosensor. *Sci. Rep.* **2018**, *8*, 11413. [CrossRef]
10. He, Y.; Xing, X.; Tang, H.; Pang, D. Graphene oxide-based fluorescent biosensor for protein detection via terminal protection of small-molecule-linked DNA. *Small* **2013**, *9*, 2097–2101. [CrossRef]
11. Zhang, Z.; Xia, X.; Xiang, X.; Huang, F.; Han, L. Conjugated cationic polymer-assisted amplified fluorescent biosensor for protein detection via terminal protection of small molecule-linked DNA and graphene oxide. *Sens. Actuators B Chem.* **2017**, *249*, 8–13. [CrossRef]
12. Girigoswami, K.; Akhtar, N. Nanobiosensors and fluorescence based biosensors: An overview. *Int. J. Nano Dimens.* **2019**, *10*, 1–17.
13. Rabiee, N.; Bagherzadeh, M.; Ghadiri, A.M.; Salehi, G. ZnAl nano layered double hydroxides for dual functional CRISPR/Cas9 delivery and enhanced green fluorescence protein biosensor. *Sci. Rep.* **2020**, *10*, 20672. [CrossRef]
14. Morales, M.A.; Halpern, J.M. Guide to Selecting a Biorecognition Element for Biosensors. *Bioconjug. Chem.* **2018**, *29*, 3231–3239. [CrossRef]
15. Sang, S.; Zhang, W.; Zhao, Y. Review on the Design Art of Biosensors. *State Art Biosens.-Gen. Asp.* **2013**. [CrossRef]

16. Zhou, L.; Mao, H.; Wu, C.; Tang, L.; Wu, Z.; Sun, H.; Zhang, H.; Zhou, H.; Jia, C.; Jin, Q.; et al. Label-free graphene biosensor targeting cancer molecules based on non-covalent modification. *Biosens. Bioelectron.* **2017**, *87*, 701–707. [CrossRef]
17. Shirani, M.; Kalantari, H.; Khodayar, M.J.; Kouchak, M.; Rahbar, N. A novel strategy for detection of small molecules based on aptamer/gold nanoparticles/graphitic carbon nitride nanosheets as fluorescent biosensor. *Talanta* **2020**, *219*, 121235. [CrossRef] [PubMed]
18. Zhang, Z.; Liu, J. An engineered one-site aptamer with higher sensitivity for label-free detection of adenosine on graphene oxide. *Can. J. Chem.* **2018**, *96*, 957–963. [CrossRef]
19. Damborský, P.; Švitel, J.; Katrlík, J. Optical biosensors. *Essays Biochem.* **2016**, *60*, 91–100. [CrossRef] [PubMed]
20. Mehrotra, P. ScienceDirect Biosensors and their applications—A review. *J. Oral Biol. Craniofacial Res.* **2016**, *6*, 153–159. [CrossRef] [PubMed]
21. Meyer, D.; Hagemann, A.; Kruss, S. Kinetic Requirements for Spatiotemporal Chemical Imaging with Fluorescent Nanosensors. *ACS Nano* **2017**, *11*, 4017–4027. [CrossRef]
22. Ehtesabi, H. Application of carbon nanomaterials in human virus detection. *J. Sci. Adv. Mater. Devices* **2020**, *5*, 436–450. [CrossRef]
23. Devi, M. Application of 2D Nanomaterials as Fluorescent Biosensors. *ACS Symp. Ser.* **2020**, *1353*, 117–141. [CrossRef]
24. Hwang, H.S.; Jeong, J.W.; Kim, Y.A.; Chang, M. Carbon nanomaterials as versatile platforms for biosensing applications. *Micromachines* **2020**, *11*, 814. [CrossRef]
25. Ramnani, P.; Saucedo, N.M.; Mulchandani, A. Carbon nanomaterial-based electrochemical biosensors for label-free sensing of environmental pollutants. *Chemosphere* **2016**, *143*, 85–98. [CrossRef]
26. Rajakumar, G.; Zhang, X.H.; Gomathi, T.; Wang, S.F.; Ansari, M.A.; Mydhili, G.; Nirmala, G.; Alzohairy, M.A.; Chung, I.M. Current use of carbon-based materials for biomedical applications-A prospective and review. *Processes* **2020**, *8*, 355. [CrossRef]
27. Kour, R.; Arya, S.; Young, S.-J.; Gupta, V.; Bandhoria, P.; Khosla, A. Review—Recent Advances in Carbon Nanomaterials as Electrochemical Biosensors. *J. Electrochem. Soc.* **2020**, *167*, 037555. [CrossRef]
28. Zhu, C.; Du, D.; Lin, Y. Graphene-like 2D nanomaterial-based biointerfaces for biosensing applications. *Biosens. Bioelectron.* **2017**, *89*, 43–55. [CrossRef]
29. Vermisoglou, E.; Panáček, D.; Jayaramulu, K.; Pykal, M.; Frébort, I.; Kolář, M.; Hajdúch, M.; Zbořil, R.; Otyepka, M. Human virus detection with graphene-based materials. *Biosens. Bioelectron.* **2020**, *166*, 112436. [CrossRef]
30. Tan, J.; Lai, Z.; Zhang, Z.; Zheng, R.; Su, J.; Huang, Y.; Huang, P.; Song, H.; Yang, N.; Zhou, S.; et al. A Graphene Oxide-Based Fluorescent Aptasensor for the Turn-on Detection of CCRF-CEM. *Nanoscale Res. Lett.* **2018**, *13*, 66. [CrossRef]
31. Liu, F.; Choi, J.Y.; Seo, T.S. Biosensors and Bioelectronics Graphene oxide arrays for detecting specific DNA hybridization by fluorescence resonance energy transfer. *Biosens. Bioelectron.* **2010**, *25*, 2361–2365. [CrossRef]
32. Tehrani, Z.; Burwell, G.; Mohd Azmi, M.A.; Castaing, A.; Rickman, R.; Almarashi, J.; Dunstan, P.; Miran Beigi, A.; Doak, S.H.; Guy, O.J. Generic epitaxial graphene biosensors for ultrasensitive detection of cancer risk biomarker. *2D Mater.* **2014**, *1*, 025004. [CrossRef]
33. Cui, F.; Ji, J.; Sun, J.; Wang, J.; Wang, H.; Zhang, Y.; Ding, H.; Lu, Y.; Xu, D.; Sun, X. A novel magnetic fluorescent biosensor based on graphene quantum dots for rapid, efficient, and sensitive separation and detection of circulating tumor cells. *Anal. Bioanal. Chem.* **2019**, *411*, 985–995. [CrossRef] [PubMed]
34. Liu, B.; Huang, P.J.J.; Kelly, E.Y.; Liu, J. Graphene oxide surface blocking agents can increase the DNA biosensor sensitivity. *Biotechnol. J.* **2016**, *11*, 780–787. [CrossRef] [PubMed]
35. Wang, L.; Wu, A.; Wei, G. Graphene-based aptasensors: From molecule-interface interactions to sensor design and biomedical diagnostics. *Analyst* **2018**, *143*, 1526–1543. [CrossRef]
36. Waiwijit, U.; Phokaratkul, D.; Kampeera, J.; Lomas, T.; Wisitsoraat, A.; Kiatpathomchai, W.; Tuantranont, A. Graphene oxide based fluorescence resonance energy transfer and loop-mediated isothermal amplification for white spot syndrome virus detection. *J. Biotechnol.* **2015**, *212*, 44–49. [CrossRef]
37. Bayraktutan, T.; Meral, K. Merocyanine 540 adsorbed on polyethylenimine-functionalized graphene oxide nanocomposites as a turn-on fluorescent sensor for bovine serum albumin. *Phys. Chem. Chem. Phys.* **2016**, *18*, 23400–23406. [CrossRef]
38. Jeong, S.; Kim, D.; Kim, D.H.; Kim, D. Fluorometric detection of in fluenza viral RNA using graphene oxide. *Anal. Biochem.* **2018**, *561–562*, 66–69. [CrossRef]
39. Hochreiter, B.; Garcia, A.P.; Schmid, J.A. Fluorescent proteins as genetically encoded FRET biosensors in life sciences. *Sensors* **2015**, *15*, 26281–26314. [CrossRef]
40. Salihoglu, O.; Kakenov, N.; Balci, O.; Balci, S.; Kocabas, C. Graphene as a Reversible and Spectrally Selective Fluorescence Quencher. *Sci. Rep.* **2016**, *6*, srep33911. [CrossRef]
41. Huang, H.; Su, S.; Wu, N.; Wan, H.; Wan, S.; Bi, H.; Sun, L. Graphene-based sensors for human health monitoring. *Front. Chem.* **2019**, *7*, 399. [CrossRef]
42. Kalkal, A.; Pradhan, R.; Kadian, S.; Manik, G.; Packirisamy, G. Biofunctionalized Graphene Quantum Dots Based Fluorescent Biosensor toward Efficient Detection of Small Cell Lung Cancer. *ACS Appl. Bio Mater.* **2020**, *3*, 4922–4932. [CrossRef]
43. Zhang, H.; Zhang, H.; Aldalbahi, A.; Zuo, X.; Fan, C.; Mi, X. Fluorescent biosensors enabled by graphene and graphene oxide. *Biosens. Bioelectron.* **2017**, *89*, 96–106. [CrossRef]
44. Alhazmi, H.A.; Ahsan, W.; Mangla, B.; Javed, S.; Hassan, M.Z. Graphene-based biosensors for disease theranostics: Development, applications, and recent advancements. *Nanotechnol. Rev.* **2022**, *11*, 96–116. [CrossRef]

45. Wang, W.; Su, H.; Wu, Y.; Zhou, T.; Li, T. Review—Biosensing and Biomedical Applications of Graphene: A Review of Current Progress and Future Prospect. *J. Electrochem. Soc.* **2019**, *166*, B505–B520. [CrossRef]
46. Di Pietrantonio, F.; Cannatà, D.; Benetti, M. *Biosensor Technologies Based on Nanomaterials*; Elsevier Inc.: Amsterdam, The Netherlands, 2019.
47. Peltomaa, R.; Glahn-Martínez, B.; Benito-Peña, E.; Moreno-Bondi, M.C. Optical Biosensors for Label-Free Detection of Small Molecules. *Sensors* **2018**, *18*, 4126. [CrossRef]
48. Gosai, A.; Khondakar, K.R.; Ma, X.; Ali, M.A. Application of functionalized graphene oxide based biosensors for health monitoring: Simple graphene derivatives to 3D printed platforms. *Biosensors* **2021**, *11*, 384. [CrossRef]
49. Kim, T.H.; Lee, D.; Choi, J.W. Live cell biosensing platforms using graphene-based hybrid nanomaterials. *Biosens. Bioelectron.* **2017**, *94*, 485–499. [CrossRef]
50. Zhao, H.; Ding, R.; Zhao, X.; Li, Y.; Qu, L.; Pei, H.; Yildirimer, L.; Wu, Z.; Zhang, W. Graphene-based nanomaterials for drug and/or gene delivery, bioimaging, and tissue engineering. *Drug Discov. Today* **2017**, *22*, 1302–1317. [CrossRef]
51. Jiang, Z.; Feng, B.; Xu, J.; Qing, T.; Zhang, P.; Qing, Z. Graphene biosensors for bacterial and viral pathogens. *Biosens. Bioelectron.* **2020**, *166*, 112471. [CrossRef]
52. Ozcelik, D.; Parks, J.W.; Wall, T.A.; Stott, M.A.; Cai, H.; Parks, J.W.; Hawkins, A.R.; Schmidt, H. Optofluidic wavelength division multiplexing for single-virus detection. *Proc. Natl. Acad. Sci. USA* **2015**, *112*, 12933–12937. [CrossRef]
53. Torrente-Rodríguez, R.M.; Lukas, H.; Tu, J.; Min, J.; Yang, Y.; Xu, C.; Rossiter, H.B.; Gao, W. SARS-CoV-2 RapidPlex: A Graphene-Based Multiplexed Telemedicine Platform for Rapid and Low-Cost COVID-19 Diagnosis and Monitoring. *Matter* **2020**, *3*, 1981–1998. [CrossRef]
54. Tymm, C.; Zhou, J.; Tadimety, A.; Burklund, A.; Zhang, J.X.J. Scalable COVID-19 Detection Enabled by Lab-on-Chip Biosensors. *Cell. Mol. Bioeng.* **2020**, *13*, 313–329. [CrossRef]
55. Hao, X.; Liu, R.; He, Y.; Xiao, X.; Xiao, W.; Zheng, Q.; Lin, X.; Tao, P.; Zhou, P.; Li, S. Multiplex PCR methods for detection of several viruses associated with canine respiratory and enteric diseases. *PLoS ONE* **2019**, *14*, e0213295. [CrossRef]
56. Zhang, X.; Hu, Y.; Yang, X.; Tang, Y.; Han, S.; Kang, A.; Deng, H.; Chi, Y.; Zhu, D.; Lu, Y. FÖrster resonance energy transfer (FRET)-based biosensors for biological applications. *Biosens. Bioelectron.* **2019**, *138*, 111314. [CrossRef]
57. Zhang, Z.; Tang, Z.; Farokhzad, N.; Chen, T.; Tao, W. Sensitive, Rapid, Low-Cost, and Multiplexed COVID-19 Monitoring by the Wireless Telemedicine Platform. *Matter* **2020**, *3*, 1818–1820. [CrossRef]
58. Li, Z.; Yin, X.; Sun, Y.; Qu, L.; Du, D.; Lin, Y. Functionalized Two-Dimensional Nanomaterials for Biosensing and Bioimaging. *ACS Symp. Ser.* **2020**, *1353*, 143–165. [CrossRef]
59. Xiao, Y.; Sheng, Y.; Zhou, J.; Chen, M.; Wen, W.; Zhang, X.; Wang, S. A novel label-free strategy for pathogenic DNA detection based on metal ions binding-induced fluorescence quenching of graphitic carbon nitride nanosheets. *Analyst* **2017**, *142*, 2617–2623. [CrossRef]
60. Bányai, K.; Estes, M.K.; Martella, V.; Parashar, U.D. Viral gastroenteritis. *Lancet* **2018**, *392*, 175–186. [CrossRef]
61. Tian, F.; Lyu, J.; Shi, J.; Yang, M. Graphene and graphene-like two-denominational materials based fluorescence resonance energy transfer (FRET) assays for biological applications. *Biosens. Bioelectron.* **2017**, *89*, 123–135. [CrossRef]
62. Chang, X.H.; Zhang, J.; Wu, L.H.; Peng, Y.K.; Yang, X.Y.; Li, X.L.; Ma, A.J.; Ma, J.C.; Chen, G.Q. Research progress of near-infrared fluorescence immunoassay. *Micromachines* **2019**, *10*, 422. [CrossRef]
63. Shi, J.; Tian, F.; Lyu, J.; Yang, M. Nanoparticle based fluorescence resonance energy transfer (FRET) for biosensing applications. *J. Mater. Chem. B* **2015**, *3*, 6989–7005. [CrossRef]
64. Morales-Narváez, E.; Merkoçi, A. Graphene oxide as an optical biosensing platform. *Adv. Mater.* **2012**, *24*, 3298–3308. [CrossRef]
65. Jung, J.H.; Cheon, D.S.; Liu, F.; Lee, K.B.; Seo, T.S. A Graphene Oxide Based Immuno-biosensor for Pathogen Detection. *Angew. Chem.* **2010**, *49*, 5708–5711. [CrossRef]
66. James, A.S.; Todd, S.; Pollak, N.M.; Marsh, G.A.; Macdonald, J. Ebolavirus diagnosis made simple, comparable and faster than molecular detection methods: Preparing for the future. *Virol. J.* **2018**, *15*, 75. [CrossRef]
67. Wen, J.; Li, W.; Li, J.; Tao, B.; Xu, Y.; Li, H.; Lu, A.; Sun, S. Study on rolling circle amplification of Ebola virus and fluorescence detection based on graphene oxide. *Sens. Actuators B Chem.* **2016**, *227*, 655–659. [CrossRef]
68. Krammer, F.; Smith, G.J.D.; Fouchier, R.A.M.; Peiris, M.; Kedzierska, K.; Doherty, P.C.; Palese, P.; Shaw, M.L.; Treanor, J.; Webster, R.G.; et al. Influenza. *Nat. Rev. Dis. Primers* **2018**, *4*, 4. [CrossRef]
69. Srivastava, S.; Singh, P.K.; Vatsalya, V.; Karch, R.C. Developments in the Diagnostic Techniques of Infectious Diseases: Rural and Urban Prospective. *Adv. Infect. Dis.* **2018**, *8*, 121–138. [CrossRef]
70. Kabir, M.A.; Zilouchian, H.; Caputi, M.; Asghar, W. Advances in HIV diagnosis and monitoring. *Crit. Rev. Biotechnol.* **2020**, *40*, 623–638. [CrossRef]
71. Qaddare, S.H.; Salimi, A. Amplified fluorescent sensing of DNA using luminescent carbon dots and AuNPs/GO as a sensing platform: A novel coupling of FRET and DNA hybridization for homogeneous HIV-1 gene detection at femtomolar level. *Biosens. Bioelectron.* **2017**, *89*, 773–780. [CrossRef]
72. Wu, Y.; Cen, Y.; Huang, L.; Yu, R.; Chu, X. Upconversion fluorescence resonance energy transfer biosensor for sensitive detection of human immunodeficiency virus antibodies in human serum. *Chem. Commun.* **2014**, *50*, 4759–4762. [CrossRef]

73. Zhang, S.; Wang, K.; Li KBin Shi, W.; Jia, W.P.; Chen, X.; Sun, T.; Han, D.M. A DNA-stabilized silver nanoclusters/graphene oxide-based platform for the sensitive detection of DNA through hybridization chain reaction. *Biosens. Bioelectron.* **2017**, *91*, 374–379. [CrossRef]
74. Arora, K. Recent Biosensing Applications of Graphene-Based Nanomaterials. *Handb. Graphene* **2019**, *6*, 297–348.
75. Zhang, Y.; Chen, X.; Roozbahani, G.M.; Guan, X. Graphene oxide-based biosensing platform for rapid and sensitive detection of HIV-1 protease. *Anal. Bioanal. Chem.* **2018**, *410*, 6177–6185. [CrossRef]
76. Abd Muain, M.F.; Cheo, K.H.; Omar, M.N.; Amir Hamzah, A.S.; Lim, H.N.; Salleh, A.B.; Tan, W.S.; Ahmad Tajudin, A. Gold nanoparticle-decorated reduced-graphene oxide targeting anti hepatitis B virus core antigen. *Bioelectrochemistry* **2018**, *122*, 199–205. [CrossRef]
77. Xavier, M.M.; Nair, P.R.; Mathew, S. Emerging trends in sensors based on carbon nitride materials. *Analyst* **2019**, *144*, 1475–1491. [CrossRef]
78. Fan, J.; Yuan, L.; Liu, Q.; Tong, C. An ultrasensitive and simple assay for the Hepatitis C virus using a reduced graphene oxide-assisted hybridization chain reaction. *Analyst* **2019**, *144*, 3972–3979. [CrossRef]
79. Ozcelik, D.; Jain, A.; Stambaugh, A.; Stott, M.A.; Parks, J.W.; Hawkins, A.; Schmidt, H. Scalable Spatial-Spectral Multiplexing of Single-Virus Detection Using Multimode Interference Waveguides. *Sci. Rep.* **2017**, *7*, 12199. [CrossRef]
80. Liu, X.; Wang, F.; Aizen, R.; Yehezkeli, O.; Willner, I. Graphene oxide/nucleic-acid-stabilized silver nanoclusters: Functional hybrid materials for optical aptamer sensing and multiplexed analysis of pathogenic DNAs. *J. Am. Chem. Soc.* **2013**, *135*, 11832–11839. [CrossRef]
81. Gingrich, J.C.; Davis, D.R.; Nguyen, Q. Multiplex detection and quantitation of proteins on Western blots using fluorescent probes. *Biotechniques* **2000**, *29*, 636–642. [CrossRef]
82. Hsu, C.C.; Franklin, C.; Riley, L.K. Multiplex Fluorescent Immunoassay for the simultaneous detection of serum antibodies to multiple rodent pathogens. *Lab Anim.* **2007**, *36*, 36–38. [CrossRef]
83. Chen, L.; Song, L.; Zhang, Y.; Wang, P.; Xiao, Z.; Guo, Y.; Cao, F. Nitrogen and Sulfur Codoped Reduced Graphene Oxide as a General Platform for Rapid and Sensitive Fluorescent Detection of Biological Species. *ACS Appl. Mater. Interfaces* **2016**, *8*, 11255–11261. [CrossRef]
84. Jin, Z.; Geißler, D.; Qiu, X.; Wegner, K.D.; Hildebrandt, N. A Rapid, Amplification-Free, and Sensitive Diagnostic Assay for Single-Step Multiplexed Fluorescence Detection of MicroRNA. *Angew. Chem. Int. Ed.* **2015**, *54*, 10024–10029. [CrossRef]
85. Zheng, Q.; Wu, H.; Wang, N.; Yan, R.; Ma, Y.; Guang, W.; Wang, J.; Ding, K. Graphene-based Biosensors for Biomolecules Detection. *Curr. Nanosci.* **2014**, *10*, 627–637. [CrossRef]
86. Perumal, V.; Hashim, U. ScienceDirect Advances in biosensors: Principle, architecture and. *J. Econ. Financ. Adm. Sci.* **2013**, *12*, 1–15. [CrossRef]
87. Kasry, A.; Ardakani, A.A.; Tulevski, G.S.; Menges, B.; Copel, M. Highly Efficient Fluorescence Quenching with Graphene. *J. Phys. Chem. C* **2012**, *4*, 2858–2862. [CrossRef]
88. Liu, M.; Zhang, Q.; Brennan, J.D.; Li, Y. Graphene-DNAzyme-based fluorescent biosensor for Escherichia coli detection. *MRS Commun.* **2018**, *8*, 687–694. [CrossRef]
89. Bitounis, D.; Ali-Boucetta, H.; Hong, B.H.; Min, D.H.; Kostarelos, K. Prospects and challenges of graphene in biomedical applications. *Adv. Mater.* **2013**, *25*, 2258–2268. [CrossRef]
90. Banerjee, A.N. Graphene and its derivatives as biomedical materials: Future prospects and challenges. *Interface Focus* **2018**, *8*, 20170056. [CrossRef]
91. Agarwal, S.; Zhou, X.; Ye, F.; He, Q.; Chen, G.C.K.; Soo, J.; Boey, F.; Zhang, H.; Chen, P. Interfacing live cells with nanocarbon substrates. *Langmuir* **2010**, *26*, 2244–2247. [CrossRef]
92. Zhang, L.; Xia, J.; Zhao, Q.; Liu, L.; Zhang, Z. Functional graphene oxide as a nanocarrier for controlled loading and targeted delivery of mixed anticancer drugs. *Small* **2010**, *6*, 537–544. [CrossRef]
93. Lammel, T.; Boisseaux, P.; Fernández-Cruz, M.L.; Navas, J.M. Internalization and cytotoxicity of graphene oxide and carboxyl graphene nanoplatelets in the human hepatocellular carcinoma cell line Hep G2. *Part. Fibre Toxicol.* **2013**, *10*, 27. [CrossRef]
94. Liao, C.; Li, Y.; Tjong, S.C. Graphene nanomaterials: Synthesis, biocompatibility, and cytotoxicity. *Int. J. Mol. Sci.* **2018**, *19*, 3564. [CrossRef]
95. Guo, S.; Dong, S. Graphene and its derivative-based sensing materials for analytical devices. *J. Mater. Chem.* **2011**, *21*, 18503–18516. [CrossRef]
96. Sengupta, J.; Hussain, C.M. Graphene and its derivatives for Analytical Lab on Chip platforms. *TrAC-Trends Anal. Chem.* **2019**, *114*, 326–337. [CrossRef]
97. Yu, X.; Cheng, H.; Zhang, M.; Zhao, Y.; Qu, L.; Shi, G. Graphene-based smart materials. *Nat. Rev. Mater.* **2017**, *2*, 17046. [CrossRef]
98. Chircov, C.; Grumezescu, A.M.; Andronescu, E. Biosensors-on-Chip: An Up-to-Date Review. *Molecules* **2020**, *25*, 6013. [CrossRef]
99. Huang, G.; Huang, Q.; Xie, L.; Xiang, G.; Wang, L.; Xu, H.; Ma, L. A rapid, low-cost, and microfluidic chip-based system for parallel identification of multiple pathogens related to clinical pneumonia. *Sci. Rep.* **2017**, *7*, 6441. [CrossRef]
100. Pang, B.; Fu, K.; Liu, Y.; Ding, X.; Hu, J.; Wu, W.; Xu, K.; Song, X.; Wang, J.; Mu, Y.; et al. Development of a self-priming PDMS/paper hybrid microfluidic chip using mixed-dye-loaded loop-mediated isothermal amplification assay for multiplex foodborne pathogens detection. *Anal. Chim. Acta* **2018**, *1040*, 81–89. [CrossRef]

101. Sengupta, P.; Khanra, K.; Chowdhury, A.R.; Datta, P. *Lab-on-a-Chip Sensing Devices for Biomedical Applications*; Elsevier Ltd.: Amsterdam, The Netherlands, 2019.
102. Kujawska, M.; Bhardwaj, S.K.; Mishra, Y.K.; Kaushik, A. Using graphene-based biosensors to detect dopamine for efficient parkinson's disease diagnostics. *Biosensors* **2021**, *11*, 433. [CrossRef]
103. Kaushik, A.; Khan, R.; Solanki, P.; Gandhi, S.; Gohel, H.; Mishra, Y.K. From nanosystems to a biosensing prototype for an efficient diagnostic: A special issue in honor of professor Bansi D. Malhotra. *Biosensors* **2021**, *11*, 359. [CrossRef]
104. Sengupta, J.; Adhikari, A.; Hussain, C.M. Graphene-based analytical lab-on-chip devices for detection of viruses: A review. *Carbon Trends* **2021**, *4*, 100072. [CrossRef]
105. Soomro, R.A. Development of Biosensors for Drug Detection Applications. *Nanobiosensors* **2020**, 203–222. [CrossRef]
106. Xiang, X.; Luo, M.; Shi, L.; Ji, X.; He, Z. Droplet-based microscale colorimetric biosensor for multiplexed DNA analysis via a graphene nanoprobe. *Anal. Chim. Acta* **2012**, *751*, 155–160. [CrossRef]
107. Kenry Yeo, J.C.; Yu, J.; Shang, M.; Loh, K.P.; Lim, C.T. Highly Flexible Graphene Oxide Nanosuspension Liquid-Based Microfluidic Tactile Sensor. *Small* **2016**, *12*, 1593–1604. [CrossRef]
108. Hashmi, A.; Nayak, V.; Rb, K.; Jain, B.; Baid, M.; Alexis, F.; Kumar, A. Potentialities of graphene and its allied derivatives to combat against SARS-CoV-2 infection. *Mater. Today Adv.* **2022**, *13*, 100208. [CrossRef]
109. Roberts, A.; Chauhan, N.; Islam, S.; Mahari, S.; Ghawri, B. Graphene functionalized field-effect transistors for ultrasensitive detection of Japanese encephalitis and Avian influenza virus. *Sci. Rep.* **2020**, *10*, 14546. [CrossRef]
110. Jin, X.; Zhang, H.; Li, Y.T.; Xiao, M.M.; Zhang, Z.L.; Pang, D.W.; Wong, G.; Zhang, Z.Y.; Zhang, G.J. A field effect transistor modified with reduced graphene oxide for immunodetection of Ebola virus. *Microchim. Acta* **2019**, *186*, 223. [CrossRef]
111. Simone, G. Graphene-based optical sensors for the pre- vention of SARS-CoV-2 viral dissemination. *arXiv* **2020**, arXiv:2011.02181.

biosensors

MDPI

Review

Recent Advances in Electrical Impedance Sensing Technology for Single-Cell Analysis

Zhao Zhang [1,†], Xiaowen Huang [2,†], Ke Liu [1], Tiancong Lan [1], Zixin Wang [3] and Zhen Zhu [1,*]

1 Key Laboratory of MEMS of Ministry of Education, Southeast University, Sipailou 2, Nanjing 210018, China; zhangzhao98@seu.edu.cn (Z.Z.); liuke@seu.edu.cn (K.L.); lantiancong@seu.edu.cn (T.L.)
2 The First Affiliated Hospital of Nanjing Medical University (Jiangsu Province Hospital), Department of Orthopedics, Nanjing 210029, China; jiangsuaaa@aliyun.com
3 School of Electronics and Information Technology, Sun Yat-Sen University, Xingang Xi Road 135, Guangzhou 510275, China; wangzix@mail.sysu.edu.cn
* Correspondence: zhuzhen@seu.edu.cn; Tel.: +86-25-83792632 (ext. 8830)
† Both authors contribute equally to this manuscript.

Abstract: Cellular heterogeneity is of significance in cell-based assays for life science, biomedicine and clinical diagnostics. Electrical impedance sensing technology has become a powerful tool, allowing for rapid, non-invasive, and label-free acquisition of electrical parameters of single cells. These electrical parameters, i.e., equivalent cell resistance, membrane capacitance and cytoplasm conductivity, are closely related to cellular biophysical properties and dynamic activities, such as size, morphology, membrane intactness, growth state, and proliferation. This review summarizes basic principles, analytical models and design concepts of single-cell impedance sensing devices, including impedance flow cytometry (IFC) to detect flow-through single cells and electrical impedance spectroscopy (EIS) to monitor immobilized single cells. Then, recent advances of both electrical impedance sensing systems applied in cell recognition, cell counting, viability detection, phenotypic assay, cell screening, and other cell detection are presented. Finally, prospects of impedance sensing technology in single-cell analysis are discussed.

Keywords: electrical impedance spectroscopy; impedance flow cytometry; single cell analysis; microfluidics

Citation: Zhang, Z.; Huang, X.; Liu, K.; Lan, T.; Wang, Z.; Zhu, Z. Recent Advances in Electrical Impedance Sensing Technology for Single-Cell Analysis. *Biosensors* **2021**, *11*, 470. https://doi.org/10.3390/bios11110470

Received: 26 October 2021
Accepted: 17 November 2021
Published: 22 November 2021

1. Introduction

Cellular biophysical properties provide signals for abnormalities in tissues and organs [1,2]. Due to the heterogeneity presented in any isogeneic cell population, conventional population-averaged approaches neglect differences among individuals in gene expression and cell processes, leading to the loss of significant information [3,4]. Studies of cellular heterogeneity allow for exploring the cause, diagnosis and targeted therapy of diseases and the discovery of drugs [5,6].

Microfluidics, which provides manipulation and analytical methods at single-cell level, has emerged as a powerful tool for single-cell analysis [7,8]. Microfluidic devices have the advantages of miniaturization, low cost, comparable geometric dimension to cell sizes and flexible structural design [9–11]. Various single-cell manipulation strategies have been developed and introduced in microfluidic devices for cell-based studies [12]. These strategies to manipulate cells could be active, such as dielectrophoresis [13], acoustophoresis [14] and optical tweezers [15], or passive, such as microwells [16], hydrodynamic traps [17] and inertial focusing in curved channels [18]. To characterize the diverse biophysical properties of single cells, analytical methods integrated with microfluidic devices have been widely expanded, such as spectroscopy [19], fluorometry [20], mass spectroscopy [21] and electrochemical probes [22]. To study cellular heterogeneity, optical characterization methods, such as optical flow cytometry and laser confocal microscopy, are most widely used to acquire biological information in single-cell resolution [20]. However, these methods require fluorescent labels in cells to characterize cell and subcellular structures.

Electrical impedance properties of single cells could be used as biophysical markers that provide important information to uncover the complex physiological states of cells [23–25]. Biosensors based on single-cell electrical impedance measurements have the advantage of probing multiple biological parameters without fluorescent labeling. The electrical impedance at different frequency ranges refers multiple properties of cells: size information at a low frequency (from 100 kHz to 1 MHz), membrane capacitance at a higher frequency (about a few MHz), and intracellular organelles, such as the conductivity of cytoplasm, at even higher frequencies [26,27]. In addition, the non-invasive and label-free impedance sensing techniques are easy to be integrated into microfluidic devices for quantitative and real-time detection of single cells. Due to the advantages mentioned above, electrical impedance integrated microfluidic devices have been widely utilized for cell-based assays in single-cell resolution.

This review first presents a brief overview of basic principles, analytical models and design concepts of electrical impedance sensing devices for single-cell analysis. Next, applications of two essential microfluidic systems for single-cell impedance measurement are focused: impedance flow cytometry for mobile cell detection, such as cell counting, identification, and classification, and electrical impedance spectroscopy for immobilized cell monitoring, such as cell differentiation, division, and proliferation. In the end, advances and prospects on electrical impedance sensing technology for single-cell analysis are discussed.

2. Theory and Modeling

In electrical impedance sensing, a frequency-dependent excitation signal in the form of an alternating current (AC) voltage $\widetilde{U}(j\omega)$ is applied to a pair of electrodes and the response current $\widetilde{I}(j\omega)$ is measured. Impedance of the measured object is defined as the ratio between excitation voltage and response current:

$$\widetilde{Z}(j\omega) = \frac{\widetilde{U}(j\omega)}{\widetilde{I}(j\omega)} = Z_{RE} + jZ_{IM} \tag{1}$$

where $\widetilde{Z}(j\omega)$ is the complex impedance. Z_{RE} and Z_{IM} are the real and imaginary part of the complex impedance, respectively. j^2 is -1 and ω is the angular frequency ($\omega = 2\pi f$). The absolute value and phase shift of the complex impedance are given by:

$$\left|\widetilde{Z}\right| = \sqrt{(Z_{RE})^2 + (Z_{IM})^2} \tag{2}$$

$$\theta = \arctan\left(\frac{Z_{IM}}{Z_{RE}}\right) \tag{3}$$

The common theory and model of cell impedance sensing is the electrical model of a spherical cell in an aqueous medium. The complex impedance of cell-medium system is given by:

$$\widetilde{Z}_{mix} = \frac{1}{j\omega\widetilde{C}_{mix}} \tag{4}$$

where $\widetilde{C}_{mix}$ is the complex capacitance of the system and is determined by $\widetilde{\varepsilon}_{mix}$:

$$\widetilde{C}_{mix} = \widetilde{\varepsilon}_{mix}\frac{S}{4\pi kd} \tag{5}$$

The well-known Maxwell's mixture theory (MMT) [26,28] is widely used to derive the complex permittivity of mixture of cell and medium as:

$$\widetilde{\varepsilon}_{mix} = \widetilde{\varepsilon}_m \frac{1 + 2\varphi \widetilde{f}_{CM}}{1 - \varphi \widetilde{f}_{CM}} \tag{6}$$

In this equation, $\widetilde{\varepsilon}_{mix}$, $\widetilde{\varepsilon}_m$ and $\widetilde{\varepsilon}_c$ represent the complex permittivity of the medium-cell mixture, suspending medium and the cell, respectively. φ represents the volume fraction of the cell to the suspending medium and $\widetilde{f}_{CM}$ represents the Clausius-Mossotti factor:

$$\widetilde{f}_{CM} = \frac{\widetilde{\varepsilon}_c - \widetilde{\varepsilon}_m}{\widetilde{\varepsilon}_c + 2\widetilde{\varepsilon}_m} \tag{7}$$

Because MMT only works well in a uniform field with low volume fractions, the volume fraction should be replaced with a corrected value in the cases of high volume fraction (usually >40%) and non-uniform field [27]. Besides, the derivation of MMT is only applicable under homogenous external electric fields and depends on configuration mode of electrodes in specific devices. Considering the geometric parameters of electrode configurations, the impedance of a mixture system can be described as:

$$\widetilde{Z}_{mix} = \frac{1}{j\omega \widetilde{\varepsilon}_{mix} G_f} \tag{8}$$

where G_f is a geometrical constant. Calculation methods of G_f under typical electrode configurations were proposed in previous literatures [28–30].

The simplest electrical model of a biological cell is the "single-shell model", which consists of an insulating shell (i.e., plasma membrane) and a conducting sphere (i.e., cytosol, assumed to be homogeneous) [23,27,28]. In addition, plant cells, Gram-negative bacteria like *Escherichia coli* (*E. coli*), and fungi like *Saccharomyces cerevisiae* (*S. cerevisiae*) have outer cell walls. The electrical model of these cells could be represented by "multi-shell model" [31]. Based on MMT, these electrical models could be simplified step by step until it becomes a homogeneous sphere according to following equation:

$$\widetilde{\varepsilon}_{ei} = \widetilde{\varepsilon}_{i+1} \frac{\left(\frac{R_{i+1}}{R_i}\right)^3 + 2 \times \frac{\widetilde{\varepsilon}_i - \widetilde{\varepsilon}_{i+1}}{\widetilde{\varepsilon}_i + 2\widetilde{\varepsilon}_{i+1}}}{\left(\frac{R_{i+1}}{R_i}\right)^3 - \frac{\widetilde{\varepsilon}_i - \widetilde{\varepsilon}_{i+1}}{\widetilde{\varepsilon}_i + 2\widetilde{\varepsilon}_{i+1}}} \tag{9}$$

where R_i (i = 1, 2, 3, 4) and $\widetilde{\varepsilon}_i$ (i = 1, 2, 3, 4) are the radius and the complex permittivity of each shell. $\widetilde{\varepsilon}_{ei}$ (i = 1, 2, 3) stands for the complex permittivity of the homogeneous sphere after i-th simplification (Figure 1A). After the third simplification, the multi-shell model becomes homogeneous and its complex permittivity ($\widetilde{\varepsilon}_{e3}$) could be substituted into Equation (7) as the complex permittivity of the cell ($\widetilde{\varepsilon}_c$).

Typically, a single cell in liquid culture is in either of two conditions: suspended in medium or adhered to the substrate/electrode. Cells in a flow system and cells captured in a fixed position are classified as suspended cells. The ECMs of impedance sensing system with suspended and adherent cells have been provided for theoretically analysis (Figure 1B) [24,26]. In ECMs, cell and medium are simply equivalent to resistors and capacitors in series and parallel according to their electrical parameters. Besides, the electrical double layers (EDLs), formed by the contract between metal electrodes and electrolyte solutions, are modelled as capacitors (C_{dl}). Ignoring the double layer, the impedance of a mixture system can be described as [24]:

$$\widetilde{Z}_{mix} = \frac{R_m(1 + j\omega R_c C_{mem})}{j\omega R_m C_{mem} + (1 + j\omega R_c C_{mem})(1 + j\omega R_m C_m)} \tag{10}$$

where R_m and R_c are the resistance of medium and cytosol, respectively. C_m, C_{mem} are the capacitance of medium and cell membrane, respectively. Optimizing the dielectric properties of suspension medium can improve the dominance of cell impedance in complex impedance of the mixture. The PBS solution, which has the best conductivity while maintaining the activity of biological cells, is therefore widely used in cell impedance sensing applications.

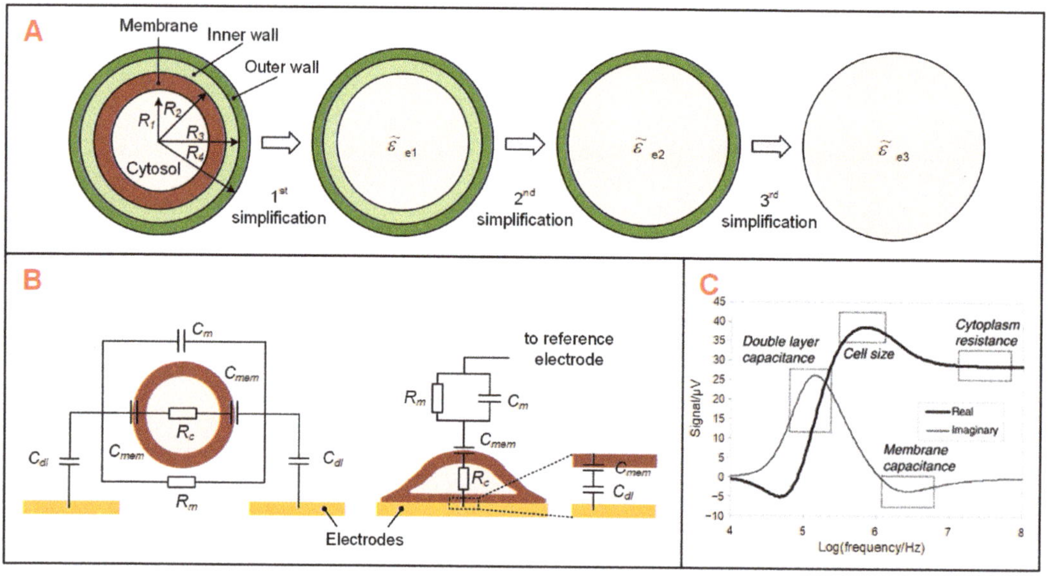

Figure 1. Electrical model and equivalent circuit models (ECMs) of a single cell. (**A**) Multi-shell model of a single cell simplified into a homogeneous sphere based on MMT. (**B**) ECMs of a cell suspended between a pair of sensing electrodes and a cell adhered on a sensing electrode. (**C**) Simulation results of cell impedance sensing using an ECM model, which presented various frequency domains corresponding to different biophysical parameters. Reproduced from [32] with the permission from Royal Society of Chemistry.

The frequency range of cell impedance spectra is from 1 Hz to 10 GHz and could be distinguished into three distinct dispersions (or relaxations), which are the well-known α, β and γ dispersions [28]. The α-dispersion, occurs below several kHz, is attributed to the polarization of cell membrane. However, it is difficult to measure due to the domination of impedance by EDLs at low frequencies. At higher frequencies (more than few tens of kHz), the β-dispersion exhibits characteristic sub-dispersions dominated by double layer capacitance, cell size, cell membrane and cytoplasm. Therefore, it is the most widely applied in the electrical impedance sensing of biological samples. The frequency ranges corresponding each sub-dispersion has been given by simulation result using ECMs (Figure 1C) [26]. In GHz range, the γ-dispersion arises from reorientation of water molecules.

3. Device Designs for Sensing Single-Cell Impedance

In this section, common designs of single-cell impedance sensing devices are reviewed first, including impedance flow cytometry to sense flowing cells and electrical impedance spectroscopy to sense immobilized cells. Then, instruments and portable platforms to implement impedance sensing are summarized. In the end, complementary metal-oxide-semiconductor (CMOS) based electrical impedance devices are described.

3.1. Impedance Flow Cytometry (IFC)

Impedance flow cytometry is a high-throughput methedology for single-cell analysis, analogous to micro Coulter particle counter (μCPC) [32]. Compared with the μCPC, IFC excels in miniaturization, less requirement for peripherals, and flexible integration of interrogation units [33]. IFC measures the variation of response current caused by single cells passing over patterned electrodes in a microfluidic channel. The sensitivity of IFC devices is mainly dependent on the distribution of AC electric field in the channel. Therefore, electrode configurations, namely the geometric setting of electrode pairs, must be considered in particular [34]. Besides, the consistency of detection results can be enhanced by utilizing particle positioning methods to ensure that suspension samples roughly pass through the same cross-sectional position of the sensing region [35,36]. Following subsections will describe several designs of IFC devices, as well as exemplify optimized systems reported recently.

3.1.1. Electrode Configurations

Coplanar electrodes. In 2001, Gawad et al. proposed the first microfluidic IFC device used for high-throughput single-cell impedance measurement at multiple frequencies [32]. As shown in Figure 2A, the basic detection unit in coplanar electrode configurations consists of two or three electrodes positioned in the bottom of microchannel. The absolute measurement scheme facilitates two electrodes to measure the impedance changes in the detection space of the microfluidic channel, which has been discussed in detail in Section 2. The current pulse caused by the passage of a flowing cell is recorded. This scheme is typically applicable for cell counting which has no requirement of detecting small signal changes [37]. In a differential measurement scheme, which has three electrodes, excitation signal was applied to the intermediate electrodes and the differential current was measured at the electrodes on both sides to provide a higher signal to noise ratio (SNR) [32]. Such a configuration enabled automatically switching between the measurement and reference electrode pairs when a single cell passing through the sensing region. The signal waveform of the differential current was a symmetric bipolar Gaussian shape and gave the information about cell size and electric parameters upon frequencies. Compared to the absolute measurement scheme, differential measurement is widely used since it corrects uneven drift of electrode properties and enables the calculation of the flow rate of cells [38–40]. The devices with coplanar electrodes can be easily fabricated with tiny deviations due to the single-step alignment between the channel and metal [32].

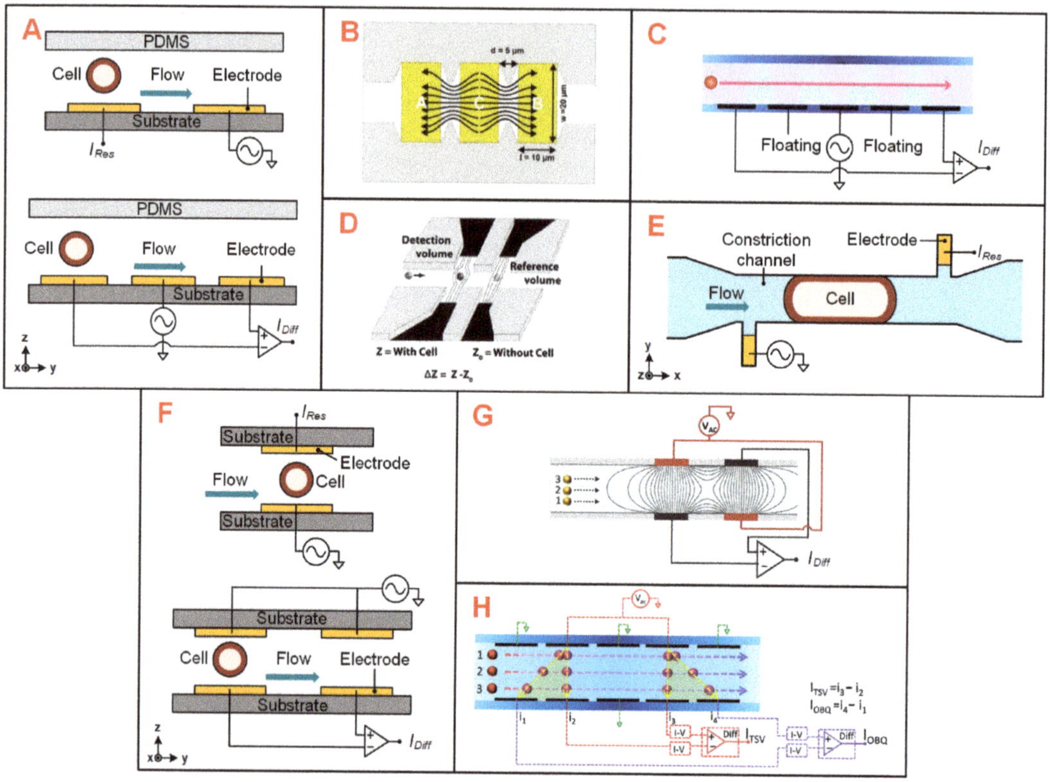

Figure 2. Different designs of IFC devices. (**A**) Schematics of absolute and differential measurement schemes of coplanar electrode configuration. I_{Res} refers to response current and I_{Diff} refers to differential current. (**B**) Coplanar electrodes with larger electrode area exposed to the medium. Reproduced from [39] with the permission from MDPI. (**C**) Coplanar electrodes with two extra floating electrodes. Reproduced from [41] with the permission from Royal Society of Chemistry. (**D**) Liquid electrodes. Reproduced from [42] with the permission from Royal Society of Chemistry. (**E**) Asymmetrical liquid electrodes with the constriction channel for cell flowing [43]. (**F**) Schematics of absolute and differential measurement schemes of facing electrode configuration. (**G**) Facing electrodes with asymmetric wiring scheme. Reproduced from [44] with the permission from Elsevier. (**H**) Five pairs of facing electrodes. Reproduced from [45] with the permission from Royal Society of Chemistry.

However, IFC devices with coplanar electrodes are sensitive to the vertical position of samples because of the non-uniform distribution of electric field. Hence, several optimizations of coplanar electrode configurations were proposed to improve the sensing sensitivity. As shown in Figure 2B, Clausen et al. proposed an optimized chip with larger electrode area exposed to the medium [39]. This design allowed for more current and greater current density between the electrodes, and thus attained an improved SNR. Xie et al. adopted a similar design and their IFC device thus had a higher SNR (23.5–32.6 dB) [46]. Besides, they proposed that this design scheme enabled lower amplitude of excitation signal to reduce the potential damage to cells. Rather than seeking ways to increase current intensity or local electric intensity, De Ninno et al. proposed a five-electrode design combining a conventional chip layout with compensation strategy which enabled accurate size measurement of particles without the need for focusing methods [41]. In this device, one floating electrode was placed between each pair of the detection electrodes to obtain more information with respect to the height of the particles (Figure 2C). As a result, the proposed compensation procedure made the "electric" diameters (the estimated diameter of particle) closer to the actual data. Liquid electrodes proposed by Demierre et al. were designed to address the non-uniform electric field in conventional coplanar electrode configuration (Figure 2D) [42,47]. In these devices, electrodes were located at both ends of short channels perpendicular to the main channel. The electrodes were far enough to generate almost homogeneous electric field over the channel height. However, this design reduces the sensitivity due to the larger detection volume. Besides, the liquid electrode designs are also used for focusing particle stream based on the principle of dielectrophoresis [48,49].

In order to compare the performance of different coplanar electrode designs, Cottet et al. evaluated the vertical and lateral sensitivity of four typical layouts [50]. As a result, the design with a longer constriction channel was considered as the best candidate since it was relatively insensitive to the particle height or longitudinal misalignment in the fabrication process.

Facing electrodes. Facing electrode design was firstly proposed by Cheung et al. [51]. In this configuration, the electrodes are positioned at the top and bottom of the channel and thus eliminate the electric field non-uniformity to some extent (Figure 2F). As the electric field is limited to smaller volume, this design has improved sensitivity largely. Absolute measurement schemes are usually used for simple cell counting and therefore require less sensitivity. Thus, absolute measurement schemes are rarely used in practical applications, while differential measurement scheme is widely used in many prototypes [52–54]. However, the fabrication process of facing electrode design needs accurate alignment of electrodes patterned at different substrates, leading to higher fabrication difficulty.

In order to obtain impedance information related to the vertical height of the single cells, Caselli et al. proposed an asymmetric wiring scheme (Figure 2G) [44]. This solution can improve the precision level of particle diameter measurement to a certain extent and simply compensate the signal impedance of eccentric particles. Besides, in order to obtain impedance data with high accuracy, Spencer et al. proposed a new design of multiple pairs of facing electrodes with an impedance signal processing algorithm, minimizing the influence of the vertical height of the particle on the impedance signal (Figure 2H) [45].

3.1.2. Particle Positioning

The relative position of particles passing through the detection area between electrode pairs has great effects on the detection stability and repeatability, especially in coplanar electrode design [37,55,56]. Except for reducing the cross-sectional area of the channel, constriction channels and various methods of focusing particles have been proposed in order to obtain higher impedance signal quality.

Constriction channel. Due to the lack of close contact between cells and electrodes when cell passing through the detection area, the electric current tends to bypass the cells by flowing through the surrounding medium. In order to solve this problem, Chen et al. applied the constriction channel design in a μCPC device [57]. In this design, cells were pressed and elongated when flowing into the constriction channel. Constriction channel allows for more accurate detection of specific membrane capacitance and cytoplasm conductivity, thus increasing the classification success rate of different cell populations [58]. Hence, it has been used for screening various types of single cells, such as blood cells [59], tumor cells [43,58] and plant cells [60]. Due to the cross-sectional area of the constriction channel must be smaller than the size of the interested cells, this design has higher risk of channel blockage and lower throughput. To this end, Zhao et al. proposed the crossing constriction channel design with bypass outlet, which allowed large particles passing through to address the possible blockage of the constriction channel [58]. Furthermore, in order to improve the accuracy of cell classification, they introduced asymmetrical liquid electrodes to obtain cell diameter (Figure 2E) [43]. In this method, the measured impedance is proportional to the elongated length of the cell, so that the relative volume of the cell can be calculated. The passage time for cells to pass through the constriction channel is related to cellular mechanical properties. In order to obtain the passage time, Han et al. introduced a pairs of coplanar electrodes at the inlet and outlet of the constriction channel [60]. The time point when a cell passes through each sensing unit is recorded, and then the passage time can be calculated according to the time interval for a cell from a sensing unit to another.

Particle focusing. Although particle focusing system sometimes increases the device complexity, it is commonly integrated in IFC devices for eliminating the influence of sectional position on impedance signals. Various particle focusing systems based on different technologies have been reported such as hydrodynamic focusing [37,61,62], acoustophoretic focusing [63–65], dielectrophoretic focusing [66,67] and inertial focusing [68]. Hydrodynamic focusing enables the sample flow to be coated by sheath flow and focused into a narrow stream either horizontally or vertically. This conventional method minimized the potential for two or more particles to enter the detection region simultaneously and ensures a uniform particles velocity. Acoustophoretic focusing methods are based on either traveling surface acoustic waves (TSAWs) [64] or standing surface acoustic waves (SSAWs) [65] to manipulate particles. The principle of dielectrophoresis (DEP) focusing is that a neutral but polarizable particle is subjected to DEP force in a nonuniform electric field. Since the DEP forces depend on the size and dielectric properties of the particles, it can also be used for single-cell trapping and separating. Inertial focusing, as a passive focusing method, depends on special channel structure and high particle flow velocity.

3.2. *Electrical Impedance Spectroscopy (EIS) Sensing Devices*

EIS sensing is suitable for real-time monitoring and tracking of a limited number of cells simultaneously. For *in-situ* EIS measurements, frequency sweeping takes several seconds every time, which requires the stable capture of single cells as a prerequisite [24]. Therefore, various trapping methods have been proposed for positioning single cells in microfluidic channel. Alternatively, single cells can be directly adhered onto the electrode or substrate for electric cell-substrate impedance sensing. Additional progress has been also made to increase the throughput of EIS devices.

3.2.1. Trapping of Suspended Single Cells

The methods to trap suspended single cells include hydrodynamic traps [69–73], negative pressure traps [74–76], DEP trapping [77–80] and optical manipulation [81]. Hydrodynamic traps are the most common design to position single cells in microfluidic devices, and usually consist of simple microstructures (such as U-shaped or three-pillar traps [69,71,73]) or special fluid channels (such as μ-fluidic traps [70,72]) to dock numerous single cells synchronously in a short time. Di Carlo et al. captured approximately 100 isolated HeLa cells by a U-shaped trap array and achieved the cell maintenance over 85% after 24 h [69]. Tang et al. applied the μ-fluidic microstructures to achieve a high cell-trapping rate of 95% (Figure 3A) [82]. In negative pressure trapping system, one or more suction channels are connected to the side or bottom of the cell-perfusion channel [74–76]. By applying negative pressure to the suction channel, single cells could be trapped in vias or slits which are the connection points of cell-perfusion channel and suction channels. Although such devices are complex in design and fabrication, negative pressure traps allow for selective capture and release of redundant or unwanted cells [76]. Han et al. demonstrated a system with backside suction channels to capture HeLa cells in cavity pores [75]. Since the diameter of the pore is between that of HeLa cells and blood cells, most of the HeLa cells could be separated from the spiked blood samples. DEP trapping could be used to perform accurate and selective capture of specific samples, as the DEP force depends on the dielectric properties of single cells and the frequency of applied electric field [79]. Besides, DEP force was also used to release redundant cells to achieve a uniform single-cell trapping [83]. Electrode configuration schemes for DEP trapping include quadrupole electrode array [77,78], microwells [79,80], and 'ring-dot' electrode structures [84]. As a representative, the quadrupole electrode array, proposed by Heida et al., could generate DEP force to repel cells away from electrodes and towards the array center (Figure 3B) [78]. Optical methods were used to manipulate single cells with a high precision. The liquid resin containing target cells could be rapidly polymerized under laser exposure and formed traps of specific shape. As such, Xu et al. performed real-time two-photon-lithography to capture single cells and achieved a high capture efficiency of 100% on a one-bead-to-one-trap basis [81].

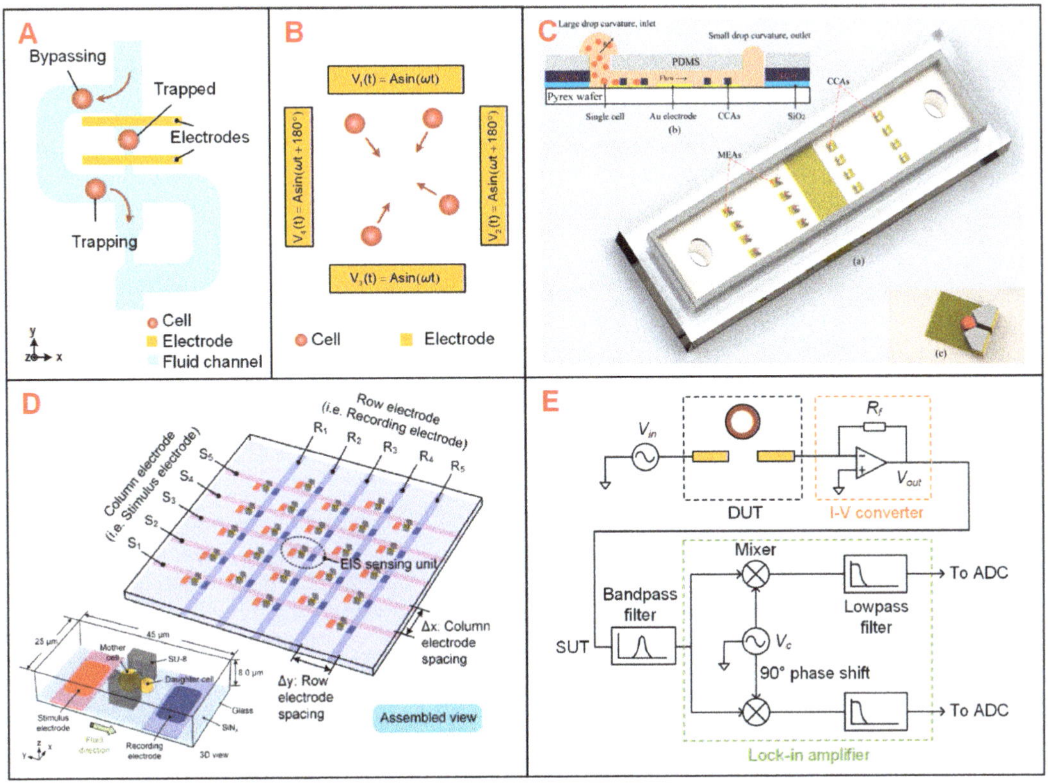

Figure 3. Different designs of EIS devices and schematics of single-cell EIS sensing system with lock-in amplifier. (**A**) Schematics of μ-fluidic traps to immobilize single cells in a cellular EIS sensing device [82]. (**B**) Quadrupole electrode array to gather cells towards the center. The phase difference of applied AC signals between adjacent electrodes is 180° [78]. (**C**) A three-dimensional (3D) single-cell culturing device to detect HeLa cell migration. Reproduced from [85] with the permission from American Chemical Society. (**D**) An addressable microelectrode array to perform single-cell immobilization and localized EIS measurement. Reproduced from [86] with the permission from John Wiley and Sons. (**E**) Schematics of single-cell impedance sensing system with lock-in amplifier.

3.2.2. Electrical Cell-Substrate Impedance Sensing (ECIS)

ECIS is a well-established technology developed to assess cellular behavior or responses to drug candidates by measuring the impedance of live cells adhered on the electrode surface [87,88]. As cells proliferate and spread on the surface of sensing electrodes, electrical current is interfered immediately and thus resulting in a drastic change in the measured impedance. Besides, ECIS could combine with a variety of single-cell manipulation techniques to perform single-cell impedance measurement [85,89–92]. Tsai et al. [91] used micro pillars to trap single HeLa cells in a microfluidic system and monitor their adhesion and spreading Zhang et al. applied DEP trapping to HeLa cells and monitored their adhesion, response to drugs, steady growth and differentiation by ECIS [92]. In addition, surface modification could be used to promote cell adhesion on metal electrodes rather than glass or plastic substrate [87,93]. Shah et al. modified the surface of the SU-8 substrate sequentially with mNH_2 linked PEG (Amino functionalized methoxyl polyethylene glycol) to avoid any nonspecific cell adhesion near the sensing electrode and, thus, eliminate the unnecessary cell crosstalk [93]. Nguyen et al. proposed a microfluidic chip with removable PDMS cover lid which enabled building up a two-dimensional or three-dimensional microenvironment for investigating single cancer cell migration (Figure 3C) [85].

3.2.3. Advanced Design to Increase the Throughput of EIS Devices

Typically, conventional EIS devices can monitor less than 20 cells simultaneously [94,95]. Although simple devices are useful for low-cost assays, they result in tedious repetition of experiments and reduced data reliability in applications, which usually require long-term monitoring of a large population of samples. In these devices, on-chip impedance sensors are connected with respective bond pads for external interconnection [83,85], where the dimension of sensors is limited by the chip margin. To solve this problem, individually addressable microelectrode arrays (MEAs) are proposed to be incorporated into EIS sensing devices [86,96]. In these devices, two sets of n microelectrode bars are arranged orthogonally to form a sensing array of $n \times n$ sites with only $2n$ bond pads (Figure 3D) [86]. Each sensing unit at the crossing position can be addressed individually by external multiplexers. Guo et al. reported a microarray chip integrated with two MEAs for cell positioning and impedance monitoring, respectively [96]. Geng et al. proposed the design of a MEA chip to measure the replicative lifespan (RLS) of budding yeasts and assessed the influence of neighbor samples upon the site-specific impedance measurement [86]. Alternatively, CMOS integrated circuit could be utilized to overcome the limitation of the number of EIS sensing units by integrating electrodes and addressing circuit (see Section 3.4).

3.3. *Instruments and Portable Platforms for Electrical Impedance Sensing Technology*

Impedance converter and lock-in amplifier (LIA) are commonly used to measure the impedance of single cells. The impedance converter is an AC self-balancing bridge, which consists of a simple op-amplifier and a feedback resistor (Figure 3E) [97]. In this approach, an AC excitation signal V_{in} is input to one port of the device under test (DUT), and the feedback resistor R_f shares the same current flowing through the DUT. Ideally, without any phase shift of the op-amplifier, the current flowing through the DUT is proportional to the voltage on R_f. Then the DUT complex impedance can be calculated from the output voltage V_{out}. LIA, also known as the phase-sensitive detector, is capable of extracting weak signals from noisy background (Figure 3E) [98]. The output voltage V_{out}, termed as signal under test (SUT), is multiplied by in-phase and quadrature carrier signal V_c, whose frequency is the same as that of V_{in}. Then, the amplitude and phase of SUT are extracted with correlative demodulation by carrier signal at the same reference frequency. Benchtop instruments, including several versions of LIAs developed by Stanford Research Systems Inc. (Sunnyvale, CA, USA), Zurich Instruments AG (Zurich, Switzerland), NF corp. (Yokohama, Japan), SBT instruments (Copenhagen, Denmark), Sine Scientific Instruments (Guangzhou, China) and Liquid Instruments (Lyneham, Australia), respectively, are usually interconnected directly or through custom printed circuit boards (PCBs) to the microfluidic devices [99–104]. Here, PCBs are typically functionalized with control modules such as multiplexers to activate target sensing electrodes on the devices.

Device integration leads to cost saving, while portable system serves for more situations. Motivated by this, researchers began to integrate impedance measuring circuit on custom PCBs. As a representative, the integrated circuit chip AD5933, a low-cost impedance analyzer system, has been introduced in embedded portable systems [105]. Notably, the frequency range of this system is only up to 100 kHz, and its accuracy is lower than that of benchtop instruments. Huang et al. developed a wide-band digital lock-in amplifier (DLIA), which features a low input noise of 4.4 $nV/\sqrt{Hz}$, 120 dB dynamic reserve and a phase deviation of less than 0.02° through the whole frequency range up to 65 MHz [106]. This portable EIS system has been demonstrated by impedance measurements of three sets of micro beads with different diameters.

3.4. CMOS-Based Impedance Sensing Devices

The miniaturization and portability of single-cell impedance sensing devices can be achieved by harnessing CMOS integrated circuit (IC) technology. CMOS IC chips have been reported as an alternative to measure biological impedance at single-cell level, showing the trend of impedance sensing system in the integration level [107,108]. However, the aluminum (Al) microelectrodes fabricated by standard CMOS processes are not biocompatible due to the biological toxicity and chemical activity of Al. Therefore, the microelectrodes are covered with gold (Au) or platinum (Pt) layer to obtain better biocompatibility and chemical inertness [107]. Since its substrate is replaced by large scale integrated circuits, such devices enable tens of thousands of sensing units featuring individually addressable microelectrodes. Chen et al. developed a high-throughput EIS sensing platform consisting of a 96 × 96 microelectrode array for tumor cell counting and analysis [107]. This CMOS chip was packaged with a PCB that contains multiplexers and the EIS measurement of cells was implemented by a LCR meter. The size of microelectrodes approximates to the scale of target cells for better sensitivity, resulting in a weak current of about 100 pA that is easily submerged in the noise. To overcome this problem, Gamo et al. introduced a current integrator acting as a I-V converter to effectively suppress noise and measure weak current signals [108]. More efforts are required to achieve the complete integration of impedance sensing circuitry on CMOS chip. Visvam et al. reported a high-density CMOS MEA system, including a programmable waveform generator, 59,760 platinum microelectrodes and 32 on-chip lock-in amplifiers for impedance sensing [109]. This improved integration level contributes to superior sensing and actuation capabilities and high signal quality.

4. Applications of Single-Cell Impedance Sensing Technology

4.1. IFC to Detect Flowing Single Cells

Recent IFC devices applied in single-cell analysis are summarized in Table 1. These applications, discussed in this subsection, are simply classified according to cell species, including blood cells [110–114], tumor cells [43,52,115–122], stem cells [123–127], plant cells [60,128–132] and microbes [53,62,133–141]. In terms of blood cells, researchers focused on the identification and counting of normal or diseased blood cells. Studies showed the capability of IFC devices in recognition of different types of dissociated tumor cells (DTCs) [43,118,120,121] or circulating tumor cells (CTCs) [52,115,119,122]. As for stem cells, the main focus is the impedance measurement of their long-term differentiation process. Studies of plant cells include the detection of pollen viability [128–132] and cell screening [60]. Besides, impedance measurements of microbes are further classified and discussed.

Table 1. Applications of IFC for single-cell analysis.

Category	First Author (Year)	Electrode and Fluidic Layouts	Frequency	Target Cells	Application	Ref.
Blood cells	Holmes (2010)	2 coplanar electrode pairs	503 kHz and 10 MHz	CD4 T-cells	Cell counting	[112]
	Du (2013)	1 coplanar electrode pair	2 MHz	Red blood cells	Detection of malaria-infected cells	[113]
	Hassan (2016)	2 coplanar electrode pairs	303 kHz and 1.7 MHz	CD4 and CD8 T-cells	Cell counting	[111]
	Liu (2018)	2 coplanar electrode pairs	156 kHz, 500 kHz and 3 MHz	Red blood cells	Detection of sickle cells	[110]
	Honrdo (2018)	2 facing electrode pairs, fluorescence detection	2–8 MHz	Red blood cells	Detection of malaria-infected cells	[114]

Table 1. *Cont.*

Category	First Author (Year)	Electrode and Fluidic Layouts	Frequency	Target Cells	Application	Ref.
Tumor cells	Choi (2013)	Two polyelectrolyte gel electrodes	DC	OVCAR-3 cells	Cell recognition	[115]
	Spencer (2014)	2 facing electrode pairs	0.5 MHz and 2 MHz	MCF-7 cells	Cell recognition	[52]
	Han (2015)	2 facing electrode pairs	500 kHz and 10 MHz	DLD-1 cells	Cell recognition	[116]
	Zhao (2016)	μCPC with constriction channel	1 kHz and 100 kHz	A549 and H1299 cells	Cell screening	[117]
	Desai (2019)	2 coplanar electrode pairs, sheath flow focusing	250 kHz	Thyroid, breast, lung, and ovarian cancer cells	Cell recognition	[118]
	Ren (2019)	1 coplanar electrode pair, 2 constriction channels	1 kHz, 10 kHz, 100 kHz, and 1 MHz	MDA-MB-231 cells	Cell recognition	[119]
	McGrath (2020)	5 facing electrode pairs	500 kHz–50 MHz	Six types of pancreatic ductal adenocarcinoma cell	Cell screening	[120]
	Ostermann (2020)	2 facing electrode pairs	6 MHz	U937 cells	Viability assay	[121]
	Zhang (2020)	1 coplanar electrode pair, asymmetrical constriction channel	100 kHz and 250 kHz	A549 and Hep G2 cells	Cell screening	[43]
Stem cells	Song (2016)	C-shaped arranged coplanar electrodes	500 kHz and 3 MHz	Mesenchymal stem cells	Monitoring differentiation process	[125]
	Xavier (2017)	2 facing electrode pairs, fluorescence detection	500 Hz and 2MHz	Skeletal stem cells	Monitoring differentiation process	[127]
Plant cells	Heidmann (2016)	2 facing electrode pairs	500 Hz and 12 MHz	Tobacco pollen	Viability assay	[128]
	Heidmann (2017)	2 facing electrode pairs	500 kHz, 3 MHz and 12 MHz	Tomato, pepper, potato and wind pollinators pollen	Viability assay	[129]
	Impe (2019)	2 facing electrode pairs	1 MHz	Wheat pollen	Viability assay	[130]
	Ascari (2020)	2 facing electrode pairs	2 MHz and 8 MHz	Hazelnut pollen	Viability assay	[131]
	Canonge (2020)	2 facing electrode pairs	500 kHz and 12 MHz	Wheat microspore	Monitoring androgenesis process	[132]
	Han (2020)	2 coplanar electrode pairs, constriction channel	500 kHz and 5 MHz	Herbaceous *Arabidopsis thaliana* and woody *Populus trichocarpa*	Cell screening	[60]
Microbes	Choi (2014)	2 polyelectrolytic gel electrodes, sheath focusing	DC	*F. tularensis* and *E. coli*	Cell recognition	[62]
	Mcgrath (2017)	2 facing electrode pairs	250 kHz, 18.3 MHz and 50 MHz	*C. parvum*	Viability assay	[136]
	Guler (2018)	1 coplanar electrode pairs	2 MHz	*E. coli*	Cell recognition	[135]
	Clausen (2018)	2 coplanar electrode pairs 2 facing electrode pairs	200 kHz and 7 MHz	*E. coli*	Cell recognition	[53]
	Chawla (2018)	1 coplanar electrode pairs	1.12 MHz and 1.5 MHz	*S. cerevisiae* cells	Monitoring cell growth rate	[137]
	Xie (2019)	1 coplanar electrode pairs	1 MHz	*S. cerevisiae* cells	Reproductive performance assessment	[139]
	Opitz (2019)	2 facing electrode pairs	0.5 MHz, 10 MHz and 12 MHz	*S. cerevisiae* cells	Viability assay	[138]
	Bertelsen (2020)	2 facing electrode pairs	366 kHz and 6.9 MHz	*E. coli*	Determination of the viability of *E. coli*	[140]
	Spencer (2020)	4 facing electrode pairs	5 MHz and 40 MHz	*K. pneumoniae*	Antimicrobial susceptibility tests	[141]

4.1.1. Blood Cells

Sickle cell disease (SCD), which causes sclerosis and membrane distortion in red blood cells (RBCs), brings about variation in cellular electrical properties [142]. Liu et al. combined on-chip oxygen control onto a single IFC chip for sickle cell disease diagnosis and monitoring [110]. They measured the electrical impedance of normal cells and sick cells at three different frequencies under normoxic and hypoxic conditions, respectively. As shown in Figure 4A, normal RBCs and sickle cells were separated clearly according to the measured impedance amplitude and phase value at 156 kHz under the normoxic condition. The results suggested that electrical impedance could serve as a new parameter to diagnose sickle cell disease. Parasite invasion can alter the dielectric properties of RBCs [113,143]. Du et al. demonstrated the discrimination of normal RBCs and *P. falciparum*-infected RBCs through analyzing the changes in the impedance signal amplitude and phase [113]. Honrado et al. developed an IFC device integrated with fluorescence interrogation to detect the dielectric properties of RBCs infected by malaria (Figure 4(Bi)) [114]. As a result, for early-stage infection (6 h), infected cells and normal cells were not distinguishable according to their impedance signals. However, as parasite growth progressed, the membrane capacitance and cytoplasmic conductivity of infected RBCs increased and thus the discrimination between two cell populations gradually became detectable (Figure 4(Bii)).

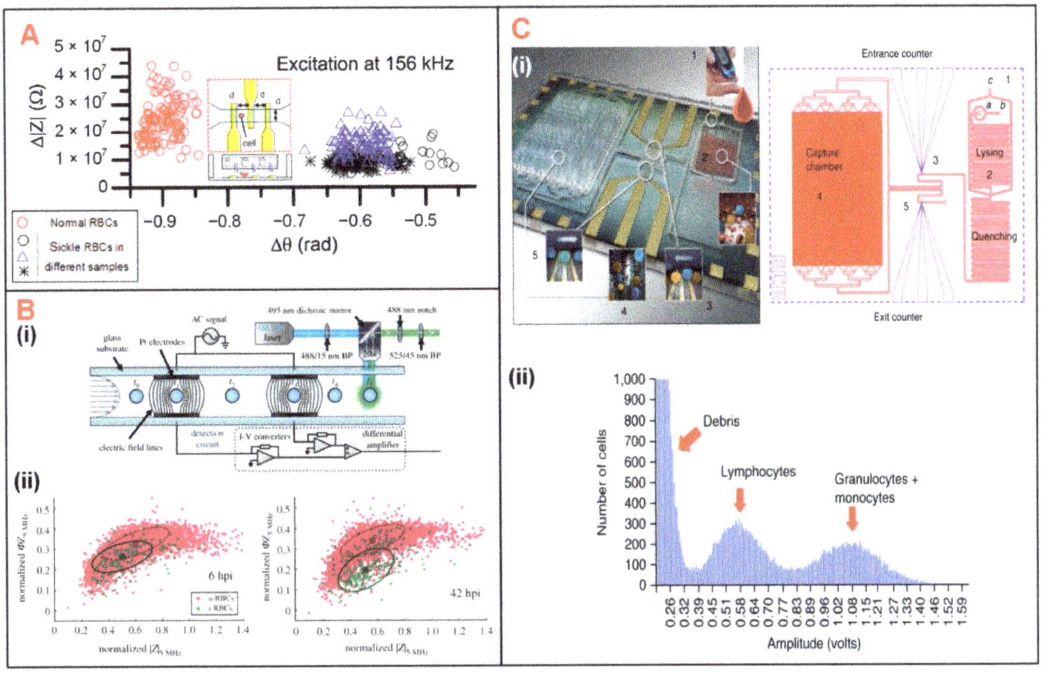

Figure 4. Blood cell analysis using IFC devices. (**A**) Measurement of $\Delta|Z|$ vs. $\Delta\theta$ for normal and sickle RBCs at 156 kHz. Reproduced from [110] with the permission from Elsevier. (**B**) (**i**) Schematics of a IFC device integrated fluorescence detection. (**ii**) Measurement of normalized phase (ΦZ_{5MHz}) vs. amplitude ($|Z_{5MHz}|$) at 6 h and 42 h after RBC infection. u-RBCs and i-RBCs stand for uninfected and infected RBCs. Reproduced from [114] with the permission from Royal Society. (**C**) (**i**) Photograph and layout of the differential immunocapture biochip. (**ii**) Pulse amplitudes of recorded impedance signals showing the size distribution of cells. Lymphocytes and granulocytes + monocytes are two groups of distinct populations of leukocytes. Reproduced from [111] with the permission from Springer Nature.

Holmes et al. discriminated and enumerated CD4 T-cells based on impedance cytometry and immune capture [112,144]. In their study, CD4 T-cells were labeled with small antibody conjugated beads, which changed the electrical properties of target T-cells. Hence, CD4 T-cells could be identified from their corresponding subpopulations based on impedance opacity ($|Z_{10MHz}|/|Z_{503kHz}|$). Recently, Hassan et al. reported an impedance biosensor based on differential immunocapture technology to perform cell counting on CD4 and CD8 T-cells with high accuracy (Figure 4(Ci)) [111,145,146]. In this device, antibodies specific to CD4 T-cells were initially adsorbed on a chamber between two conventional IFC modules. As the leukocytes flowed into the chamber, CD4 T-cells were captured and immobilized on the antibodies (Figure 4(Cii)). The cell number of each population can be calculated according to the impedance pulses caused by the passage of cells through the chamber. This protocol can be used to enumerate specific cell types with their corresponding antibodies immobilized in the capture chamber.

4.1.2. Tumor Cells

Tumor diagnosis is underpinned by determining which cells are malignant in acquired biopsy, leading to the need to accurately distinguish DTCs from normal cells in tissue [2]. Zhao et al. classified two tumor cell lines (A549 and H1299) based on different cellular membrane capacitance (C_m) and cytoplasm conductivity (σ_p) [117]. Desai et al. separated lung cancer DTCs (LC-DTCs) from RBCs, peripheral blood mononucleated cells (PBMCs) and normal lung cells based on impedance amplitude [118]. When LC-DTCs pass through the coplanar electrodes, impedance amplitude signal generates more significant pulse compared to that of normal cells. They also determined five major cancer types (lung, thyroid, breast, ovarian, and kidney cancers) from their corresponding counterpart target cells. Pancreatic ductal adenocarcinoma (PDAC) is an aggressive cancer lacking specific biomarkers. Aimed at this situation, McGrath et al. reported an IFC device to separate single PDAC tumor cells against xenografts [120]. They found that the phase of impedance signal of six PDAC cell types showed some correlations to specific gene expression, especially the *KRAS* mutations that led to higher phase variation. T188 and T738 are primary stage tumors with unknown *KRAS* mutations showing lower impedance phase contrast than other PDAC samples. Zhang et al. developed a microfluidic IFC platform with asymmetrical constriction channel to better detect the dielectric properties and diameters of different types of single tumor cells (Figure 2E) [43]. The classification accuracy between two tumor cell lines, A459 and HEP G2 cells, could be significantly improved with the combination of the individual intrinsic bioelectrical markers of membrane capacitance, cytoplasm conductivity and cell diameter. Besides, Ostermann et al. reported that necrotic and viable U937 human lymphoma cells could be clearly discriminated based on the phase of impedance signals by using a commercial IFC device [121]. Dead and viable cells can be discriminated by impedance signals at high frequency as the imaginary component of cell impedance depends on the membrane integrity of the cell (Figure 5A).

Identification and characterization of CTCs in blood stream is key to monitor the progression of cancer metastasis [122]. Choi et al. proposed a simple DC impedance microcytometer for identifying CTCs according to the cell volume [115]. Ren et al. reported an IFC device featuring parallel cyclic deformability channels and coplanar electrodes, to collect both biomechanical and bioelectrical properties for tumor cell analysis [119]. The deformation and transition time of tumor cells could be obtained from the time points when impedance amplitude changes abruptly (Figure 5B). In clinical application, due to the very small amount of CTCs in a blood sample, it is necessary to pre-enrich CTCs before measurement [52,116]. According to the different membrane capacitance between tumor cells and normal PBMCs, Spencer et al. measured the dielectric properties of MCF-7 cells (a representative of CTCs) at 500 kHz and 2 MHz and distinguished them from leukocytes when mixed in the whole blood [52]. Compared with optical approaches, electrical impedance measurement shows better performance in separating MCF-7 cells from other blood cells (Figure 5C). Besides, Han et al. reported a microfluidic system integrated with both enrichment and impedance detection units to discriminate CTCs [116]. In this study, immunomagnetic nanobeads (MNBs) and highly-conductive graphene nanoplates (GNPs) were bonded to the surface of DLD-1 cells (a representative colorectal cancer cell line). Compared with normal blood cells, the impedance signal of DLD-1 cells coated with GNPs shows a phase shift of 100 degrees for identification (Figure 5D), successfully.

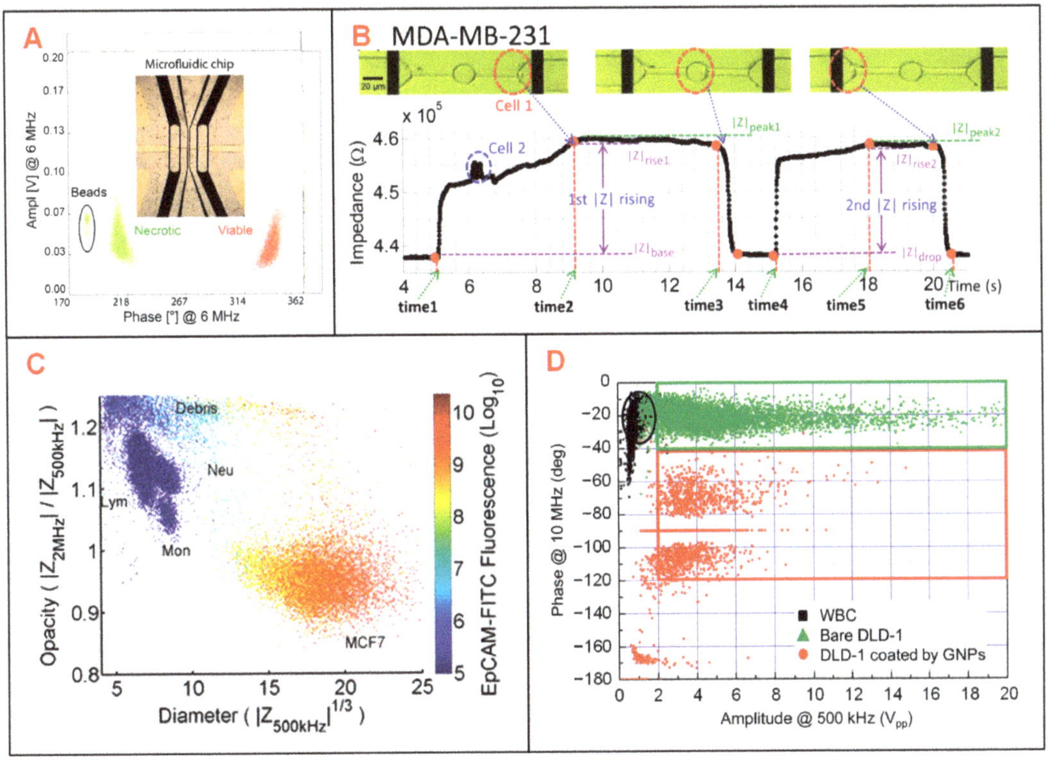

Figure 5. Tumor cell analysis using IFC devices. (**A**) Scatter plot of amplitude and phase values at 6 MHz for necrotic and viable U937 human lymphoma cells and 10-µm beads. Reproduced from [121] with the permission from Springer Nature. (**B**) Recording of the impedance variation when a breast cancer cell (MDA-MB-231) passing through the constriction channel. Reproduced from [119] with the permission from American Institute of Physics. (**C**) Scatter plot of opacity ($|Z_{2MHz}|/|Z_{500kHz}|$) and electric diameter ($|Z_{500kHz}|^{1/3}$) for MCF-7 cells and other blood cells. Reproduced from [52] with the permission from AIP Publishing. (**D**) Scatter plot of impedance amplitude ($|Z_{10MHz}|$) and phase (ΦZ_{500kHz}) to classify white blood cells, bare DLD-1 cells and DLD-1 cells coated by GNPs. Reproduced from [116] with the permission from American Chemical Society.

4.1.3. Stem Cells

Hildebrandt et al. demonstrated that the osteogenic differentiation process of human mesenchymal stem cells (hMSCs) could be monitored by tracking their impedance variation [123]. Song et al. was the first to propose a dual-micropore microfluidic IFC device to monitor the same differentiation process (Figure 6(Ai)) [124,125]. In this device, when MSCs or osteoblasts passed through micropores, a pulse in impedance amplitude was recorded to determine the proportion of differentiated cells at each stage [125]. Moreover, a support vector machine (SVM) algorithm was employed in data analysis to reach a classification accuracy of 87%. It is notable that the training data set of SVM included a total number of 1028 impedance signals combining both relative phase at 3 MHz and impedance opacity ($|Z_{2MHz}|/|Z_{500kHz}|$). The optimal SVM-based model was also used to characterize the differentiation process (from MSCs into osteoblasts), in which the proportion of osteoblasts was increasing while that of MSCs was decreasing (Figure 6(Aii)). Zhao et al. reported that the intercellular electrical markers, such as specific membrane capacitance ($C_{specific\ membrane}$) and cytoplasm conductivity ($\sigma_{cytoplasm}$) of neural stem cells could be used to evaluate their differentiation processes [126]. During different stages of differentiation, the distribution difference of $C_{specific\ membrane}$ differs a lot from that of $\sigma_{cytoplasm}$ (Figure 6B). Besides, Xavier et al. developed an IFC device equipped with two pairs of facing electrodes combined with confocal microscopic monitoring of the osteogenic differentiation of skeletal stem cells (SSCs) (Figure 6(Ci)) [127]. The sample of human bone marrow mononuclear cells (hBMMNCs) extracted from human bone marrow (BM) were cultured in vitro and was injected into a microfluidic chip for impedance detection every two weeks. Changes in the opacity ($|Z_{2MHz}|/|Z_{500kHz}|$) of impedance data could characterize the osteoblast differentiation process of SSCs. (Figure 6(Cii)). In the first two weeks (from BM to P0), the decreased opacity corresponded to the increased cell size and membrane capacitance during SSCs osteogenic differentiation. After P0, there was no significant change of opacity indicating the completion of osteogenic differentiation.

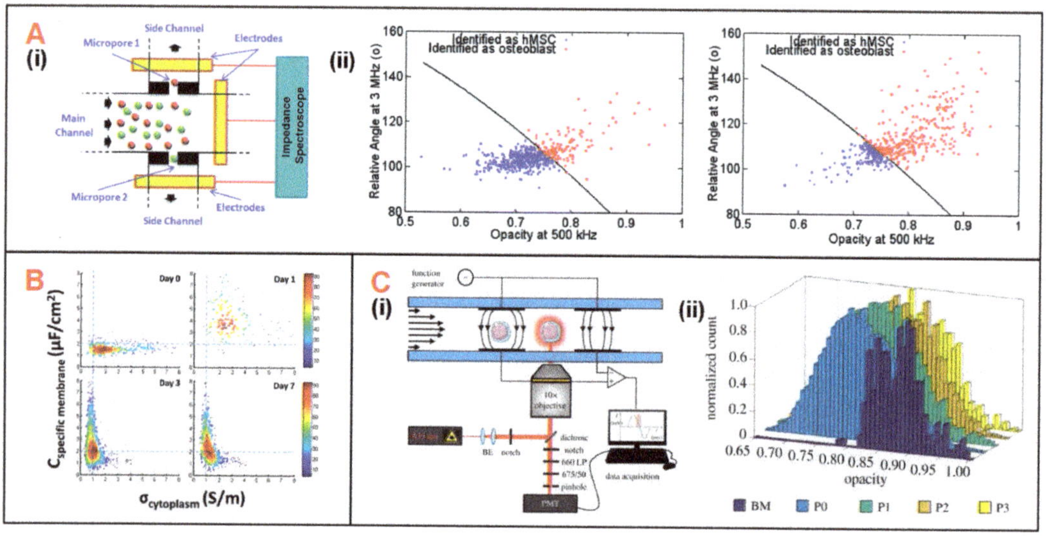

Figure 6. Stem cell analysis using IFC devices. (**A**) (**i**) Schematics of a dual-micropore based IFC device consisting of a main channel and two deputy channels through micropores. (**ii**) Scatter plot of signal phase at 3 MHz vs. opacity ($|Z_{3MHz}|/|Z_{500kHz}|$) for hMSCs and osteoblasts at 7 days (on the left) and 14 days (on the right) after post-induction. Reproduced from [125] with the permission from Royal Society of Chemistry. (**B**) $C_{specific\ membrane}$ and $\sigma_{cytoplasm}$ variations of rat neural stem cells within the differentiation process of 7 days. Reproduced from [126] with the permission from Public Library of Science. (**C**) (**i**) Schematics of an IFC device integrated fluorescence detection. (**ii**) Changes of impedance signal opacity ($|Z_{2MHz}|/|Z_{500kHz}|$) within 56 days SSCs differentiation process. Reproduced from [127] with the permission from Royal Society.

4.1.4. Plant Cells

Qualifying pollen or spores, especially their viability and germination capacity, is important for industrial production and plant breeding [128]. Heidmann et al. measured the viability of pollen samples by using a commercial impedance device [128,129]. In one of these studies, pollen samples were measured before and after heat treatment [128]. As a result, larger phase of the impedance signal corresponding to viable samples was no longer presented after heat-inactivation, which suggested that heat treatment inactivated the pollen samples and destroyed the integrity of cell membrane. Furthermore, Heidmann et al. predicted the germination rate of tomato pollen population by measuring the amount of viable and dead pollens [129]. Impe et al. [130] and Ascari et al. [131] assessed pollen viability of hazelnut and wheat, and further identified various factors (sugar, H_3BO_3, $CaCl_2 \cdot 2H_2O/Ca(NO_3)_2 \cdot 2H_2O$ concentration and pH) affecting pollen viability with the same commercial device. Canonge et al. utilized the IFC device to track and characterize the developmental process of wheat (*Triticum aestivum* L.) genotype Pavon microspores in gametogenesis and anrogenesis [132]. According to this study, throughout all sporophytic developmental stages, some of the viable microspores showed a continuous increase in both impedance amplitude and phase. As a result, electrical impedance could serve as a fast and reliable reactivity marker for tracking wheat microspores in androgenesis.

The biological and physiological properties of cell wall unique to plant cells offer the potential to increase phenotyping resolution and identify nonanatomic markers [147]. Han et al. developed an IFC device to characterize the biophysical properties of two model plant species, herbaceous *Arabidopsis thaliana* and woody *Populus trichocarpa* [60]. In the regeneration process of primary cell wall (PCW), plant cells are gradually covered by the fibrillary network, which becomes thick and interlaced, resulting in the decrease of capacitance of cell membrane and PCW [148]. Thus, the researchers found that the *Arabidopsis* cells with regenerated PCW were less deformable and electrically conductive than that without PCW.

4.1.5. Microbes

IFC devices have been utilized extensively in detection, separation and viability analysis of unicellular microbes, including bacteria [53,62,133–135,140,141], protozoa [136,149] and fungi [137–139].

Bernabini et al. demonstrated the feasibility to detect bacteria according to cell size in an IFC device [133]. This device features small cross-sectional area of the flow channel and narrow width of electrodes, since the size of bacteria is usually smaller than that of mammalian cells. Without the measurement of signal phase, *E. coli* could be identified by volume rather than membrane capacitance, and thus *E. coli* was indistinguishable from particles in similar size [133]. To solve this problem, phase metric was introduced and thus the viability and species of bacteria could be determined [134]. In order to precisely measure the diameter of different bacterium, Choi reported an IFC device with position-adjustable virtual wall [62]. The movement of virtual wall is modulated by adjusting the flow rate of sample suspension focused by low conductive sheath flow. The cross-sectional area of sample flow could be adjusted to approximate that of bacteria, making the impedance changes caused by the bacteria passage more significant (Figure 7A). Recently, Guler et al. merged the amplitude and phase information of impedance signals to achieve higher size sensitivity and detection throughput of bacteria (Figure 7B) [135]. Besides, Clausen et al. used two simple IFC devices with coplanar and facing electrodes, respectively, to detect different types of bacteria [53]. These IFC devices could be used to accurately measure any change in bacteria concentration and distinguish methicillin-sensitive *Staphylococcus aureus* (MSSA) from *E. coli* according to the impedance phase signal at 8 MHz (Figure 7C). Moreover, Bertelsen et al. detected and characterized *E. coli* inactivated by ethanol, heat and autoclaving, respectively [140]. The population of ethanol-treated bacteria showed a similar amplitude to 1.5-μm polystyrene beads, which was consistent with the hypothesis of membrane disruption. Supported by the experimental data, ethanol treatment caused membrane disruption while heat process did no obvious harm to cell membrane. In detail, the loss of membrane integrity corresponded to changes in impedance signal amplitude and phase (Figure 7D). Notably, Spencer et al. developed a method to optimize the prescription of antibiotic by an impedance-based fast antimicrobial susceptibility test (iFAST) [141]. By applying microfluidic impedance cytometry with differential electrode configuration, the phenotype response (electrical opacity and electrical diameter) of *Klebsiella pneumoniae* (*K. pneumoniae*) to specific antibiotic was accurately analyzed.

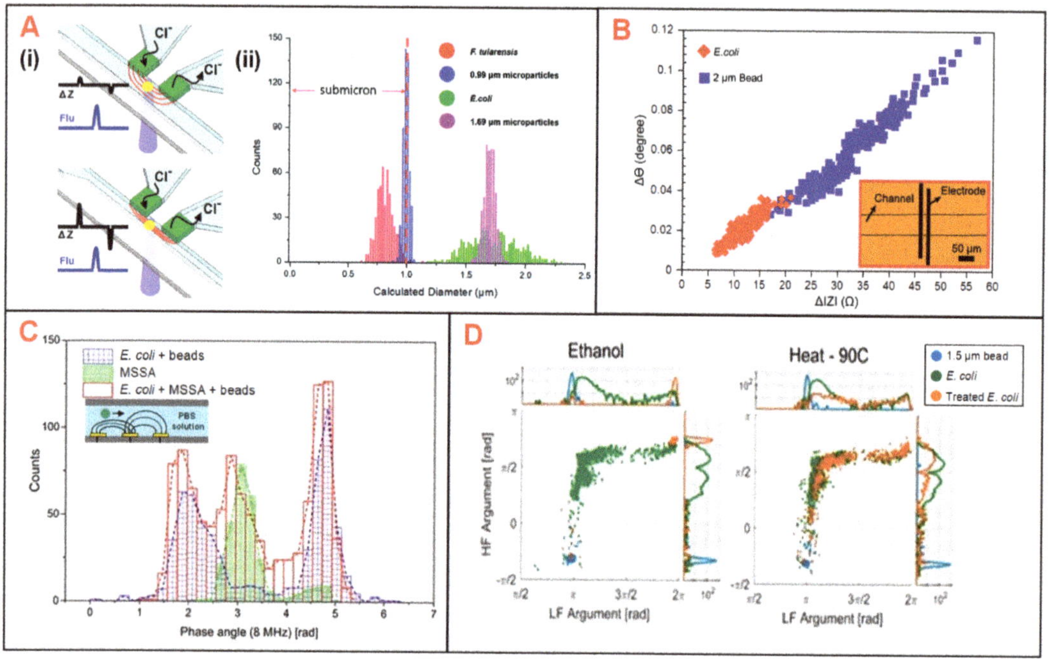

Figure 7. Bacteria analysis using IFC devices. (**A**) (**i**) Schematics of an IFC device with position-adjustable virtual wall. (**ii**) Histograms of bacteria and particle counts in distribution of calculated diameter derived from impedance signals. Reproduced from [62] with the permission from Royal Society of Chemistry. (**B**) Scatter plots of impedance signal amplitude and phase changes at 2 MHz for *E. coli* and 2-µm beads. Reproduced from [135] with the permission from Elsevier. (**C**) Histograms of cell counts in distribution of impedance signal phase at 8 MHz for *E. coli*, MSSA and *E. coli* + MSSA + beads. Reproduced from [53] with the permission from MDPI. (**D**) Scatter plots of impedance signal amplitude and phase at low (366 kHz) and high frequencies (6.9 MHz) for *E. coli* with ethanol and heat treatment. Reproduced from [140] with the permission from MDPI.

Besides, accurate recognition and viability analysis of protozoan pathogens have advanced in parasitic diseases diagnosis of human and livestock [149]. To this end, Mcgrath et al. detected single protozoan oocysts utilizing a continuous IFC system [136]. The heat treatment performed on *Cryptosporidium parvum* (*C. parvum*) lowered the impedance signal amplitude and phase especially at high frequency representing the internal properties of the oocyst (50 MHz) (Figure 8(Ai)). The difference of viable and inactive populations can be enhanced by increasing the conductivity of medium suspension. In addition, according to the amplitude at 250 kHz and phase at 18.3 MHz, the major human-pathogenic species (*C. parvum*, *Cryptosporidium muris* (*C. muris*) and *Giardia Lamblia* (*G. lamblia*)) were discriminated from other parasite species that posed little or no risk to human health (Figure 8(Aii)).

Yeast cells, easily accessible and culturable, have been widely used as an important model organism to study cell growth and division in cell cycle progression [94]. Xie et al. optimized an IFC device with a constriction channel to detect the size of single budding yeast (*S. cerevisiae*) cells and calculate the late-budding rates of populations [139]. As shown in Figure 8B, due to the impact of velocity gradient near the constriction channel, rod particles are aligned with the electric field lines, so that the length of rod particles could be assessed by pulse width of impedance amplitude at 1 MHz. In this way, rod and spherical particles could be clearly discriminated. Moreover, late-budding yeast, namely mother cell with a daughter cell that is nearly mature, can be viewed as a rod-shaped cell, while other yeast is approximately oval. Using the same principle, the shape of target cells as well as the budding stages can be obtained. In another work, Chawla et al. developed a microfluidic platform allowing for long-term culturing and independent monitoring of growth rate of budding yeast (Figure 8(Ci)) [137]. In this device, multiple cell populations were anchored to pads and their daughter cells were then washed away, flowing through the impedance sensing unit. By analyzing the impedance signal phase at 1.5 MHz, passages of cells through electrodes were recorded. Then by counting the flowing cells in unit time, the growth rate of cell population can be calculated. As shown in Figure 8(Cii), the phase fluctuated drastically as the cell passes through the electrodes, and each phase pulse corresponds to a single cell flowing through the sensing area. This device enabled culturing and monitoring of various groups of budding yeast simultaneously. At the same time, cell populations can be exposed to different medium solution and their growth rates can be calculated indirectly from impedance signal phase. In addition, Opitz et al. focused on monitoring and analyzing of yeast population under different culturing conditions [138]. In their study, impedance signals at 12 MHz were analyzed to characterize cell viability in a three-day brewing process. On the first day, the high phase indicated that the cell population had high viability and they began to breed by large numbers (Figure 8D). By the end of the third day, the cell population showed lower viability. The cell loss could be ascribed to the depletion of oxygen and the accumulated ethanol.

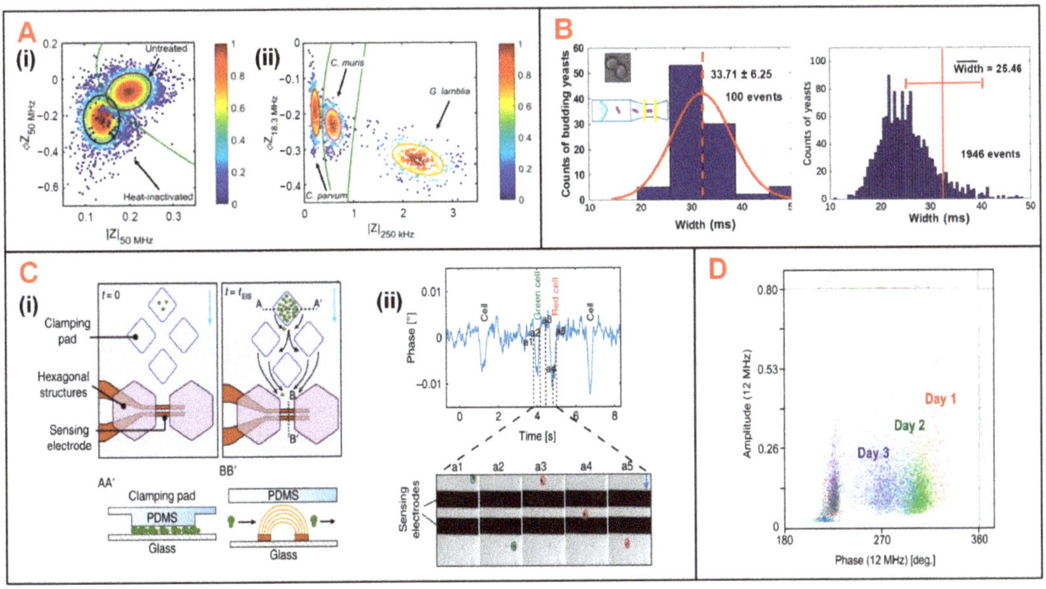

Figure 8. Microbial analysis using IFC devices. (**A**) (**i**) Scatter plot of phase (ΦZ_{50MHz}) vs. amplitude ($|Z_{50MHz}|$) for heat-inactivated and untreated *C. parvum*. (**ii**) Scatter plot of phase ($\Phi Z_{18.3MHz}$) vs. amplitude ($|Z_{250kHz}|$) for *C. parvum*, *C. muris* and *G. lamblia*. Reproduced from [136] with the permission from Nature. (**B**) Histograms of particle counts in distribution of signal pulse width measured at 1 MHz. Width refers to the time that yeast cells take to pass through the sensing electrodes. Reproduced from [139] with the permission from American Chemical Society. (**C**) (**i**) Schematics of an IFC device used for long-term budding yeast culturing and growth-rate measurement. (**ii**) Signal phase changes corresponding to five events that yeast cells passing through the impedance sensing electrodes. Reproduced from [137] with the permission from Nature. (**D**) Scatter plot of impedance signal amplitude vs. phase at 12 MHz for yeast population during three-day brewing process. Reproduced from [138] with the permission from Springer.

4.2. EIS to Detect Suspended or Adherent Single Cells

Different from IFC devices that are commonly used for cell recognition and screening with high throughput, EIS sensing devices are capable of extracting broadband impedance information and tracking dynamic variations of single cells. Recent EIS sensing devices applied in single-cell analysis are summarized in Table 2. These devices are classified into two categories: one is to determine the optimal frequency at which the impedance of different cell lines or cell states is most sensitive [82,150] and the other is to continuously monitor the dynamic cell process or cell behavior and phenotypic changes [83,85,91,92,94,95,148,151–153].

Table 2. Applications of EIS measurement for single cells. OT: Observation time. Throughput: Maximum number of single cells that can be simultaneously measured.

First Author (Year)	Techniques	Frequency Range	Throughput	OT	Target Cells	Application	Ref.
Primiceri (2011)	ECIS	1 Hz to 1 MHz	/	4 h	Hepatocellular carcinoma cells	Monitoring cell migration	[152]
Hong (2012)	DEP traps	20 kHz to 101 kHz	/	/	A549, MDA-MB-231, MCF-7, and HeLa cells	Electrical characteristics analysis of cancer cells	[151]
Nguyen (2013)	Hydrodynamic traps and ECIS	100 Hz to 1 MHz	16	/	MDA-MB-231 and MCF-7 cells	Monitoring cell capture, adhesion, and spreading process	[85]
Zhu (2014)	Negative pressure traps	10 kHz to 10 MHz	10	42 min	*S. cerevisiae* cells	Monitoring bud growth and cell motion	[83]
Zhu (2015)	Negative pressure traps	10 kHz to 10 MHz	10	120 min	*S. pombe* cells	Cell cycle determination	[94]
Zhou (2016)	Hydrodynamic traps	100 Hz to 20 MHz	10	48 h	Mouse embryonic stem cells	Monitoring the differentiation process	[95]
Park (2016)	Negative pressure traps	5 kHz to 1 MHz	5	/	Cancerous human urothelial cells (TCCSUP)	Cell recognition	[150]
Tsai (2016)	Hydrodynamic traps	10 kHz to 100 kHz	3	24 h	HeLa cells	Monitoring electrical characteristics	[91]
Tang (2017)	Hydrodynamic traps	1.953 kHz to 1 MHz	10	/	MCF-7 cells	Monitoring the capture process and cell screening	[82]
Chen (2020)	Hydrodynamic traps	100 kHz to 2 MHz	/	24 h	*Arabidopsis* mesophyll cells	Monitoring the regeneration process of primary cell wall	[148]
Zhang (2020)	DEP traps and ECIS	100 kHz	32	5 min	HeLa, MCF-7, and 293T cells	Monitoring the recovery process after electroporation	[153]
Zhang (2020)	DEP traps and ECIS	100 kHz	32	21 days	Mesenchymal stem cells	Monitoring differentiation process	[92]

EIS sensing technology has been used to investigate the optimal frequency at which the characteristic parameters extracted from EIS signals are most prominent in measuring specific dielectric properties of cells [82,150]. Park et al. proposed two types of devices to distinguish cancerous from normal human urothelial cell lines (Figure 9(Ai)) [150]. In one device, single cells were captured at 3D traps by applying negative pressure underneath. Then, the impedance of immobilized single cells was individually measured at frequencies from 5 kHz to 1 MHz. According to the EIS signals in Figure 9(Aii) plot, 119 kHz was supposed to be the optimal frequency, at which the impedance of two types of cells had the greatest divergence. The real-time impedance of the cell lines was measured at 119 kHz in the other device (an IFC device) to identify cancerous cells. These two devices potentially provide a supplementary platform to detect urothelial cancer of the bladder (UCB). In another study, Tang et al. developed a portable single-cell analytical system combining hydrodynamic traps and EIS measurement to accurately detect the sizes of MCF-7 cells [82]. Under the hydrodynamic forces, MCF-7 cells could be initially captured at the entrance of the narrow channel and then squeezed into it. Impedance signals were collected from three groups, among which one is the control group of PBS solution without cells, another one is the trapped cells in suspension, and the third one is the squeezed cells (Figure 9B). According to the sweep-frequency measurement of EIS, the frequency was optimized to 500 kHz, at which, cellular trapping-releasing-squeezing manipulation and cell size could be detected more accurately.

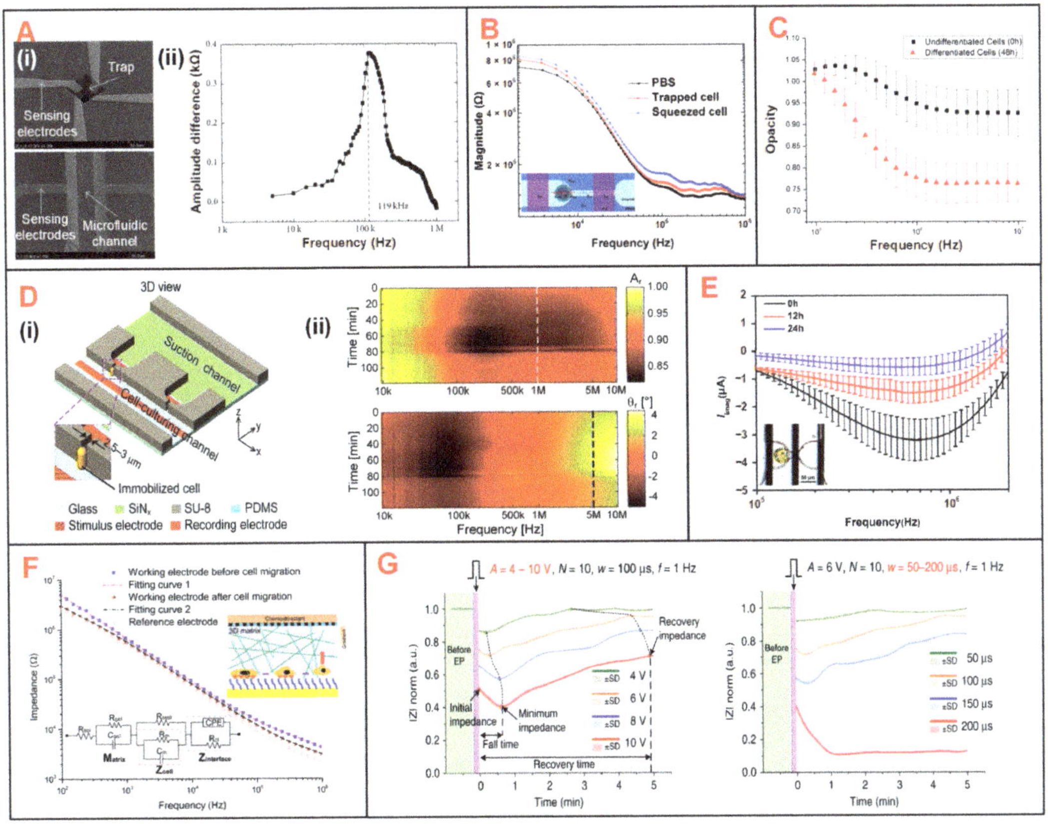

Figure 9. Cell-based assay using EIS sensing devices. (**A**) (**i**) SEM images of the two devices used to detect cancerous urothelial cells. Left one is an EIS sensing device with a negative pressure trap used to investigate the optimal frequency. Right one is an IFC device to perform high-throughput electrical impedance measurement of normal and cancerous urothelial cells. (**ii**) Measurement of the amplitude difference between normal and cancerous urothelial cells in the frequency range of 5 kHz to 1 MHz. Reproduced from [150] with the permission from Hindawi. (**B**) Schematics of a EIS sensing device to measure the amplitude and phase signal of MCF-7 cells under three typical conditions: PBS solution without cells, cell trapped and cell squeezed. Reproduced from [82] with the permission from Springer. (**C**) Using a EIS sensing device with microfluidic traps to distinguish the undifferentiated and differentiated cells by measuring the impedance over the frequency range from 100 kHz to 10 MHz. Reproduced from [95] with the permission from Elsevier. (**D**) (**i**) Schematics of an EIS-integrated single-cell culturing device for immobilization and impedance recording of *Schizosaccharomyces pombe* (*S. pombe*) cells. (**ii**) Recorded EIS amplitude and phase signals over the frequency range from 10 kHz to 10 MHz showing the growth and division of single *S. pombe* cells. Reproduced from [94] with the permission from Nature. (**E**) Imaginary part of current response for *Arabidopsis* mesophyll cells at different status (0 h, 12 h and 24 h after incubation, respectively). Reproduced from [148] with the permission from Elsevier. (**F**) The Bode impedance spectra measured on working electrode before and after cell migration, as well as on reference electrode without cells over the frequency range from 100 Hz to 1 MHz. Reproduced from [85] with the permission from American Chemical Society. (**G**) Recording of $|Z|_{norm}$ for HeLa cells in the recovery process under different conditions of electroporation. *A*, *N*, *w* and *f* stand for pulse amplitude, number, width and frequency, respectively. Reproduced from [153] with the permission from Nature.

EIS sensing technology has been used to monitor cell behavior and phenotypic changes, including differentiation of stem cells [92,148,154], cell growth and division [83,94,155], formation of cell wall [95], migration of tumor cells [85,152] and recovery process after electroporation [153].

In order to characterize the differentiation process of stem cells, Zhou et al. analyzed the impedance data from mouse embryonic stem cells (mESCs) at different time points in a cell differentiation cycle [95]. In this study, impedance opacity ($|Z_{1MHz}|/|Z_{50kHz}|$) was increasing during the 48-h cell differentiation process, and was significant at above 1 MHz (Figure 9C). Based on this finding, they observed the metastable transition state, from which stem cells could either differentiate irreversibly or return to pre-differentiation state at 24 h. Zhang et al. proposed a multifunctional microfluidic chip, which featured DEP trapping, electrical stimulation and real-time impedance monitoring of single cells [92,153,154]. They recorded the real-time impedance changes of two groups of MSCs with (OM group) or without electrical stimulation (OM + ES group) [92]. The results showed that electrical stimulation could accelerate the response to drug and advance the differentiation of MSCs. Besides, this device provided additional phenotypic indicators that were not available in cell traction force sensor and contributed to multimodal characterization of long-term physiological variations in the cell differentiation process [154].

Ghenim et al. were the first to monitor the impedance variation in the mitosis of a single mammalian cell [155]. Zhu et al. presented a microfluidic cell-culturing chip to trap, cultivate and selectively release individual yeast cells [156]. Then, this device was used to monitor the cell dynamics in a cell cycle of yeast cells (Figure 9(Di)) [83,94]. As an example, electrodes originally used to generate DEP forces were used to measure the electrical impedance spectrum of rod-shaped *S. pombe* cells, which were immobilized in an upright position at the traps [94]. Cell growth, nuclear division and cytokinesis in a cell cycle were sensitively characterized by EIS signal amplitude at 1 MHz and phase at 5 MHz (Figure 9(Dii)).

Chen et al. investigated the formation process of primary cell wall of *Arabidopsis* mesophyll cells [148]. As discussed in Section 4.1.4, the formation of the cell wall reduced the capacitance of entire plant cell and thus led to an increase in the imaginary part of impedance signal [60]. In support of this hypothesis, they measured the differential current response of *Arabidopsis* mesophyll cells at three status of cell wall formation (Figure 9E).

Cell migration, which serves as the initiation of cancer metastasis, could be recorded by ECIS technology [157]. Primiceri et al. demonstrated that cell migration could be monitored and automatically analyzed by a EIS biochip [152]. Nguyen et al. proposed a microfluidic chip with ECIS for monitoring the migration of single cancer cells in 3D matrixes (Figure 3C) [85]. In this study, the impedance measurements were performed with a voltage of 10 mV over the frequency range from 100 Hz to 1 MHz and showed the significant decrease of EIS amplitude after cell migration (Figure 9F). The real-time EIS recording was carried out at 4 kHz and demonstrated that MCF-7 cells were less metastatic than MDA-MB-231 cells. Zhang et al. monitored the recovery processes of HeLa cells after electroporation by using impedance measurement (Figure 9G) [153]. HeLa cells were trapped and electroporated with different working modes of center electrodes. Within 5 min after electroporation, normalized amplitude curves were slowly rising corresponded to the reversible EP processes, while those stabilizing at the minimum values indicated the irreversible EP and cell death.

5. Conclusions

Electrical impedance sensing technology, as a rapid and non-invasive method to probe cellular biophysical information, has become appealing in single-cell study. The basic theories and modeling methods of single-cell impedance sensing have been re-viewed herein and recent advances in this field have been highlighted with respect to the device design and applications.

Generally, the way to implement electrical impedance measurement in a microfluidic device is categorized into IFC and EIS sensing. IFC features measuring impedance of single frequencies for large number of cells, while EIS sensing is capable of re-al-time monitoring of a few cells over a wide frequency range. A variety of optimal electrode layouts, fluidic channel configurations, hydrodynamic focusing systems have been proposed to improve the sensitivity and consistency of IFC in the measurement of cellular electrical parameters. With various trapping methods, suspended single cells could be stably immobilized for *in-situ* EIS sensing. ECIS can recognize cellular behavior sensitively in response to defined stimulus. Besides, individually addressable MEAs have been incorporated into EIS sensing devices in order to overcome the limitation of throughput. Impedance converters and LIAs, as the basic electronic components to measure impedance, have been further integrated and miniaturized from instruments to portable platforms. As an alternative, CMOS-based impedance sensing devices have been developed to increase the integration level of impedance sensing system.

Since IFC devices have the merits of rapid measurement and high throughput, they have been widely used in the identification and classification of various species of single cells and determination of cell viability. The impedance signals at different frequencies reveal dielectric characteristics of different cellular structures. Hence, a combination of multiple impedance parameters, such as amplitude at high frequency and low frequency or amplitude and phase at the same frequency, has been commonly used to identify single cells with various phenotypes, in different life stages or under multiple external conditions. Applications of the IFC devices in plant cell analysis, especially pollen screening, have been proposed and promoted, and the strategy of combining mechanical characterization and impedance measurement has been also developed. In addition, machine learning has been used in impedance data analysis to improve the performance of IFC devices.

EIS sensing has been used to choose the most sensitive frequency for subsequent high-speed analysis or long-term monitoring of cell behavior and phenotypes. The various cellular physiological processes, including adhesion, growth, division, differentiation, proliferation and cellular structure formation, have been characterized by the measured electrical impedance spectra.

From this review, insights into challenges and prospects of electrical impedance sensing technology for single-cell analysis could be provided as follows. Different cell subpopulations are hard to be accurately classified based on impedance information at specific frequencies. To this end, multi-frequency impedance signals and the combination of multiple biophysical parameters could be used to enrich characteristic information of different cells, and thus could favor the improvement of phenotyping resolution. Besides, machine learning algorithm, such as SVM and neural networks, in data analysis could help to correspond broadband impedance signals to cell phenotypic characteristics or single-cell physiological processes. On the other hand, the high cost of chip fabrication and benchtop instruments hinders the popularization of single-cell impedance sensing technology. It can be supposed that single-cell impedance measurement devices could soon appear in the clinical laboratory in a more user-friendly format, with the help of development and promotion of commercial equipment and portable platforms. By then, a quantum leap will appear in the fields of rapid diagnosis, smart healthcare and personalized medicine.

Author Contributions: Conceptualization, Z.Z. (Zhao Zhang), X.H. and Z.Z. (Zhen Zhu); writing—original draft preparation, Z.Z. (Zhao Zhang) and X.H.; writing—review and editing, K.L., T.L. and Z.Z. (Zhen Zhu); supervision, Z.W. and Z.Z. (Zhen Zhu); project administration, Z.W. and Z.Z. (Zhen Zhu); funding acquisition, Z.Z. (Zhen Zhu). All authors have read and agreed to the published version of the manuscript.

Funding: This research was funded by National Natural Science Foundation of China, grant number 61774036; the National Key R&D Program of China, grant numbers 2018YFF01012100; the Fundamental Research Funds for the Central Universities; the Open Research Fund of State Key Laboratory of Bioelectronics, Southeast University.

Institutional Review Board Statement: Not applicable.

Informed Consent Statement: Not applicable.

Data Availability Statement: Not applicable.

Conflicts of Interest: The authors declare no conflict of interest.

References

1. El-Ali, J.; Sorger, P.K.; Jensen, K.F. Cells on chips. *Nature* **2006**, *442*, 403–411. [CrossRef] [PubMed]
2. Suresh, S. Biomechanics and biophysics of cancer cells. *Acta Biomater.* **2007**, *3*, 413–438. [CrossRef]
3. Graf, T.; Stadtfeld, M. Heterogeneity of embryonic and adult stem cells. *Cell Stem Cell* **2008**, *3*, 480–483. [CrossRef]
4. Hunter, K. Host genetics influence tumour metastasis. *Nat. Rev. Cancer.* **2006**, *6*, 141–146. [CrossRef] [PubMed]
5. Wang, D.; Bodovitz, S. Single cell analysis: The new frontier in 'omics'. *Trends Biotechnol.* **2010**, *28*, 281–290. [CrossRef] [PubMed]
6. Armbrecht, L.; Dittrich, P.S. Recent advances in the analysis of single cells. *Anal. Chem.* **2017**, *89*, 2–21. [CrossRef] [PubMed]
7. Zare, R.N.; Kim, S. Microfluidic platforms for single-cell analysis. *Annu. Rev. Biomed. Eng.* **2010**, *12*, 187–201. [CrossRef]
8. Kim, S.; Song, H.; Ahn, H.; Kim, T.; Jung, J.; Cho, S.K.; Shin, D.-M.; Choi, J.-R.; Hwang, Y.-H.; Kim, K. A Review of Advanced Impedance Biosensors with Microfluidic Chips for Single-Cell Analysis. *Biosensors* **2021**, *11*, 412. [CrossRef]
9. Li, W.; Zhang, L.; Ge, X.; Xu, B.; Zhang, W.; Qu, L.; Choi, C.H.; Xu, J.; Zhang, A.; Lee, H.; et al. Microfluidic fabrication of microparticles for biomedical applications. *Chem. Soc. Rev.* **2018**, *47*, 5646–5683. [CrossRef]
10. Marcus, J.S.; Anderson, W.F.; Quake, S.R. Microfluidic single-cell mRNA isolation and analysis. *Anal. Chem.* **2006**, *78*, 3084–3089. [CrossRef] [PubMed]
11. Andersson, H.; van den Berg, A. Microfluidic devices for cellomics: A review. *Sens. Actuators B Chem.* **2003**, *92*, 315–325. [CrossRef]
12. Gao, D.; Jin, F.; Zhou, M.; Jiang, Y. Recent advances in single cell manipulation and biochemical analysis on microfluidics. *Analyst* **2019**, *144*, 766–781. [CrossRef] [PubMed]
13. Hunt, T.P.; Westervelt, R.M. Dielectrophoresis tweezers for single cell manipulation. *Biomed. Microdevices* **2006**, *8*, 227–230. [CrossRef]
14. Lin, S.C.; Mao, X.; Huang, T.J. Surface acoustic wave (SAW) acoustophoresis: Now and beyond. *Lab Chip* **2012**, *12*, 2766–2770. [CrossRef] [PubMed]
15. Ramser, K.; Hanstorp, D. Optical manipulation for single-cell studies. *J. Biophotonics* **2010**, *3*, 187–206. [CrossRef] [PubMed]
16. Rettig, J.R.; Folch, A. Large-scale single-cell trapping and imaging using microwell arrays. *Anal. Chem.* **2005**, *77*, 5628–5634. [CrossRef] [PubMed]
17. Ryley, J.; Pereira-Smith, O.M. Microfluidics device for single cell gene expression analysis in Saccharomyces cerevisiae. *Yeast* **2006**, *23*, 1065–1073. [CrossRef] [PubMed]
18. Ozkumur, E.; Shah, A.M.; Ciciliano, J.C.; Emmink, B.L.; Miyamoto, D.T.; Brachtel, E.; Yu, M.; Chen, P.I.; Morgan, B.; Trautwein, J.; et al. Inertial focusing for tumor antigen–dependent and –independent sorting of rare circulating tumor cells. *Sci. Transl. Med.* **2013**, *5*, 179ra147. [CrossRef] [PubMed]
19. Wu, H.; Volponi, J.V.; Oliver, A.E.; Parikh, A.N.; Simmons, B.A.; Singh, S. In vivo lipidomics using single-cell Raman spectroscopy. *Proc. Natl. Acad. Sci. USA* **2011**, *108*, 3809–3814. [CrossRef] [PubMed]
20. Amann, R.; Fuchs, B.M. Single-cell identification in microbial communities by improved fluorescence in situ hybridization techniques. *Nat. Rev. Microbiol.* **2008**, *6*, 339–348. [CrossRef] [PubMed]
21. He, X.; Chen, Q.; Zhang, Y.; Lin, J.M. Recent advances in microchip-mass spectrometry for biological analysis. *Trends Analyt. Chem.* **2014**, *53*, 84–97. [CrossRef]
22. Actis, P.; Tokar, S.; Clausmeyer, J.; Babakinejad, B. Electrochemical nanoprobes for single-cell analysis. *ACS Nano* **2014**, *8*, 875–884. [CrossRef]
23. Sun, T.; Morgan, H. Single-cell microfluidic impedance cytometry: A review. *Microfluid. Nanofluid.* **2010**, *8*, 423–443. [CrossRef]
24. Morgan, H.; Sun, T.; Holmes, D.; Gawad, S.; Green, N.G. Single cell dielectric spectroscopy. *J. Phys. D Appl. Phys.* **2007**, *40*, 61–70. [CrossRef]
25. Zhang, X.; Hatamie, A.; Ewing, A.G. Nanoelectrochemical analysis inside a single living cell. *Curr. Opin. Electrochem.* **2020**, *22*, 94–101. [CrossRef]
26. Sun, T.; Green, N.G.; Nicolas, G.G.; Hywel, M. Analysis and numerical modeling methods for impedance analysis of single cells on-chip. *Nano* **2008**, *3*, 55–63. [CrossRef]
27. Gawad, S.; Cheung, K.; Seger, U.; Bertsch, A.; Renaud, P. Dielectric spectroscopy in a micromachined flow cytometer: Theoretical and practical considerations. *Lab Chip* **2004**, *4*, 241–251. [CrossRef] [PubMed]
28. Asami, K. Characterization of heterogeneous systems by dielectric spectroscopy. *Prog. Polym. Sci.* **2002**, *27*, 1617–1659. [CrossRef]
29. Sun, T.; Green, N.G.; Gawad, S.; Morgan, H. Analytical electric field and sensitivity analysis for two microfluidic impedance cytometer designs. *IET Nanobiotechnol.* **2007**, *1*, 69–79. [CrossRef] [PubMed]
30. Sun, T.; Morgan, H.; Green, N.G. Analytical solutions of ac electrokinetics in interdigitated electrode arrays: Electric field, dielectrophoretic and traveling-wave dielectrophoretic forces. *Phys. Rev. E* **2007**, *76*, 046610. [CrossRef] [PubMed]
31. Franco, J.L.S.; Otero, A.S.; Madronero, J.R.; Martin, S.M.S. Dielectric characterization of the yeast cell budding cycle. *Prog. Electromagn. Res.* **2013**, *134*, 1–22. [CrossRef]
32. Gawad, S.; Schild, L.; Renaud, P.H. Micromachined impedance spectroscopy flow cytometer for cell analysis and particle sizing. *Lab Chip* **2001**, *1*, 76–82. [CrossRef] [PubMed]
33. Cheung, K.C.; Di Berardino, M.; Schade-Kampmann, G.; Hebeisen, M.; Pierzchalski, A.; Bocsi, J.; Mittag, A.; Tarnok, A. Microfluidic impedance-based flow cytometry. *Cytometry A* **2010**, *77*, 648–666. [CrossRef] [PubMed]

34. Xu, Y.; Xie, X.; Duan, Y.; Wang, L.; Cheng, Z.; Cheng, J. A review of impedance measurements of whole cells. *Biosens. Bioelectron.* **2016**, *77*, 824–836. [CrossRef] [PubMed]
35. Yan, S.; Yuan, D. Continuous microfluidic 3D focusing enabling microflow cytometry for single-cell analysis. *Talanta* **2021**, *221*, 121401. [CrossRef] [PubMed]
36. Daguerre, H.; Solsona, M.; Cottet, J.; Gauthier, M.; Renaud, P.; Bolopion, A. Positional dependence of particles and cells in microfluidic electrical impedance flow cytometry origin, challenges and opportunities. *Lab Chip* **2020**, *20*, 3665–3689. [CrossRef] [PubMed]
37. Rodriguez-Trujillo, R.; Castillo-Fernandez, O.; Garrido, M.; Arundell, M.; Valencia, A.; Gomila, G. High-speed particle detection in a micro-Coulter counter with two-dimensional adjustable aperture. *Biosens. Bioelectron.* **2008**, *24*, 290–296. [CrossRef] [PubMed]
38. Kirkegaard, J.; Clausen, C.H.; Rodriguez-Trujillo, R.; Svendsen, W.E. Study of paclitaxel-treated HeLa cells by differential electrical impedance flow cytometry. *Biosensors* **2014**, *4*, 257–272. [CrossRef] [PubMed]
39. Clausen, C.; Skands, G.; Bertelsen, C.; Svendsen, W. Coplanar electrode layout optimized for increased sensitivity for electrical impedance spectroscopy. *Micromachines* **2014**, *6*, 110–120. [CrossRef]
40. Carminati, M.; Ferrari, G.; Vahey, M.D.; Voldman, J.; Sampietro, M. Miniaturized impedance flow cytometer: Design rules and integrated readout. *IEEE Trans. Biomed. Circuits Syst.* **2017**, *11*, 1438–1449. [CrossRef] [PubMed]
41. De Ninno, A.; Errico, V.; Bertani, F.R.; Businaro, L.; Bisegna, P.; Caselli, F. Coplanar electrode microfluidic chip enabling accurate sheathless impedance cytometry. *Lab Chip* **2017**, *17*, 1158–1166. [CrossRef] [PubMed]
42. Valero, A.; Braschler, T.; Renaud, P. A unified approach to dielectric single cell analysis: Impedance and dielectrophoretic force spectroscopy. *Lab Chip* **2010**, *10*, 2216–2225. [CrossRef] [PubMed]
43. Zhang, Y.; Liang, H.; Tan, H.; Chen, D.; Wang, Y.; Xu, Y.; Wang, J.; Chen, J. Development of microfluidic platform to high-throughput quantify single-cell intrinsic bioelectrical markers of tumor cell lines, subtypes and patient tumor cells. *Sens. Actuators B Chem.* **2020**, *317*, 128231. [CrossRef]
44. Caselli, F.; De Ninno, A.; Reale, R.; Businaro, L.; Bisegna, P. A novel wiring scheme for standard chips enabling high-accuracy impedance cytometry. *Sens. Actuators B Chem.* **2018**, *256*, 580–589. [CrossRef]
45. Spencer, D.; Caselli, F.; Bisegna, P.; Morgan, H. High accuracy particle analysis using sheathless microfluidic impedance cytometry. *Lab Chip* **2016**, *16*, 2467–2473. [CrossRef]
46. Xie, X.; Cheng, Z.; Xu, Y.; Liu, R.; Li, Q.; Cheng, J. A sheath-less electric impedance micro-flow cytometry device for rapid label-free cell classification and viability testing. *Anal. Methods* **2017**, *9*, 1201–1212. [CrossRef]
47. Demierre, N.; Braschler, T.; Muller, R.; Renaud, P. Focusing and continuous separation of cells in a microfluidic device using lateral dielectrophoresis. *Sens. Actuators B Chem.* **2008**, *132*, 388–396. [CrossRef]
48. Demierre, N.; Braschler, T.; Linderholm, P.; Seger, U.; van Lintel, H.; Renaud, P. Characterization and optimization of liquid electrodes for lateral dielectrophoresis. *Lab Chip* **2007**, *7*, 355–365. [CrossRef]
49. Shaker, M.; Colella, L.; Caselli, F.; Bisegna, P.; Renaud, P. An impedance-based flow microcytometer for single cell morphology discrimination. *Lab Chip* **2014**, *14*, 2548–2555. [CrossRef] [PubMed]
50. Cottet, J.; Kehren, A.; van Lintel, H.; Buret, F.; Frénéa-Robin, M.; Renaud, P. How to improve the sensitivity of coplanar electrodes and micro channel design in electrical impedance flow cytometry: A study. *Microfluid. Nanofluid.* **2019**, *23*, 11. [CrossRef]
51. Cheung, K.; Gawad, S.; Renaud, P. Impedance spectroscopy flow cytometry: On-chip label-free cell differentiation. *Cytometry A* **2005**, *65*, 124–132. [CrossRef]
52. Spencer, D.; Hollis, V.; Morgan, H. Microfluidic impedance cytometry of tumour cells in blood. *Biomicrofluidics* **2014**, *8*, 064124. [CrossRef]
53. Clausen, C.H.; Dimaki, M.; Bertelsen, C.V.; Skands, G.E.; Rodriguez-Trujillo, R.; Thomsen, J.D.; Svendsen, W.E. Bacteria detection and differentiation using impedance flow cytometry. *Sensors* **2018**, *18*, 3496. [CrossRef] [PubMed]
54. Bürgel, S.C.; Escobedo, C.; Haandbæk, N.; Hierlemann, A. On-chip electroporation and impedance spectroscopy of single-cells. *Sens. Actuators B Chem.* **2015**, *210*, 82–90. [CrossRef]
55. Spencer, D.; Morgan, H. Positional dependence of particles in microfludic impedance cytometry. *Lab Chip* **2011**, *11*, 1234–1239. [CrossRef] [PubMed]
56. Wan, J.; Yang, L.; Hu, S.; Liu, K.; Jia, J.; Yao, J. Position Calibration of a Single Cell Measurement with Electrochemical Impedance Spectroscopy. *IEEE Sens. J.* **2021**, 1. [CrossRef]
57. Chen, J.; Zheng, Y.; Tan, Q.; Shojaei-Baghini, E.; Zhang, Y.L.; Li, J.; Prasad, P.; You, L.; Wu, X.Y.; Sun, Y. Classification of cell types using a microfluidic device for mechanical and electrical measurement on single cells. *Lab Chip* **2011**, *11*, 3174–3181. [CrossRef] [PubMed]
58. Zhao, Y.; Wang, K.; Chen, D.; Fan, B.; Xu, Y.; Ye, Y.; Wang, J.; Chen, J.; Huang, C. Development of microfluidic impedance cytometry enabling the quantification of specific membrane capacitance and cytoplasm conductivity from 100,000 single cells. *Biosens. Bioelectron.* **2018**, *111*, 138–143. [CrossRef] [PubMed]
59. Zheng, Y.; Shojaei-Baghini, E.; Azad, A.; Wang, C.; Sun, Y. High-throughput biophysical measurement of human red blood cells. *Lab Chip* **2012**, *12*, 2560–2567. [CrossRef] [PubMed]
60. Han, Z.; Chen, L.; Zhang, S.; Wang, J.; Duan, X. Label-free and simultaneous mechanical and electrical characterization of single plant cells using microfluidic impedance flow cytometry. *Anal. Chem.* **2020**, *92*, 14568–14575. [CrossRef]

61. Watkins, N.; Venkatesan, B.M.; Toner, M.; Rodriguez, W.; Bashir, R. A robust electrical microcytometer with 3-dimensional hydrofocusing. *Lab Chip* **2009**, *9*, 3177–3184. [CrossRef] [PubMed]
62. Choi, H.; Jeon, C.S.; Hwang, I.; Ko, J.; Lee, S.; Choo, J.; Boo, J.H.; Kim, H.C.; Chung, T.D. A flow cytometry-based submicron-sized bacterial detection system using a movable virtual wall. *Lab Chip* **2014**, *14*, 2327–2333. [CrossRef] [PubMed]
63. Grenvall, C.; Antfolk, C.; Bisgaard, C.Z.; Laurell, T. Two-dimensional acoustic particle focusing enables sheathless chip Coulter counter with planar electrode configuration. *Lab Chip* **2014**, *14*, 4629–4637. [CrossRef]
64. Ng, J.W.; Collins, D.J.; Devendran, C.; Ai, Y.; Neild, A. Flow-rate-insensitive deterministic particle sorting using a combination of travelling and standing surface acoustic waves. *Microfluid. Nanofluid.* **2016**, *20*, 151. [CrossRef]
65. Sriphutkiat, Y.; Zhou, Y. Particle accumulation in a microchannel and its reduction by a standing surface acoustic wave (SSAW). *Sensors* **2017**, *17*, 106. [CrossRef] [PubMed]
66. Mernier, G.; Duqi, E.; Renaud, P. Characterization of a novel impedance cytometer design and its integration with lateral focusing by dielectrophoresis. *Lab Chip* **2012**, *12*, 4344–4349. [CrossRef] [PubMed]
67. Evander, M.; Ricco, A.J.; Morser, J.; Kovacs, G.T.; Leung, L.L.; Giovangrandi, L. Microfluidic impedance cytometer for platelet analysis. *Lab Chip* **2013**, *13*, 722–729. [CrossRef]
68. Carlo, D.D.; Irimia, D.; Tompkins, R.G.; Toner, M. Continuous inertial focusing, ordering, and separation of particles in microchannels. *Proc. Natl. Acad. Sci. USA* **2007**, *104*, 18892–18897. [CrossRef] [PubMed]
69. Di Carlo, D.; Wu, L.Y.; Lee, L.P. Dynamic single cell culture array. *Lab Chip* **2006**, *6*, 1445–1449. [CrossRef]
70. Tan, W.H.; Takeuchi, S. A trap-and-release integrated microfluidic system for dynamic microarray applications. *Proc. Natl. Acad. Sci. USA* **2006**, *104*, 1146–1151. [CrossRef]
71. Jang, L.S.; Wang, M.H. Microfluidic device for cell capture and impedance measurement. *Biomed. Microdevices* **2007**, *9*, 737–743. [CrossRef] [PubMed]
72. Tan, W.H.; Takeuchi, S. Dynamic microarray system with gentle retrieval mechanism for cell-encapsulating hydrogel beads. *Lab Chip* **2008**, *8*, 259–266. [CrossRef] [PubMed]
73. Malleo, D.; Nevill, J.T.; Lee, L.P.; Morgan, H. Continuous differential impedance spectroscopy of single cells. *Microfluid. Nanofluid.* **2010**, *9*, 191–198. [CrossRef] [PubMed]
74. Cho, Y.H.; Yamamoto, T.; Sakai, Y.; Fujii, T.; Kim, B. Development of microfluidic device for electrical/physical characterization of single cell. *J. Microelectromech. Syst.* **2006**, *15*, 287–295. [CrossRef]
75. Han, K.H.; Han, A.; Frazier, A.B. Microsystems for isolation and electrophysiological analysis of breast cancer cells from blood. *Biosens. Bioelectron.* **2006**, *21*, 1907–1914. [CrossRef]
76. Younghak, C.; Hyun Soo, K.; Frazier, A.B.; Chen, Z.G.; Dong Moon, S.; Han, A. Whole-cell impedance analysis for highly and poorly metastatic cancer cells. *J. Microelectromech. Syst.* **2009**, *18*, 808–817. [CrossRef]
77. Lan, K.C.; Jang, L.S. Integration of single-cell trapping and impedance measurement utilizing microwell electrodes. *Biosens. Bioelectron.* **2011**, *26*, 2025–2031. [CrossRef] [PubMed]
78. Heida, T.; Rutten, W.L.C.; Marani, E. Understanding dielectrophoretic trapping of neuronal cells: Modelling electric field, electrodeliquid interface and fluid flow. *J. Phys. D Appl. Phys.* **2002**, *35*, 1592–1602. [CrossRef]
79. Rosenthal, A.; Voldman, J. Dielectrophoretic traps for single-particle patterning. *Biophys. J.* **2005**, *88*, 2193–2205. [CrossRef]
80. Jang, L.S.; Huang, P.H.; Lan, K.C. Single-cell trapping utilizing negative dielectrophoretic quadrupole and microwell electrodes. *Biosens. Bioelectron.* **2009**, *24*, 3637–3644. [CrossRef] [PubMed]
81. Xu, B.; Shi, Y.; Lao, Z.; Ni, J.; Li, G.; Hu, Y.; Li, J.; Chu, J.; Wu, D.; Sugioka, K. Real-time two-photon lithography in controlled flow to create a single-microparticle array and particle-cluster array for optofluidic imaging. *Lab Chip* **2018**, *18*, 442–450. [CrossRef]
82. Tang, W.; Tang, D.; Ni, Z.; Xiang, N.; Yi, H. A portable single-cell analysis system integrating hydrodynamic trapping with broadband impedance spectroscopy. *Sci. China Technol. Sci.* **2017**, *60*, 1707–1715. [CrossRef]
83. Zhu, Z.; Frey, O.; Franke, F.; Haandbaek, N.; Hierlemann, A. Real-time monitoring of immobilized single yeast cells through multifrequency electrical impedance spectroscopy. *Anal. Bioanal. Chem.* **2014**, *406*, 7015–7025. [CrossRef]
84. Taff, B.M.; Voldman, J. A scalable addressable positive-dielectrophoretic cell-sorting array. *Anal. Chem.* **2005**, *77*, 7976–7983. [CrossRef]
85. Nguyen, T.A.; Yin, T.-I.; Reyes, D.; Urban, G.A. Microfluidic chip with integrated electrical cell-Impedance sensing for monitoring single cancer cell migration in three-dimensional matrixes. *Anal. Chem.* **2013**, *85*, 11068–11076. [CrossRef] [PubMed]
86. Geng, Y.; Zhu, Z.; Zhang, Z.; Xu, F.; Marchisio, M.A.; Wang, Z.; Pan, D.; Zhao, X.; Huang, Q.A. Design and 3D modeling investigation of a microfluidic electrode array for electrical impedance measurement of single yeast cells. *Electrophoresis* **2021**, *42*, 1996–2009. [CrossRef]
87. Asphahani, F.; Zhang, M. Cellular impedance biosensors for drug screening and toxin detection. *Analyst* **2007**, *132*, 835–841. [CrossRef]
88. Kovacs, G.T.A. Electronic sensors with living cellular components. *Proc. IEEE* **2003**, *91*, 915–929. [CrossRef]
89. Seriburi, P.; McGuire, S.; Shastry, A.; Böhringer, K.F.; Meldrum, D.R. Measurement of the cell-substrate separation and the projected area of an individual adherent cell using electric cell-substrate impedance sensing. *Anal. Chem.* **2008**, *80*, 3677–3683. [CrossRef] [PubMed]
90. Thein, M.; Asphahani, F.; Cheng, A.; Buckmaster, R.; Zhang, M.; Xu, J. Response characteristics of single-cell impedance sensors employed with surface-modified microelectrodes. *Biosens. Bioelectron.* **2010**, *25*, 1963–1969. [CrossRef] [PubMed]

91. Tsai, S.L.; Wang, M.H. 24 h observation of a single HeLa cell by impedance measurement and numerical modeling. *Sens. Actuators B Chem.* **2016**, *229*, 225–231. [CrossRef]
92. Zhang, Z.; Zheng, T.; Zhu, R. Microchip with single-cell impedance measurements for monitoring osteogenic differentiation of mesenchymal stem cells under electrical stimulation. *Anal. Chem.* **2020**, *92*, 12579–12587. [CrossRef] [PubMed]
93. Shah, P.; Zhu, X.; Zhang, X.; He, J.; Li, C.Z. Microelectromechanical system-based sensing arrays for comparative in vitro nanotoxicity assessment at single cell and small cell-population using electrochemical impedance spectroscopy. *ACS Appl. Mater. Interfaces* **2016**, *8*, 5804–5812. [CrossRef]
94. Zhu, Z.; Frey, O.; Haandbaek, N.; Franke, F.; Rudolf, F.; Hierlemann, A. Time-lapse electrical impedance spectroscopy for monitoring the cell cycle of single immobilized S. pombe cells. *Sci. Rep.* **2015**, *5*, 17180. [CrossRef]
95. Zhou, Y.; Basu, S.; Laue, E.; Seshia, A.A. Single cell studies of mouse embryonic stem cell (mESC) differentiation by electrical impedance measurements in a microfluidic device. *Biosens. Bioelectron.* **2016**, *81*, 249–258. [CrossRef] [PubMed]
96. Guo, X.; Zhu, R. Controllable in-situ cell electroporation with cell positioning and impedance monitoring using micro electrode array. *Sci. Rep.* **2016**, *6*, 31392. [CrossRef] [PubMed]
97. Sun, S.; Xu, L.; Cao, Z.; Zhou, H.; Yang, W. A high-speed electrical impedance measurement circuit based on information-filtering demodulation. *Meas. Sci. Technol.* **2014**, *25*, 075010. [CrossRef]
98. Kishore, K.; Akbar, S.A. Evolution of lock-in amplifier as portable sensor interface platform: A review. *IEEE Sens. J.* **2020**, *20*, 10345–10354. [CrossRef]
99. Zurich Instruments. Available online: https://www.zhinst.com/others/en (accessed on 14 September 2021).
100. Liquid Instruments. Available online: https://www.liquidinstruments.com/company/ (accessed on 14 September 2021).
101. NF Corporation. Available online: https://www.nfcorp.co.jp/english/index.html (accessed on 14 September 2021).
102. SRS Home Page. Available online: https://thinksrs.com/index.html (accessed on 14 September 2021).
103. SBT Instruments. Available online: https://sbtinstruments.com/ (accessed on 12 November 2021).
104. Sine Scientific Instruments. Available online: https://www.ssi-instrument.com/ (accessed on 21 November 2021).
105. Grossi, M.; Riccò, B. Electrical impedance spectroscopy (EIS) for biological analysis and food characterization: A review. *J. Sens. Sens. Syst.* **2017**, *6*, 303–325. [CrossRef]
106. Huang, K.; Geng, Y.; Zhang, X.; Chen, D.; Cai, Z.; Wang, M.; Zhu, Z.; Wang, Z. A wide-band digital lock-in amplifier and its application in microfluidic impedance measurement. *Sensors* **2019**, *19*, 3519. [CrossRef] [PubMed]
107. Chen, Y.; Wong, C.C.; Pui, T.S.; Nadipalli, R.; Weerasekera, R.; Chandran, J.; Yu, H.; Rahman, A.R.A. CMOS high density electrical impedance biosensor array for tumor cell detection. *Sens. Actuators B Chem.* **2012**, *173*, 903–907. [CrossRef]
108. Gamo, K.; Nakazato, K.; Niitsu, K. Design, theoretical analysis, and experimental verification of a CMOS current integrator with $1.2 \times 2.05\ \mu m^2$ microelectrode array for high-sensitivity bacterial counting. *Jpn. J. Appl. Phys.* **2017**, *56*, 01AH01. [CrossRef]
109. Viswam, V.; Bounik, R.; Shadmani, A.; Dragas, J.; Urwyler, C.; Boos, J.A.; Obien, M.E.J.; Muller, J.; Chen, Y.; Hierlemann, A. Impedance spectroscopy and electrophysiological imaging of cells with a high-density CMOS microelectrode array system. *IEEE Trans. Biomed. Circuits Syst.* **2018**, *12*, 1356–1368. [CrossRef] [PubMed]
110. Liu, J.; Qiang, Y.; Alvarez, O.; Du, E. Electrical impedance microflow cytometry with oxygen control for detection of sickle cells. *Sens. Actuators B Chem.* **2018**, *255*, 2392–2398. [CrossRef]
111. Hassan, U.; Watkins, N.N.; Reddy, B., Jr.; Damhorst, G.; Bashir, R. Microfluidic differential immunocapture biochip for specific leukocyte counting. *Nat. Protoc.* **2016**, *11*, 714–726. [CrossRef] [PubMed]
112. Holmes, D.; Morgan, H. Single cell impedance cytometry for identification and counting of CD4 T-cells in human blood using impedance labels. *Anal. Chem.* **2010**, *82*, 1455–1461. [CrossRef] [PubMed]
113. Du, E.; Ha, S.; Diez-Silva, M.; Dao, M.; Suresh, S.; Chandrakasan, A.P. Electric impedance microflow cytometry for characterization of cell disease states. *Lab Chip* **2013**, *13*, 3903–3909. [CrossRef]
114. Honrado, C.; Ciuffreda, L.; Spencer, D.; Ranford-Cartwright, L.; Morgan, H. Dielectric characterization of Plasmodium falciparum-infected red blood cells using microfluidic impedance cytometry. *J. R. Soc. Interface* **2018**, *15*, 20180416. [CrossRef]
115. Choi, H.; Kim, K.B.; Jeon, C.S.; Hwang, I.; Lee, S.; Kim, H.K.; Kim, H.C.; Chung, T.D. A label-free DC impedance-based microcytometer for circulating rare cancer cell counting. *Lab Chip* **2013**, *13*, 970–977. [CrossRef]
116. Han, S.I.; Han, K.H. Electrical detection method for circulating tumor cells using graphene nanoplates. *Anal. Chem.* **2015**, *87*, 10585–10592. [CrossRef] [PubMed]
117. Zhao, Y.; Jiang, M.; Chen, D.; Zhao, X.; Xue, C.; Hao, R.; Yue, W.; Wang, J.; Chen, J. Single-cell electrical phenotyping enabling the classification of mouse tumor samples. *Sci. Rep.* **2016**, *6*, 19487. [CrossRef] [PubMed]
118. Desai, S.P.; Coston, A.; Berlin, A. Micro-electrical impedance spectroscopy and identification of patient-derived, dissociated tumor cells. *IEEE Trans. Nanobiosci.* **2019**, *18*, 369–372. [CrossRef] [PubMed]
119. Ren, X.; Ghassemi, P.; Strobl, J.S.; Agah, M. Biophysical phenotyping of cells via impedance spectroscopy in parallel cyclic deformability channels. *Biomicrofluidics* **2019**, *13*, 044103. [CrossRef]
120. McGrath, J.S.; Honrado, C.; Moore, J.H.; Adair, S.J.; Varhue, W.B.; Salahi, A.; Farmehini, V.; Goudreau, B.J.; Nagdas, S.; Blais, E.M.; et al. Electrophysiology-based stratification of pancreatic tumorigenicity by label-free single-cell impedance cytometry. *Anal. Chim. Acta* **2020**, *1101*, 90–98. [CrossRef]
121. Ostermann, M.; Sauter, A.; Xue, Y.; Birkeland, E.; Schoelermann, J.; Holst, B.; Cimpan, M.R. Label-free impedance flow cytometry for nanotoxicity screening. *Sci. Rep.* **2020**, *10*, 142. [CrossRef] [PubMed]

122. Krebs, M.G.; Hou, J.M.; Ward, T.H.; Blackhall, F.H.; Dive, C. Circulating tumour cells: Their utility in cancer management and predicting outcomes. *Ther. Adv. Med. Oncol.* **2010**, *2*, 351–365. [CrossRef] [PubMed]
123. Hildebrandt, C.; Buth, H.; Cho, S.; Impidjati; Thielecke, H. Detection of the osteogenic differentiation of mesenchymal stem cells in 2D and 3D cultures by electrochemical impedance spectroscopy. *J. Biotechnol.* **2010**, *148*, 83–90. [CrossRef] [PubMed]
124. Song, H.; Wang, Y.; Rosano, J.M.; Prabhakarpandian, B.; Garson, C.; Pant, K.; Lai, E. A microfluidic impedance flow cytometer for identification of differentiation state of stem cells. *Lab Chip* **2013**, *13*, 2300–2310. [CrossRef]
125. Song, H.; Rosano, J.M.; Wang, Y.; Garson, C.J.; Prabhakarpandian, B.; Pant, K.; Klarmann, G.J.; Perantoni, A.; Alvarez, L.M.; Lai, E. Identification of mesenchymal stem cell differentiation state using dual-micropore microfluidic impedance flow cytometry. *Anal. Methods* **2016**, *8*, 7437–7444. [CrossRef]
126. Zhao, Y.; Liu, Q.; Sun, H.; Chen, D.; Li, Z.; Fan, B.; George, J.; Xue, C.; Cui, Z.; Wang, J.; et al. Electrical property characterization of neural stem cells in differentiation. *PLoS ONE* **2016**, *11*, e0158044. [CrossRef]
127. Xavier, M.; de Andres, M.C.; Spencer, D.; Oreffo, R.O.C.; Morgan, H. Size and dielectric properties of skeletal stem cells change critically after enrichment and expansion from human bone marrow: Consequences for microfluidic cell sorting. *J. R. Soc. Interface* **2017**, *14*, 20170233. [CrossRef]
128. Heidmann, I.; Schade-Kampmann, G.; Lambalk, J.; Ottiger, M.; Di Berardino, M. Impedance flow cytometry: A novel technique in pollen analysis. *PLoS ONE* **2016**, *11*, e0165531. [CrossRef] [PubMed]
129. Heidmann, I.; Di Berardino, M. Impedance flow cytometry as a tool to analyze microspore and pollen quality. *Methods Mol. Biol.* **2017**, *1669*, 339–354.
130. Impe, D.; Reitz, J.; Kopnick, C.; Rolletschek, H.; Borner, A.; Senula, A.; Nagel, M. Assessment of pollen viability for wheat. *Front. Plant Sci.* **2019**, *10*, 1588. [CrossRef] [PubMed]
131. Ascari, L.; Cristofori, V.; Macri, F.; Botta, R.; Silvestri, C.; De Gregorio, T.; Huerta, E.S.; Di Berardino, M.; Kaufmann, S.; Siniscalco, C. Hazelnut pollen phenotyping using label-free impedance flow cytometry. *Front. Plant Sci.* **2020**, *11*, 615922. [CrossRef] [PubMed]
132. Canonge, J.; Philippot, M.; Leblanc, C.; Potin, P.; Bodin, M. Impedance flow cytometry allows the early prediction of embryo yields in wheat (*Triticum aestivum* L.) microspore cultures. *Plant Sci.* **2020**, *300*, 110586. [CrossRef] [PubMed]
133. Bernabini, C.; Holmes, D.; Morgan, H. Micro-impedance cytometry for detection and analysis of micron-sized particles and bacteria. *Lab Chip* **2011**, *11*, 407–412. [CrossRef] [PubMed]
134. David, F.; Hebeisen, M.; Schade, G.; Franco-Lara, E.; Di Berardino, M. Viability and membrane potential analysis of Bacillus megaterium cells by impedance flow cytometry. *Biotechnol. Bioeng.* **2012**, *109*, 483–492. [CrossRef] [PubMed]
135. Guler, M.T.; Bilican, I. Capacitive detection of single bacterium from drinking water with a detailed investigation of electrical flow cytometry. *Sens. Actuator A Phys.* **2018**, *269*, 454–463. [CrossRef]
136. McGrath, J.S.; Honrado, C.; Spencer, D.; Horton, B.; Bridle, H.L.; Morgan, H. Analysis of parasitic protozoa at the single-cell level using microfluidic impedance cytometry. *Sci. Rep.* **2017**, *7*, 2601. [CrossRef]
137. Chawla, K.; Burgel, S.C.; Schmidt, G.W.; Kaltenbach, H.M.; Rudolf, F.; Frey, O.; Hierlemann, A. Integrating impedance-based growth-rate monitoring into a microfluidic cell culture platform for live-cell microscopy. *Microsyst. Nanoeng.* **2018**, *4*, 8. [CrossRef] [PubMed]
138. Opitz, C.; Schade, G.; Kaufmann, S.; Berardino, M.D. Rapid determination of general cell status, cell viability, and optimal harvest time in eukaryotic cell cultures by impedance flow cytometry. *Appl. Microbiol. Biotechnol.* **2019**, *103*, 8619–8629. [CrossRef] [PubMed]
139. Xie, X.; Zhang, Z.; Ge, X.; Zhao, X.; Hao, L.; Cheng, Z.; Zhou, W.; Du, Y.; Wang, L.; Tian, F.; et al. Particle self-aligning, focusing, and electric impedance microcytometer device for label-free single cell morphology discrimination and yeast budding analysis. *Anal. Chem.* **2019**, *91*, 13398–13406. [CrossRef] [PubMed]
140. Bertelsen, C.V.; Franco, J.C.; Skands, G.E.; Dimaki, M.; Svendsen, W.E. Investigating the use of impedance flow cytometry for classifying the viability state of E. coli. *Sensors* **2020**, *20*, 6339. [CrossRef]
141. Spencer, D.C.; Paton, T.F.; Mulroney, K.T.; Inglis, T.J.J.; Sutton, J.M.; Morgan, H. A fast impedance-based antimicrobial susceptibility test. *Nat. Commun.* **2020**, *11*, 5328. [CrossRef] [PubMed]
142. van Beers, E.J.; Samsel, L.; Mendelsohn, L.; Saiyed, R.; Fertrin, K.Y.; Brantner, C.A.; Daniels, M.P.; Nichols, J.; McCoy, J.P.; Kato, G.J. Imaging flow cytometry for automated detection of hypoxia-induced erythrocyte shape change in sickle cell disease. *Am. J. Hematol.* **2014**, *89*, 598–603. [CrossRef] [PubMed]
143. Küttel, C.; Nascimento, E.; Demierre, N.; Silva, T.; Braschler, T.; Renaud, P.; Oliva, A.G. Label-free detection of Babesia bovis infected red blood cells using impedance spectroscopy on a microfabricated flow cytometer. *Acta Trop.* **2007**, *102*, 63–68. [CrossRef] [PubMed]
144. Holmes, D.; Pettigrew, D.; Reccius, C.H.; Gwyer, J.D.; van Berkel, C.; Holloway, J.; Davies, D.E.; Morgan, H. Leukocyte analysis and differentiation using high speed microfluidic single cell impedance cytometry. *Lab Chip* **2009**, *9*, 2881–2889. [CrossRef]
145. Hassan, U.; Reddy, B., Jr.; Damhorst, G.; Sonoiki, O.; Ghonge, T.; Yang, C.; Bashir, R. A microfluidic biochip for complete blood cell counts at the point-of-care. *Technology* **2015**, *3*, 201–213. [CrossRef] [PubMed]
146. Watkins, N.N.; Hassan, U.; Damhorst, G.; Ni, H.; Vaid, A.; Rodriguez, W.; Bashir, R. Microfluidic $CD4^+$ and $CD8^+$ T lymphocyte counters for point-of-care HIV diagnostics using whole blood. *Sci. Transl. Med.* **2013**, *5*, 214ra170. [CrossRef]

147. Durand-Smet, P.; Chastrette, N.; Guiroy, A.; Richert, A.; Berne-Dedieu, A.; Szecsi, J.; Boudaoud, A.; Frachisse, J.-M.; Bendahmane, M.; Hamant, O.; et al. A comparative mechanical analysis of plant and animal cells reveals convergence across kingdoms. *Biophys. J.* **2014**, *107*, 2237–2244. [CrossRef] [PubMed]
148. Chen, L.; Han, Z.; Fan, X.; Zhang, S.; Wang, J.; Duan, X. An impedance-coupled microfluidic device for single-cell analysis of primary cell wall regeneration. *Biosens. Bioelectron.* **2020**, *165*, 112374. [CrossRef] [PubMed]
149. Houssin, T.; Follet, J.; Follet, A.; Dei-Cas, E.; Senez, V. Label-free analysis of water-polluting parasite by electrochemical impedance spectroscopy. *Biosens. Bioelectron.* **2010**, *25*, 1122–1129. [CrossRef]
150. Park, Y.; Kim, H.W.; Yun, J.; Seo, S.; Park, C.J.; Lee, J.Z.; Lee, J.H. Microelectrical impedance spectroscopy for the differentiation between normal and cancerous human urothelial cell lines: Real-time electrical impedance measurement at an optimal frequency. *BioMed Res. Int.* **2016**, *2016*, 8748023. [CrossRef] [PubMed]
151. Hong, J.-L.; Lan, K.-C.; Jang, L.-S. Electrical characteristics analysis of various cancer cells using a microfluidic device based on single-cell impedance measurement. *Sens. Actuators B Chem.* **2012**, *173*, 927–934. [CrossRef]
152. Primiceri, E.; Chiriaco, M.S.; Dioguardi, F.; Monteduro, A.G.; D'Amone, E.; Rinaldi, R.; Giannelli, G.; Maruccio, G. Automatic transwell assay by an EIS cell chip to monitor cell migration. *Lab Chip* **2011**, *11*, 4081–4086. [CrossRef]
153. Zhang, Z.; Zheng, T.; Zhu, R. Single-cell individualized electroporation with real-time impedance monitoring using a microelectrode array chip. *Microsyst. Nanoeng.* **2020**, *6*, 81. [CrossRef] [PubMed]
154. Zhang, Z.; Zheng, T.; Zhu, R. Long-term and label-free monitoring for osteogenic differentiation of mesenchymal stem cells using force sensor and impedance measurement. *J. Mater. Chem. B* **2020**, *8*, 9913–9920. [CrossRef]
155. Ghenim, L.; Kaji, H.; Hoshino, Y.; Ishibashi, T.; Haguet, V.; Gidrol, X.; Nishizawa, M. Monitoring impedance changes associated with motility and mitosis of a single cell. *Lab Chip* **2010**, *10*, 2546–2550. [CrossRef] [PubMed]
156. Zhu, Z.; Frey, O.; Ottoz, D.S.; Rudolf, F.; Hierlemann, A. Microfluidic single-cell cultivation chip with controllable immobilization and selective release of yeast cells. *Lab Chip* **2012**, *12*, 906–915. [CrossRef] [PubMed]
157. Crowell, L.L.; Yakisich, J.S.; Aufderheide, B.; Adams, T.N.G. Electrical impedance spectroscopy for monitoring chemoresistance of cancer cells. *Micromachines* **2020**, *11*, 832. [CrossRef] [PubMed]

biosensors

MDPI

Review

Magnetoimpedance Biosensors and Real-Time Healthcare Monitors: Progress, Opportunities, and Challenges

Valery Ortiz Jimenez [1,†], Kee Young Hwang [1,†], Dang Nguyen [1,2], Yasif Rahman [1], Claire Albrecht [1], Baylee Senator [1], Ongard Thiabgoh [3,*], Jagannath Devkota [1,4], Vinh Duc An Bui [5], Dao Son Lam [1,6], Tatiana Eggers [1,*] and Manh-Huong Phan [1,*]

1 Laboratory for Advanced Materials and Sensors, Department of Physics, University of South Florida, Tampa, FL 33620, USA; valeryortizj@usf.edu (V.O.J.); keeyoung@usf.edu (K.Y.H.); dang3@usf.edu (D.N.); yasif@usf.edu (Y.R.); clairealbrecht9@gmail.com (C.A.); senator.bn.2018@wvwc.edu (B.S.); jagannath.devkota@netl.doe.gov (J.D.); daosonlamln@gmail.com (D.S.L.)
2 Department of Biomedical Engineering, University of South Florida, Tampa, FL 33620, USA
3 Department of Physics, Faculty of Science, Ubon Ratchathani University, Warinchamrap, Ubon Ratchathani 34190, Thailand
4 National Energy Technology Laboratory, Pittsburgh, PA 15236, USA
5 Hue Central Hospital, Hue 52000, Vietnam; buiducanvinh@gmail.com
6 Institute of Materials Science, Vietnam Academy of Science and Technology, 18 Hoang Quoc Viet, Ha Noi 10072, Vietnam
* Correspondence: ongard.t@ubu.ac.th (O.T.); tatianaeggers@gmail.com (T.E.); phanm@usf.edu (M.-H.P.); Tel.: +813-974-4322 (M.-H.P.)
† These authors contributed equally to this work.

Abstract: A small DC magnetic field can induce an enormous response in the impedance of a soft magnetic conductor in various forms of wire, ribbon, and thin film. Also known as the giant magnetoimpedance (GMI) effect, this phenomenon forms the basis for the development of high-performance magnetic biosensors with magnetic field sensitivity down to the picoTesla regime at room temperature. Over the past decade, some state-of-the-art prototypes have become available for trial tests due to continuous efforts to improve the sensitivity of GMI biosensors for the ultrasensitive detection of biological entities and biomagnetic field detection of human activities through the use of magnetic nanoparticles as biomarkers. In this review, we highlight recent advances in the development of GMI biosensors and review medical devices for applications in biomedical diagnostics and healthcare monitoring, including real-time monitoring of respiratory motion in COVID-19 patients at various stages. We also discuss exciting research opportunities and existing challenges that will stimulate further study into ultrasensitive magnetic biosensors and healthcare monitors based on the GMI effect.

Keywords: magnetoimpedance; magnetic biosensors; healthcare monitors; COVID-19 detection

Citation: Jimenez, V.O.; Hwang, K.Y.; Nguyen, D.; Rahman, Y.; Albrecht, C.; Senator, B.; Thiabgoh, O.; Devkota, J.; Bui, V.D.A.; Lam, D.S.; et al. Magnetoimpedance Biosensors and Real-Time Healthcare Monitors: Progress, Opportunities, and Challenges. *Biosensors* **2022**, *12*, 517. https://doi.org/10.3390/bios12070517

Received: 13 May 2022
Accepted: 8 July 2022
Published: 12 July 2022

1. Introduction

Since the turn of the 21st century, magnetic biosensor research has steadily grown year after year [1–6]. Magnetic phenomena such as the giant magneto-resistance effect, nuclear magnetic resonance, and superconducting quantum interference have often been proposed as the transducers of magnetic biosensors [2,7,8]. Although these magnetic phenomena offer valuable precision as transducers, their complicated measurement protocols, expensive equipment, and requisite for cryogenic temperatures have prevented them from being fully harnessed in the healthcare industry. This is where the giant magnetoimpedance effect, made prominent in 1994, found its footing in the world of biosensing and healthcare. High magnetic field sensitivity at room temperature coupled with classical and easy-to-model impedance–frequency–temperature relationships make the giant magnetoimpedance effect a qualified transducer of biometric data.

The significant change in the impedance of a high (magnetic) permeability material when subjected to a small magnetic field was first reported in 1936 by Harrison et al. as an impedance magnetometer [9]. Much later, in 1991, Makhotkin et al. demonstrated a magnetic field sensor made of a soft ferromagnetic ribbon of FeCoSiB [10]. However, it was not until 1994, when two independent groups simultaneously published articles, that the phenomenological theory of the enormous impedance change of magnetic wires when experiencing a weak magnetic field was detailed [11,12]. At this point, the terminology "giant magnetoimpedance effect", or GMI effect, was coined. Owing to its ultra-high magnetic field sensitivity, the GMI effect in soft ferromagnetic materials has been extensively explored and applied to both fundamental research and industrial applications. Excellent review articles on the GMI effect and its applications have been published thus far [13,14]. It is worth noticing in Figure 1 that the number of published articles per year greatly increased from 1994 to 2000 and slightly fluctuated between 2000 and 2020, while the number of citations has rapidly increased since 1994.

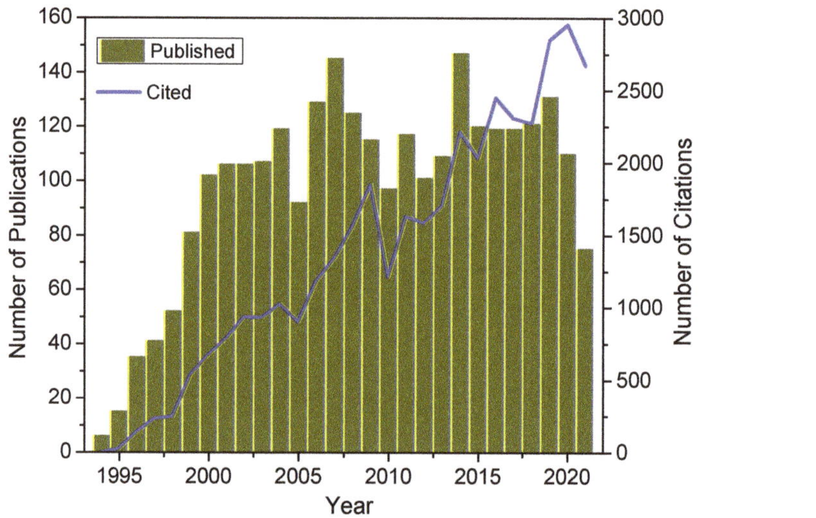

Figure 1. Number of published articles and citations per year in the field of magnetoimpedance materials and sensors. The data were collected from Web of Science with "magnetoimpedance" or "magneto-impedance" as a keyword.

Magnetoimpedance (MI) biosensors are the products of a combination of the magnetoimpedance effect and electrochemical or affinity biosensors. Brought to the mainstream in 1994 by electrical engineers, the MI effect experienced a research peak in the fields of electrical engineering and materials science. Shortly after, the use of the MI effect for biological sensing applications was proposed by Mohri and coworkers [15]. Since then, condensed matter and materials science groups, electrical engineers, and biomedical researchers alike have conducted heavy research into using the GMI effect to detect biomagnetic particles and even record magnetic biometric data from the heart and brain [8,16–22]. Indeed, a wide range of ultrasensitive MI sensor prototypes and their potential applications, including biomagnetic sensing, have been proposed and developed by Aichi Micro Intelligent Corporation (see Figure 2) [23]. This technology is not without its shortcomings. As it stands today, GMI sensors suffer from two main limitations: (a) sensitivity and (b) quantification: two issues that researchers have attempted to solve both from a sensor design and a material engineering perspective. Sensor geometry, materials, structure, and operating frequency and current have been explored in order to optimize the performance of GMI-based sensors [18].

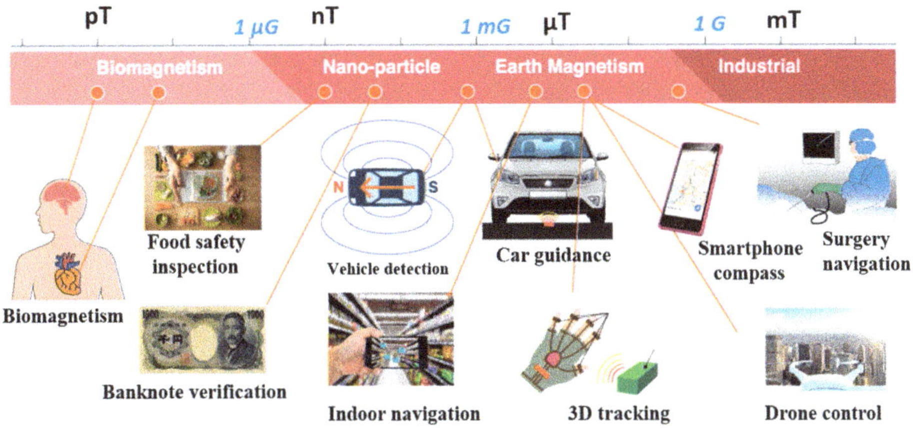

Figure 2. The potential applications of MI sensors proposed by Aichi Micro Intelligent Corporation. Reprinted with permission from Ref. [23]. Copyright 2022 Aichi Steel Corporation.

The implementation of GMI sensors in biomedical applications is largely achieved in conjunction with magnetic particles. Efforts to increase their sensitivity to the stray field emanating from magnetic particles is extensive and ongoing [24–29]. On the other hand, research into tailoring and functionalizing the properties of magnetic particles is equally important and complimentary to the development of GMI biosensors [30]. Magnetic particles can be functionalized to attach to specific molecules [24] or to carry drugs for targeted delivery in the human body [31], which requires a deep understanding of their magnetic properties and accurate detection to for applications in the field of nanomedicine.

In this review article, we focus on the short history of magnetoimpedance biosensors, how they have been improved over the past decade, and the state-of-the-art prototypes published within the past few years. We also highlight our recent developments in GMI-based medical devices for application in healthcare monitoring, including real-time monitoring of COVID-19 patients at various stages. Emerging opportunities and challenges in this rapidly expanding research field are also discussed to help guide future research and development of GMI-based biosensors for healthcare applications.

2. Basic Principles

The principle of MI-based biosensors is based on the detection of a small-magnitude (or "weak") magnetic field by a change in impedance of a soft magnetic material. Essentially, an MI-based biosensor is a transducer that converts small changes of the magnetic field experienced by the sensing element into electrical signals. The sources of these small magnetic fields can be magnetic nanoparticles, red blood cells, magnetic signals from the brain, or the motion of tiny magnetic particles in media or tissue. MI biosensor applications range from qualitatively detecting the presence of small biomagnetic fields to providing a quantitative field measurement that can be translated into, for example, a particular concentration of magnetic particles. In this section, we first provide a concise review of the magnetoimpedance effect, as many great reviews on the subject have already been published [5,6,32–34]. Then, we outline the history of the earliest prototypes of magnetoimpedance-based biosensors sorted by their unique applications to biology and healthcare.

2.1. The Giant Magnetoimpedance Phenomenon

The GMI phenomenon refers to a large change in the complex impedance of a soft ferromagnetic conductor when subjected to an external static magnetic field. The change in impedance of a conductor consists of resistive (R) and reactive (X) components. The origin of the GMI effect can be demonstrated by the skin effect in classical electrodynamics and the circumferential magnetic permeability associated with circular domain wall movements. Generally, GMI materials possess large magnetic permeability and low resistivity, and understanding the enhancements and tradeoffs between these intrinsic material properties is crucial to optimizing the GMI effect for biosensor applications.

The complex impedance of a ferromagnetic conductor can be expressed as

$$Z(\mu, f, H) = R(\mu, f, H) = +jX(\mu, f, H) \tag{1}$$

where R is resistance, X is reactance, μ is magnetic permeability, f is operating frequency, H is the external magnetic field, and j is the imaginary unit. As $Z(\mu, f, H)$ and $\mu(f, H) = \mu' - j\mu''$ both vary with f and H, the analysis of the GMI effect in a ferromagnetic conductor can be quite complex. Because of this, the GMI effect is typically categorized by frequency into three different regimes: low frequency, high frequency, and very high frequency. Many published articles have focused on the low-frequency response of the GMI effect; therefore, the scope of this section will mainly encompass the high-frequency response of soft magnetic materials.

The theory of magnetism and the dynamical response of magnetization can explain the origin of the high-frequency GMI effect. Beginning with Maxwell's equations [35], we have:

$$\nabla^2 \vec{H} - \nabla\left(\nabla \cdot \vec{H}\right) = \mu_0 \sigma \frac{\partial}{\partial t}\left(\vec{H} + \vec{M}\right) \tag{2}$$

where $\vec{H}$ and $\vec{M}$ are the external magnetic field and spontaneous magnetization vector, respectively. The dynamical response of the magnetization from an applied external magnetic field can be described by the Landau–Lifshitz–Gilbert (LLG) equation as [36–38]:

$$\frac{\partial \vec{M}}{\partial t} = -\mu_0 \gamma \left(\vec{M} \times \vec{H}\right) + \alpha \left(\vec{M} \times \frac{\partial \vec{M}}{\partial t}\right) \tag{3}$$

where γ is the gyromagnetic ratio, α is the damping parameter, and $\vec{H}$ and $\vec{M}$ are the external field and spontaneous magnetization within a domain, respectively. By applying the corresponding boundary conditions and solving the coupled Equations (2) and (3) [36–41], the longitudinal impedance of a cylindrical conductor for any frequency range can be expressed as:

$$Z(\omega) = \frac{1}{2} ka R_{dc} \frac{J_0(ka)}{J_1(ka)} \tag{4}$$

where $k = \frac{(1-j)}{\delta} = (1-j)\sqrt{\mu f \mu_\varnothing \sigma}$, $\mu_\varnothing$ is the circular permeability, and R_{dc} is the DC resistance of the magnetic wire.

Similarly, for a magnetic slab of thickness $2a$, the impedance can be expressed as:

$$Z(\omega) = R_{dc} jka \coth(jka) \tag{5}$$

where $k = \frac{(1-j)}{\delta} = (1-j)\sqrt{\mu f \mu_T \sigma}$, μ_T is the transverse permeability, and R_{dc} is the DC resistance of the magnetic slab.

The magnitude of the GMI effect of a magnetic microwire is defined by the change in Z, R, and/or X due to the external DC magnetic field. The figure-of-merit of GMI materials is the GMI ratio [14], and it is defined as follows:

$$MI\% = \frac{Z(H) - Z(H_{max})}{Z(H_{max})} \times 100\ \%, \tag{6}$$

where $Z(H)$ is the impedance at field H, and H_{max} represents the maximum value of the applied magnetic field. The magnetoresistance (MR) and magnetoreactance (MX) are defined in the same manner as in Equation (6), with R or X interchanged with Z. The magnetic field sensitivity (η) is defined as:

$$\eta = \frac{d}{dH}\left(\frac{\Delta Z}{Z}\right) \tag{7}$$

The sensitivity of the magnetoresistance (MR) and magnetoreactance (MX) are defined in the same manner as Equation (7), with R or X interchanged with Z.

For the detection of magnetic particles, the sensitivity of the biosensor is defined as the difference between the maximum value in MI, MR, or MX (i.e., corresponding to the field value H_k) of the test sample (TS) and reference sample (ref), which are calculated as:

$$\Delta\eta_{\xi} = [\xi]_{max,\ TS} - [\xi]_{max,\ ref} \tag{8}$$

where $[\xi]_{max}$ stands for $\xi = \frac{\Delta R}{R}, \frac{\Delta X}{X}, \frac{\Delta Z}{Z}$, which are the maximum values of the MR, MX, and MI ratios, respectively. These parameters are considered important figures-of-merit for assessing the sensitivity of an MI biosensor. In most studies, $\Delta\eta_R$, $\Delta\eta_X$, and $\Delta\eta_Z$ can also be denoted as MR, MX, and MI detection sensitivities, respectively.

2.2. Detection Principles

The fundamental detection principle of an MI-based biosensor is the detection of the stray magnetic field of magnetic markers attached to the biomolecules of interest. To simplify this complex scheme, we can approximate the stray field of a magnetic biomarker as one generated by a single magnetized microsphere with a magnetic moment (m) symmetric about the center of the sphere (Figure 3). Then, the magnetic induction can be expressed as [42–44]:

$$\vec{B}\left(\vec{r}\right) = \mu_0\vec{H} + \frac{\mu_0}{4\pi}\cdot\frac{3\vec{r}\left(\vec{r}\cdot\vec{m}\right) - \left(\vec{r}\cdot\vec{r}\right)\vec{m}}{r^5} \tag{9}$$

where $\vec{H}$ is the applied external magnetic field, μ_0 is the magnetic permeability of free space, and r is the radial vector in spherical coordinates. Figure 3a,b show the schematics of stray magnetic field detection of magnetic beads without an external magnetic field [43] and with an applied external magnetic field [27], respectively.

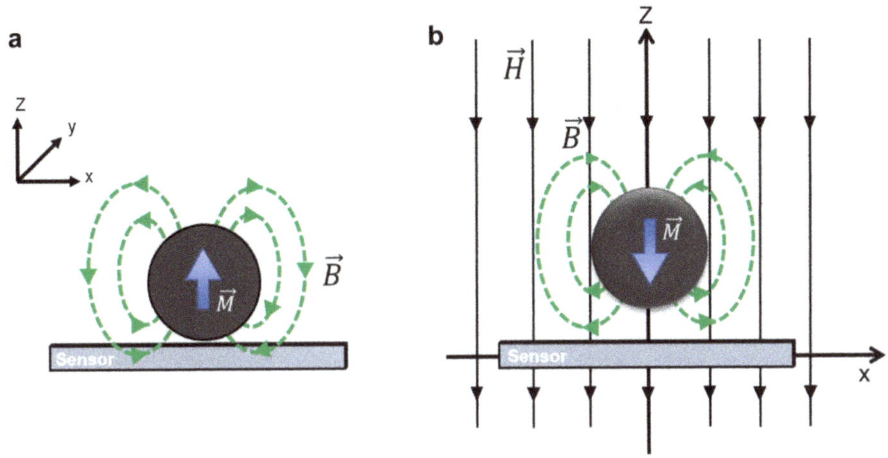

Figure 3. Schematics of stray magnetic field detection of magnetic beads (**a**) without an external magnetic field. Reprinted with permission from Ref. [43]. Copyright 2022 Elsevier, and (**b**) with an applied external magnetic field. Reprinted with permission from Ref. [27]. Copyright 2022 AIP Publishing.

From Equation (9), the in-plane magnetic field ($\overrightarrow{B}_x$) at distance (d) along the x-axis from the center of the magnetic bead radius (a) and a small distance over sensing element (t) can be expressed as [27]:

$$B_x = \mu_0 M \frac{a^3(a+t)d}{\left[(a+t)^2 + d^2\right]^{5/2}} \quad (10)$$

Similarly, the transverse magnetic field ($\overrightarrow{B}_z$) can be expressed as:

$$B_z = \mu_0 M \frac{(2z^2 - d^2)}{[z^2 + d^2]^{5/2}} \quad (11)$$

where $M = m/V$ is the magnetization of the magnetic bead.

The detection of stray magnetic fields emanating from magnetic beads or nanoparticles is the principal method behind potential applications of magnetoimpedance biosensing and medical diagnostics [16,29,44–49]. There are two main approaches to the measurement or detection of the magnetic particles: (i) detection directly on the surface of the sensor (Figure 4a) or (ii) detection from some distance away from the sensor (Figure 4b). For instance, a GMI sensor can be used as a probe of the presence of magnetic nanoparticles inside cells, which is the so-called magnetic label detection method (Figure 4a). A very thin layer of gold (~2 nm) is often coated on the surface of the sensing element to assure signal stability and biocompatibility [31].

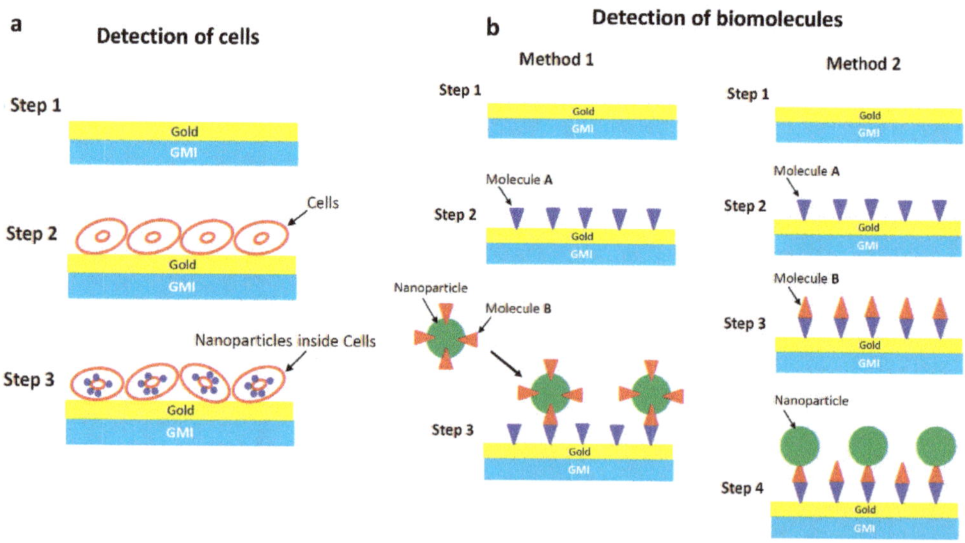

Figure 4. (**a**) Schematic illustration of the principle of detecting magnetic nanoparticles as magnetic labels inside cells; (**b**) schematic illustration of the principles of targeting and recognizing biomolecules. In Method 1, the magnetic nanoparticles are functionalized with target molecules (Molecule B), while in Method 2 the target molecules (Molecule B) are bound to the surface of the sensor.

For the detection of biomolecules using the GMI sensor and magnetic nanoparticles, two methods are often considered (Figure 4b). In Method 1, GMI measurements are first conducted on the sensitive element. Probe molecules (Molecule A) are then bound to the surface of the sensitive element, and GMI measurements are repeated. Magnetic nanoparticles, which are already functionalized with target molecules (Molecule B), will be bound to the probe molecules (Molecule A). In Method 2, target molecules (Molecule B) are first bound to the probe molecules (Molecule A), and the solution of nanoparticles then flows along the surface of the sensitive element. In both cases, the small magnetic fields generated by the nanoparticles that reach the surface of the sensitive element are detected using the GMI sensor [17].

Another new perspective is the contactless measurement of magnetic particles/biomagnetic samples. The stray magnetic fields of the magnetic beads and biomagnetic fields have recently been explored via ultra-sensitive MI-based sensors thanks to the promise of biosensing applications. Fodil et al. [29] designed an experimental setup to detect MNPs flowing through a microchannel at a distance, which is not coplanar with respect to the sensor, resulting in signal improvement. Additionally, the biomagnetic field of ileal musculature was measured ~1 mm below the sample using a gradio-magneto sensor. This detection approach can be developed for biomedical and medical diagnosis for in vivo samples [44]. Development and applications of these detection methods will be discussed in the following sections.

3. Early Sensor Prototypes (2000–2016)

3.1. Early Planar Prototypes

Amorphous ferromagnetic ribbons were first identified as potential weak magnetic field detectors in 1991 [10]. The earliest works on utilizing these magnetically soft ribbons as biosensors via the GMI effect were published a decade later by Kurlyandskya and coworkers [45–47]. In these works, rapidly quenched amorphous ribbons served as platforms for the detection of magnetic particles. Indeed, when a commercial ferrofluid or a suspension of Dynabeads was placed in close contact with the ribbons, their magnetoimpedance response increased due to their stray field interacting with the sensing element, and the field distribution of the magnetoimpedance widened.

A few years later, Yang and coworkers [48] demonstrated a Metglas ribbon as the sensing element of an MI-based biosensor for the detection and genotyping of human papilloma virus (HPV). This amorphous Metglas ribbon served as the sensing element, and it was micropatterned in a meander shape to increase the surface area for detection as well as for magnetic field sensitivity. The detection principle of this MI biosensor is to detect the stray magnetic fields of magnetic nanoclusters that label or tag HPV. The tagged HPV is then captured by specific probes on the surface of a microchannel in corresponding detection regions. When compared to the fluorescence method of HPV genotyping, the MI-based method had fewer steps, and the total assay time was significantly shortened.

The continuous flow detection of magnetic particles using an FeCoCrSiB ribbon and $[FeNi/Ti]_3/Cu/[FeNi/Ti]_3$ multilayer film was comparatively demonstrated in [49]. In this work, a 10 μL microfluidic chip was placed on top of the two sensing elements, and two different particle suspensions (Chemicell beads and Dynabeads) were pumped through the chamber as the magnetoimpedance was measured. While both the ribbon and thin-film sensor prototypes demonstrated a clear change in impedance when the particles entered the chamber, the results were not easy to reproduce. Although the use of a transmission line to measure the impedance improved the noise level in the experiments compared to prior work, the authors acknowledged the need for signal filtering to bring the sensitivity and reproducibility of these biosensor prototypes up to commercial standards.

While most efforts have been focused on developing a biosensor based on the MI effect, which has limited sensitivity, Devkota et al. [50–52] showed that by exploiting the MR and MX effects, it is possible to improve the sensitivity of the biosensor ($\Delta\eta_R, \Delta\eta_X$) by up to 50% and 100%, respectively. The increase of $\Delta\eta_R$, $\Delta\eta_X$, and $\Delta\eta_Z$ with increasing concentration of iron oxide nanoparticles (Figure 5a) can be attributed to the increase of transverse susceptibility μ_T due to the strong coupling of the magnetic fringe fields of the nanoparticles to the AC transverse magnetic field. This coupling becomes independent of iron oxide nanoparticles after the concentration of nanoparticles exceeds a critical amount, and no further increase in $\Delta\eta$ is therefore obtained. The authors also demonstrate that patterning the ribbon surface with nano/micro-sized holes is an effective way to improve the detection sensitivity of a ribbon-based MI biosensor [53]. This is particularly important as improvement in detection sensitivity can lead to highly sensitive detection of bioanalytes tagged to magnetic markers or cells that have taken up magnetic markers. Indeed, the authors fabricated a novel sensor probe by patterning four holes, each of dimension 2 μm × 2 μm, on a soft ferromagnetic ribbon using a focused ion beam (FIB) [54]. They analyzed the MI and MX responses for the probe itself and with 10 μL of the cell medium (as the control), unlabeled Lewis lung carcinoma (LLC) cancer cells, and magnetically labelled Lewis lung carcinoma (ML-LLC) cells (Figure 5b–d). The results showed that the sensor probe, cell medium, and label-free LLC cells did not have significant difference in their MI profiles (MI and MX ratios), while the ML-LLC cells had higher values. This demonstrates the possibility of using a hole-based MI biosensor to separate ML-LLC cells from unlabeled LLC cells. A GMI sensing platform could thus be developed as the new generation of diagnosis systems for reliable and quick biodetection at room temperature that can also be used as a new, low-cost, fast, and easy pre-detection method prior to magnetic resonance imaging (MRI).

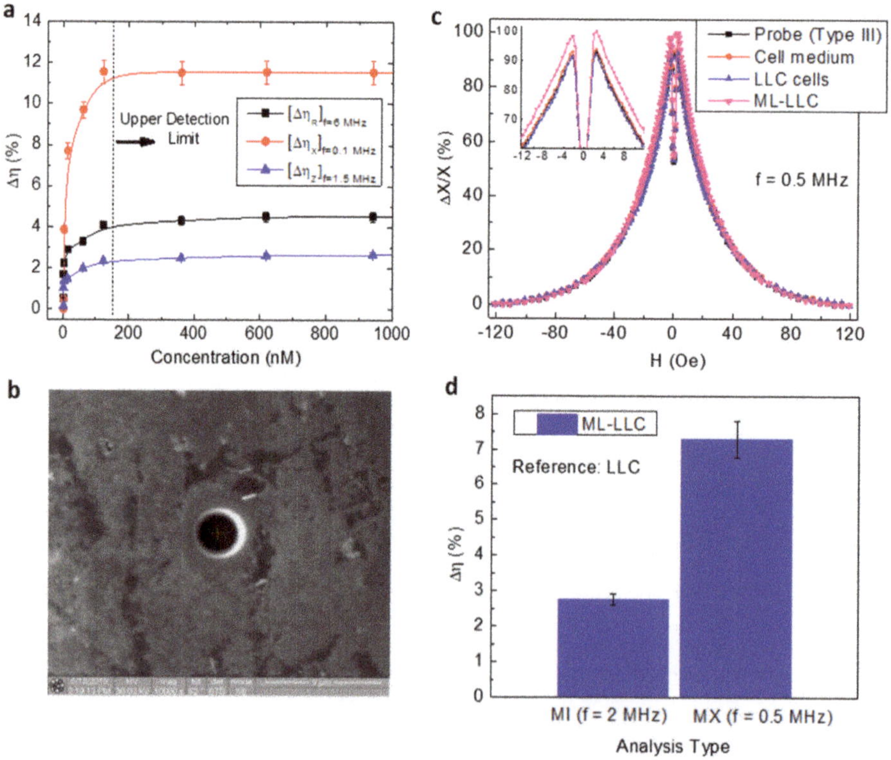

Figure 5. (**a**) Fe_3O_4 particle concentration dependence on MR, MX, and MI detection sensitivities and (**b**) SEM image of a hole on the ribbon. (**c**) Magnetic field dependence on MX ratio for a hole-based MX biosensor with cell medium, LLC cells, and magnetically labelled LLC cells (ML-LLC). (**d**) MI- and MX-based detection sensitivities of the probe for ML-LLC with reference to LLC.

Early biosensor prototypes utilizing amorphous ferromagnetic ribbons certainly encouraged further study of MI-based biosensors. However, when considering the incorporation of amorphous ribbons into magnetic biosensors for applications in the healthcare industry, some problems arise. For one, due to the rapid quenching technique used to prepare them, the physical and magnetic properties of the ribbons can vary significantly between different sections of the same batch [55]. Thus, mass-produced ribbons used as biosensing elements in commercial products would require individual characterization and adjustment due to differences in sensitivity. Furthermore, as the operating frequency increases, the formation of eddy currents leads to a sharp increase in heat losses, which can deteriorate the magnetic properties of most commercial ribbons, for which the thickness is in the range of tens of microns [56,57].

While the planar geometry of amorphous magnetic ribbons is well-suited to magnetic particle detection, which is the most popular application of magnetoimpedance biosensors, the requirement of miniaturization forces sensing elements toward smaller and smaller sizes. This is where thin-film structures exhibiting the magnetoimpedance effect begin to outshine amorphous ribbons as sensing elements. While miniaturization itself leads to a new set of challenges, ribbons are essentially disordered bulk structures and cannot meet the standards of mass reproducibility desired for commercial production. On the other hand, thin-film growth and patterning techniques have long been established to produce consistent results. It is clear that thin-film-based magnetoimpedance sensors serve as a way to overcome some of the limitations of ribbon-based sensors, and are further explored in the remainder of this section.

Some of the earliest thin-film biosensor prototypes used the GMI effect of FeCuNbSiB/Cu/FeCuNbSiB thin-film trilayer systems [58]. The composition FeCuNbSiB is of the FINEMET family, which are well known as rapidly quenched ribbons. In thin-film form, FINEMET structures were deposited via RF sputtering and annealed to induce ferromagnetic Fe–Si nanocrystalline grains within a ferromagnetic matrix that remained amorphous. Indeed, it is this microstructure that makes the magnetoimpedance effect of FINEMET films highly sensitive to weak magnetic fields and thus well-suited to MI-based biosensors. The uniformly distributed nanograins embedded in the amorphous matrix assure near-zero magnetostrictive anisotropy. Moreover, the magnetocrystalline anisotropy in nanocrystalline materials is averaged out, which also contributes to its high sensitivity to weak magnetic fields [59].

In 2009, Volchkov and coworkers [60] introduced another MI biosensor prototype based on a NiFe trilayer structure. In the world of spintronics, NiFe, or permalloy, is a well-studied and often used soft magnetic material. In this work, the authors varied the width of the sensing element, which was either a single NiFe film, a rectangular NiFe/Cu/NiFe trilayer structure, or a NiFe/Cu/NiFe trilayer structure where the center Cu conductor was narrower than the NiFe layers. The sensors were operated at frequencies in the upper hundreds of MHz, where the MI effect was found to be the largest. The authors discovered a tradeoff between the width of the sensing element and the observed MI effect; that is, when increasing the width of the sensing element to increase the detection area, the MI effect decreased. This work illustrates the importance of geometry when designing magnetoimpedance biosensors while keeping in mind the size of the particles to be detected. Wang et al. later reported a multilayered NiFe/Cu/NiFe meander film for MI-based biosensing grown using micro-electro–mechanical systems (MEMS) [61,62]. The authors noted that these films' maximum GMI ratio was observed at significantly lower frequencies than most ribbons, a desirable feature in biosensing applications. A quantitative study of the detection of Dynabeads Protein A was performed, demonstrating the accurate detection of Dynabeads down to a concentration of 0.1 $\mu g\ mL^{-1}$ at a frequency of 1.4–1.5 MHz. These studies further highlight the importance of the geometry of the sensing element, but also demonstrate the potential of thin-film-based MI biosensors for accurate quantification of magnetic biomarkers

In summary, planar geometries of magnetoimpedance-based sensors are available in a wide variety of compositions and shapes, all of which have the advantage of a large surface area. A large sensing area is ideal for the detection of magnetic particles and biomarkers, especially in contact-based measurements. While amorphous ribbons possess high sensitivity, their production methods limit mass reproducible sensors without the need to individually characterize them. Thin-film-based biosensors offer a clear alternative to mass-reproducible sensors, and the growth techniques are compatible with miniaturization standards. Two major issues persist in early MI planar sensors: sensitivity and quantification. Sensitivity is limited by the composition and operating frequency of the sensing element, and due to the limited sensitivity, quantification is a challenge, especially in an unshielded environment. Both issues are addressed in more recent works, as we discuss in Section 4. In the following subsection, we focus on MI sensor prototypes based on soft magnetic wires and compare their performance and applications with that of planar MI sensors.

3.2. Early Wire-Based Prototypes

So far, much of this review has focused on MI biosensors for the detection of stray magnetic fields emanating from magnetic particles that could be tagged to biomolecules. However, there is another equally important application of MI biosensing, and that is the detection of small-magnitude magnetic fields produced by biochemical current flow or the presence of small quantities of ferromagnetic contaminants [63]. For example, the motion of red blood cells in the body, the nervous system, and the movement of neurons within the brain all produce magnetic fields. Indeed, prototypes of wire-based MI biosensors used to detect weak biological magnetic fields predate many of the planar particle-detection prototypes mentioned in the previous section, and, additionally, wire-based MI protypes have consistently shown greater field sensitivity than planar prototypes.

The scientists responsible for constructing the first MI sensor are L. Panina, K. Mohri, T. Uchiyama, and coworkers [64]. In their pioneering 1995 work, the authors fabricated a highly sensitive magnetic field sensor using a 200 MHz resonant multi-vibrator bridge circuit that combined two CoFeSiB amorphous microwires and two field-effect transistors. It is worth noting again that this wire geometry sensor prototype predates the ribbon- and film-based sensors mentioned in the previous sections, which also came with extensive and custom integrated circuit elements to reduce noise and improve quantitative measurements.

Shortly thereafter, Uchiyama and coworkers demonstrated the use of this MI biosensor to detect the position of brain tumors in rats after injection with a solution of 25-nm magnetite nanoparticles dispersed in agarose [65]. The authors found that their MI sensor was able to detect the position and size of the tumor and produce a basic topographical map of the tumor by detecting the stray field produced by the magnetite particles embedded in it.

Chirac et al. [24] proposed an MI-based biosensor using a combination of ssDNA hybridization capture with streptavidin-covered magnetic microparticles (Figure 6a). The sensing element was a glass-covered CoFeSiB microwire, and its impedance was simply measured in a four-point configuration with a probe frequency around 10 MHz. Concentrations as low as 25–30 magnetic particles/µL could be detected in this arrangement (Figure 6b). Expanding on this work, Chirac and coworkers fabricated an MI biosensor prototype based on an array of glass-coated amorphous microwires and applied it to the detection of commercial polymer-based magnetic particles (Figure 6c,d) [25]. It was found that the number of active microwires enlarged the relative change in MI response, and this microwire arrangement could be used as an MI biosensor.

In the same year, Chiriac and coworkers published a systematic study on the detection of different sizes of magnetic particles (four ranges of sizes: 40–60 µm, 60–100 µm, 100–150 µm, and 150–300 µm) by a single CoFeSiB glass-coated microwire to predict whether these types of particles could function well as magnetic markers [26]. The results indicated that all sizes of magnetic particles in this study produced a notable and easily detectable MI response. The largest increase in MI effect, about 43%, was found with microparticles in the size range of 150–300 µm and a detection configuration of DC field parallel to the wire and measurement frequency of 10 MHz.

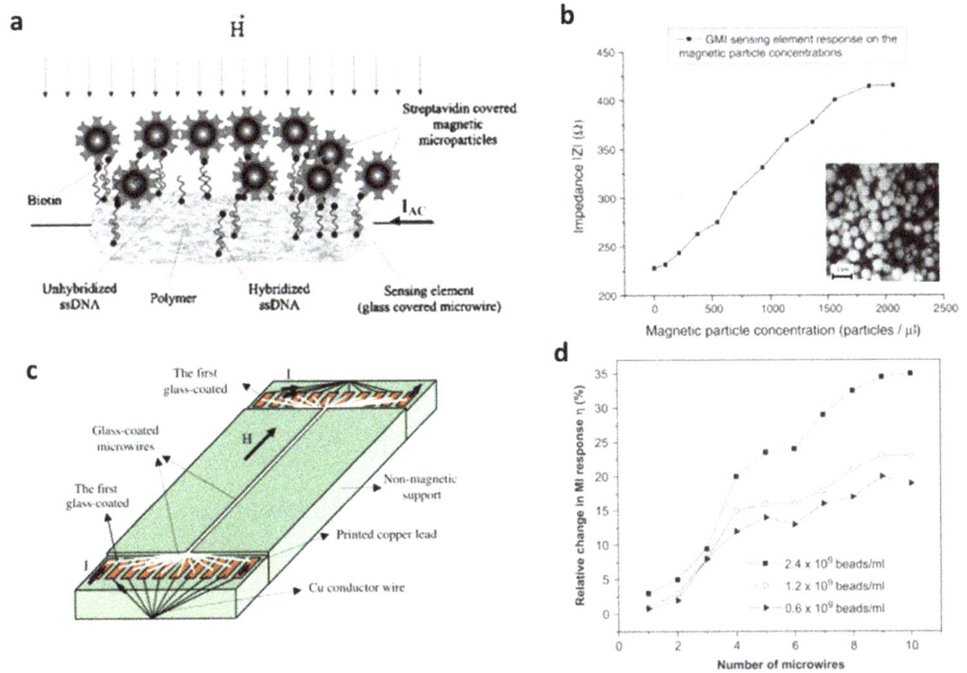

Figure 6. (**a**) The principle of a GMI-based magnetic biosensor using the ssDNA hybridization phenomenon as an example; (**b**) impedance response of the sensing element on the magnetic particles concentration. Reprinted with permission from Ref. [24]. Copyright 2022 Elsevier; (**c**) schematic design of the multiwire-based MI device; and (**d**) relative change in MI response as a function of the number of active glass-coated microwires for different bead concentrations. Reprinted with permission from Ref. [25]. Copyright 2022 Elsevier.

In 2013, Fodil et al. [29] combined microfluidics with an MI micro-magnetometer to detect 20 nm iron oxide nanoparticles, which could be functionalized for biomarkers. The micro-magnetometer was an MI biosensor based on a 40 μm CoFeSiBNb microwire sensing element at a measurement frequency of 15 MHz. In this experimental configuration, the microfluidic channel was a small distance away from the sensor. The authors showed two successive flows of MNPs measured by the MI microwire sensor, and the peak of the detected magnetic field was at a 180 nL volume of MNP flowing near the sensor at 4.3 mm/s. As a next step in the development of the sensor, Fodil and coworkers [66] reported the in-flow detection of a very low concentration of superparamagnetic nanoparticles (as small as 5.47×10^{-9} mol), which was also confirmed theoretically in a subsequent publication [67].

Wire-based sensors lack the surface-area planar geometries but make up for it in increased sensitivity for contactless measurements. Early prototypes not only detected the stray field of magnetic particles of different sizes but also detected biomagnetic fields, which ribbon and thin-film sensors cannot do. Wire-based sensor sensitivity can be further improved in combination with integrated circuit elements, novel sensing arrays, and other approaches that are discussed in the following section.

4. Current Magnetoimpedance Biosensors and Healthcare Monitors (2016–Now)

4.1. Detection of Magnetic Particles

The structure of a thin-film MI sensor significantly impacts its sensitivity. In general, two types of thin-film structures are being studied: straight line films and meander films.

In both cases, different multilayer structures have been investigated with different levels of success that are summarized below.

Presently, some of the most widely studied materials in this field are permalloy-based films, largely due to their well-established deposition techniques that allow for consistency across different films [68]. In general, the thickness of the material must match the skin depth associated with the operational frequency to obtain a significant MI ratio. For permalloys probed at a frequency in the few to tens of MHz, the corresponding skin depth are in the micrometer range, which is large compared to the minimum thickness one can achieve with sputtering techniques without degrading its magnetic properties [68].

It has been well-established in MI research that a metallic conductor placed between the two ferromagnetic layers increases the MI effect of the ferromagnetic conductor [28,69–75]. Therefore, multilayer structures are a research direction that achieves the desirable thickness for large skin-depth variations to enhance the MI effect of thin-films. A popular structure that incorporates these findings is FeNi/Cu/FeNi sandwich films, where Cu separates FeNi multilayers and serves as the central conductive spacer [76]. Including an insulator between the metallic and ferromagnetic layers has also been shown to further increase this effect as a consequence of changing the distribution of the electromagnetic field in the film to promote a change in impedance upon the application of an external field [77]. Kurlyandskaya et al. reported that the closer the conductivity of the spacer to that of the film, the more evenly distributed the electromagnetic field will be, which is conducive to a larger MI effect in the material [69]. The reported MI ratios for these structures vary with composition, thickness, length, and shape; these results are summarized in Table 1.

Table 1. Thin-film structures and MI ratios (%).

Structure (Thickness)	Dimensions	Max MI Ratio (%)	Max Sensitivity	MP Detection Applications	Ref.
		Straight line			
[Py(100 nm)/Ti(6 nm)]$_4$/ Cu(400 nm)/[Ti(6 nm)/Py(100 nm)]$_4$	10 mm × 0.5 mm 1.5 mm × 90 microns	350 220	300%/Oe 75%/Oe	N/A	[68]
[$Fe_{21}Ni_{79}$(100 nm)/ Cu(3 nm)]5/Cu(500 nm)/ [Cu(3 nm)/$Fe_{21}Ni_{79}$(100 nm)]$_5$	10 mm × 0.5 mm	160	41%/Oe	Stray field of MP in blood vessels	[28]
[$Fe_{19}Ni_{81}$(50 nm)/Ti(6 nm)]$_6$/ Cu(500 nm)/[Ti(6 nm)/$Fe_{19}Ni_{81}$ (50 nm)]$_6$	1 mm × 10 mm	~135	0.4 Ω/Oe	N/A	[69]
[FeNi(170 nm)/Ti(6 nm)]$_3$/ Cu(500 nm)/[Ti(6 nm)/ FeNi(170 nm]$_3$/Ti(6 nm)	10 mm × 0.5 mm		50%/Oe	Ferrogel detection	[70]
		Meander			
[Py(100 nm)/Ti(6 nm)]$_4$/ Cu(400 nm)/[Py(100 nm/ Ti(6 nm)]$_4$	5 mm × 4 mm, 12 strips 0.16 mm wide each	53.5	5.1 Ω/Oe	N/A	[71]
[FeNi(100 nm)/Cu(3 nm)]$_4$/ FeNi(100 nm)/Cu(500 nm) [FeNi(100 nm)/Cu(3 nm)]$_4$/ FeNi(100 nm)	200 microns wide, 14 strips 300 microns wide, 10 strips	60 165	-	Detection of polymer/MNP composites	[72]
$Fe_{17}Ni_{83}$(2 microns)/ Cu(140 microns)/ $Fe_{17}Ni_{83}$(2 microns)	5 mm long, 3 turns 6 turns	55.2 161.6	-	N/A	[73]
NiFe/Cu/NiFe	5 mm long 1.26 mm wide, 3 turns	97.54	0.1 μg mL^{-1} DPA concentration, 1 ng/mL AFP concentration	Dynabeads protein A, Alpha-fetoprotein detection	[61,74]

There are several factors that must be carefully considered when choosing the composition of such structures. Current research has shown that the choice of nonmagnetic conductor and its thickness impacts the MI response of the films [68]. The thickness and dimensions of the ferromagnetic layers also plays an important role that will set a limit to the sensitivity of the film. Figure 7a–c shows the dependence of the MI ratio on the dimensions of the different layers. This shows that the size of the central Cu layer relative to the size of the magnetic multilayers impacts the MI ratio; therefore, it should be a property that is carefully controlled to optimize the sensitivity of the film. In summary, the choice of structure should be prudently considered based on the desired application, as it will have a considerable impact on sensitivity.

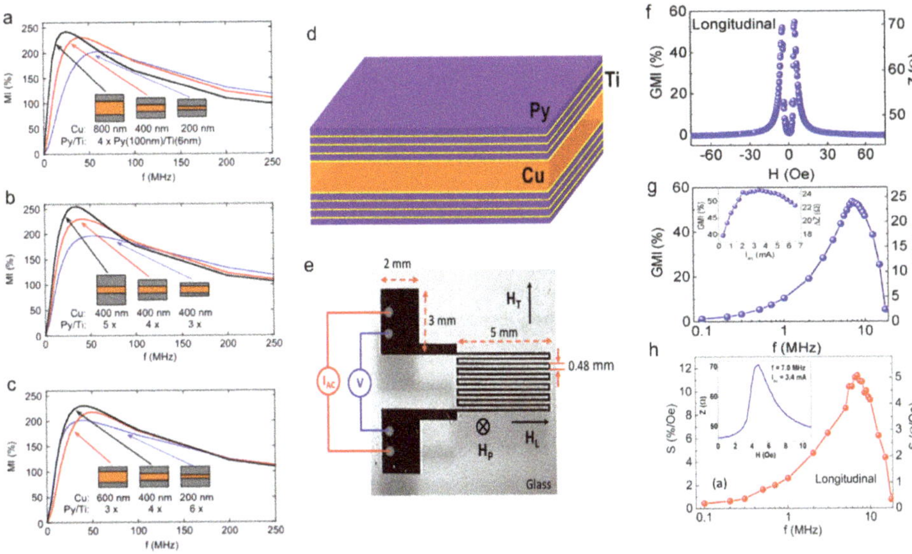

Figure 7. Compilation of results showing optimization of a multilayer structure's MI ratio as a function of frequency. (**a**–**c**) The optimization of the dimensions of a single strip [Py/Ti]/Cu/[Py/Ti] film (i.e., sensing element) in a meander structure. Reprinted with permission from Ref. [68]. Copyright 2022 Elsevier. (**d**) The film was grown in a layered structure [Py/Ti]/Cu/[Py/Ti]. Reprinted with permission from Ref. [71]. Copyright 2022 AIP Publishing. (**e**) The meander has 12 segments, each one 0.16 mm wide and 5 mm long, and the separation between segments is 0.48 mm. GMI ratio as a function of magnetic field (**f**), frequency (**g**), and ac current (inset of (**g**)). (**h**) Field sensitivity of GMI as a function of frequency, with an inset showing $Z(H)$.

Another widely studied structure for thin-film MI sensors is meander type films. While the structures of meanders are more complex, they offer an advantage over single-line strip films due to the formation of a transverse AC magnetic field on each segment of the meander [75,77] that may result in an electromagnetic coupling effect that enhances the MI effect. Meanders incorporate the same multilayer structures used in straight-line MI sensors but in a compact form ideal for applications while still maintaining a significant MI ratio (Figure 7d,e). Additionally, the AC magnetic field may prove useful in magnetizing any magnetic particles near the film. This is particularly important when working with superparamagnetic nanoparticles, where the magnetization process may allow for the detection of these particles without the necessity of an external magnetizing field [77]. An additional parameter that may also be studied is the operating current of the sensors, which has been shown to impact sensor sensitivity (Figure 7f–h).

The main application of MI-based biosensors has been for the detection of biomarkers, such as magnetic particles. Several works have been dedicated to the detection of small quantities of superparamagnetic beads [77–79] due to their potential as biomarkers that make it possible to employ MI sensing elements as biosensors. It has been shown that thin-film-based biosensors have been used to detect as few as 10 magnetic beads [77]. Wang et al. [77] demonstrated how a meander-shaped MI sensor can be used to detect alpha-fetoprotein (AFP) by conjugating superparamagnetic Dynabeads. They used a

$Cr/Cu/NiFe/Cu/NiFe/Al_2O_3/Cr/Au$ meander structure. The Al_2O_3 layer served as an insulation layer between the sensing element and the immunoplatform (Au). The authors showed that the sensor is capable of detecting concentrations of AFPs as small as 1 ng/mL although based on their previous work detecting Dynabeads [61], they reported that their detection limit was closer to 1 pg/mL.

While most of the studies attempted to demonstrate the capacity of GMI biosensors in detecting weak magnetic signals from low-concentration magnetic nanoparticles that are used as magnetic biomarkers, these biosensors are unable to provide quantitative values of the stray fields created by the magnetic nanoparticles. To address this, Phan's group developed an effective method based on the linear-field GMI response, which can detect and determine stray field values of magnetic (Fe_3O_4) nanoparticles at different concentrations (Figure 8). A single Co-rich microwire was used as a sensing element (inset of Figure 8a). To optimize the GMI biosensor's performance, they applied an external magnetic field (~2 Oe) below the anisotropy field (H_K) of the microwire, which brought the impedance change to a linear detection regime (Figure 8a). By fitting the linear region of the GMI curve (here only the R vs. H dependence was considered; Figure 8b,c), they were able to determine the minimum detection capability to be about 19 mOe from 10 mg of Fe_3O_4 nanoparticles placed 1 mm from one end of the microwire (Figure 8d). Future studies should focus on increasing the sensitivity of the microwire further to precisely detect even lower fields. This can be achieved in a myriad of ways, including optimizing the composition of the wires or modifying the domain structures with stress, heat treatment, and magnetic field annealing. Further improvements to signal processing can be achieved by carefully engineering circuits [80,81], understanding noise behavior [82], and exploring more complex sensor designs that integrate multiple sensing elements for signal filtering [20–22,46,83–87].

4.2. Biomagnetic Field Detection

Biomagnetic fields are magnetic fields produced from living systems, and they have been a hot topic of interest due to their small but significant effects, such as their direction-seeking effect on bird migration, their effect on the movement of bacteria, and as important signal sources from the human heart and brain. Therefore, biomagnetic measurement of magnetic fields is a crucial tool in investigating the functional organization of some human organs. Due to their extremely high field sensitivity down to the picoTesla level at room temperature, GMI-based sensors have been employed for biomagnetic field detection. The localized biomagnetic fields generated by smooth muscle cells, cardiogram signals, and smooth muscle tissue samples taken from a guinea pig have been detected using an MI sensor without a magnetic shield [19,20,46]. The biomagnetic signal from ileum musculature samples have been measured to be up to several nT [46]. Recently, numerous studies have suggested that the superior field sensitivity of GMI-based sensors is highly promising for magnetocardiography (MCG) and magnetoencephalography (MEG) [22,32].

Measurement of an MCG signal using an off-diagonal MI gradiometer in an unshielded environment at room temperature was reported in 2017 [83]. The sensor head of the gradio-magneto sensor was located 10 mm from the body surface of the test subject. Simultaneous measurements of electrocardiogram (ECG) and MCG signals were performed. Because of the chest movement from the test subject, the shown signal was averaged over more than 50 cycles. An active magnetic shielding system was developed for this gradiometer, which effectively reduced environmental magnetic noise around the sensor head and reduced the number of averaging cycles from 50 to 25 [84]. Moreover, Mohri et al. [22] utilized amorphous wire MI sensors integrated with CMOS to measure the back MCG from the left scapula of a test subject. The obtained signals showed excellent results when compared to simultaneous measurements of ECG and MCG. Ma and Uchiyama [85] later developed a new type of MI gradiometer consisting of a pair of CoFeSiB amorphous wires and a pick-up coil, which used a peak-to-peak voltage detector by synchronized switching. With this highly sensitive sensor, they performed MCG and MEG at room temperature (Figure 9) with an average of 10–15 cycles in an unshielded environment and achieved a noise amplitude of 100 pT [86].

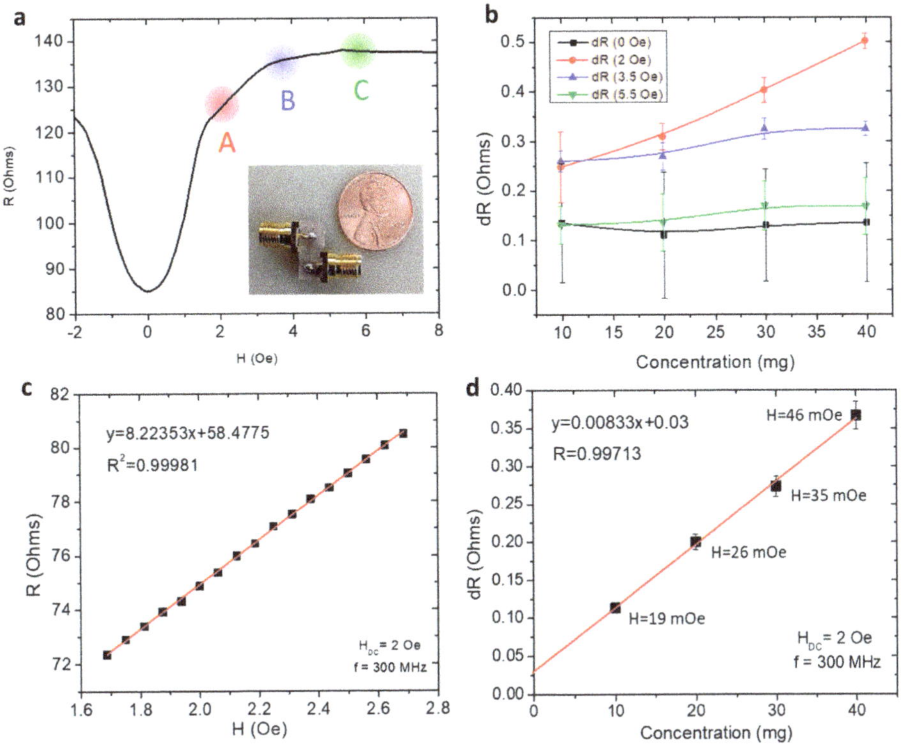

Figure 8. (**a**) The low-field (-4 Oe to 9 Oe) GMI curve demonstrating the regions measured during nanoparticle detection. 'A' corresponds to H_{DC} = 2 Oe; 'B' is H_{DC} = 3.5 Oe (~H_K: the anisotropy field); and 'C' is H_{DC} = 5.5 Oe (>H_K); (**b**) the change in resistance (dR) in a soft ferromagnetic Co-rich wire when exposed to different quantities of nanoparticles relative to the resistance with no nanoparticles present. Measurement was performed at each of the DC fields H_{DC} = 0 Oe, 2 Oe, 3.5 Oe, and 5.5 Oe; (**c**) linear fit of the low-field region of the MI curve allowed us to calculate the stray field of Fe_3O_4 nanoparticles; and (**d**) the linear change in resistance (d*R*) for H_{DC} = 2 Oe indicates that this MI-based microwire sensor can be used to detect small concentrations of magnetic nanoparticles.

Real-time brain activity measurement was carried out using a highly sensitive MI sensor without any magnetic shielding at room temperature [21,87,88]. Uchiyama and coworkers set up a GMI-based gradiometer to measure a small magnetic field 5 mm from the back left of the subject's head. The compared signals were ~500 pT in magnitude between eye opening and eye closing with respect to the background field. An analogous experiment was reported in [87], where an auditory evoked field (AEF) brainwave was probed using a highly sensitive MI sensor (picoTesla resolution) in a normal environment. The authors showed a difference in the received signals when the test subject opened or closed his/her eyes [87]. This real-time monitoring of brain activity signals supports the findings of the gradiometer measurement. Recent studies have explored alpha rhythm and visual event-related field (ERF) measurements using a high-performance MI sensor system in an unshielded environment [88]. The authors showed the difference in the measured MEG and EEG signals simultaneously when the test subject opened or closed his/her eyes. This noninvasive real-time monitoring of human biomagnetic fields using a highly sensitive MI-based sensor could be applied for brain activity measurement.

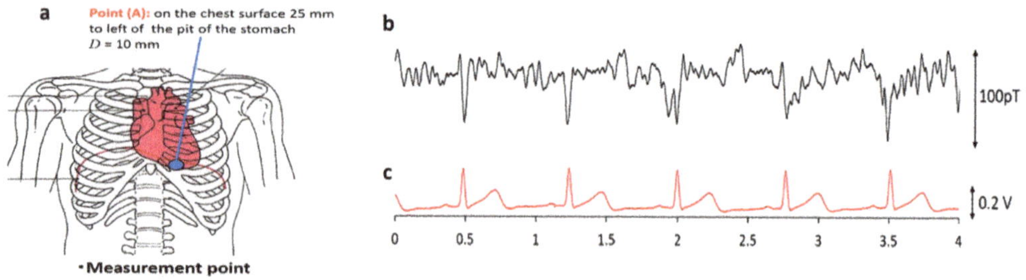

Figure 9. (**a**) Measurement location of MCG, which was 25 mm to the left of the pit of the stomach and 10 mm from the surface. (**b**,**c**) Real-time recordings of MCG and ECG signals measured by a peak-to-peak VD-type MI gradiometer without any magnetic shielding equipment. Reprinted with permission Ref. [86]. Copyright 2022 Elsevier.

4.3. Microfluidics

The superior magnetic field sensitivity of MI biosensors makes them attractive sensors in sensitive microfluidic platforms. Recently, the detection of biofunctionalized magnetic particles using a variety of MI geometries as sensing elements has been explored from the viewpoint of biomedical and clinical diagnostics. In 2016, the in-flow detection of ultra-small magnetic particles (20 nm) with a nanomolar concentration of superparamagnetic nanoparticles was achieved using a Co-rich microwire-based sensor (see Figure 10). The experimental results were theoretically validated with a model of the magnetization of a linear homogeneous isotropic material [67]. Later, an off-diagonal GMI biosensor integrated with a simple microfluidics chip was employed to detect antibody and alpha-protein (AFP) antigens labelled with magnetic beads in concentrations as low as 100 fg/ml under an external magnetic field. The experiment reveals that linear correlation with the AFP concentration is a useful approach for detecting cancer biomarkers [89].

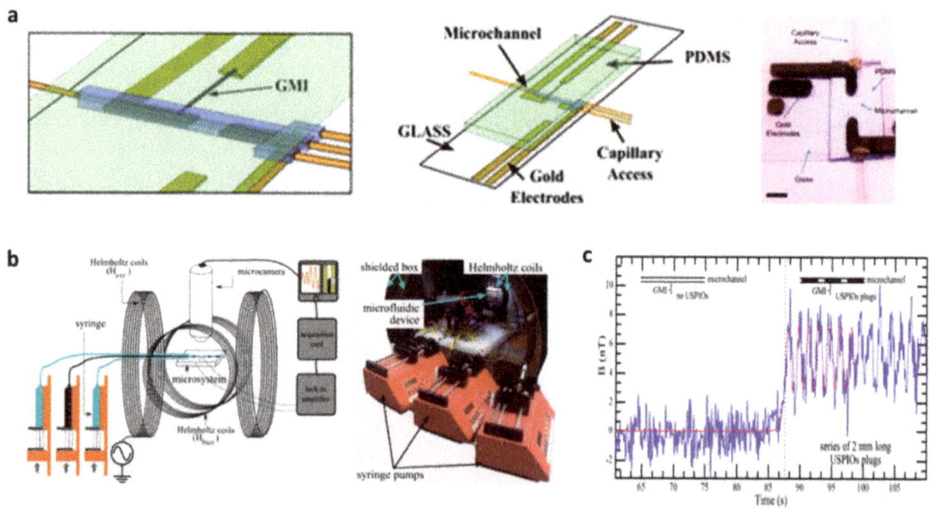

Figure 10. (**a**) Left to right: 3D schematics of the microfluidic system showing the PDMS microchannel on one side of the glass substrate, the GMI microwire with electrical connections on the other

side, and a picture of the actual device. (**b**) Picture of the experimental bench where one can see the programmable syringe pumps, the shielded box, the Helmholtz coils, and the microfluidic system. (**c**) Measured magnetic signal before and after (black dotted, t = 87.5 s) injection of the USPIOs plugs with 5.47×10^{-9} mol contents (2 mm long-20 nL volume with a molar concentration of 230 mmol/L). Reprinted with permission from Ref. [67]. Copyright 2022 AIP Publishing.

Feng et al. employed a GMI-based immunoreaction platform to detect the stray magnetic field and variation of magnetic signal from immunomagnetic beads [90]. In this configuration, a meander-line structured Co-based ribbon was integrated onto a microfluidic chip fabricated by MEMS. The integrated microfluidic chip consisted of incoming inlets and incubation and reactive chambers used as the biomarker detection system. In this detection scheme, prostate specific antigen (PSA) was detected at a concentration of 0.1 ng/mL, which is quite promising for further quantitative analysis. Recently, Melnikov et al. fabricated FeNiCu multilayer nanostructures that were used for the detection of magnetic particles in a blood-vessel-like structure [28]. The measurement of stray fields generated by iron-oxide microparticles under an external magnetic field was carried out using the longitudinal MI effect. The position of magnetic composite samples mimicking a thrombus (blood clot) was performed experimentally and modelled theoretically using COMSOL. This experiment is promising for thrombosis evaluation and therapy [28].

4.4. Real-Time Healthcare Monitoring of Patients with Respiratory Illness or COVID-19

Breathing is vital to life. Therefore, the real-time monitoring of a patient's breathing pattern is crucial to the support of respiratory rehabilitation therapies, such as magnetic resonance exams for respiratory-triggered imaging, chronic pulmonary disease treatment, and synchronized functional electrical stimulation. While several respiratory monitoring devices have already been developed [91–96], they are often in direct contact with a patient, which increases the chance of inaccurate or limited data. In this context, Thiabgoh et al. [94,96] developed a novel, noninvasive, and contactless magnetic sensing platform based on magneto-LC resonance (MLCR) technology that can precisely monitor a patient's breathing, movement, or sleep patterns, thus providing efficient monitoring at a clinic or at home. A combination of the GMI and LC-resonance effects makes the MLCR sensor extremely sensitive to small variations in magnetic field. By placing a tiny permanent magnet on a patient's chest, the MLCR sensor can precisely convert the magnetic oscillations generated by the patient's breathing into an impedance spectrum (Figure 11a), which allows deep analysis of breathing variation to help identify respiratory-related diseases. Hwang et al. recently reported that the MLCR sensor can yield a distinct breathing pattern for each person tested and reveal abnormal breathing [95]. They showed that when individuals get older, they manifest weaker breathing marked by increased respiration rate. Older individuals held their breath for shorter periods of time and often experienced more respiratory issues than younger people. They also observed that as people became more relaxed (Figure 11b), their breathing occurred more regularly, and eventually more slowly while sleeping (Figure 11c) as compared to when awake (Figure 11d). Listening to relaxing music while sleeping was also found to help the patient breathe more regularly and slowly (Figure 11e) than in the case without music (Figure 11d) [95]. Research has shown that music can slow down our breathing rates, which helps to release stresses and promote relaxation as well as treat chronic pulmonary disease [97,98]. Research also has shown that breath-training techniques can help us relax, sleep, and breathe more naturally and effectively [99]. It is therefore anticipated that the ultrasensitive MLCR monitor can provide not only valuable information on a patient's current health status, but also a novel breathing control tool for improving our health and physical performance.

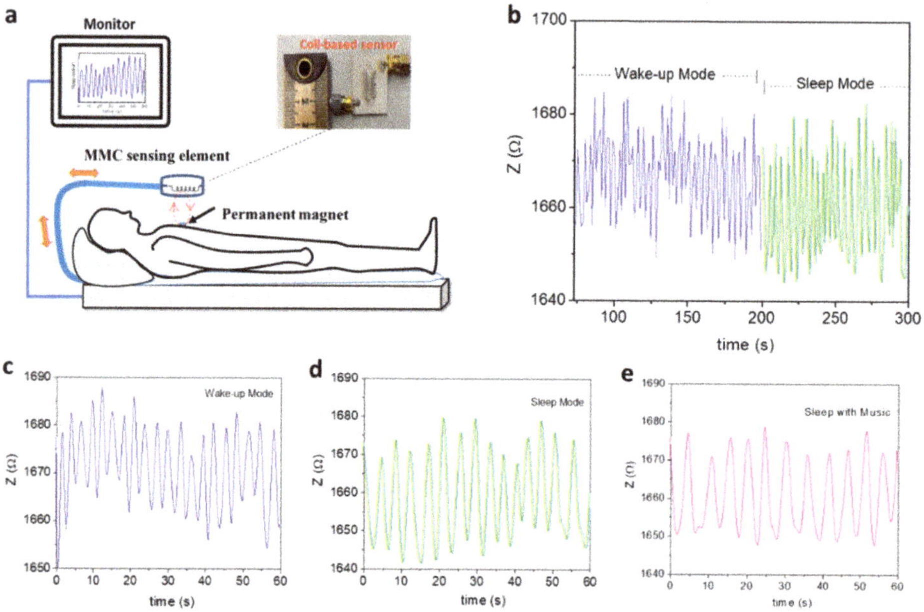

Figure 11. (**a**) Schematic for the respiratory motion test and results; inset is an image of the soft-magnetic-coil-based sensor probe; (**b**) breathing patterns of a 42-year-old patient were continuously tracked from waking to sleeping; (**c**) waking of the patient; (**d**) during sleep, the patient breathed more deeply (higher amplitude) and slowly (14 times per minute) than when awake (20 times per minute); (**e**) breathing patterns of this patient while sleeping with piano music; the person breathed more regularly and slowly (11 times per minute) than when sleeping without music (14 times per minute) (**d**). The abnormal breathing observed around 45 s almost disappeared when sleeping with music.

COVID-19 has killed more than 80 million people around the world, a number which continues to increase daily. This outbreak represents an unprecedented global public health challenge. To limit the spread of COVID-19 and to help doctors in clinical decision making, detection and real-time monitoring of symptoms during early, intermediate, and severe states is critical. However, most of the existing detection methods yield limited information with long processing time, and often require significant amounts of sample data from the subject, which requires human contact [100–102]. Therefore, there is an urgent demand for developing contactless devices that enable early and fast detection of COVID-19 and to track its growth rate in real time. Recall that common symptoms of COVID-19 include (i) shortness of breath or difficulty breathing, (ii) cough, and (iii) fever (Figure 12a). Since the MLCR sensor can detect abnormal breathing, it can be employed to distinguish breathing patterns of healthy and COVID-19-infected individuals as well as to track in real time breathing pattern variations of COVID-19 patients at different stages of illness [103]. Phan's group demonstrates the excellent capacity of using this technology to reveal shortness of breath and abnormal breathing in the breathing patterns of COVID-19 patients (Figure 12b–e) that are often absent in healthy people. It is worth mentioning that the tested COVID-19 patient lost the ability to hold his/her breath for extended periods and required a much longer time to return to a regular breathing pattern (Figure 12e) as compared to healthy people. Combining this technology with AI and machine learning, it could be possible to determine a COVID-19 patient's health status (early, intermediate, or severe) and help propose an appropriate medical treatment plan based on the available data [100].

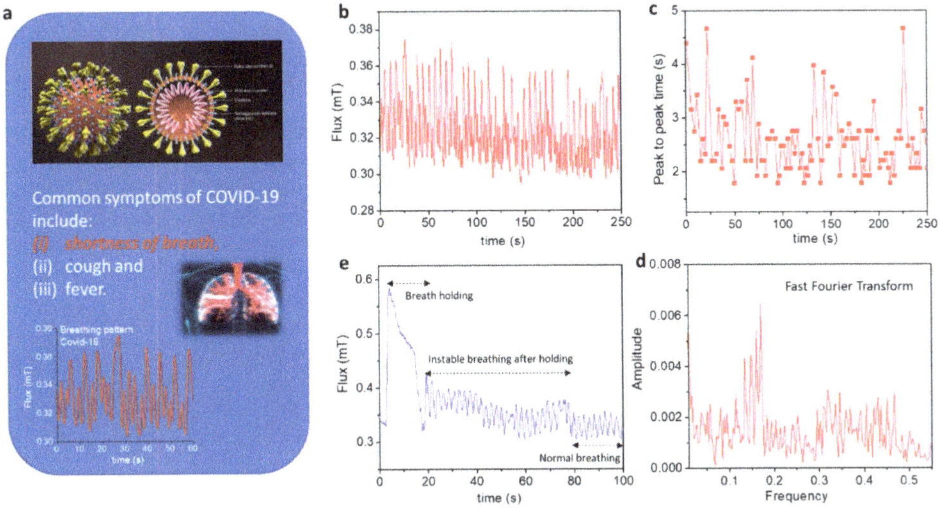

Figure 12. (**a**) Illustration of COVID-19 symptoms; top: schematic of the virus, bottom: breathing pattern of a patient with COVID-19. (**b**) Breathing pattern of the COVID-19 patient revealed irregular amplitude and several breathing anomalies. (**c**) Shortness of breath is evident from analysis of the peak-to-peak time versus measurement time. (**d**) The broad distribution of frequencies deduced from Fourier transform also reveals these features. (**e**) The COVID-19 patient lost the ability to hold his/her breath for a long time and required a much longer time to return to normal breathing.

5. Concluding Remarks and Future Outlook

While magnetoimpedance-based biosensors show remarkable sensitivity at room temperature, much of the work performed in the detection of biomarkers remains qualitative. The ability to quantify magnetic and nonmagnetic biomarkers is key for the application of magnetoimpedance sensors beyond the detection of magnetic fields. In this respect, amorphous ribbons suffer from poor reproducibility due to the rapid-quenching fabrication process. We previously discussed the work by Devkota et al. that demonstrates the potential of introducing patterned holes into the surface of ribbons for improving their sensitivity. This approach warrants further study, and we propose that even the introduction of a single micron/nano-sized hole should have a significant impact on the detection sensitivity of ribbon-based magnetoimpedance biosensors. The interaction between magnetic particles and patterned holes of comparable sizes could permit detection of a single particle. While patterned holes can enhance the sensitivity of such sensors, reproducibility remains an issue that should be addressed by future studies. Multilayered films are the clear choice for quantifiable and reproducible sensors that can be mass produced due to their well-established fabrication techniques. Further work is necessary to enhance the detection limit of these sensors, which could be achieved by improving signal filtering to reduce noise, expanding fabrication techniques, and/or exploring different sensor geometries.

Beyond the detection of magnetic biomarkers, magnetoimpedance-based sensors show ultrahigh sensitivity to ultrasmall magnetic fields. Magnetic microwires show significantly higher sensitivity than ribbons or multilayered films but lack a platform for contact detection of magnetic biomarkers. However, their ultrahigh sensitivity to stray fields allows quantification of the stray field of magnetic nanoparticles, both static and in flow, as well as the detection of biomagnetic fields. This highlights the importance of choosing an appropriate magnetoimpedance-sensing element based on the desired application and is a testament to their versatility.

We have seen an increasing number of reports of GMI sensors for human healthcare monitoring. An interesting application that stands out is the work by Hwang et al. that uses a GMI-based magnetic field sensor to measure the respiratory motion of human subjects in a completely contactless manner. The authors further proposed using this technique to track the progress of COVID-19 patients and to determine if there is a correlation between different stages of COVID-19 and respiratory patterns. AI and machine learning have proven to be powerful tools to analyze these types of data, and in conjunction with other indicators such as heart rate, blood oxygen, etc., could be used for future diagnosis of COVID-19 patients. This model could be expanded to study several other respiratory illnesses, sleeping disorders, and healthcare tracking to offer a wide range of potential applications for novel GMI biosensors for human healthcare.

Given the fact that music can help us relax and improve our breathing, the development of magnetic–musical biofeedback for breathing regulation and control appears to be a novel approach that will stimulate further studies to fully exploit its unique practicality and wide-ranging healthcare monitoring applications. This technology can also find its place among other important applications in artificial intelligence, emotion detection, and space control and management.

The fast Fourier transform (FFT) algorithm, which converts a signal from the time domain to the frequency domain and vice versa, can be used in future research to extract spectral features from breathing patterns. This would also be used in combination with machine learning and other algorithms for signal processing to provide comprehensive information on a patient's health status and physical performance.

Author Contributions: Conceptualization, V.O.J., K.Y.H., D.N., O.T., J.D., T.E. and M.-H.P.; methodology, O.T., J.D., K.Y.H., Y.R. and D.N.; software, K.Y.H. and V.O.J.; validation, V.O.J., O.T., T.E. and M.-H.P.; formal analysis, O.T., J.D., K.Y.H., D.N. C.A. and B.S.; investigation, O.T., J.D., K.Y.H., D.N., Y.R., C.A., B.S., V.D.A.B. and D.S.L.; resources, K.Y.H., D.N., and V.D.A.B.; data curation, K.Y.H., D.N. and V.D.A.B.; writing—original draft preparation, V.O.J., O.T. and T.E.; writing—review and editing, V.O.J., O.T., T.E. and M.-H.P.; visualization, O.T., T.E. and M.-H.P.; supervision, M.-H.P.; project administration, M.-H.P.; funding acquisition, M.-H.P. All authors have read and agreed to the published version of the manuscript.

Funding: Research at the University of South Florida was supported by the USF COVID-19 Research Foundation under grant number 125300.

Institutional Review Board Statement: Ethical review and approval were waived for this study because the research presents no more than minimal risk of harm to subjects and involves no procedures for which written consent is normally required outside the research context.

Informed Consent Statement: Not applicable.

Data Availability Statement: Not applicable.

Acknowledgments: The authors acknowledge Hari Srikanth of the University of South Florida, Vu Dinh Lam of the Vietnam Academy of Science and Technology, and Anh-Tuan Le of Phenikaa University for their useful discussions.

Conflicts of Interest: The authors declare no conflict of interest.

References

1. Su, D.; Wu, K.; Saha, R.; Peng, C.; Wang, J.-P. Advances in Magnetoresistive Biosensors. *Micromachines* **2020**, *11*, 34. [CrossRef] [PubMed]
2. Khan, M.A.; Sun, J.; Li, B.; Przybysz, A.; Kosel, J. Magnetic Sensors-A Review and Recent Technologies. *Eng. Res. Express* **2021**, *3*, 022005. [CrossRef]
3. Li, Y.; Cheng, H.; Alhalili, Z.; Xu, G.; Gao, G. The Progress of Magnetic Sensor Applied in Biomedicine: A Review of Non-Invasive Techniques and Sensors. *J. Chin. Chem. Soc.* **2021**, *68*, 216–227. [CrossRef]
4. Lin, G.; Makarov, D.; Schmidt, O.G. Magnetic Sensing Platform Technologies for Biomedical Applications. *Lab Chip* **2017**, *17*, 1884–1912. [CrossRef]
5. Denmark, D.J.; Bustos-Perez, X.; Swain, A.; Phan, M.-H.; Mohapatra, S.; Mohapatra, S.S. Readiness of Magnetic Nanobiosensors for Point-of-Care Commercialization. *J. Electron. Mater.* **2019**, *48*, 4749–4761. [CrossRef]
6. Nabaei, V.; Chandrawati, R.; Heidari, H. Magnetic Biosensors: Modelling and Simulation. *Biosens. Bioelectron* **2018**, *103*, 69–86. [CrossRef]

7. Ripka, P. Magnetic Sensors: Principles and Applications. In *Encyclopedia of Materials: Science and Technology*; Elsevier: Amsterdam, The Netherlands, 2007; pp. 1–11. [CrossRef]
8. Wang, S.X.; Li, G.X. Advances in Giant Magnetoresistance Biosensors with Magnetic Nanoparticle Tags: Review and Outlook. *IEEE Trans. Magn.* **2008**, *44*, 1687–1702. [CrossRef]
9. Harrison, E.P.; Turney, G.L.; Rowe, H. Electrical Properties of Wires of High Permeability. *Nature* **1935**, *135*, 961. [CrossRef]
10. Makhotkin, V.E.; Shurukhin, B.P.; Lopatin, V.A.; Marchukov, P.Y.; Levin, Y.K. Magnetic Field Sensors Based on Amorphous Ribbons. *Sens. Actuators A Phys.* **1991**, *27*, 759–762. [CrossRef]
11. Beach, R.S.; Berkowitz, A.E. Giant Magnetic Field Dependent Impedance of Amorphous FeCoSiB Wire. *Appl. Phys. Lett.* **1994**, *64*, 3652–3654. [CrossRef]
12. Panina, L.V.; Mohri, K. Magneto-impedance Effect in Amorphous Wires. *Appl. Phys. Lett.* **1994**, *65*, 1189–1191. [CrossRef]
13. Knobel, M.; Vázquez, M.; Kraus, L. Giant Magnetoimpedance. In *Handbook of Magnetic Materials*; Elsevier: Amsterdam, The Netherlands, 2003; pp. 497–563. ISBN 9780444514592.
14. Phan, M.-H.; Peng, H.-X. Giant Magnetoimpedance Materials: Fundamentals and Applications. *Pro. Mater. Sci.* **2008**, *53*, 323–420. [CrossRef]
15. Mohri, K.; Bushida, K.; Noda, M.; Yoshida, H.; Panina, L.V.; Uchiyama, T. Magneto-Impedance Element. *IEEE Trans. Magn.* **1995**, *31*, 2455–2460. [CrossRef]
16. Kurlyandskaya, G.V. Giant Magnetoimpedance for Biosensing: Advantages and Shortcomings. *J. Magn. Magn. Mater.* **2009**, *321*, 659–662. [CrossRef]
17. Wang, T.; Zhou, Y.; Lei, C.; Luo, J.; Xie, S.; Pu, H. Magnetic Impedance Biosensor: A Review. *Biosens. Bioelectron.* **2017**, *90*, 418–435. [CrossRef]
18. Sayad, A.; Skafidas, E.; Kwan, P. Magneto-Impedance Biosensor Sensitivity: Effect and Enhancement. *Sensors* **2020**, *20*, 5213. [CrossRef]
19. Uchiyama, T.; Nakayama, S. Magnetic Sensors Using Amorphous Metal Materials: Detection of Premature Ventricular Magnetic Waves. *Physiol. Rep.* **2013**, *1*, e00030. [CrossRef]
20. Uchiyama, T.; Nakayama, S.; Mohri, K.; Bushida, K. Biomagnetic Field Detection Using Very High Sensitivity Magnetoimpedance Sensors for Medical Applications. *Phys. Status Solidi A* **2009**, *206*, 639–643. [CrossRef]
21. Uchiyama, T.; Mohri, K.; Honkura, Y.; Panina, L.V. Recent Advances of Pico-Tesla Resolution Magneto-Impedance Sensor Based on Amorphous Wire CMOS IC MI Sensor. Element. *IEEE Trans. Magn.* **2012**, *48*, 3833–3839. [CrossRef]
22. Mohri, K.; Yamamoto, M.; Uchiyama, T. Application Topics of Amorphous Wire CMOS IC Magneto-Impedance Micromagnetic Sensors for I-o-T Smart Society. *J. Sens.* **2019**, *2019*, 8285240. [CrossRef]
23. Aichi Steel Corporation. Available online: https://www.aichi-mi.com/e-applications/ (accessed on 7 May 2022).
24. Chiriac, H.; Tibu, M.; Moga, A.E.; Herea, D.D. Magnetic GMI Sensor for Detection of Biomolecules. *J. Magn. Magn. Mater.* **2005**, *293*, 671–676. [CrossRef]
25. Chiriac, H.; Herea, D.D.; Corodeanu, S. Microwire Array for Giant Magneto-Impedance Detection of Magnetic Particles for Biosensor Prototype. *J. Magn. Magn. Mater.* **2007**, *311*, 425–428. [CrossRef]
26. Chiriac, H.; Herea, D.D. Magneto-Impedance Sensor for Biomedical Applications. *Int. J. Appl. Electromagn.* **2007**, *25*, 453–459. [CrossRef]
27. Huong Giang, D.T.; Dang, D.X.; Toan, N.X.; Tuan, N.V.; Phung, A.T.; Duc, N.H. Distance Magnetic Nanoparticle Detection Using a Magnetoelectric Sensor for Clinical Interventions. *Rev. Sci. Instrum.* **2017**, *88*, 015004. [CrossRef] [PubMed]
28. Melnikov, G.Y.; Lepalovskij, V.N.; Svalov, A.V.; Safronov, A.P.; Kurlyandskaya, G.V. Magnetoimpedance Thin Film Sensor for Detecting of Stray Fields of Magnetic Particles in Blood Vessel. *Sensors* **2021**, *21*, 3621. [CrossRef] [PubMed]
29. Fodil, K.; Denoual, M.; Dolabdjian, C.; Harnois, M.; Senez, V. Dynamic Sensing of Magnetic Nanoparticles in Microchannel Using GMI Technology. *IEEE Trans. Magn.* **2013**, *49*, 93–96. [CrossRef]
30. Materón, E.M.; Miyazaki, C.M.; Carr, O.; Joshi, N.; Picciani, P.H.S.; Dalmaschio, C.J.; Dacis, F.; Shimizu, F.M. Magnetic Nanoparticles in Biomedical Applications: A Review. *Appl. Surf. Sci. Adv.* **2021**, *6*, 100163. [CrossRef]
31. Anderson, S.D.; Gwenin, V.V.; Gwenin, C.D. Magnetic Functionalized Nanoparticles for Biomedical, Drug Delivery and Imaging Applications. *Nanoscale Res. Lett.* **2019**, *14*, 188. [CrossRef]
32. Murzin, D.; Mapps, D.J.; Levada, K.; Belyaev, V.; Omelyanchik, A.; Panina, L.; Rodionova, V. Ultrasensitive Magnetic Field Sensors for Biomedical Applications. *Sensors* **2020**, *20*, 1569. [CrossRef]
33. Stupin, D.D.; Kuzina, E.A.; Abelit, A.A.; Emelyanov, A.K.; Nikolaev, D.M.; Ryazantsev, M.N.; Koniakhin, S.V.; Dubina, M.V. Bioimpedance Spectroscopy: Basics and Applications. *ACS Biomater. Sci. Eng.* **2021**, *7*, 1962–1986. [CrossRef]
34. Peng, H.X.; Qin, F.; Phan, M.-H. *Ferromagnetic Microwire Composites: From Sensors to Microwave Applications*; Springer: Berlin/Heidelberg, Germany, 2016; ISBN 9783319292762.
35. Landau, L.D.; Lifshitz, E.M.; King, A.L. Electrodynamics of Continuous Media. *Am. J. Phys.* **1961**, *29*, 647–648. [CrossRef]
36. Melo, L.G.C..; Ciureanu, P.; Yelon, A. Permeability Deduced from Impedance Measurements at Microwave Frequencies. *J. Magn. Magn. Mater.* **2002**, *249*, 337–341. [CrossRef]
37. Ménard, D.; Frankland, D.; Ciureanu, P.; Yelon, A.; Rouabhi, M.; Cochrane, R.W.; Chiriac, H.; Óvári, T.A. Modeling of Domain Structure and Anisotropy in Glass-Covered Amorphous Wires. *J. Appl. Phys.* **1998**, *83*, 6566–6568. [CrossRef]
38. Kraus, L. GMI Modeling and Material Optimization. *Sens. Actuators A Phys.* **2003**, *106*, 187–194. [CrossRef]

39. Díaz-Rubio, A.; García-Miquel, H.; García-Chocano, V.M. In-Plane Omnidirectional Magnetic Field Sensor Based on Giant Magneto Impedance (GMI). *J. Magn. Magn. Mater.* **2017**, *444*, 249–255. [CrossRef]
40. Sommer, R.L.; Chien, C.L. Role of Magnetic Anisotropy in the Magnetoimpedance Effect in Amorphous Alloys. *Appl. Phys. Lett.* **1995**, *67*, 857–859. [CrossRef]
41. Tannous, C.; Gieraltowski, J. Giant Magneto-Impedance and Its Applications. *J. Mater. Sci.* **2004**, *15*, 125–133. [CrossRef]
42. Dressel, M.; Grüner, G. *Electrodynamics of Solids: Optical Properties of Electrons in Matter*; Cambridge University Press: Cambridge, UK, 2002; ISBN 9780521592536.
43. Rife, J.C.; Miller, M.M.; Sheehan, P.E.; Tamanaha, C.R.; Tondra, M.; Whitman, L.J. Design and Performance of GMR Sensors for the Detection of Magnetic Microbeads in Biosensors. *Sens. Actuators A Phys.* **2003**, *107*, 209–218. [CrossRef]
44. Nakayama, S.; Uchiyama, T. Real-Time Measurement of Biomagnetic Vector Fields in Functional Syncytium Using Amorphous Metal. *Sci. Rep.* **2015**, *5*, 8837. [CrossRef]
45. Kurlyandskaya, G.V.; Sánchez, M.L.; Hernando, B.; Prida, V.M.; Gorria, P.; Tejedor, M. Giant-Magnetoimpedance-Based Sensitive Element as a Model for Biosensors. *Appl. Phys. Lett.* **2003**, *82*, 3053–3055. [CrossRef]
46. Kurlyandskaya, G.V.; Levit, V. Magnetic Dynabeads®Detection by Sensitive Element Based on Giant Magnetoimpedance. *Biosens. Bioelectron.* **2005**, *20*, 1611–1616. [CrossRef] [PubMed]
47. Kurlyandskaya, G.V.; Fal Miyar, V. Surface Modified Amorphous Ribbon Based Magnetoimpedance Biosensor. *Biosens. Bioelectron.* **2007**, *22*, 2341–2345. [CrossRef] [PubMed]
48. Yang, H.; Chen, L.; Lei, C.; Zhang, J.; Li, D.; Zhou, Z.M.; Bao, C.C.; Hu, H.Y.; Chen, X.; Cui, F.; et al. Giant Magnetoimpedance-Based Microchannel System for Quick and Parallel Genotyping of Human Papilloma Virus Type 16/18. *Appl. Phys. Lett.* **2010**, *97*, 043702. [CrossRef]
49. García-Arribas, A.; Martínez, F.; Fernández, E.; Ozaeta, I.; Kurlyandskaya, G.V.; Svalov, A.V.; Berganzo, J.; Barandiaran, J.M. GMI Detection of Magnetic-Particle Concentration in Continuous Flow. *Sens. Actuators A Phys.* **2011**, *172*, 103–108. [CrossRef]
50. Devkota, J.; Wang, C.; Ruiz, A.; Mohapatra, S.; Mukherjee, P.; Srikanth, H.; Phan, M.-H. Detection of Low-Concentration Superparamagnetic Nanoparticles Using an Integrated Radio Frequency Magnetic Biosensor. *J. Appl. Phys.* **2013**, *113*, 104701. [CrossRef]
51. Devkota, J.; Mai, T.T.T.; Stojak, K.; Ha, P.T.; Pham, H.N.; Nguyen, X.P.; Mukherjee, P.; Srikanth, H.; Phan, M.-H. Synthesis, Inductive Heating, and Magnetoimpedance-Based Detection of Multifunctional Fe_3O_4 Nanoconjugates. *Sens. Actuators B Chem.* **2014**, *190*, 715–722. [CrossRef]
52. Devkota, J.; Wingo, J.; Mai, T.T.T.; Nguyen, X.P.; Huong, N.T.; Mukherjee, P.; Srikanth, H.; Phan, M.-H. A Highly Sensitive Magnetic Biosensor for Detection and Quantification of Anticancer Drugs Tagged to Superparamagnetic Nanoparticles. *J. Appl. Phys.* **2014**, *115*, 178503. [CrossRef]
53. Devkota, J.; Ruiz, A.; Mukherjee, P.; Srikanth, H.; Phan, M.-H. Magneto-Impedance Biosensor with Enhanced Sensitivity for Highly Sensitive Detection of Nanomag-D Beads. *IEEE Trans. Magn.* **2013**, *49*, 4060–4063. [CrossRef]
54. Devkota, J. Enhanced Magnetoimpedance and Microwave Absorption Responses of Soft Ferromagnetic Materials for Biodetection and Energy Sensing. Ph.D. Dissertation, University of South Florida, Tampa, FL, USA, 7 April 2015. Available online: https://digitalcommons.usf.edu/etd/5862/ (accessed on 1 March 2022).
55. Sato, T.; Fujine, T.; Miyazaki, J. Variation of Magnetic Properties along ribbon length in an amorphous $Fe_{80.5}Si_{6.5}B_{12}C_1$ Alloy. *J. Magn. Magn. Mater* **1988**, *71*, 255–262. [CrossRef]
56. Masood, A.; Baghbaderani, H.A.; Strom, V.; Stamenov, P.; McCloskey, P.; Mathuna, C.O.; Kulkarni, S. Facbrication and Soft Magnetic Properties of Rapidly Quenched Co-Fe-B-Si-Nb Ultra-Thin Amorphous Ribbons. *J. Magn. Magn. Mater* **2019**, *483*, 54–58. [CrossRef]
57. Kulkarni, S.; Li, D.; Wang, N.; Roy, S.; Mathuna, C.O.; Young, G.; McCloskey, P. Low Loss Magnetic Thin Films for Off-Line Power Conversion. *IEEE Trans. Magn.* **2014**, *50*, 2004404. [CrossRef]
58. Bethke, C.; Yakabchuk, H.; Tarasenko, V.; Hammer, H.; Kisker, E.; Koppers, E.; Christoph, S.; Zirwes, R.; Müller, J. Detektion Superparamagnetischer Marker Mittels GMI-Sensorik. *TM Tech. Mess.* **2003**, *70*, 574–576. [CrossRef]
59. Moulin, J.; Woytasik, M.; Shahosseini, I.; Alves, F. Micropatterning of Sandwiched FeCuNbSiB/Cu/FeCuNbSiB for the Realization of Magneto-Impedance Microsensors. *Microsyst. Technol.* **2011**, *17*, 637–644. [CrossRef]
60. Volchkov, S.O.; Svalov, A.V.; Kurlyandskaya, G.V. Giant Magnetic Impedance of Film Nanostructures Adapted for Biodetection. *Russ. Phys. J.* **2009**, *52*, 769–776. [CrossRef]
61. Wang, T.; Zhou, Y.; Lei, C.; Lei, J.; Yang, Z. Development of an Ingenious Method for Determination of Dynabeads Protein A Based on a Giant Magnetoimpedance Sensor. *Sens. Actuators B Chem.* **2013**, *186*, 727–733. [CrossRef]
62. Wang, T.; Zhou, Y.; Lei, C.; Lei, J.; Yang, Z. Ultrasensitive Detection of Dynabeads Protein A Using the Giant Magnetoimpedance Effect. *Microchim. Acta* **2013**, *180*, 1211–1216. [CrossRef]
63. Schwarzschild, B.M. Weak Magnetic Fields in Human Body. *Phys. Today* **1979**, *32*, 18–21. [CrossRef]
64. Panina, L.V.; Mohri, K.; Uchiyama, T.; Noda, M.; Bushida, K. Giant Magneto-Impedance in Co-Rich Amorphous Wires and Films. *IEEE Trans. Magn.* **1995**, *31*, 1249–1260. [CrossRef]
65. Uchiyama, T.; Mohri, K.; Shinkai, M.; Ohshima, A.; Honda, H.; Kobayashi, T.; Wakabayashi, T.; Yoshida, J. Position Sensing of Magnetite Gel Using MI Sensor for Brain Tumor Detection. *IEEE Trans. Magn.* **1997**, *33*, 4266–4268. [CrossRef]

66. Fodil, K.; Denoual, M.; Dolabdjian, C.; Treizebre, A.; Senez, V. Model Calculation of the Magnetic Induction Generated by Magnetic Nanoparticles Flowing into a Microfluidic System: Performance Analysis of the Detection. *IEEE Trans. Magn.* **2014**, *50*, 1–8. [CrossRef]
67. Fodil, K.; Denoual, M.; Dolabdjian, C.; Treizebre, A.; Senez, V. In-Flow Detection of Ultra-Small Magnetic Particles by an Integrated Giant Magnetic Impedance Sensor. *Appl. Phys. Lett.* **2016**, *108*, 173701. [CrossRef]
68. García-Arribas, A.; Fernández, E.; Svalov, A.; Kurlyandskaya, G.v.; Barandiaran, J.M. Thin-Film Magneto-Impedance Structures with Very Large Sensitivity. *J. Magn. Magn. Mater.* **2016**, *400*, 321–326. [CrossRef]
69. Kurlyandskaya, G.V.; Shcherbinin, S.V.; Buznikov, N.A.; Chlenova, A.A.; Svalov, A.V. Magnetic Materials for Thin Film Based Magnetoimpedance Biosensing. *Phys. Met. Metallogr.* **2019**, *120*, 1243–1251. [CrossRef]
70. Kurlyandskaya, G.V.; Fernández, E.; Safronov, A.P.; Svalov, A.V.; Beketov, I.; Beitia, A.B.; García-Arribas, A.; Blyakhman, F.A. Giant Magnetoimpedance Biosensor for Ferrogel Detection: Model System to Evaluate Properties of Natural Tissue. *Appl. Phys. Lett.* **2015**, *106*, 193702. [CrossRef]
71. Vilela, G.L.S.; Monsalve, J.G.; Rodrigues, A.R.; Azevedo, A.; Machado, F.L.A. Giant Magnetoimpedance Effect in a Thin-Film Multilayer Meander-like Sensor. *J. Appl. Phys.* **2017**, *121*, 124501. [CrossRef]
72. Lodewijk, K.J.; Fernandez, E.; Garcia-Arribas, A.; Kurlyandskaya, G.V.; Lepalovskij, V.N.; Safronov, A.P.; Kooi, B.J. Magnetoimpedance of Thin Film Meander with Composite Coating Layer Containing Ni Nanoparticles. *J. Appl. Phys.* **2014**, *115*, 17A323. [CrossRef]
73. Wang, T.; Lei, C.; Lei, J.; Yang, Z.; Zhou, Y. Preparation of Meander Thin-Film Microsensor and Investigation the Influence of Structural Parameters on the Giant Magnetoimpedance Effect. *Appl. Phys. A* **2012**, *109*, 205–211. [CrossRef]
74. Wang, T.; Yang, Z.; Lei, C.; Lei, J.; Zhou, Y. An Integrated Giant Magnetoimpedance Biosensor for Detection of Biomarker. *Biosens. Bioelectron* **2014**, *58*, 338–344. [CrossRef]
75. Zhou, Z.; Zhou, Y.; Chen, L.; Lei, C. Transverse, Longitudinal and Perpendicular Giant Magnetoimpedance Effects in a Compact Multiturn Meander NiFe/Cu/NiFe Trilayer Film Sensor. *Meas. Sci. Technol.* **2011**, *22*, 035202. [CrossRef]
76. Buznikov, N.A.; Safronov, A.P.; Orue, I.; Golubeva, E.V.; Lepalovskij, V.N.; Svalov, A.V.; Chlenova, A.A.; Kurlyandskaya, G.V. Modelling of Magnetoimpedance Response of Thin Film Sensitive Element in the Presence of Ferrogel: Next Step toward Development of Biosensor for in-Tissue Embedded Magnetic Nanoparticles Detection. *Biosens. Bioelectron* **2018**, *117*, 366–372. [CrossRef]
77. Wang, T.; Chen, Y.; Wang, B.; He, Y.; Li, H.; Liu, M.; Rao, J.; Wu, Z.; Xie, S.; Luo, J. A Giant Magnetoimpedance-Based Separable-Type Method for Supersensitive Detection of 10 Magnetic Beads at High Frequency. *Sens. Actuators A Phys.* **2019**, *300*, 111656. [CrossRef]
78. García-Arribas, A. The Performance of the Magneto-Impedance Effect for the Detection of Superparamagnetic Particles. *Sensors* **2020**, *20*, 1961. [CrossRef] [PubMed]
79. Alonso, J.; Khurshid, H.; Devkota, J.; Nemati, Z.; Khadka, N.K.; Srikanth, H.; Pan, J.; Phan, M.-H. Superparamagnetic Nanoparticles Encapsulated in Lipid Vesicles for Advanced Magnetic Hyperthermia and Biodetection. *J. Appl. Phys.* **2016**, *119*, 083904. [CrossRef]
80. Costa Silva, E.; Hall Barbosa, C.R.; Gusmao, L.A.P.; Leipner, Y.; Fortaleza, L.G.S.; Costa Monteiro, E. Point Matching: A New Electronic Method for Homogenizing the Phase Characteristics of Giant Magnetoimpedance Sensors. *Rev. Sci. Instrum* **2014**, *85*, 084708. [CrossRef]
81. Zhang, D.; Pan, Z.; Zhuo, H.; Zhang, W. Magnetic Sensor Based on Giant Magneto-Impedance Effect Using the Self-Regulating Technology on the Bias Magnetic Field. *Sens. Actuator A Phys.* **2016**, *249*, 225–230. [CrossRef]
82. Jin, F.; Tu, X.; Wang, J.; Yang, B.; Dong, K.; Mo, W.; Hui, Y.; Peng, J.; Jiang, J.; Xu, L.; et al. Noise Modeling and Simulation of Giant Magnetic Impedance (GMI) Magnetic Sensor. *Sensors* **2020**, *20*, 960. [CrossRef]
83. Uchiyama, T.; Takiya, T. Development of Precise Off-Diagonal Magnetoimpedance Gradiometer for Magnetocardiography. *AIP Adv.* **2017**, *7*, 056644. [CrossRef]
84. Takiya, T.; Uchiyama, T. Development of an Active Shielding-Type MI Gradiometer: Its Application for Magnetocardiography. *IEEE Trans. Magn.* **2017**, *53*, 4002804. [CrossRef]
85. Ma, J.; Uchiyama, T. Development of Peak-to-Peak Voltage Detector-Type MI Gradiometer for Magnetocardiography. *IEEE Trans. Magn.* **2018**, *54*, 5000605. [CrossRef]
86. Uchiyama, T.; Ma, J. Development of Pico Tesla Resolution Amorphous Wire Magneto-Impedance Sensor for Bio-Magnetic Field Measurements. *J. Magn. Magn. Mater.* **2020**, *514*, 167148. [CrossRef]
87. Wang, K.; Cai, C.; Yamamoto, M.; Uchiyama, T. Real-Time Brain Activity Measurement and Signal Processing System Using Highly Sensitive MI Sensor. *AIP Adv.* **2017**, *7*, 056635. [CrossRef]
88. Ma, J.; Uchiyama, T. Alpha Rhythm and Visual Event-Related Fields Measurements at Room Temperature Using Magneto-Impedance Sensor System. *IEEE Trans. Magn.* **2019**, *55*, 4002706. [CrossRef]
89. Zhu, Y.; Zhang, Q.; Li, X.; Pan, H.; Wang, J.; Zhao, Z. Detection of AFP with an Ultra-Sensitive Giant Magnetoimpedance Biosensor. *Sens. Actuators B Chem.* **2019**, *293*, 53–58. [CrossRef]
90. Feng, Z.; Zhi, S.; Guo, L.; Zhou, Y.; Lei, C. An Integrated Magnetic Microfluidic Chip for Rapid Immunodetection of the Prostate Specific Antigen Using Immunomagnetic Beads. *Microchim. Acta* **2019**, *186*, 252. [CrossRef]

91. Al-Khalidi, F.Q.; Saatchi, R.; Burke, D.; Elphick, H.; Tan, S. Respiration Rate Monitoring Methods: A Review. *Pediatr. Pulmonol.* **2011**, *46*, 523–529. [CrossRef] [PubMed]
92. Padasdao, B.; Shahhaidar, E.; Stickley, C.; Boric-Lubecke, O. Electromagnetic Biosensing of Respiratory Rate. *IEEE Sens. J.* **2013**, *13*, 4204–4211. [CrossRef]
93. Fekr, A.R.; Janidarmian, M.; Radecka, K.; Zilic, Z. A Medical Cloud-Based Platform for Respiration Rate Measurement and Hierarchical Classification of Breath Disorders. *Sensors* **2014**, *14*, 11204–11224. [CrossRef]
94. Thiabgoh, O.; Eggers, T.; Phan, M.H. A New Contactless Magneto-LC Resonance Technology for Real-Time Respiratory Motion Monitoring. *Sens. Actuators A Phys.* **2017**, *265*, 120–126. [CrossRef]
95. Hwang, K.Y.; Jimenez, V.O.; Muchharla, B.; Eggers, T.; Le, A.T.; Lam, V.D.; Phan, M.H. A Novel Magnetic Respiratory Sensor for Human Healthcare. *Appl. Sci.* **2021**, *11*, 3585. [CrossRef]
96. Thiabgoh, O. Novel Magneto-LC Resonance Sensors for Industrial and Bioengineering Applications. Ph.D. Dissertation, University of South Florida, Tampa, FL, USA, 26 March 2018. Available online: https://digitalcommons.usf.edu/etd/7650/ (accessed on 1 March 2022).
97. Canga, B.; Azoulay, R.; Raskin, J.; Loewy, J. AIR: Advances in Respiration—Music Therapy in the Treatment of Chronic Pulmonary Disease. *Respir. Med.* **2015**, *109*, 1532–1539. [CrossRef]
98. Sakaguchi, Y.; Aiba, E. Relationship between Musical Characteristics and Temporal Breathing Pattern in Piano Performance. *Front. Hum. Neurosci.* **2016**, *10*, 381. [CrossRef] [PubMed]
99. Costa, T.; Vara, M.; Cristino, C.; Zanella, T.; Nogueira-Neto, G.; Nohama, P. Breathing Monitoring and Pattern Recognition with Wearable Sensors. In *Wearable Devices—The Big Wave of Innovation*; IntechOpen: London, UK, 2019. [CrossRef]
100. Spackman, E.; Senne, D.A.; Myers, T.J.; Bulaga, L.L.; Garber, L.P.; Perdue, M.L.; Lohman, K.; Daum, L.T.; Suarez, D.L. Development of a Real-Time Reverse Transcriptase PCR Assay for Type A Influenza Virus and the Avian H5 and H7 Hemagglutinin Subtypes. *J. Clin. Microbiol.* **2002**, *40*, 3256–3260. [CrossRef] [PubMed]
101. Marras, S.A.E. Interactive Fluorophore and Quencher Pairs for Labeling Fluorescent Nucleic Acid Hybridization Probes. *Mol. Biotechnol.* **2008**, *38*, 247–255. [CrossRef] [PubMed]
102. Wu, J.; Liu, J.; Li, S.; Peng, Z.; Xiao, Z.; Wang, X.; Yan, R.; Luo, J. Detection and Analysis of Nucleic Acid in Various Biological Samples of COVID-19 Patients. *Travel Med. Infect. Dis.* **2020**, *37*, 101673. [CrossRef] [PubMed]
103. Hwang, K.Y.; Chapel, W.; Jimenez, V.O.; Muchharla, B. Data, Real-Time Monitoring of Coronavirus Progress Using Magnetic Sensing and Machine Learning. US20200029862A1, 2021. Available online: https://patents.google.com/patent/US20210369137A1/en (accessed on 1 March 2022).

biosensors

MDPI

Review

Microfluidic-Based Droplets for Advanced Regenerative Medicine: Current Challenges and Future Trends

Hojjatollah Nazari [1], Asieh Heirani-Tabasi [2,3], Sadegh Ghorbani [4], Hossein Eyni [5,6], Sajad Razavi Bazaz [1], Maryam Khayati [7], Fatemeh Gheidari [8], Keyvan Moradpour [9], Mousa Kehtari [10], Seyed Mohsen Ahmadi Tafti [11], Seyed Hossein Ahmadi Tafti [2] and Majid Ebrahimi Warkiani [1,12,*]

1 School of Biomedical Engineering, University of Technology Sydney, Sydney, NSW 2007, Australia; hojjatollah.nazari@student.uts.edu.au (H.N.); sajad.razavibazaz@student.uts.edu.au (S.R.B.)
2 Research Center for Advanced Technologies in Cardiovascular Medicine, Tehran Heart Center Hospital, Tehran University of Medical Sciences, Tehran 14535, Iran; asieh.heirani@gmail.com (A.H.-T.); hosseinahmaditafti@yahoo.com (S.H.A.T.)
3 Department of Cell Therapy and Hematology, Faculty of Medical Sciences, Tarbiat Modares University, Tehran 14535, Iran
4 Interdisciplinary Nanoscience Center (iNANO), Aarhus University, 8000 Aarhus, Denmark; sadeghqorbani@gmail.com
5 Cellular and Molecular Research Center, School of Medicine, Iran University of Medical Sciences, Tehran 14535, Iran; h.eyni1990@gmail.com
6 Department of Anatomical Sciences, School of Medicine, Iran University of Medical Sciences, Tehran 14535, Iran
7 Department of Pharmaceutical Nanotechnology, School of Pharmacy, Zanjan University of Medical Sciences, Zanjan 45371, Iran; Khayati293@yahoo.com
8 Department of Biotechnology, University of Tehran, Tehran 14535, Iran; fatemeh.gheidari@gmail.com
9 Department of Chemical Engineering, Sharif University of Technology, Tehran 14535, Iran; k1moradpour@gmail.com
10 Department of Biology, Faculty of Science, University of Tehran, Tehran 14535, Iran; Mousakehtari@gmail.com
11 Colorectal Surgery Research Center, Imam Hospital Complex, Tehran University of Medical Sciences, Tehran 14535, Iran; smohsenahmadi1364@gmail.com
12 Institute of Molecular Medicine, Sechenov University, 119991 Moscow, Russia
* Correspondence: majid.warkiani@uts.edu.au; Tel.: +61-2951-43141

Citation: Nazari, H.; Heirani-Tabasi, A.; Ghorbani, S.; Eyni, H.; Razavi Bazaz, S.; Khayati, M.; Gheidari, F.; Moradpour, K.; Kehtari, M.; Ahmadi Tafti, S.M.; et al. Microfluidic-Based Droplets for Advanced Regenerative Medicine: Current Challenges and Future Trends. *Biosensors* **2022**, *12*, 20. https://doi.org/10.3390/bios12010020

Received: 5 December 2021
Accepted: 29 December 2021
Published: 31 December 2021

Publisher's Note: MDPI stays neutral with regard to jurisdictional claims in published maps and institutional affiliations.

Abstract: Microfluidics is a promising approach for the facile and large-scale fabrication of monodispersed droplets for various applications in biomedicine. This technology has demonstrated great potential to address the limitations of regenerative medicine. Microfluidics provides safe, accurate, reliable, and cost-effective methods for encapsulating different stem cells, gametes, biomaterials, biomolecules, reagents, genes, and nanoparticles inside picoliter-sized droplets or droplet-derived microgels for different applications. Moreover, microenvironments made using such droplets can mimic niches of stem cells for cell therapy purposes, simulate native extracellular matrix (ECM) for tissue engineering applications, and remove challenges in cell encapsulation and three-dimensional (3D) culture methods. The fabrication of droplets using microfluidics also provides controllable microenvironments for manipulating gametes, fertilization, and embryo cultures for reproductive medicine. This review focuses on the relevant studies, and the latest progress in applying droplets in stem cell therapy, tissue engineering, reproductive biology, and gene therapy are separately evaluated. In the end, we discuss the challenges ahead in the field of microfluidics-based droplets for advanced regenerative medicine.

Keywords: droplet; microfluidics; regenerative medicine; stem cell therapy; tissue engineering

1. Introduction

Regenerative medicine is an emerging therapeutic method to reconstruct the human body's damaged organs and tissues [1]. This growing interdisciplinary science combines different approaches and technologies, such as medicine, stem cell biology, genetics, biomaterials, and chemistry. Regenerative medicine's primary field is cell therapy that directly implements stem cells and their secretions—such as exosomes and extracellular vesicles (EVs)—for tissue repair [1,2]. The accompanying of stem cell technology with biomaterial sciences, nanotechnology, and manufacturing techniques has led to the advent of tissue engineering, trying to improve cells' efficacy and functionality in injured tissues [3–5]. Moreover, the promising area of gene therapy resulting from tremendous improvements in genetics and molecular biology enables researchers to manipulate genes in cells [6]. The efficacy of all of these fields in regenerative medicine could not be improved without the provision of the stem cells being encapsulated safely in three-dimensional (3D) microenvironments mimicking their native home, or being confined in small rooms that prepare micro/nano-scale necessities and conditions for cellular manipulation and analysis [7,8]. Meeting this objective could significantly enhance the efficacy and success of stem-cell-based therapeutic methods, achievable by implementing droplets made by using microfluidic systems.

Droplet-based microfluidics is the technology of generating and manipulating small volumes of fluids (in the range of nL to pL) in immiscible phases. The droplets made using this technique can be produced at high frequency with desirable monodispersed sizes and well-defined volumes [9]. Moreover, well-controlled sequences of reproducible droplets can be obtained by implementing different geometries and patterns in microfluidic devices [10], which prepare desirable microenvironments for the isolation and encapsulation, culturing, gene editing, analysis, manipulation, and fusion of cells [11]. Therefore, they can provide a proper environment for cellular and molecular studies in different fields of regenerative medicine, including stem cell therapy, tissue engineering [12], gene therapy, and reproductive biology [13].

Microfluidic systems can provide safe and high-efficacy encapsulation methods for capturing stem cells inside aqueous droplets [14]. The droplets can be filled with specific culturing media for growing stem cells, or be filled with 3D hydrogels to mimic the native microenvironment of tissues for stem cells to proliferate and differentiate toward the desired adult cells [15]. Therefore, these droplets can mimic stem cells' niches and improve complicated cellular complexes [16]. In addition to mimicking the stem cells' niche, encapsulation of stem cells inside droplets can reveal their behavior and fates in different microenvironmental indications, such as cell–cell signals and possible cell–biomaterial interactions [17]. Moreover, droplets can simulate the biological and mechanical conditions inside niches of stem cells to aggregate and form mono/multicellular spheroids. Furthermore, the microenvironment created by droplets enables the development of intricate cellular clusters such as organoids via self-assembly and tissue formation. Droplets can also be appropriate platforms for performing high-throughput analysis and screening stem cells at a single-cell level to provide valuable information required for cell therapy and tissue engineering [17]. For instance, this technology enables us to study single stem cells' gene expression profiles in different stages of differentiation. Moreover, the microenvironments made by droplets can prepare conditions necessary for other high-throughput analyses—such as RNA sequencing—more efficiently and cost-effectively [18].

Implementing microfluidics-based droplets with assisted reproductive technology (ART) techniques can also improve the efficacy of generating artificial uteri, stem-cell-derived gametes, and human cloning [19,20]. Furthermore, droplets can also prepare desired genetic manipulation conditions in gene-therapy-based therapeutic methods [21]. Droplets act as nano-scale platforms to improve the efficacy of correcting genetic disorders via the insertion of exogenous genetic materials into cells based on non-viral vector systems. Therefore, the side effects caused by viral infections during gene editing of cells can be reduced by encapsulating cells and desired genomic materials inside droplets and implementing physical forces such as electric, optic, and hydrodynamic forces.

In this review, we present the state-of-the-art microfluidics-based droplet technologies for preparing microenvironments needed for the main domains of regenerative medicine, including stem cell therapy, tissue engineering, reproductive biology, and gene therapy. This review is organized into microfluidics-based droplet generation methods and the latest applications of droplets in each regenerative medicine domain (Figure 1). Finally, our conclusion and future remarks for implementing microfluidics-based droplets to bridge the gaps between the regenerative medicine lab and clinical practice are discussed.

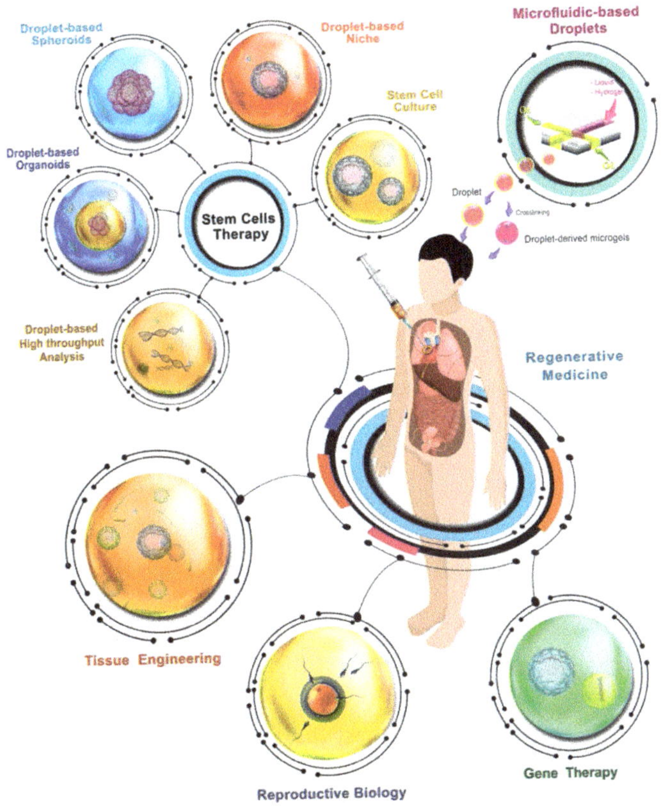

Figure 1. Microfluidics-based droplets for advanced regenerative medicine: Their applications in stem cell research (cell culture, niche engineering, spheroid forming, organoid generation, and high-throughput analysis), tissue engineering, reproductive biology, and gene therapy.

2. Microfluidics-Based Droplets

Microfluidics technology has undergone outstanding progress after its debut in the 1990s, and has shown massive functionality for biomedical applications [22–24]. Furthermore, the capacity of microfluidics to manipulate multiphase systems and generate monodispersions of polymer particles, emulsions, bubbles, and droplets has increased due to vast advances in microfabrication over the past decade [25]. Microfluidics-based systems implement the fundamentals of two-phase dynamics in microchannels for fabricating droplets for regenerative medicine purposes. These platforms facilitate droplet fabrication based on emulsification methods within their microchannels, and generate monodispersed droplets. Moreover, these systems can eliminate some common challenges in droplet generation using bulk techniques [26]. One of the biggest challenges in non-microfluidic methods is mixing two fluids in the bulk volume to enhance the turbulence, causing the break-up of droplets due to shear flows. Microfluidic devices can precisely control individual droplets, reduce sample consumption, and decrease polydispersity [27].

Furthermore, the unique convective flow profile within individual droplets enhances the mixing, leading to significant improvement of heat and mass transfer and acceleration of reactions [28]. For the aim of droplet generation, a variety of geometries and patterns can be implemented. These systems can also be integrated with automation systems and computers for versatile manipulation of the droplet formation process.

Droplet formation and manipulation in microfluidic devices can be either passive or active [29]. Passive droplet formations include co-flow (coaxial), crossflow (mainly in T-junction), and flow focusing [30]. In co-flow systems, samples are introduced via coaxial microchannels. The inner channel is assigned for the dispersed fluid phase, while the outer channel is responsible for the continuous phase. For the crossflow designs, the continuous and the dispersed fluid meet at an angle, where the most common case is a T-junction channel at an angle of 90 degrees. In the flow-focusing category, the continuous and the dispersed flow streams are introduced coaxially inside a confined region, which leads to the generation of droplets. The interfacial tension, viscosity, and flow rate ratio of continuous and dispersed flows define the gap, size, and generation rate of droplets. Active droplet formation and manipulation also involves electric, magnetic, and/or acoustic forces. Within the electric control, both alternating and direct current can be used (see the EWOD concept in Section 5). The droplet size can be precisely controlled by adjusting the electric field strength in the electrical methods. In dielectrophoresis (DEP), through the use of an electric field, uniform droplets can be generated by pulling the droplets from the reservoir of fluid. For the magnetic fields, the magnetic control occurs via ferrofluids, which are magnetic nanoparticle suspensions in an aqueous or oil-based carrier [31]. In droplet microfluidics, ferrofluids can be considered the dispersed- or continuous-phase flow. Droplets in magnetic-assisted droplet microfluidics can be characterized using the magnetic bond number, which is defined as the ratio of the magnetic force to the interfacial tension strength. Acoustic forces can also be helpful in droplet generation. For instance, surface acoustic wave devices can be utilized to induce mechanical vibration, assisting in droplet generation [32].

Many factors affect droplet formation. Therefore, proper comprehension of droplet formation and its dynamics is enlightening for utilizing its applications [33]. In passive droplet microfluidics, immiscible flow phases exist, generating both individual volumes of fluids and moving interfaces. Within each individual droplet, the convective flow profile eases the mixing. The linear Stokes equations govern the microfluidic droplet dynamics [34]. However, certain nonlinearities exist because of the two-phase flow interface and the variable interfacial tension. In passive microfluidics and pressure-driven flow, the interface deformation and the associated droplet breakup are affected by the channel junction geometry design and the local fluid flow. In droplet microfluidics, dimensionless numbers characterize the fluid behavior. The first crucial dimensionless number is the Reynolds number, defining the ratio of inertial to viscous forces. However, the Reynolds number is usually small; therefore, inertial forces become negligible. Another substantial dimensionless number is the capillary number, Ca, which identifies the ratio of viscous strength to interfacial forces. Indeed, two competing effects of the interface extension and deformation, which are mainly caused by deformation resistance and shear stress, affect the droplet formation. The other important dimensionless number is the Weber number, which defines the ratio of inertial to interfacial tension. At high flow rates, the transition of individual droplets to continuous jets is affected by inertial effects. Although the dimensionless numbers mentioned above are the most critical numbers in droplet microfluidics, other factors such as buoyancy, gravity, or elastic effects also exist under certain circumstances, and must be evaluated through the definition of proper dimensionless numbers [33].

The emulsions used for droplet generation inside microfluidic devices are classified into four system structures: oil-in-water, water-in-oil, water-in-oil-in-water, and oil-in-water-in-oil emulsions. The oil-in-water emulsions are systems consisting of oil droplets dispersed in an aqueous phase; these emulsions have been used to encapsulate and deliver various bioactive lipids, fatty acids in food products, and oil-soluble drugs for pharmaceutics [35,36]. The second category is water-in-oil emulsions—systems of water droplets dispersed in an oil phase; these emulsions have been utilized in several industries, including foods, cosmetics, chemical synthesis, and printers. These suspended aqueous droplets can encapsulate water-soluble ingredients (such as polymers and hydrogels) or living cells (including stem cells and progenitor cells) after crosslinking, and are highly useful in the fields of tissue engineering and regenerative medicine [37]. There are also water-in-oil-in-water and oil-in-water-in-oil emulsions, which are known as multiple emulsions consisting of water and oil droplets dispersed within larger opposite-phase droplets, e.g., the water-in-oil-in-water emulsions are double emulsions in which oil droplets enclosing water droplets are dispersed in water (Figure 2). These multiple emulsions are widely used in the food industry [38], cosmetic production, pharmaceutical research, and chemical separation [39,40]. Furthermore, many methods exist for droplet manipulation.

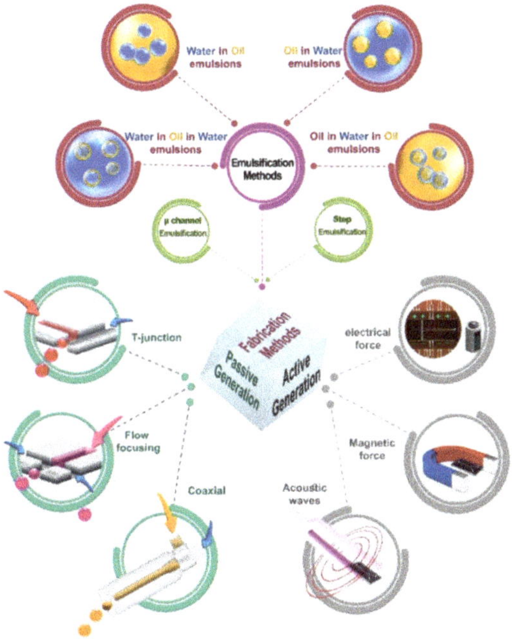

Figure 2. Microfluidics-based methods for droplet generation. The emulsification methods, types of droplets, geometries used in the passive generation, and forces can be implemented for the active generation of droplets.

3. Microfluidics-Based Droplets for Stem Cell Therapy

Stem-cell-based therapy is a regenerative medicine division in which stem cells are implemented for repairing tissues and organs [41]. Due to the tremendous advances in stem cell technologies over the past decade, this therapeutic approach has extended to the clinical trial stages to treat various diseases [42]. The advent of different types of stem cells—including pluripotent stem cells (PSCs), mesenchymal stem cells (MSCs), embryonic stem cells (ESCs), and induced pluripotent stem cells (iPSCs)—has made advanced stem cell therapy a game-changer in regenerative medicine [43]. The setup of precise derivation methods and treatment of cells with specific culturing media is necessary in order to differentiate these stem cells into the desired functional adult cells. Such environmental conditions can be created inside droplets made via microfluidic technology. In this section, the utilization of droplets for culturing and differentiation of stem cells, simulating stem cell niches' microenvironments, and the formation of spheroids and organoids is reviewed.

In the end, the droplet-assisted methods for isolation and single-cell analysis of stem cells are described.

3.1. Droplet-Based Stem Cell Culture

The technologies associated with the expansion, editing, and transplantation of stem cells and progenitor cells play an essential role in cell therapy and regenerative medicine [44]. Three significant characteristics identify stem cells: first, they have self-renewal capacity; second, they are unspecialized cells; third, they can differentiate into specialized subpopulations [45]. These cells reside in specific dynamic 3D microenvironments in the body called niches, which provide complicated biochemical and biophysical cues that aid in these cells' survival and determine their fate [46]. The advancement of microfluidics-based droplets can enhance our perception of the regulating functions involved in microenvironmental signaling, significantly improving cell therapy's chance of success. The generation of monodispersed micro-scale droplets via microfluidic systems provides stem cell researchers with facilities to encapsulate stem cells inside 3D microgels with the desirable microenvironment, removing some challenges in stem cell therapy. For instance, the low retention and homing of transplanted stem cells after transplantation dramatically reduce their therapeutic effects on patients. In other words, it is necessary to support transplanted cells' retention and chemotaxis in order to achieve long-term and efficient cell therapy [47,48]. Therefore, we need approaches such as droplet carriers or droplet-derived microgels that prolong cell culture time before and after transplantation. Another challenge in regenerating damaged tissue is the simulation of the 3D structure of tissue, which provides a range of physiologically more significant features than 2D, such as culturing stem cells or sphere formation to form multicellular tissue cultures in vitro for further implantation [49].

Droplets can enhance the efficacy of one of the primary stem cell sources in cell therapy—MSCs [50]. These cells can be separated from the bone marrow, adipose tissue, amniotic fluid, amniotic membrane, umbilical cord, and placenta, which secrete different anti-inflammatory, angiogenic, and anti-fibrotic factors, and cause immunomodulation without noticeable activation of any immune responses [51]. The low homing capabilities in targeting of tissues by MSCs can be addressed by encapsulating MSCs inside 3D droplet-derived hydrogels [52,53]. Moreover, these elements are suitable for the simultaneous carrying of sensitive biological cargoes, including nucleic acids, proteins, drugs, nanoparticles, and living cells [54,55]. By controlling the droplet-derived microgels' porosity, the presence of immunostimulatory agents produced by allogeneic or xenogeneic cells transplanted to the immune system in surrounding tissues and capillaries could be reduced. At the same time, the diffusion of oxygen and cellular waste products is preserved. Microgel porosity can be easily controlled by different hydrogel biomaterials and concentrations [56]. Many studies have investigated the impact of encapsulating MSCs on the secretion of biochemical agents and signaling proteins made by cells inside hydrogels. For instance, Headen et al. used a two-layer parallel microfluidic system to encapsulate human MSCs in synthetic microspheres with less than 100 µm diameter; they observed that these 3D microspheres increased cells' viability and enhanced the secretion of vascular endothelial growth factor (VEGF), which is necessary for angiogenesis [57]. VEGF is an exclusive mitogen for vascular endothelial cells, which induces proliferation, promotes migration, and inhibits apoptosis of endothelial cells [58]. In addition to isolating transplanted cells from the host immune system, 3D gels allow engineering of the cell-laden microenvironment, which presents cells with the stiffness of matrix–adhesive ligand interactions. These factors significantly influence the secretory function and differentiation of encapsulated stem cells [59,60].

Human embryonic stem cells (hESCs) are also popular stem cell sources in clinical trials; they can be endlessly expanded in culture and differentiated into any given cell type in the body [61]. The differentiation of hESCs is generally performed either in 3D aggregates known as embryoid bodies (EBs), or in two 2D monolayer cultures. Droplets can also assist in the encapsulation of hESCs inside biomaterials and microenvironments [62]. For instance, Chayosumrit et al. established a 3D model to expand and differentiate hESCs and encapsulate them in calcium alginate microcapsules. The microgels induced cellular interactions, which are essential for preserving both pluripotency and differentiation of hESCs, allowing separation of hESCs from fibroblasts and providing immune isolation during transplantation (Figure 3A) [63].

3.2. Droplet-Based Spheroids

When stem cells settle in their native 3D niche microenvironment, they assemble with other cells and form aggregates, called cell spheroids [51]. These stem cells inside spheroids can prepare complicated interactions with other types of stem cells, progenitor cells, stromal cells (via their secretary molecules), and extracellular matrix (ECM) molecules, leading to their differentiation, activation, and performance. The in vitro encapsulation and culturing of stem cells inside droplets with similar physiological and mechanical properties to the native niche can enable us to address such challenges [64].

Despite developing a variety of techniques for in vitro spheroid formation—including hanging drop, spinner flask, and suspension culture—over the past decades, some limitations decrease the efficacy of spheroids for research and clinical purposes, including the laborious procedure, and the limited diffusion of oxygen and nutrients to the spheroids [65–67]. The microfluidics-based droplets can act as platforms for high-throughput culturing and precise size control of spheroids inside 3D microenvironments to address these issues. Furthermore, the fabrication of stem cell multicellular spheroids with controlled dimensions, which are in high demand in regenerative medicine, is achievable using droplets because they can provide different cells inside spheroids with the ability to interact, regulating biological processes such as hemostasis and disease development. In vitro formation of multicellular spheroids enables the intercellular interactions required for cell viability, phenotyping, and function maintenance of stem cells [68].

In terms of multicellular spheroids, Chan et al. developed a method for generating double-emulsion water-in-oil-in-water ($w/o/w$) droplets for high-throughput production of spheroids via droplets encapsulating different cell types, such as human mesenchymal stem cells (hMSCs); they applied these droplets as bioreactors to accelerate the growth and formation rates in spheroids; moreover, they extracted these spheroids and encapsulated them in alginate and alginate/RGD (arginine-glycine-aspartic acid) 3D microenvironments, and found that the incorporation of modified alginate enhanced the osteogenic differentiation of spheroids (Figure 3B) [69]. The liquid microenvironment of droplets can also create cellular spheroids via their cell–cell interactions in a pre-hatching embryo. Agarwal et al. constructed a core–shell microcapsule to imitate the 3D structure of pre-hatching embryos with a hydrogel alginate shell and an aqueous core of embryonic cells; their results demonstrated a survival rate of over 92% in encapsulated embryonic cells. Inside these microfluidic-derived droplets, single cells could turn into embryonic bodies (EBs) consisting of almost 20 cells, while hundreds of cells are required to form embryonic bodies via the hanging drop method. Agarwal et al. also observed a significant increase in pluripotency gene expression compared to in 3D culture (Figure 3C). Finally, they differentiated these EBs into beating cardiomyocytes with the aid of only a small molecule instead of a complex combination of growth factors [70].

One of the most crucial requirements to vastly employ spheroids in regenerative medicine, drug discovery, and pharmaceutics is a production process that is cost-effective, easy to implement, and not prone to human error. A variety of research groups are attempting to advance robotized droplet microfluidic platforms to produce practical microspheres for cell encapsulation and culture on a large scale. For instance, Langer and Joensson developed an automated microfluidics-based droplet generation system for spheroid formation; they fabricated scaffold-free cell spheroids with highly monodisperse droplets, at a high production rate of 85,000 spheroids per microfluidic device per hour; furthermore, their platform consisted of main steps, including droplet generation, formation, and recovery, as well as dispensing of spheroids [71]. In another study, Cedillo-Alcantar et al. developed an automated microfluidic system that produced droplets to cultivate as well as biosensing hepatic spheroids. The droplets were able to mimic hormones' physiological microenvironment in the hepatic portal circulation and support the long-term preservation of primary hepatocytes. Additionally, spheroids' biochemical responses inside droplets (secretion of glucose and albumin) were monitored in a real-time manner using an enzymatic assay [72].

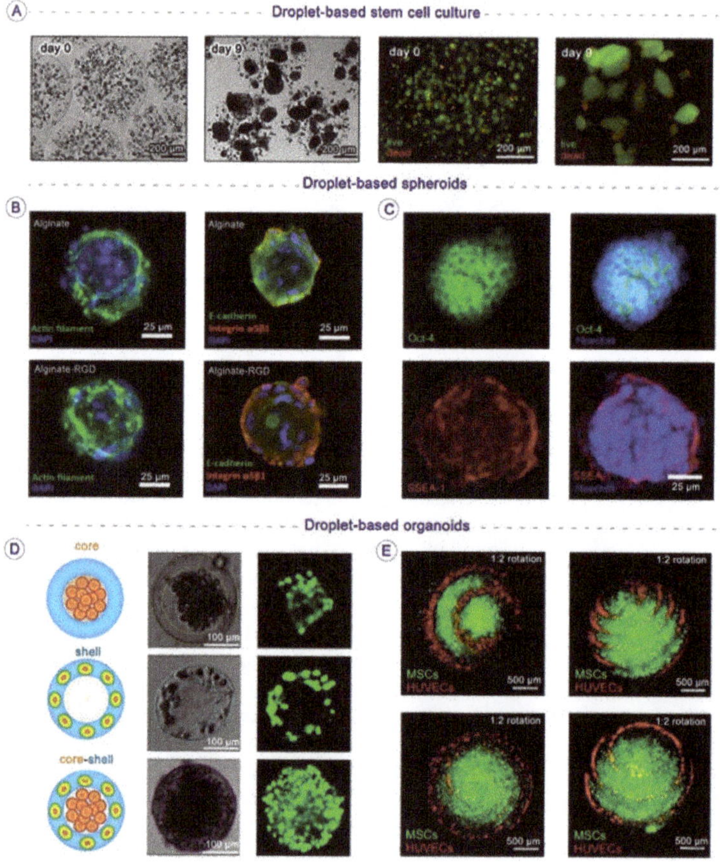

Figure 3. Microfluidics-based droplets for generating 3D culture conditions, niches, spheroids, and organoids for stem cell therapy applications: (**A**) Viability and morphology of encapsulated hESCs at different timepoints after encapsulation in alginate microcapsules (viable cells: green GFP; non-viable cells: red PI) [63]; Copyright, Elsevier. (**B**) Immunostaining for E-cadherin and integrin for spheroids encapsulated in alginate or alginate/RGD microgel after three days [69]; Copyright, Nature. (**C**) Immunohistochemical staining of pluripotency marker proteins (Oct-4, green; SSEA-1, red; Hoechst, blue) in the embryonic spheroids [70]; Copyright, The Royal Society of Chemistry. (**D**) Spatial assembly of different cells in the 3D core–shell scaffolds. Up: HepG2 cells are confined in the core by the hydrogel shell. Middle: NIH-3T3fibroblasts immobilized by the crosslinked alginate network in the shell. Down: Simultaneous assembly of hepatocytes in the core and fibroblasts in the shell, forming an artificial liver in a drop [73]; Copyright, The Royal Society of Chemistry. (**E**) Construction of organoids through the co-culture of HUVECs and hMSCs in spiral-based microspheroids. HUVECs were spirally distributed on the surface of the spheroids, with a composition ratio (QHUVECs:QhMSCs) of 1:2 [74]; Copyright, Wiley Online Library.

3.3. Droplet-Based Organoids

Organoids are intricate cell clusters made of different stem/progenitor cells of specific organs that form complex tissue-specific structures in the presence of 3D ECM microenvironments via self-assembly. Improving the in vitro 3D culture systems to mimic the native microenvironment of organoids in the body and generate organoids that can produce vessel-like structures for fluid transportation is a demanding concern that can be addressed using droplet-derived microgels. Moreover, the challenge of deficient nutritional supply in organoids can cause limitations of their size, survival, and functionality, which can be addressed via a microfluidics-based droplet-encapsulating approach [75].

Safe encapsulation of different types of cells in liquid microenvironments or 3D microscale constructions made of biocompatible ECM is a great challenge in the development of functional organoid models, achievable using droplet-based microfluidics [76] (Figure 4A). For instance, the liver comprises various cell types, including primary hepatocytes, hepatic stellate cells, Kupffer cells, endothelial cells, and fibroblasts. The proper liver-specific functions are achievable when these cells are cultured together in 3D cell co-culture systems. Chen et al. utilized a flow-focusing microfluidic device to produce droplets containing an aqueous core and an alginate hydrogel shell. To generate the human liver model in each droplet, they encapsulated hepatocytes and fibroblasts in the core and hydrogel shells, respectively. The co-culture of hepatocytes and fibroblasts increased homotypic and heterotypic cell–cell interactions, followed by high liver-specific functions inside droplets. Additionally, the alginate shell's high permeability resulted in the high cell viability of micro artificial organs after more than 10 days (Figure 4D) [73]. In another study to produce heterogeneous human organoids, Zhao et al. utilized a microfluidic nozzle to improve the efficacy of an airflow-assisted 3D bioprinter for the printing of cell-laden spiral microarchitectures; they encapsulated human umbilical vein endothelial cells (HUVECs) and hMSCs inside spiral-based microspheroids to establish a complicated co-culture 3D microenvironment for osteogenesis and angiogenesis in these organoids [74]. These cell-laden spheres supported in vitro production of osteogenic nodules in functional organoids (Figures 4E and 5C) [74].

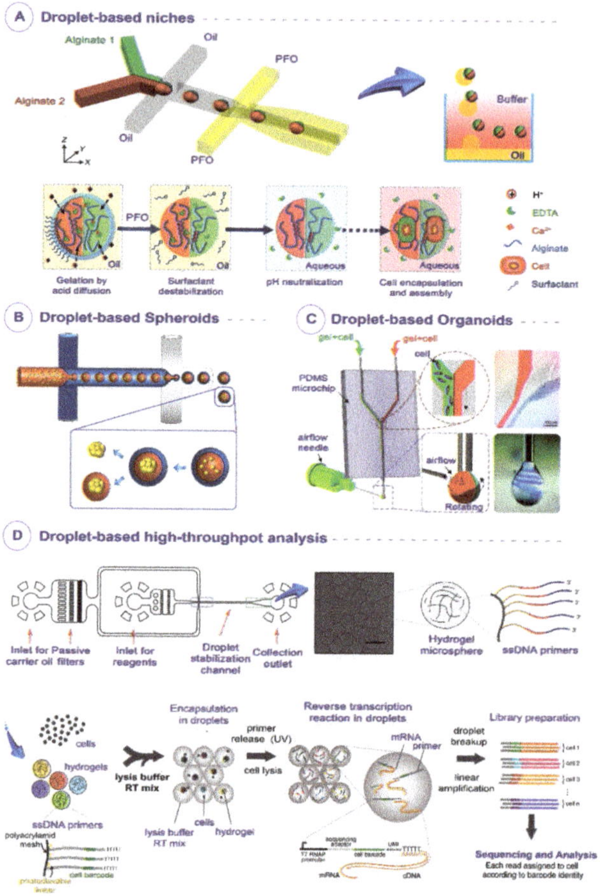

Figure 4. Microfluidics-based methods for droplet fabrication for stem cell therapies: (**A**) Up: Design of a microfluidic device to produce single emulsion drops consisting of multiple aqueous phases; the alginate droplets are solidified inside the chip and are immediately collected inside cell media. Down: The detailed mechanism to form multi-compartment alginate microgels [69]; Copyright, Wiley Online Library. (**B**) Rapid formation of multicellular spheroids after assembly of encapsulated cells. The spheroids can be released with or without microgel encapsulation [76]; Copyright, Nature. (**C**) Design of an airflow-assisted 3D bioprinter with a microfluidic nozzle to manufacture 3D spiral-based cell-laden spheroids. Sodium alginate solutions are extruded from the microchip and rotated by the airflow [74]; Copyright, Wiley Online Library. (**D**) Design of a droplet-based method for single-cell sequencing. Hydrogel microspheres are made containing barcoded oligo(dT) primers and sequencing adaptors. Each barcoded hydrogel microsphere is encapsulated with a single cell into a droplet with lysis buffer and reverse-transcription mix. Primers are released regarding UV exposure, and cDNA synthesis is performed in each droplet. Droplets are then broken, and barcoded cDNA of each cell is linearly amplified, followed by sequencing [76]; Copyright, Elsevier.

3.4. High-Throughput Analysis of Stem Cells Using Droplet-Based Techniques

The high-throughput analysis of each stem cell in a heterogeneous population can provide valuable information about their biological systems' complex behavior during clinical applications. The confining of single cells inside droplets can enhance the efficacy of laborious, expensive, and low-throughput molecular techniques such as Sanger sequencing, real-time PCR, blotting techniques, and fluorescent microscopy. Moreover, the implementation of droplets in available high-throughput cell analysis techniques—such as polymerase chain reaction [77], DNA/RNA sequencing [78], enzyme kinetics [79] and assays [80], protein crystallization [81,82], protein expression quantification [83], protein quantification mass spectrometry (LC–MS), and GC–MS metabolomics—makes it more important to be able to analyze single cells. Therefore, the high potency of droplet-based microfluidic techniques makes them excellent platforms for single-cell analysis and small samples, such as stem cells, tumors, and micro-biopsies [84]. For instance, Klein et al. designed a method for barcoding, capturing, and profiling transcriptomes, with quick collection, low technical noise, and no limitations regarding the number of cells; they encapsulated single ESCs, which are highly heterogeneous due to their pluripotency inside droplets, and then performed all of the RNA sequencing steps on an individual basis [85].

The accurate isolation and analysis of distinct cells can provide critical information on various biological parameters and processes in stem cell research—especially gene- and cell-based therapies. Conventional methods of cell separation to date—including fluorescence-activated cell sorting (FACS), magnetic-activated cell sorting (MACS), density centrifugation, limiting dilution, and immunoaffinity—are not cost-effective, need a large amount of input sample, have low throughput, and are limited to specific applications [86]. The integration of microfluidic systems producing droplets can improve cell sorting and analysis via the above techniques. Furthermore, these picoliter-sized droplets provide an isolated environment for the distinct reactions of many biological and chemical assays.

The expression of both intracellular and extracellular proteins is among the critical characteristics of stem cells that can be screened or quantified at the single-cell level using droplets. The proteins of stem cells should be analyzed, because they play a crucial role in stem cell viability, metabolism, differentiation stages, generation of recombinant proteins or antibodies by recombinant or hybridoma cells, and finding protein-producer colonies for immunological and pharmacological purposes. The routine techniques for this aim—including fluorophore–antibody detection of proteins via flow cytometry, immunohistochemistry, Western blotting, immunocytochemistry, and enzyme-linked immunosorbent assay (ELISA)—are expensive, time-consuming, and need a large number of cell samples and reagents; moreover, they cannot enable us to screen rare cell phenotypes that may be of high importance as colonies of the expensive drug–protein producer, or as specifically characterized and efficient stem cells or progenitor cells. Microfluidics-based droplets make the screening and detection of rare cell phenotypes at a single-cell level possible. The small volumes of liquids inside the cell-laden droplets increase the concentrations of secreted biomolecules such as monoclonal antibodies to levels that are easily detectable using diagnostic methods. For instance, Koster et al. designed a droplet-producing microfluidic system that could detect a low amount of released antibodies secreted from single hybridoma cells just 6 h after secretion using the kinetic ELISA method [87].

Next-generation sequencing (NGS), by providing the possibility of whole-transcriptome sequencing at a single-cell level, has emerged as a novel technology in the field of stem cells; it reveals the heterogeneity of gene expression profiles in cell populations related to different stages of cellular differentiation or function. Whole-exome sequencing gives us a wide range of data on gene expression profiles, compared to high-throughput DNA/RNA arrays that are not whole-genome-wide. Microfluidics has helped scientists to perform next-generation sequencing at the single-cell level in order to help analyze heterogeneous cell populations and stem cells. For instance, Streets et al. developed a microfluidic device to capture and lyse single cells and synthesize cDNA from mRNAs via a next-generation sequencing platform. Using their device, they prepared 94 libraries from single mouse ECSs, and improved RNA detection sensitivity made it possible to reconstruct a majority of the bulk transcriptome from 10 single cells with 0.2 M reads per cell. Enhanced measurement precision and technical throughput helped to distinguish variations in expression between and within different types of murine embryonic cells [88]. Moreover, microfluidics-based droplets have provided scientists with numerous achievements in single-cell analysis. For instance, Klein et al. reported developing a high-throughput system that can encapsulate single cells and lysis buffer, reverse-transcription reaction mix, and hydrogels containing barcoded sequencing primers into droplets. After encapsulation and lysis of cells, ultraviolet (UV) treatment of droplets releases barcoded primers from hydrogels for cDNA synthesis. Following linear amplification and breakup of droplets, next-generation sequencing (NGS) makes specific reads from each cell's genetic content regarding sequencing adaptors and unique barcodes on the cDNA. Therefore, Klein et al. applied this approach to murine ESCs after withdrawal of leukemia inhibitory factor (LIF), showing the heterogeneous onset of differentiation in this cell population and revealing gene expression relationships; their method showed acceptable reproducibility of data and a low noise baseline (Figure 4D) [85].

Zilionis et al. established a droplet microfluidics platform called inDrops for whole transcriptomics or genomics analysis of more than 15,000 individual cells in one hour, with minimal reagent use. More than 75% of cells in a sample are encapsulated into nanoliter aqueous droplets with hydrogel bead cores that bear barcoding primers. Following lysis in droplets, barcoded cDNA of cells is synthesized from mRNA via barcoding primers by reverse transcription, and then an RNA-Seq library is prepared. Zilionis et al. have used inDrops for murine primary immune cells and murine ESCs, and showed that this is the best method of single-cell transcriptomics for limiting samples consisting of less than 200,000 cells, such as rare stem cell populations or tumor biopsies [89]. Another microfluidics-based droplet technique called Drop-Seq, which Macosko et al. implemented for transcriptome analysis of murine retinal cells, had a lower rate of 2–4% of cells barcoded [90]. Applying such a high-throughput genome-wide expression profiling technique in a single cell enables scientists to precisely search for gene expression variations during stem cell culture and differentiation.

4. Microfluidics-Based Droplets for Tissue Engineering

Tissue engineering is defined as an interdisciplinary field—a combination of engineering, nanotechnology, biomaterials, and stem cell science to develop functional tissue-like structures for the replacement, repair, and rehabilitation of damaged tissues/organs [91]. During the past decade, various scaffolds with different shapes, morphologies, and mechanical properties have been developed using different materials for engineering a variety of tissues, among which cell-laden scaffolds were more successful in mimicking the 3D environments of desired tissues. These types of scaffolds should be highly biocompatible, with controlled degradability, mechanical stability, uniform size, and suitable mass transport functions [92].

To date, numerous materials have been applied for encapsulating cells in the cell-laden scaffold, among which hydrogels were more successful in mimicking the native microenvironment of human tissues for regulating stem cell proliferation and differentiation, synthesis of ECM proteins, and cell–matrix and cell–cell biochemical and mechanical interactions. Cell-laden hydrogels are usually produced by incorporating cells in precursor solution, followed by gelation using bulk- or microfabrication techniques. The encapsulation of cells within macroscopic hydrogels leads to low survival, incomplete cell–cell and cell–matrix interactions, and poor nutrient exchange because of the low diffusion rate in the scaffold [93]. One of the difficulties in the macro-scale 3D culture systems is that they are not suitable for long-term cell culture, because the maximum oxygen diffusion inside scaffolds is usually limited to 400 μm. On the other hand, the size of scaffolds made by routine methods is much bigger than this limit, and most of the encapsulated cells lose their functionality and viability in a short time. The cells encapsulated in microgels with a diameter of 100–400 μm can survive and proliferate safely. The number of cells encapsulated in the microgel and total cell encapsulation efficiency can be controlled via microfluidic techniques. In contrast, cell-laden microgels produced via microfabrication methods have a large surface-area-to-volume ratio, and can improve the efficient mass transfer and different nano-scale interactions between cells and their surrounding ECM [54]. These microgels can assemble inside damaged parts after injection and form engineered tissue-like constructs with similar morphology and functions to native tissue. The cell-laden microparticles can be manufactured through numerous techniques, one of which is droplet microfluidics.

Droplets made by microfluidic systems can prepare aqueous microenvironments to encapsulate cells and monomers of hydrogel solutions to be polymerized and form cell-laden microgels. These spherical cell-laden 3D scaffolds mimic the microenvironments needed for the proliferation and differentiation of stem cells, and generate the desirable mechanical, chemical, and physiological properties required for supporting damaged organs [94]. These systems can also encapsulate different types of cells, as well as biomolecules such as growth hormones, exosomes, biological stimulators, and various nanoparticles [95]. Therefore, the fabricated microgels can ultimately serve as functional building blocks for injection to the tissue constructs. In addition to the cell-encapsulating capacity of microfluidic systems, cells that can be cultured on the surface of the microfluidic microgels have several advantages: First, cells grown on the surface are not restricted, and can proliferate quickly within a short time. Second, detachment of cells culturing on the microgel surface is relatively easy without enzymatic operations, via mechanical separation methods or thermally mediated degradation. In the following section, we will briefly overview the applications of microparticles made using droplet technology in bone, cartilage, skin, pancreas, liver, and cardiovascular tissue engineering.

4.1. Bone Tissue Engineering

The employment of advanced technologies such as tissue engineering and stem cell therapy for advancing bone regeneration is highly demanding [96]. Despite the vast progress in bone tissue engineering, researchers and clinicians face serious challenges, such as low retention and low survival and transplanted stem cells' metabolic activity in the host body. Microfluidics-derived droplets can help to prepare acceptable microenvironments enriched with biological cues of native bone tissue, and enhance the survival, proliferation, and osteogenic differentiation of stem cells. For instance, Moshaverinia et al. developed injectable hydrogel microspheres made of RGD (as a biological cue) and alginate for the encapsulation of periodontal ligament stem cells (PDLSCs) and gingival mesenchymal stem cells (GMSCs); their results showed that adding RGD to microspheres' structure improved the viability and osteogenic differentiation of encapsulated cells and the expression of osteogenic markers such as RUNX family transcription factor 2 (Runx2), alkaline phosphatase (ALP), and osteocalcin in both in vitro and in vitro conditions. These cell-laden microspheres, which were less than 500 μm in diameter, were half the size of the ones obtained via bulk methods. This approach improved the transport of nutrients, oxygen, and waste products through the microspheres, resulting in high cell viability [97].

In addition to implementing biochemical cues in spheres, the rapid generation of injectable osteogenic tissue constructs is another critical issue in bone stem cell therapy. Regarding this, Zhao et al. encapsulated bone-marrow-derived mesenchymal stem cells (BMSCs) and bone morphogenetic protein-2 (BMP-2, as an osteogenic growth factor) inside photo-crosslinkable gelatin methacryloyl (GelMA) microspheres with a high-speed production rate using a capillary microfluidic device. Incorporating BMP-2 into GelMA microgels improved stem cell proliferation, osteogenesis, and mineralization in both in vitro and in vivo studies. These cell-laden structures supported the spreading of stem cells within the microspheres and their migration out of the microspheres [98]. In another study, Hou et al. prepared injectable, degradable poly(vinyl alcohol) (PVA) microgels (100–200 μm in diameter) loaded with hMSCs and BPM-2 via high-throughput microfluidics-based systems. The joining of this growth factor to PVA microgels increased ALP activity, calcium content, and Runx2 expression in encapsulated cells. The mild crosslinking and cell-compatible conditions enabled the encapsulation of hMSCs with prolonged survival, differentiation, and migration [99]. Other studies encapsulated single stem cells inside microspheres for the purpose of osteogenic differentiation. For example, An et al. encapsulated rat MSCs at a single-cell level inside alginate microspheres. The immune-shielding properties of he alginate shell supported bone formation after transplantation to the rat tibial ablation model (Figure 5A) [100].

4.2. Cartilage Tissue Engineering

The direct injection of stem cells with hydrogels to the damaged articular cartilage is considered a potential therapeutic method [101]. The hydrogels prepared based on bulk fabrication methods face encapsulated cells with significant challenges, including high apoptosis rate, low retention in the host cartilage, and inadequate chondrogenic differentiation and functionality. A variety of microfabrication methods have been used to encapsulate cells for cartilage regeneration, among which a limited number are based on droplets. For instance, Li et al. encapsulated BMSCs inside visible light-cured microgels composed of gelatin norbornene (GelNB) and polyethylene glycol (PEG), and differentiated them into chondrocytes in chondroinductive media. The encapsulation process using droplets increased the chondrogenesis of cells compared to the bulk hydrogel and 2D culture [102]. In an in vivo study, Feng et al. designed and fabricated injectable cell-laden microgels to self-assemble in situ inside the injury point and from a 3D porous scaffold without external stimulation; they encapsulated BMSCs inside these hybrid microspheres composed of gelatin and hyaluronic acid (HA) microgels via a thiol-Michael addition reaction. These cell-laden microgels self-assembled into cartilage-like structures via cell–cell interconnectivity, and supported proliferation and chondrogenic differentiation of encapsulated BMSCs. Furthermore, the occurrence of vascularization and hypertrophy—which are both considered huge barriers in cartilage regeneration—decreased significantly after transplantation (Figure 5B) [103].

In addition to the cell-encapsulating role of droplets, these microstructures can be employed for the sustained release of biological cues inside the injury point. For instance, the transforming growth factor beta-3 (TGF-β3)—a cytokine—plays an essential role in increasing the viability and ECM secretion of chondrocytes, and in enhancing the chondrogenic differentiation of stem cells. In this regard, Qasim et al. loaded poly(L-lactic-co-glycolic acid) (PLGA) particles with TGF-β3 using a droplet-based microfluidic system, and incorporated them into polycaprolactone (PCL) nanofibers using the electrospinning method. The fabricated microspheres improved the loading efficiency of TGF-β3 by up to 80%. They found that these microparticles blended with nanofibrous scaffolds promoted the proliferation and chondrogenic differentiation of hMSCs [104].

4.3. Wound Healing

As the largest organ in terms of surface area in the body, skin protects internal tissues from various damaging elements, including mechanical injuries, infections, ultraviolet radiation, and temperature. Severe skin-related injuries endanger the patient's life and negatively impact the healthcare economy [105]. Even though various therapeutic methods for wound healing have been developed in the clinic over the past decade, these approaches are practical only for moderate injuries. The arrival of advanced skin tissue engineering methods helps to regenerate damaged parts [106]. For this aim, researchers developed scaffolds using droplet-based microfabrication techniques for severe skin damage. For instance, Yu et al. fabricated hollow microspheres of bacterial cellulose using a droplet-based fabrication process, and applied them as wound-healing materials for the regeneration of significantly damaged skin. To this end, they encapsulated *Gluconacetobacter xylinus* bacteria in core–shell microspheres with an alginate core and an agarose shell. The bacteria inside the core secrete and tangles cellulose in the shell part. The alginate core and bacteria are removed, and hollow cellulose microspheres remain in the structure. These porous microspheres showed acceptable mechanical properties that supported the proliferation and migration of PC-9 cells. Furthermore, in collaboration with human primary epidermal keratinocytes, these scaffolds enhanced wound healing in male Sprague Dawley rats [107].

The fabrication of hydrogel microparticles permits direct delivery of cells through needles to the injured tissue for wound healing with the least clinical invasiveness. In this regard, Griffin et al. synthesized a new class of injectable microporous annealed-particle gels that enabled a stable, interconnected network of micropores for cell migration. They prepared complex three-dimensional networks based on self-assembly of monodisperse microgel building blocks made using a microfluidic device (Figure 5C). The annealing of these building blocks created an interconnected microporous scaffold, in which cells could be placed in the interconnected pores. Furthermore, the building blocks were easily injected into the mice's bodies, and caused faster skin regeneration and lower immune response than non-porous scaffolds after one week. The nature of these droplet-based building blocks allows the combination of an extensive range of materials, signals, and cell populations [108].

4.4. Pancreas Regeneration

Diabetes mellitus is one of the most prevalent metabolic diseases all over the world. The in vitro culture and transplantation of pancreatic islets is considered an effective therapeutic method to aid patients with type I diabetes. Droplet-based encapsulation methods can increase the safety of the encapsulation process, control the size of scaffolds, permit long-term culture periods, and prepare immunological protection of these pancreatic islets in the engineering of pancreatic tissue [109].

Fabrication of core–shell scaffold structures using microfluidics-derived droplets can provide islet cells with more immunoprotection after injection into the host. For instance, Ma et al. encapsulated rat pancreatic islets with high viability in alginate core–shell microcapsules and injected them into the type I diabetic mouse model. The transplantation of cell-laden core–shell structures decreased blood glucose to the normal range without stimulating the immune system [110]. In another study, Headen et al. encapsulated human pancreatic islets in scaffolds made of maleimide-functionalized PEG (PEG-4MAL) via a flow-focusing microfluidic system. The encapsulated cells inside scaffolds showed viability of 90% after eight days of culture (Figure 5D). On the other hand, capsules showed a selective permeability to biomolecules such as glucose, insulin, and antibodies [111]. The presence of oils and surfactants in droplet fabrication procedures can endanger the viability and functionality of encapsulated cells. In this regard, Lue et al. encapsulated rat pancreatic islets (β-TC6) inside calcium alginate hydrogels using a water-in-water system; they used a pneumatic valve combined with a crossflow microfluidic device to assist in the production of droplets and improve controllability. The encapsulated cells responded to the stimulation of glucose during cultivation for seven days [112].

4.5. Liver Regeneration

More than half a billion people are suffering from various liver diseases worldwide [113]. Liver transplantation is still the only therapeutic method for severe hepatic disorders, and patients face a severe shortage of donors and a long waiting list. Thus, regenerative-medicine-based methods are in high demand in the field of liver regeneration [114]. The morphology of hepatic cells, their cell–cell and cell–scaffold interactions, and the capability of spheroid formation, albumin secretion, and urea synthesis are the main challenges in liver regeneration, and are highly dependent on the microenvironment in which the cells are encapsulated.

The hepatic spheroid-laden hydrogels are of great interest in liver regeneration. One of the greatest limitations of the conventional microencapsulation method is that a constant number of spheroids cannot be encapsulated and produced per microgel. Chan et al. encapsulated hepatocyte spheroids inside alginate microgels using a droplet-based system. The gelation time of hydrogel permitted hepatocytes to improve their assembly within the liquid core. Additionally, their study showed that the co-culturing of hepatocytes with endothelial progenitor cells (5:1) improved the hepatic functionality, differentiation, angiogenesis, and long-term performance of cells (Figure 5E) [115].

The cell-laden spheres simulate the niche of hepatic cells, and also protect them against the immune system. In this regard, Wang et al. fabricated core–shell cell-laden microgels made of methylcellulose and GelMA. The device incorporated HepG2 cells with and without HUVECs through a methylcellulose core. This core was encapsulated in a GelMA shell, and microgels were cultured for a two-week in vitro study. The presence of HUVECs enhanced the functionality of hepatocytes, which can be related to homotypic and heterotypic cell–cell interactions in such microenvironments [116].

4.6. Cardiovascular Regeneration

Heart engineering using various scaffolds has been investigated as an alternative for managing cardiovascular diseases in the future [117]. Droplet-based encapsulation of cells can provide controllable biocompatible microenvironments to fabricate scaffolds with desirable mechanical properties for cardiovascular tissue engineering. For instance, Gal et al. encapsulated cardiac cells in personalized spherical hydrogels made of human acellular omentum and transplanted them to murine muscle. After transplantation, the encapsulated cells were spread at the injection site, exhibiting striation of actinin and connexin-43 (Figure 5F) [118]. Droplets can also serve as a temporary shield for preserving cells during severe processes such as electrospinning. Weidenbacher et al. fabricated droplets to temporarily protect murine myoblast cells via their encapsulation in gelatin microspheres for the electrospinning process. These microcapsules were electrosprayed onto the surface of nanofibrous polyvinylidene fluoride-co-hexafluoropropylene sheets to make a hybrid scaffold. The encapsulated cells were protected from the toxicity of dimethylformamide during electrospinning [119].

A variety of native biomolecules can be used to enhance cellular attachments. For instance, RGD is an integrin-binding peptide found as an adhesion motif in many ECM proteins—including fibronectin, fibrinogen, and von Willebrand factor (VWF)—commonly used to increase cellular regeneration [120]. Therefore, controlled functionalization of cell-encapsulated microparticles with adhesive peptides using cytocompatible crosslinking agents affords an environment capable of long-term cell viability. Such a microenvironment may be practical for either cell encapsulation or directing stem cell behavior and fate. In another study by Cha et al., a microfluidic flow-focusing device was utilized to fabricate GelMA microgels by photo-crosslinking UV light as a highly promising injectable tissue construct. The variable ratios of flow rates of aqueous and oil phases could control the droplet size. Then, the cardiac population cells were cultured on the GelMA microgel's surface. The adhered cells on the microgel surface proliferated over time, while maintaining high viability (~90%) and migrating to their cell-conductive surrounding areas. Furthermore, a thin biocompatible and biodegradable silica hydrogel was coated on the cell-seeded microgels' surface via the sol–gel method, as a protective shell against peroxide-induced oxidative stress during and after implantation in host tissues. The silica hydrogel shell degraded over time, without affecting cellular activities [121].

The endothelial tissue plays a crucial role in regulating homeostasis in the cardiovascular system. The monolayer of endothelial cells is a kind of biological barrier between blood and tissues; this tissue regulates different immunohematology processes in the body, including secretion of vasomotor and growth factors, and starting mechanisms of inflammation and clotting in vascular injuries [122]. This semi-permeable barrier dynamically regulates the transportation of different elements between blood and the underlying tissue. For measuring endothelial cells' functions in a microenvironment close to in vivo conditions, Crampton et al. generated collagen microparticles using a microfluidics-based droplet system and coated the surfaces with endothelial cells. They found that endothelial cells on the surface of microspheres showed typical morphology and produced tight junction proteins; their results also revealed that these cell layers had a permeability similar to that observed in vivo, and were responsive to modulators of endothelial permeability such as tumor necrosis factor-alpha (TNF-α) and TGF-β [123].

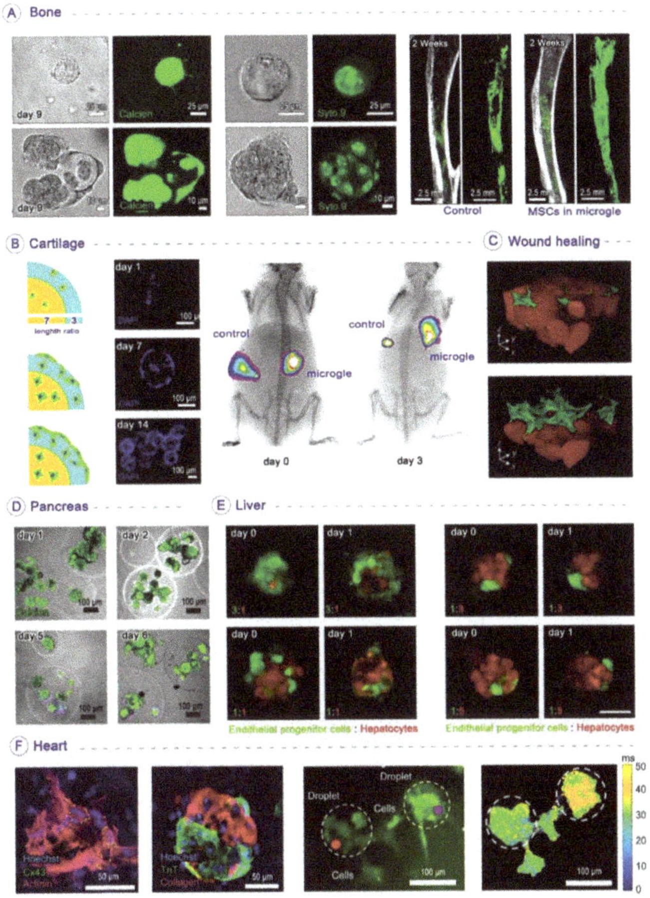

Figure 5. Microfluidics-based droplets for the fabrication of microgels in tissue engineering applications: (**A**) Left: Confocal microscopic images showing MSCs encapsulated in RGD-alginate microgels cultured in vitro for bone regeneration. Due to alginate microgels' stiffness, cells remained spherical even though they proliferated within the matrix. Middle: The confocal microscopic images of MSCs stained by SYTO 9 nuclei staining of the microgels. Right: 2D and 3D reconstructed micro-CT images of a tibia's medullary cavity after two weeks of in vivo transplantation [100]; Copyright, Elsevier. (**B**) Left: In vitro biological characterization of self-assembled BMSC-laden gelatin–hyaluronic acid microgels after 14 days. The gelatin–hyaluronic acid microgels were divided into core and shell parts with length ratios of 7:3. (DAPI-stained cell nuclei). Right: BMSC-laden microgels were hypodermically implanted into the right of the nude mouse, and pure BMSCs were hypodermically implanted into the left of the nude mouse as a control [103]; Copyright, Wiley Online Library. (**C**) Accelerated wound healing by fabrication of injectable microporous gel scaffolds assembled from annealed building blocks [108]; Copyright, Nature. (**D**) Microfluidics-based PEG-4MAL microgels retain the viability and function of encapsulated human islets after microencapsulation [111]; Copyright, Wiley Online Library. (**E**) Microfluidics-based production of microencapsulated functional hepatocyte spheroids with different cell portions [115]; Copyright, Wiley Online Library. (**F**) The morphology and contractility of the encapsulated cardiac cells. Left: Immunostaining of α-sarcomeric actinin (red), connexin-43 (green), and nuclei (blue). Immunostaining of collagen (red), troponin T (green), and nuclei (blue). Middle: In vitro migration and function of encapsulated cardiac cells. Right: Calcium imaging between two adjacent droplets [118]; Copyright, Wiley Online Library.

5. Microfluidics-Based Droplets for Assisted Reproductive Technology

Infertility is an ever-increasing public health concern worldwide, leading to the concern that more than 70 million couples worldwide have experienced infertility issues at least once in their life [124]. A variety of therapeutic methods have been developed to assist patients in the clinic. For instance, the development of mammalian reproductive technologies—most notably ART—has been one of the most outstanding achievements in the treatment of infertility [125]. ART and its different subcategories—including ovulation induction, intrauterine insemination (IUI), and in vitro fertilization (IVF)—provide an extraordinary chance for infertile couples to have offspring. Combining ART with stem cell therapy and tissue engineering approaches—such as scaffold fabrication techniques, cell encapsulation, 3D culture microenvironments, and microfluidics—could lead to the development of artificial reproductive organs, stem-cell-derived gametes, or even human cloning [125–127].

Microfluidic platforms based on single-phase continuous flow have been successfully used in various ART fields, including infertility diagnosis, sperm selection, sperm guidance, oocyte analysis, insemination, embryo culture, embryo selection, and cryopreservation. Recent research interest has been shifted toward microfluidics-based droplets, since they can offer well-controlled environments for the encapsulation of cells and other elements necessary for ART operations inside micro-scale droplets [128]. The fabrication and handling techniques of droplet-based ART systems can be divided into emulsion-based droplet microfluidics and digital microfluidics.

In emulsion-based droplet microfluidics, the droplet can automatically be formed by the interaction of two immiscible fluids, automatically generating individual micro-bioreactors from the cell suspension. For ART applications, this microfluidics category is very similar to what embryologists perform in clinical IVF, such as encapsulating the gamete in a microdroplet covered with an oil overlay. These droplets can also act as platforms for culturing embryos in the early stages. For instance, Agarwal et al. devised a core–shell encapsulation technique for 3D culture of embryonic cells) [70]. Moreover, this technique has been implemented for in vitro culturing of ESCs and ovarian preantral follicles with alginate and type I collagen as shell and core structures, respectively [129]. Moreover, the micro-segmented flow systems can be used for gamete encapsulation and culture. The most relevant work in this field was reported by Funfak et al., who used the microfluid segment technique to encapsulate the eggs of zebrafish until hatching time in a polytetrafluoroethylene (PTFE) tube, with perfluoromethyldecalin (PP9) as the carrier liquid [130].

Digital microfluidics is another promising field for precisely generating and manipulating droplets by implementing noncontact forces such as electricity, magnetism, or heat for ART and IVF applications, such as gamete manipulation and embryo culture. In this regard, two standard digital microfluidics techniques have been used in the literature: electrowetting on dielectric (EWOD), and liquid marbles (LMs). EWOD is a facile method to manipulate discretized droplets on a solid surface in a programmable manner, using electrodes made of a conductor such as indium tin oxide (ITO). In a recent study, murine embryos were encapsulated in a culture medium (3.5 μL) as a core and an oil droplet (1.5 μL) as a shell at the bottom of an EWOD plate. The EWOD technique can also be used for slow-freezing cryopreservation (SFC) and vitrification of cells, tissues, and embryos. For instance, Park et al. showed that EWOD digital microfluidics could be successfully used to select the optimal mixture of cryoprotective agents for efficient SFC [131].

Liquid marbles can also be regarded as another emerging digital microfluidics technique for the straightforward and facile manipulation of water droplets on solid surfaces [132]. A liquid marble can be directly prepared by rolling water droplets on hydrophobic materials such as PTFE or lycopodium and manipulating them with various external forces [133]. Furthermore, they can be used as bioreactors for 3D spheroid formation. For instance, Sarvi et al. demonstrated the feasibility of forming uniform EBs from Oct4B2-ESCs within a three-day culture inside liquid marbles; they also showed that these microenvironments could facilitate the differentiation of embryo bodies into cardiomyocytes. Furthermore, these droplets can be implemented as an efficient micro-bioreactor for in vitro oocyte maturation in ART [134].

6. Microfluidics-Based Droplets for Gene Therapy/Delivery

Gene therapy/delivery is considered one of the most promising methods in advanced regenerative medicine [135]. Mutated or missing genes result in the expression of dysfunctional or abnormal intra- and extracellular proteins that eventually lead to many diseases. As a valuable method, gene therapy was developed to correct genetic disorders by inserting genetic materials into cells. The exogenous genetic materials must be delivered across the cell membrane without any influences on the ability of cells to integrate successfully into the innate cell genome [136]. Regarding the continuous lipid bilayer structure of the spherical cell membrane, the transfer of foreign genetic materials through the plasma membrane is a stumbling block in gene therapy. Several physical techniques—such as sonoporation and electroporation—create temporary pores in the cell membrane using ultrasound and electric pulses. Although electroporation is the most popular method of intracellular delivery, the efficiency of sonoporation systems increases when attached to microbubbles. Concerning the downsides of conventional gene transfection systems—such as low transfection efficiency, intricate procedure, and low cell viability—the advent of microfluidics has opened up new avenues for gene delivery [21].

Gene delivery through microfluidic systems is subject to high precision and outstanding control that can finally cause the efficient generation of various vectors and other components used in gene therapy. Several genes in bulk solution have been explored by implementing a microfluidic environment combined conventionally with an electric pulse, optical energy, and hydrodynamic force. Moreover, most microfluidics-based electroporation systems require much lower voltages for gene transport, resulting in higher efficiency and cell viability than traditional electroporation methods [137]. Several microfluidic geometries, diverse materials, electrodes, and microfluidic channels have been reported to increase the insertion of gene materials into cells via electroporation. However, further development for increased efficiency is required.

Recently, new methods with a combination of microfluidics-derived droplets and conventional gene transfection techniques have emerged, with the ability to enhance the accuracy of transportation of exogenous genetic factors at a single-cell level [138]. For instance, using these methods, Zhan et al. delivered an enhanced green fluorescent protein (EGFP) plasmid vector into Chinese hamster ovary (CHO) cells; they encapsulated CHO and EGFP plasmids inside droplets, which then passed through a pair of microelectrodes with a constant voltage. The electroporated cells' GFP expression proved high-throughput insertion of functional genetic materials based on droplet microfluidics (Figure 6A) [139]. In another study, nanocomplexes were synthesized using microfluidics-assisted confinement (MAC) for non-viral vector delivery of genes. A complicated microfluidics-based droplet generator was applied to introduce various materials—such as cationic gene carriers, plasmid DNA, buffers, and oils—into each channel to generate monodisperse droplets. To form DNA nanocomplexes, the pDNA and polycation solutions were confined to discrete droplets and successfully self-assembled electrostatically (Figure 6B) [140]. Upon incubation with human embryonic kidney cells (HEK293), these homogeneous MAC-generated nanocomplexes exhibited lower cytotoxicity and higher transfection efficiency than their bulk-synthesized counterparts. To further explore the merits of the MAC system for non-viral gene transfer, another experiment in 2013 revealed that the emulsion-based droplets' microfluidic environment offers greater control of the preparation of polyplexes to generate more constant potential gene delivery systems (Figure 6C) [141]. By operating such a system, both plasmid DNA and messenger RNA payloads are consistently delivered into primary cells, stem cells, and human cell lines. The cellular unpacking of polymer–DNA nanocomplexes quantified by a flow cytometric quantum dot Förster resonance energy transfer (QD-FRET) nanosensor system demonstrated that transfection was significantly enhanced in a broad series of cell types in terms of both uncultured transgene expression and the number of cells transfected. Yang et al. also demonstrated that polymer-based nanoparticles using the dielectrophoresis (DEP) effect could be generated inside droplets within a roughly nanoliter-scale volume. They used a sort of nanomaterial for tumor-targeted gene therapy—PEI600-CyD-FA (H1)—and then mixed it with DNA plasmids to produce polyplex products that can be self-assembled. These nano-scale polymer-based particles were examined in HUVECs, and the results demonstrated that the gene transfection efficiency increased fourfold compared to the control group [142].

The picoliter microfluidic reactor and incubator (PMRI) system is the other droplet-based gene transfection method that can control the cationic lipid and DNA complex (CL-DNA) by tuning of the mixture incubation time and the order and rate of mixing of vectors. Hsieh et al. used human osteosarcoma U2OS cells (ATCC) with the pEGFP-C1 DNA vector to characterize the transfection efficiency of the prepared CL-DNA via the PMRI technique (Figure 6D); their results demonstrate that PMRI consistently mixes cationic lipids and DNA simultaneously to create a narrower lipoplex scattering. The latter, however, makes it essential to recognize the optimal lipoplex size for extreme transfection efficiency [143]. Droplet-based microfluidics may also provide a platform for viral vector delivery. In a study reported by Madrigal et al., alginate and LentiVector were incorporated into microfluidic technology to create LentiVector-loaded microgels using different gelation methods. Comparisons of three alginate gelation strategies revealed that internal gelation with $CaCO_3$/D-glucono-δ-lactone (GDL) and external gelation with $CaCl_2$ are suitable for creating LentiVector-compatible microgels. In contrast, alginate gelation with chelated calcium confirmed low utility for gene delivery due to a loss of LentiVector function under acidic gelation conditions (Figure 6E). These results demonstrate alginate microgels' capacity to successfully encapsulate and release functional LentiVector for gene delivery in vitro [144].

In conclusion, microfluidics-assisted gene transfection systems have achieved higher transfection efficiency and cell viability than conventional gene transfection processes. The microfluidics-based droplet environment allows for better spatiotemporal control over the environment. The strength and duration of the transfection stimulus eliminate the randomness of traditional transfection methods in the target cells. A further improvement would cause applied droplet-based microfluidics technology for long-term single-cell culturing, homogenous transfection, and single-cell sorting to replace the complicated operation procedures involved in gene transfection in conventional benchtop systems.

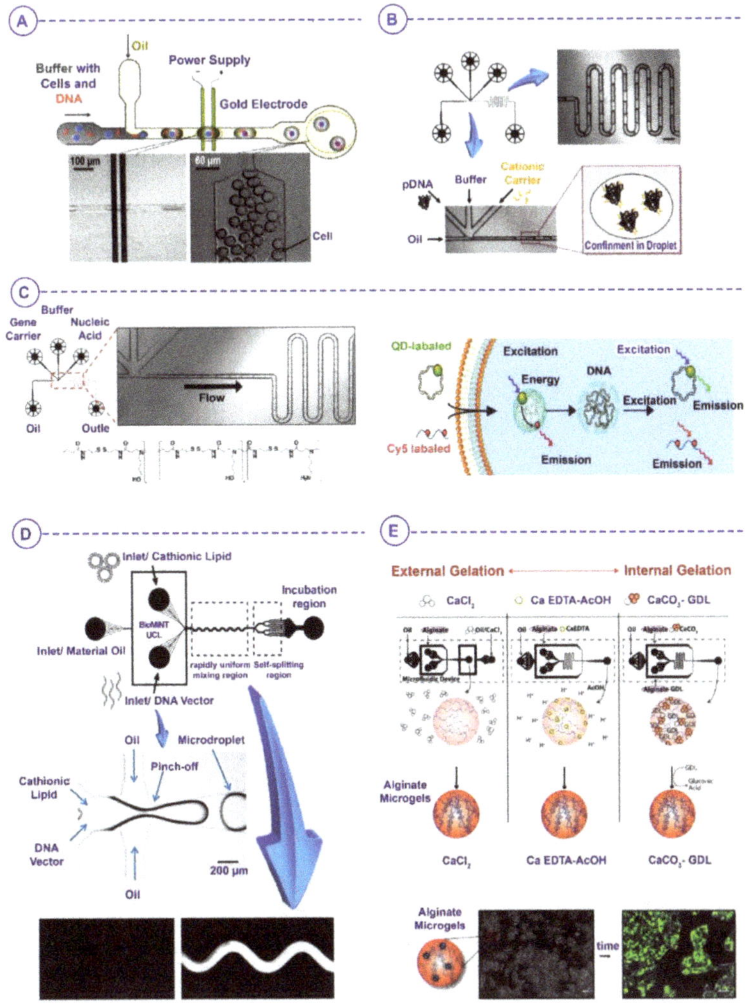

Figure 6. Microfluidics-based droplets for gene therapy/delivery: (**A**) Design of a microfluidic device for electroporation. Cell- and DNA-containing droplets rapidly flow through the two microelectrodes on the substrate and droplets with encapsulated cells after electroporation at the device's exit [139]; Copyright 2009, Wiley Online Library. (**B**) Design of microfluidics-assisted self-assembly for fabricating picoliter-sized droplets. The plasmid DNA, buffer, cationic gene carrier, and oil are introduced into each channel to generate monodisperse water-in-oil droplets (scale bar: 200 μm) [140]; Copyright, American Chemical Society. (**C**) Design of a microfluidic chip for non-viral transport of genes, as well as the cellular internalization and intracellular unpacking of genes. The microfluidic system uses crossflow geometry to produce emulsified aqueous droplets containing the polymeric gene carrier and nucleic acids [141]; Copyright, Nature. (**D**) Design of the picoliter microfluidic reactor and incubator system for controlled formulation of non-viral vectors (CL-DNA) [143]; Copyright 2009, The Royal Society of Chemistry. (**E**) Representative phase-contrast photomicrographs of alginate microgels fabricated using three different gelation techniques [144]; Copyright, The Royal Society of Chemistry.

7. Conclusions and Perspectives

The translation of regenerative medicine to clinical practice is highly dependent on advanced technologies, manufacturing methods, clinical trial success, and commercial support. Microfluidics-based droplets are a promising technology that enables the simultaneous encapsulation of various stem cells, biological agents, drugs, and nanoparticles inside specifically designed droplets for advancing integrated aspects of regenerative medicine. This review article evaluated the state-of-the-art microfluidics-based strategies for encapsulating stem cells inside droplets for stem cell therapy, tissue engineering, reproductive biology, and gene therapy (Table 1).

In droplet-based stem cell therapy, preserving the viability of stem cells during the manufacturing process is a challenge that can be addressed by implementing nontoxic biomaterials, increasing automation, and reducing manufacturing time. The challenges in stem-cell-based therapy—such as the risk of tumorigenicity and immunogenicity—have not been widely investigated enough (during in vivo and clinical trial studies) for droplets carrying stem cells, spheroids, and organoids. Furthermore, there are no established standards or procedures for droplet-based stem cell therapy. Recent studies show that the regeneration caused by stem cells is mainly caused by secretome proteins, exosomes, and EVs generated by transplanted stem cells [145]. This knowledge can be implemented to generate secretome-enriched droplets or droplet-derived microgels for cell-free regenerative medicine, thus reducing the possible side effects caused by the transplantation of stem cells.

The human body's tissues, organs, and systems are developed based on a complicated innate bottom-up organization [146]. It is essential to mimic the target tissue's conditions and build repair-tissue-like building blocks in the same manner in order to regenerate the tissues. Despite the several reports and advances in this area, the scaffolds and microgels made by droplets still cannot mimic many aspects of tissues, such as transport of fluids, transport of electrophysiological signals, angiogenesis, and neovascularization. The available polymer options are limited to some hydrogels for safely encapsulating stem cells via this approach, and need to be expanded and evaluated in future studies [76]. Moreover, scaffolds' mechanical strength plays an essential role in tissue engineering, while the natural hydrogels used in reported microgels have weak mechanical properties that are not desirable for hard tissues [92]. The possible adverse effects of crosslinkers, oils, and mechanical forces during the encapsulation process and injection should be investigated during long-term in vitro and in vivo studies. In addition to the fact that the retention of the cells in the tissue is a critical challenge, the retention of cell-laden microgels and their integration and unification with the damaged tissue after injection is also essential, and can be addressed using surface modification methods of microgels for self-assembly purposes [140]. The number of in vitro and in vivo studies implementing droplet-derived microgels for tissue engineering is considerably fewer than those using conventional scaffolds. In reproductive biology, the automation of droplet-based systems and improvement of digital microfluidic systems with programmable hardware is essential for removing the human errors during complicated droplet formation and manipulation procedures, gamete isolation, manipulation, culture, fertilization, and embryo culture [131]. The high sensitivity, accuracy, and control in delivering specific genes to the individual target cells inside droplets using advanced operations and instruments are still highly expensive, and the cost needs to be reduced for commercialization purposes. The studies related to these fields should also be expanded to in vivo and clinical trials.

Here, the applications of microfluidics-based droplets in the fields of stem cell therapies, tissue engineering, reproductive biology, and gene therapy have been reviewed and discussed. Despite the considerable achievements in droplet-based regenerative medicine, the field is expected to develop significantly in the future, removing the gaps between lab and clinic in each area. In the first step, for exploiting these droplets or their products in clinical procedures, significant issues should be considered, including automation and medical costs, preparing hygienic conditions, clean rooms, advanced manufacturing instruments, and Food and Drug Administration (FDA)-approved materials, and following regulatory necessities and ethical issues. Soon, droplets will be able to revive the dream of preparing off-the-shelf products for curing patients with mild or severe dysfunctionalities.

Table 1. Applications of droplet-based microfluidics for stem cell therapy, tissue engineering, assisted reproductive biology, and gene therapy.

Field	Application	Generation Technique	Droplet Size	Morphology/ Materials	Cell Type	Ref
Stem Cell Therapy	Stem cell culture	Flow-focusing; two-layer parallel system	64–65 µm	Microspheres	MSCs	[57]
		Air-driven droplet generator	400–500 µm	Alginate microcapsules	hESCs	[63]
	Spheroid culture	Flow-focusing	65–90 µm 150–210 µm	Alginate and alginate/RGD	hMSCs; spheroids	[69]
		Flow-focusing	190–260 µm	Alginate microcapsules	Ambryonic bodies	[70]
		Flow-focusing: - Negative pressure-driven - Micropipette-based	290 µm	Droplets	HEK293; RT4 cells; A-431 cells	[71]
		Flow-focusing	-	Droplets	Primary rat hepatocytes	[72]
	Organoid culture	Flow-focusing	169 ± 6 µm	Alginate capsules	Hepatocytes fibroblasts	[73]
		Airflow-assisted 3D bioprinting; flow-focusing	-	Spiral alginate: - Spherical - Rose-like - Tai-chi-like	HUVECs; hMSCs	[74]
	High-throughput analysis	Flow-focusing	-	Droplets for RNA sequencing	ESCs	[85]
		Flow-focusing	10–150 µm	Droplets for ELISA	Hybridoma cells	[87]
Tissue Engineering	Bone	Flow-focusing	427 µm	RGD; alginate	PDLSCs; GMSCs	[97]
		Capillary-based	90–230 µm	BMP-2; GelMA	BMSCs	[98]
		Flow-focusing	100–200 µm	BMP-2; PVA	hMSCs	[99]
		Flow-focusing	61.2–50 µm	Alginate	Single MSCs	[100]
	Cartilage	Capillary-based	320 ± 9 µm 574 ± 9 µm	Gelatine norbornene; PEG	BMSCs	[102]
		T-junction with Y-shaped inlets	-	Gelatine; hyaluronic acid	BMSCs	[103]
		Flow-focusing	10–30 µm	TGF-β3/PLGA droplets for PCL fibres	hMSCs	[104]
	Wound	Flow-focusing	20, 40, 50 µm	Alginate core, cellulose shell	Gluconacetobacter xylinus PC-9 cells	[107]
		Flow-focusing	30–150 µm	4-arm PEG vinyl	HDF; hMSCs	[108]
	Pancreas	Coaxial electrojetting	500 µm	Core–shell alginate	Pancreatic islets	[110]
		Flow-focusing	300–800 µm	PEG-4MAL	Human pancreatic islets	[111]
		Eight flow-focusing orifices	90.4 ± 3.0 µm	Alginate	β-TC6	[112]
	Liver	Flow-focusing	≈200 µm	Alginate	Hepatocytes; endothelial cells	[115]
		Flow-focusing	289.7 ± 8.3 µm	Methylcellulose GelMA	HepG2; HUVECs	[116]
	Cardiac /vascular	Flow-focusing	100–200 µm	Acellular omentum	iPSCs-derived cardiac cells	[118]
		Flow-focusing	40–140 µm	Gelatine	Murine myoblast cells	[119]
		Flow-focusing	35–150 µm	GelMA; silica hydrogel	Cardiac cells	[121]
		Flow-focusing	300 µm	Collagen	Endothelial cells	[123]

Table 1. *Cont.*

Field	Application	Generation Technique	Droplet Size	Morphology/ Materials	Cell Type	Ref
Assisted Reproductive Technology		Flow-focusing	380 μm 285 μm	Core–shell collagen alginate microcapsules	Ovarian preantral follicles; ESCs	[129]
		-	1.2 mm	PP9	Zebrafish eggs	[130]
		Electrowetting on dielectric (EWOD)	-	f DMSO-PBS mixture	MCF-7	[134]
Gene Therapy/ Delivery	Electroporation	T-junction	92 μm	EGFP plasmid vector droplets	Chinese hamster ovary cells	[139]
	Microfluidics-assisted confinement	Crossflow	0.5–1 μm	Plasmid DNA droplets	-	[140]
	Gene transfer	Crossflow	-	Plasmid DNA and messenger RNA droplets	Primary cells; stem cells; human cell lines	[141]
	Gene delivery	T-junction	-	PEI600-CyD-FA (H1) DNA plasmids; polyplexes; droplets	HUVECs	[142]
	Gene transfection	Flow-focusing	39 μm	pEGFP-C1 DNA vector	Human U2OS cells	[143]
	Viral gene transfection	Flow-focusing	100–130 μm	Alginate; gelation; LentiVector	HEK-293T	[144]
	CRISPR/Cas9 transfection	Flow-focusing	80–87 μm		hiPSCs	[138]

Funding: This research received no external funding.

Institutional Review Board Statement: Not applicable.

Informed Consent Statement: Not applicable.

Conflicts of Interest: The authors declare no conflict of interest.

References

1. Mason, C.; Dunnill, P. *A Brief Definition of Regenerative Medicine*; Future Medicine Ltd.: London, UK, 2008.
2. Jing, H.; He, X.; Zheng, J. Exosomes and regenerative medicine: State of the art and perspectives. *Transl. Res.* **2018**, *196*, 1–16. [CrossRef] [PubMed]
3. Bianco, P.; Robey, P.G. Stem cells in tissue engineering. *Nature* **2001**, *414*, 118–121. [CrossRef] [PubMed]
4. Hubbell, J.A. Biomaterials in tissue engineering. *Biotechnology* **1995**, *13*, 565–576. [CrossRef] [PubMed]
5. Hasan, A.; Morshed, M.; Memic, A.; Hassan, S.; Webster, T.J.; Marei, H.E.-S. Nanoparticles in tissue engineering: Applications, challenges and prospects. *Int. J. Nanomed.* **2018**, *13*, 5637. [CrossRef]
6. Ruiz, M.M.; Regueiro, J.R. New tools in regenerative medicine: Gene therapy. In *Stem Cell Transplantation*; Springer: Berlin/Heidelberg, Germany, 2012; pp. 254–275.
7. Lund, A.W.; Yener, B.; Stegemann, J.P.; Plopper, G.E. The natural and engineered 3D microenvironment as a regulatory cue during stem cell fate determination. *Tissue Eng. Part B Rev.* **2009**, *15*, 371–380. [CrossRef]
8. Giobbe, G.G.; Zagallo, M.; Riello, M.; Serena, E.; Masi, G.; Barzon, L.; Di Camillo, B.; Elvassore, N. Confined 3D microenvironment regulates early differentiation in human pluripotent stem cells. *Biotechnol. Bioeng.* **2012**, *109*, 3119–3132. [CrossRef]
9. Orellana, L.A.C.; Baret, J.-C. Rapid stabilization of droplets by particles in microfluidics: Role of droplet formation. *arXiv* **2019**, arXiv:1906.08517.
10. Guzey, D.; McClements, D.J. Formation, stability and properties of multilayer emulsions for application in the food industry. *Adv. Colloid Interface Sci.* **2006**, *128*, 227–248. [CrossRef]
11. Xu, S.; Nie, Z.; Seo, M.; Lewis, P.; Kumacheva, E.; Stone, H.A.; Garstecki, P.; Weibel, D.B.; Gitlin, I.; Whitesides, G.M. Generation of monodisperse particles by using microfluidics: Control over size, shape, and composition. *Angew. Chem. Int. Ed.* **2005**, *117*, 734–738. [CrossRef]
12. Mashaghi, S.; Abbaspourrad, A.; Weitz, D.A.; van Oijen, A.M. Droplet microfluidics: A tool for biology, chemistry and nanotechnology. *TrAC Trends Anal. Chem.* **2016**, *82*, 118–125. [CrossRef]
13. Yin, H.; Marshall, D. Microfluidics for single cell analysis. *Curr. Opin. Biotechnol.* **2012**, *23*, 110–119. [CrossRef]

14. Allazetta, S.; Lutolf, M.P. Stem cell niche engineering through droplet microfluidics. *Curr. Opin. Biotechnol.* **2015**, *35*, 86–93. [CrossRef]
15. Guan, Z.-Y.; Chen, L.-Y.; Li, X.-L.; Cui, Y.-R.; Liu, R.-H. Molecular mechanism of quercitrin on osteogenic differentiation and adipogenic differentiation of rat bone marrow stromal stem cells (rBMSCs). *Chin. Herb. Med.* **2018**, *10*, 184–190. [CrossRef]
16. Au, S.H.; Chamberlain, M.D.; Mahesh, S.; Sefton, M.V.; Wheeler, A.R. Hepatic organoids for microfluidic drug screening. *Lab Chip* **2014**, *14*, 3290–3299. [CrossRef]
17. Joensson, H.N.; Andersson Svahn, H. Droplet microfluidics—A tool for single-cell analysis. *Angew. Chem. Int. Ed.* **2012**, *51*, 12176–12192. [CrossRef]
18. Lun, A.T.; Riesenfeld, S.; Andrews, T.; Gomes, T.; Marioni, J.C. EmptyDrops: Distinguishing cells from empty droplets in droplet-based single-cell RNA sequencing data. *Genome Biol.* **2019**, *20*, 1–9. [CrossRef]
19. Cai, Y.; Wu, F.; Yu, Y.; Liu, Y.; Shao, C.; Gu, H.; Li, M.; Zhao, Y. Porous scaffolds from droplet microfluidics for prevention of intrauterine adhesion. *Acta Biomater.* **2019**, *84*, 222–230. [CrossRef]
20. Huang, H.-Y.; Shen, H.-H.; Chung, L.-Y.; Chung, Y.-H.; Chen, C.-C.; Hsu, C.-H.; Fan, S.-K.; Yao, D.-J. Fertilization of mouse gametes in vitro using a digital microfluidic system. *IEEE Trans. NanoBiosci.* **2015**, *14*, 857–863. [CrossRef]
21. Kim, J.; Hwang, I.; Britain, D.; Chung, T.D.; Sun, Y.; Kim, D.-H. Microfluidic approaches for gene delivery and gene therapy. *Lab Chip* **2011**, *11*, 3941–3948. [CrossRef]
22. Whitesides, G.M. The origins and the future of microfluidics. *Nature* **2006**, *442*, 368–373. [CrossRef]
23. Razavi Bazaz, S.; Mashhadian, A.; Ehsani, A.; Saha, S.C.; Krüger, T.; Ebrahimi Warkiani, M. Computational inertial microfluidics: A review. *Lab Chip* **2020**, *20*, 1023–1048. [CrossRef]
24. Mahmoudi, Z.; Mohammadnejad, J.; Razavi Bazaz, S.; Abouei Mehrizi, A.; Saidijam, M.; Dinarvand, R.; Ebrahimi Warkiani, M.; Soleimani, M. Promoted chondrogenesis of hMCSs with controlled release of TGF-β3 via microfluidics synthesized alginate nanogels. *Carbohydr. Polym.* **2019**, *229*, 115551. [CrossRef]
25. Garstecki, P.; Gitlin, I.; DiLuzio, W.; Whitesides, G.M.; Kumacheva, E.; Stone, H.A. Formation of monodisperse bubbles in a microfluidic flow-focusing device. *Appl. Phys. Lett.* **2004**, *85*, 2649–2651. [CrossRef]
26. Tan, Y.-C.; Fisher, J.S.; Lee, A.I.; Cristini, V.; Lee, A.P. Design of microfluidic channel geometries for the control of droplet volume, chemical concentration, and sorting. *Lab Chip* **2004**, *4*, 292–298. [CrossRef]
27. Teh, S.-Y.; Lin, R.; Hung, L.-H.; Lee, A.P. Droplet microfluidics. *Lab Chip* **2008**, *8*, 198–220. [CrossRef]
28. Mugele, F.; Baret, J.-C. Electrowetting: From basics to applications. *J. Condens. Matter Phys.* **2005**, *17*, R705. [CrossRef]
29. Zhu, P.; Wang, L. Passive and active droplet generation with microfluidics: A review. *Lab Chip* **2017**, *17*, 34–75. [CrossRef]
30. Christopher, G.F.; Anna, S.L. Microfluidic methods for generating continuous droplet streams. *J. Phys. D Appl. Phys.* **2007**, *40*, R319–R336. [CrossRef]
31. Liu, J.; Tan, S.-H.; Yap, Y.F.; Ng, M.Y.; Nguyen, N.-T. Numerical and experimental investigations of the formation process of ferrofluid droplets. *Microfluid. Nanofluid.* **2011**, *11*, 177–187. [CrossRef]
32. Collins, D.J.; Alan, T.; Helmerson, K.; Neild, A. Surface acoustic waves for on-demand production of picoliter droplets and particle encapsulation. *Lab Chip* **2013**, *13*, 3225–3231. [CrossRef]
33. Shang, L.; Cheng, Y.; Zhao, Y. Emerging Droplet Microfluidics. *Chem. Rev.* **2017**, *117*, 7964–8040. [CrossRef] [PubMed]
34. Lashkaripour, A.; Mehrizi, A.A.; Goharimanesh, M.; Rasouli, M.; Bazaz, S.R. Size-controlled droplet generation in a microfluidic device for rare dna amplification by optimizing its effective parameters. *J. Mech. Med. Biol.* **2018**, *18*, 1850002. [CrossRef]
35. Gutiérrez, J.; González, C.; Maestro, A.; Solè, I.; Pey, C.; Nolla, J. Nano-emulsions: New applications and optimization of their preparation. *Curr. Opin. Colloid Interface Sci.* **2008**, *13*, 245–251. [CrossRef]
36. McClements, D.; Decker, E.; Weiss, J. Emulsion-based delivery systems for lipophilic bioactive components. *J. Food Sci.* **2007**, *72*, R109–R124. [CrossRef]
37. Clausell-Tormos, J.; Lieber, D.; Baret, J.C.; El-Harrak, A.; Miller, O.J.; Frenz, L.; Blouwolff, J.; Humphry, K.J.; Koster, S.; Duan, H.; et al. Droplet-based microfluidic platforms for the encapsulation and screening of Mammalian cells and multicellular organisms. *Chem. Biol.* **2008**, *15*, 427–437. [CrossRef]
38. Muschiolik, G. Multiple emulsions for food use. *Curr. Opin. Colloid Interface Sci.* **2007**, *12*, 213–220. [CrossRef]
39. Tirnaksiz, F.; Kalsin, O. A topical $w/o/w$ multiple emulsions prepared with Tetronic 908 as a hydrophilic surfactant: Formulation, characterization and release study. *J. Pharm. Pharm. Sci.* **2005**, *8*, 299–315.
40. Chakraborty, M.; Ivanova-Mitseva, P.; Bart, H.J. Selective Separation of Toluene from n-Heptane via Emulsion Liquid Membranes Containing Substituted Cyclodextrins as Carrier. *Sep. Sci. Technol.* **2006**, *41*, 3539–3552. [CrossRef]
41. Kim, S.U.; De Vellis, J. Stem cell-based cell therapy in neurological diseases: A review. *J. Neurosci. Res.* **2009**, *87*, 2183–2200. [CrossRef]
42. Trounson, A.; McDonald, C. Stem cell therapies in clinical trials: Progress and challenges. *Cell Stem Cell* **2015**, *17*, 11–22. [CrossRef]
43. Zakrzewski, W.; Dobrzyński, M.; Szymonowicz, M.; Rybak, Z. Stem cells: Past, present, and future. *Stem Cell Res. Ther.* **2019**, *10*, 1–22. [CrossRef]
44. Strauer, B.E.; Kornowski, R. Stem cell therapy in perspective. *Circulation* **2003**, *107*, 929–934. [CrossRef]
45. Weissman, I.L. Stem cells: Units of development, units of regeneration, and units in evolution. *Cell* **2000**, *100*, 157–168. [CrossRef]
46. Discher, D.E.; Mooney, D.J.; Zandstra, P.W. Growth factors, matrices, and forces combine and control stem cells. *Science* **2009**, *324*, 1673–1677. [CrossRef]

47. Karp, J.M.; Teo, G.S.L. Mesenchymal stem cell homing: The devil is in the details. *Cell Stem Cell* **2009**, *4*, 206–216. [CrossRef]
48. Bidkhori, H.R.; Ahmadiankia, N.; Matin, M.M.; Heirani-Tabasi, A.; Farshchian, M.; Naderi-Meshkin, H.; Shahriyari, M.; Dastpak, M.; Bahrami, A.R. Chemically primed bone-marrow derived mesenchymal stem cells show enhanced expression of chemokine receptors contributed to their migration capability. *Iran. J. Basic Med. Sci.* **2016**, *19*, 14.
49. Andersen, T.; Auk-Emblem, P.; Dornish, M. 3D cell culture in alginate hydrogels. *Microarrays* **2015**, *4*, 133–161. [CrossRef]
50. Lee, Y.B.; Kim, E.M.; Byun, H.; Chang, H.-K.; Jeong, K.; Aman, Z.M.; Choi, Y.S.; Park, J.; Shin, H. Engineering spheroids potentiating cell-cell and cell-ECM interactions by self-assembly of stem cell microlayer. *Biomaterials* **2018**, *165*, 105–120. [CrossRef]
51. Cesarz, Z.; Tamama, K. Spheroid culture of mesenchymal stem cells. *Stem Cells Int.* **2016**, *2016*, 9176357. [CrossRef]
52. Chamberlain, G.; Fox, J.; Ashton, B.; Middleton, J. Concise review: Mesenchymal stem cells: Their phenotype, differentiation capacity, immunological features, and potential for homing. *Stem Cells* **2007**, *25*, 2739–2749. [CrossRef]
53. Murphy, M.B.; Moncivais, K.; Caplan, A.I. Mesenchymal stem cells: Environmentally responsive therapeutics for regenerative medicine. *Exp. Mol. Med.* **2013**, *45*, e54. [CrossRef]
54. Velasco, D.; Tumarkin, E.; Kumacheva, E. Microfluidic encapsulation of cells in polymer microgels. *Small* **2012**, *8*, 1633–1642. [CrossRef]
55. Seiffert, S. Microgel capsules tailored by droplet-based microfluidics. *ChemPhysChem* **2013**, *14*, 295–304. [CrossRef]
56. Nemir, S.; Hayenga, H.N.; West, J.L. PEGDA hydrogels with patterned elasticity: Novel tools for the study of cell response to substrate rigidity. *Biotechnol. Bioeng.* **2010**, *105*, 636–644. [CrossRef]
57. Headen, D.M.; García, J.R.; García, A.J. Parallel droplet microfluidics for high throughput cell encapsulation and synthetic microgel generation. *Microsyst. Nanoeng.* **2018**, *4*, 1–9. [CrossRef]
58. Neufeld, G.; Cohen, T.; Gengrinovitch, S.; Poltorak, Z. Vascular endothelial growth factor (VEGF) and its receptors. *FASEB J.* **1999**, *13*, 9–22. [CrossRef]
59. Park, J.S.; Chu, J.S.; Tsou, A.D.; Diop, R.; Tang, Z.; Wang, A.; Li, S. The effect of matrix stiffness on the differentiation of mesenchymal stem cells in response to TGF-β. *Biomaterials* **2011**, *32*, 3921–3930. [CrossRef]
60. García, J.R.; Clark, A.Y.; García, A.J. Integrin-specific hydrogels functionalized with VEGF for vascularization and bone regeneration of critical-size bone defects. *J. Biomed. Mater. Res. A* **2016**, *104*, 889–900. [CrossRef]
61. Manley, N.C.; Priest, C.A.; Denham, J.; Wirth III, E.D.; Lebkowski, J.S. Human embryonic stem cell-derived oligodendrocyte progenitor cells: Preclinical efficacy and safety in cervical spinal cord injury. *Stem Cells Transl. Med.* **2017**, *6*, 1917–1929. [CrossRef]
62. Dawson, E.; Mapili, G.; Erickson, K.; Taqvi, S.; Roy, K. Biomaterials for stem cell differentiation. *Adv. Drug Deliv. Rev.* **2008**, *60*, 215–228. [CrossRef]
63. Chayosumrit, M.; Tuch, B.; Sidhu, K. Alginate microcapsule for propagation and directed differentiation of hESCs to definitive endoderm. *Biomaterials* **2010**, *31*, 505–514. [CrossRef] [PubMed]
64. Kleine-Brüggeney, H.; Van Vliet, L.D.; Mulas, C.; Gielen, F.; Agley, C.C.; Silva, J.C.; Smith, A.; Chalut, K.; Hollfelder, F. Long-Term Perfusion Culture of Monoclonal Embryonic Stem Cells in 3D Hydrogel Beads for Continuous Optical Analysis of Differentiation. *Small* **2019**, *15*, 1804576. [CrossRef] [PubMed]
65. Tsai, A.-C.; Liu, Y.; Yuan, X.; Ma, T. Compaction, fusion, and functional activation of three-dimensional human mesenchymal stem cell aggregate. *Tissue Eng. Part A* **2015**, *21*, 1705–1719. [CrossRef] [PubMed]
66. Utech, S.; Prodanovic, R.; Mao, A.S.; Ostafe, R.; Mooney, D.J.; Weitz, D.A. Microfluidic generation of monodisperse, structurally homogeneous alginate microgels for cell encapsulation and 3D cell culture. *Adv. Healthc. Mater* **2015**, *4*, 1628–1633. [CrossRef]
67. Lin, R.Z.; Chang, H.Y. Recent advances in three-dimensional multicellular spheroid culture for biomedical research. *Biotechnol. J. Healthc. Nutr. Technol.* **2008**, *3*, 1172–1184.
68. Bosenberg, M.W.; Massagué, J. Juxtacrine cell signaling molecules. *Curr. Opin. Cell Biol.* **1993**, *5*, 832–838. [CrossRef]
69. Chan, H.F.; Zhang, Y.; Ho, Y.-P.; Chiu, Y.-L.; Jung, Y.; Leong, K.W. Rapid formation of multicellular spheroids in double-emulsion droplets with controllable microenvironment. *Sci. Rep.* **2013**, *3*, 3462. [CrossRef]
70. Agarwal, P.; Zhao, S.; Bielecki, P.; Rao, W.; Choi, J.K.; Zhao, Y.; Yu, J.; Zhang, W.; He, X. One-step microfluidic generation of pre-hatching embryo-like core-shell microcapsules for miniaturized 3D culture of pluripotent stem cells. *Lab Chip* **2013**, *13*, 4525–4533. [CrossRef]
71. Langer, K.; Joensson, H.N. Rapid production and recovery of cell spheroids by automated droplet microfluidics. *Slas Technol. Transl. Life Sci. Innov.* **2020**, *25*, 111–122. [CrossRef]
72. Cedillo-Alcántar, D.F.; Stybayeva, G.; de Hoyos-Vega, J.M.; Han, Y.D.; Choi, J.; García-Cordero, J.L.; Revzin, A. A Microfluidic System Combining Valve Automation and Spheroid Cultures to Characterize Hepatic Glucose Metabolism during Hormonal Stimulation. In Proceedings of the 2019 20th International Conference on Solid-State Sensors, Actuators and Microsystems & Eurosensors XXXIII (TRANSDUCERS & EUROSENSORS XXXIII), Berlin, Germany, 23–27 June 2019; pp. 488–491.
73. Chen, Q.; Utech, S.; Chen, D.; Prodanovic, R.; Lin, J.-M.; Weitz, D.A. Controlled assembly of heterotypic cells in a core–shell scaffold: Organ in a droplet. *Lab Chip* **2016**, *16*, 1346–1349. [CrossRef]
74. Zhao, H.; Chen, Y.; Shao, L.; Xie, M.; Nie, J.; Qiu, J.; Zhao, P.; Ramezani, H.; Fu, J.; Ouyang, H. Airflow-assisted 3D bioprinting of human heterogeneous microspheroidal organoids with microfluidic nozzle. *Small* **2018**, *14*, 1802630. [CrossRef]
75. Takeo, M.; Tsuji, T. Organ regeneration based on developmental biology: Past and future. *Curr. Opin. Genet. Dev.* **2018**, *52*, 42–47. [CrossRef]

76. Zhang, L.; Chen, K.; Zhang, H.; Pang, B.; Choi, C.H.; Mao, A.S.; Liao, H.; Utech, S.; Mooney, D.J.; Wang, H. Microfluidic templated multicompartment microgels for 3D encapsulation and pairing of single cells. *Small* **2018**, *14*, 1702955. [CrossRef]
77. White, A.; Heyries, K.; Doolin, C.; Vaninsberghe, M.; Hansen, C. High-throughput microfluidic single-cell digital polymerase chain reaction. *Anal. Chem.* **2013**, *85*, 7182–7190. [CrossRef]
78. Lan, F.; Demaree, B.; Ahmed, N.; Abate, A.R. Single-cell genome sequencing at ultra-high-throughput with microfluidic droplet barcoding. *Nat. Biotechnol.* **2017**, *35*, 640. [CrossRef]
79. Song, H.; Ismagilov, R.F. Millisecond kinetics on a microfluidic chip using nanoliters of reagents. *J. Am. Chem. Soc.* **2003**, *125*, 14613–14619. [CrossRef]
80. Huebner, A.; Olguin, L.F.; Bratton, D.; Whyte, G.; Huck, W.T.; De Mello, A.J.; Edel, J.B.; Abell, C.; Hollfelder, F. Development of quantitative cell-based enzyme assays in microdroplets. *Anal. Chem.* **2008**, *80*, 3890–3896. [CrossRef]
81. Zheng, B.; Roach, L.S.; Ismagilov, R.F. Screening of protein crystallization conditions on a microfluidic chip using nanoliter-size droplets. *J. Am. Chem. Soc.* **2003**, *125*, 11170–11171. [CrossRef]
82. Yadav, M.K.; Gerdts, C.J.; Sanishvili, R.; Smith, W.W.; Roach, L.S.; Ismagilov, R.F.; Kuhn, P.; Stevens, R.C. In situ data collection and structure refinement from microcapillary protein crystallization. *J. Appl. Crystallogr.* **2005**, *38*, 900–905. [CrossRef]
83. Huebner, A.; Srisa-Art, M.; Holt, D.; Abell, C.; Hollfelder, F.; Demello, A.; Edel, J. Quantitative detection of protein expression in single cells using droplet microfluidics. *Chem. Commun.* **2007**, *12*, 1218–1220. [CrossRef]
84. Guo, M.T.; Rotem, A.; Heyman, J.A.; Weitz, D.A. Droplet microfluidics for high-throughput biological assays. *Lab Chip* **2012**, *12*, 2146–2155. [CrossRef]
85. Klein, A.M.; Mazutis, L.; Akartuna, I.; Tallapragada, N.; Veres, A.; Li, V.; Peshkin, L.; Weitz, D.A.; Kirschner, M.W. Droplet barcoding for single-cell transcriptomics applied to embryonic stem cells. *Cell* **2015**, *161*, 1187–1201. [CrossRef]
86. Srisa-Art, M.; Bonzani, I.C.; Williams, A.; Stevens, M.M.; Edel, J.B. Identification of rare progenitor cells from human periosteal tissue using droplet microfluidics. *Analyst* **2009**, *134*, 2239–2245. [CrossRef]
87. Köster, S.; Angile, F.E.; Duan, H.; Agresti, J.J.; Wintner, A.; Schmitz, C.; Rowat, A.C.; Merten, C.A.; Pisignano, D.; Griffiths, A.D. Drop-based microfluidic devices for encapsulation of single cells. *Lab Chip* **2008**, *8*, 1110–1115. [CrossRef]
88. Streets, A.M.; Zhang, X.; Cao, C.; Pang, Y.; Wu, X.; Xiong, L.; Yang, L.; Fu, Y.; Zhao, L.; Tang, F. Microfluidic single-cell whole-transcriptome sequencing. *Proc. Natl. Acad. Sci. USA* **2014**, *111*, 7048–7053. [CrossRef]
89. Zilionis, R.; Nainys, J.; Veres, A.; Savova, V.; Zemmour, D.; Klein, A.M.; Mazutis, L. Single-cell barcoding and sequencing using droplet microfluidics. *Nat. Protoc.* **2017**, *12*, 44. [CrossRef]
90. Macosko, E.Z.; Basu, A.; Satija, R.; Nemesh, J.; Shekhar, K.; Goldman, M.; Tirosh, I.; Bialas, A.R.; Kamitaki, N.; Martersteck, E.M. Highly parallel genome-wide expression profiling of individual cells using nanoliter droplets. *Cell* **2015**, *161*, 1202–1214. [CrossRef]
91. Lanza, R.; Langer, R.; Vacanti, J.P.; Atala, A. *Principles of Tissue Engineering*; Academic Press: Cambridge, MA, USA, 2020.
92. Jiang, W.; Li, M.; Chen, Z.; Leong, K.W. Cell-laden microfluidic microgels for tissue regeneration. *Lab Chip* **2016**, *16*, 4482–4506. [CrossRef]
93. Hidalgo San Jose, L.; Stephens, P.; Song, B.; Barrow, D. Microfluidic encapsulation supports stem cell viability, proliferation, and neuronal differentiation. *Tissue Eng. Part C Methods* **2018**, *24*, 158–170. [CrossRef]
94. Zhao, P.; Gu, H.; Mi, H.; Rao, C.; Fu, J.; Turng, L.-S. Fabrication of scaffolds in tissue engineering: A review. *Front. Mech. Eng.* **2018**, *13*, 107–119. [CrossRef]
95. Zhao, Z.; Wang, Z.; Li, G.; Cai, Z.; Wu, J.; Wang, L.; Deng, L.; Cai, M.; Cui, W. Injectable Microfluidic Hydrogel Microspheres for Cell and Drug Delivery. *Adv. Funct. Mater.* **2021**, *31*, 2103339. [CrossRef]
96. Madrid, A.P.M.; Vrech, S.M.; Sanchez, M.A.; Rodriguez, A.P. Advances in additive manufacturing for bone tissue engineering scaffolds. *Mater. Sci. Eng. C* **2019**, *100*, 631–644. [CrossRef] [PubMed]
97. Moshaverinia, A.; Chen, C.; Xu, X.; Akiyama, K.; Ansari, S.; Zadeh, H.H.; Shi, S. Bone regeneration potential of stem cells derived from periodontal ligament or gingival tissue sources encapsulated in RGD-modified alginate scaffold. *Tissue Eng. Part A* **2013**, *20*, 611–621. [CrossRef] [PubMed]
98. Zhao, X.; Liu, S.; Yildirimer, L.; Zhao, H.; Ding, R.; Wang, H.; Cui, W.; Weitz, D. Injectable stem cell-laden photocrosslinkable microspheres fabricated using microfluidics for rapid generation of osteogenic tissue constructs. *Adv. Funct. Mater.* **2016**, *26*, 2809–2819. [CrossRef]
99. Hou, Y.; Xie, W.; Achazi, K.; Cuellar-Camacho, J.L.; Melzig, M.F.; Chen, W.; Haag, R. Injectable degradable PVA microgels prepared by microfluidic technology for controlled osteogenic differentiation of mesenchymal stem cells. *Acta Biomater.* **2018**, *77*, 28–37. [CrossRef]
100. An, C.; Liu, W.; Zhang, Y.; Pang, B.; Liu, H.; Zhang, Y.; Zhang, H.; Zhang, L.; Liao, H.; Ren, C. Continuous Microfluidic Encapsulation of Single Mesenchymal Stem Cells Using Alginate Microgels as Injectable Fillers for Bone Regeneration. *Acta Biomater.* **2020**, *111*, 181–196. [CrossRef]
101. Liu, M.; Zeng, X.; Ma, C.; Yi, H.; Ali, Z.; Mou, X.; Li, S.; Deng, Y.; He, N. Injectable hydrogels for cartilage and bone tissue engineering. *Bone Res.* **2017**, *5*, 1–20. [CrossRef]
102. Li, F.; Truong, V.X.; Thissen, H.; Frith, J.E.; Forsythe, J.S. Microfluidic encapsulation of human mesenchymal stem cells for articular cartilage tissue regeneration. *ACS Appl. Mater. Interfaces* **2017**, *9*, 8589–8601. [CrossRef]

103. Feng, Q.; Li, Q.; Wen, H.; Chen, J.; Liang, M.; Huang, H.; Lan, D.; Dong, H.; Cao, X. Injection and Self-Assembly of Bioinspired Stem Cell-Laden Gelatin/Hyaluronic Acid Hybrid Microgels Promote Cartilage Repair In Vivo. *Adv. Funct. Mater.* **2019**, *29*, 1906690. [CrossRef]
104. Qasim, M.; Le, N.X.T.; Nguyen, T.P.T.; Chae, D.S.; Park, S.-J.; Lee, N.Y. Nanohybrid biodegradable scaffolds for TGF-β3 release for the chondrogenic differentiation of human mesenchymal stem cells. *Int. J. Pharm.* **2020**, *581*, 119248. [CrossRef]
105. Rodrigues, M.; Kosaric, N.; Bonham, C.A.; Gurtner, G.C. Wound healing: A cellular perspective. *Physiol. Rev.* **2019**, *99*, 665–706. [CrossRef]
106. Beheshtizadeh, N.; Lotfibakhshaiesh, N.; Pazhouhnia, Z.; Hoseinpour, M.; Nafari, M. A review of 3D bio-printing for bone and skin tissue engineering: A commercial approach. *J. Mater. Sci.* **2020**, *55*, 3729–3749. [CrossRef]
107. Yu, J.; Huang, T.R.; Lim, Z.H.; Luo, R.; Pasula, R.R.; Liao, L.D.; Lim, S.; Chen, C.H. Production of hollow bacterial cellulose microspheres using microfluidics to form an injectable porous scaffold for wound healing. *Adv. Healthc. Mater.* **2016**, *5*, 2983–2992. [CrossRef]
108. Griffin, D.R.; Weaver, W.M.; Scumpia, P.O.; Di Carlo, D.; Segura, T. Accelerated wound healing by injectable microporous gel scaffolds assembled from annealed building blocks. *Nat. Mater.* **2015**, *14*, 737–744. [CrossRef]
109. Moberg, L.; Johansson, H.; Lukinius, A.; Berne, C.; Foss, A.; Källén, R.; Østraat, Ø.; Salmela, K.; Tibell, A.; Tufveson, G. Production of tissue factor by pancreatic islet cells as a trigger of detrimental thrombotic reactions in clinical islet transplantation. *Lancet* **2002**, *360*, 2039–2045. [CrossRef]
110. Ma, M.; Chiu, A.; Sahay, G.; Doloff, J.C.; Dholakia, N.; Thakrar, R.; Cohen, J.; Vegas, A.; Chen, D.; Bratlie, K.M. Core–shell hydrogel microcapsules for improved islets encapsulation. *Adv. Healthc. Mater.* **2013**, *2*, 667–672. [CrossRef]
111. Headen, D.M.; Aubry, G.; Lu, H.; García, A.J. Microfluidic-based generation of size-controlled, biofunctionalized synthetic polymer microgels for cell encapsulation. *Adv. Mater.* **2014**, *26*, 3003–3008. [CrossRef]
112. Liu, H.T.; Wang, H.; Wei, W.B.; Liu, H.; Jiang, L.; Qin, J.H. A Microfluidic Strategy for Controllable Generation of Water-in-Water Droplets as Biocompatible Microcarriers. *Small* **2018**, *14*, 1801095. [CrossRef]
113. Murray, C.J.; Vos, T.; Lozano, R.; Naghavi, M.; Flaxman, A.D.; Michaud, C.; Ezzati, M.; Shibuya, K.; Salomon, J.A.; Abdalla, S. Disability-adjusted life years (DALYs) for 291 diseases and injuries in 21 regions, 1990–2010: A systematic analysis for the Global Burden of Disease Study 2010. *Lancet* **2012**, *380*, 2197–2223. [CrossRef]
114. Dutkowski, P.; Oberkofler, C.E.; Béchir, M.; Müllhaupt, B.; Geier, A.; Raptis, D.A.; Clavien, P.A. The model for end-stage liver disease allocation system for liver transplantation saves lives, but increases morbidity and cost: A prospective outcome analysis. *Liver Transpl.* **2011**, *17*, 674–684. [CrossRef]
115. Chan, H.F.; Zhang, Y.; Leong, K.W. Efficient One-Step Production of Microencapsulated Hepatocyte Spheroids with Enhanced Functions. *Small* **2016**, *12*, 2720–2730. [CrossRef]
116. Wang, H.; Liu, H.; Liu, H.; Su, W.; Chen, W.; Qin, J. One-Step Generation of Core–Shell Gelatin Methacrylate (GelMA) Microgels Using a Droplet Microfluidic System. *Adv. Mater. Technol.* **2019**, *4*, 1800632. [CrossRef]
117. Ashtari, K.; Nazari, H.; Ko, H.; Tebon, P.; Akhshik, M.; Akbari, M.; Alhosseini, S.N.; Mozafari, M.; Mehravi, B.; Soleimani, M. Electrically conductive nanomaterials for cardiac tissue engineering. *Adv. Drug Deliv. Rev.* **2019**, *144*, 162–179. [CrossRef]
118. Gal, I.; Edri, R.; Noor, N.; Rotenberg, M.; Namestnikov, M.; Cabilly, I.; Shapira, A.; Dvir, T. Injectable Cardiac Cell Microdroplets for Tissue Regeneration. *Small* **2020**, *16*, 1904806. [CrossRef] [PubMed]
119. Weidenbacher, L.; Abrishamkar, A.; Rottmar, M.; Guex, A.G.; Maniura-Weber, K.; Ferguson, S.; Rossi, R.; Fortunato, G. Electrospraying of microfluidic encapsulated cells for the fabrication of cell-laden electrospun hybrid tissue constructs. *Acta Biomater.* **2017**, *64*, 137–147. [CrossRef] [PubMed]
120. Balaoing, L.R.; Post, A.D.; Lin, A.Y.; Tseng, H.; Moake, J.L.; Grande-Allen, K.J. Laminin peptide-immobilized hydrogels modulate valve endothelial cell hemostatic regulation. *PLoS ONE* **2015**, *10*, e0130749. [CrossRef] [PubMed]
121. Cha, C.; Oh, J.; Kim, K.; Qiu, Y.; Joh, M.; Shin, S.R.; Wang, X.; Camci-Unal, G.; Wan, K.-T.; Liao, R. Microfluidics-assisted fabrication of gelatin-silica core–shell microgels for injectable tissue constructs. *Biomacromolecules* **2014**, *15*, 283–290. [CrossRef]
122. Michiels, C. Endothelial cell functions. *J. Cell. Physiol.* **2003**, *196*, 430–443. [CrossRef]
123. Crampton, A.L.; Cummins, K.A.; Wood, D.K. A high-throughput microtissue platform to probe endothelial function in vitro. *Integr. Biol.* **2018**, *10*, 555–565. [CrossRef]
124. Nosrati, R.; Graham, P.J.; Zhang, B.; Riordon, J.; Lagunov, A.; Hannam, T.G.; Escobedo, C.; Jarvi, K.; Sinton, D. Microfluidics for sperm analysis and selection. *Nat. Rev. Urol.* **2017**, *14*, 707–730. [CrossRef]
125. Kashaninejad, N.; Shiddiky, M.J.A.; Nguyen, N.T. Advances in microfluidics-based assisted reproductive technology: From sperm sorter to reproductive system-on-a-chip. *Adv. Biosyst.* **2018**, *2*, 1700197. [CrossRef]
126. Shams, A.; Eslahi, N.; Movahedin, M.; Izadyar, F.; Asgari, H.; Koruji, M. Future of spermatogonial stem cell culture: Application of nanofiber scaffolds. *Curr. Stem Cell Res. Ther.* **2017**, *12*, 544–553. [CrossRef]
127. Eyni, H.; Ghorbani, S.; Shirazi, R.; Salari Asl, L.; P Beiranvand, S.; Soleimani, M. Three-dimensional wet-electrospun poly (lactic acid)/multi-wall carbon nanotubes scaffold induces differentiation of human menstrual blood-derived stem cells into germ-like cells. *J. Biomater. Appl.* **2017**, *32*, 373–383. [CrossRef]
128. Maleki, M.A.; Soltani, M.; Kashaninejad, N.; Nguyen, N.-T. Effects of magnetic nanoparticles on mixing in droplet-based microfluidics. *Phys. Fluids* **2019**, *31*, 32001. [CrossRef]

129. Agarwal, P.; Choi, J.K.; Huang, H.; Zhao, S.; Dumbleton, J.; Li, J.; He, X. A Biomimetic Core-Shell Platform for Miniaturized 3D Cell and Tissue Engineering. *Part. Part. Syst. Charact.* **2015**, *32*, 809–816. [CrossRef]
130. Funfak, A.; Brosing, A.; Brand, M.; Kohler, J.M. Micro fluid segment technique for screening and development studies on Danio rerio embryos. *Lab Chip* **2007**, *7*, 1132–1138. [CrossRef]
131. Park, S.; Wijethunga, P.A.L.; Moon, H.; Han, B. On-chip characterization of cryoprotective agent mixtures using an EWOD-based digital microfluidic device. *Lab Chip* **2011**, *11*, 2212–2221. [CrossRef]
132. Aussillous, P.; Quere, D. Liquid marbles. *Nature* **2001**, *411*, 924–927. [CrossRef]
133. Khaw, M.K.; Ooi, C.H.; Mohd-Yasin, F.; Vadivelu, R.; St John, J.; Nguyen, N.-T. Digital microfluidics with a magnetically actuated floating liquid marble. *Lab Chip* **2016**, *16*, 2211–2218. [CrossRef]
134. Ledda, S.; Idda, A.; Kelly, J.; Ariu, F.; Bogliolo, L.; Bebbere, D. A novel technique for in vitro maturation of sheep oocytes in a liquid marble microbioreactor. *J. Assist. Reprod. Genet.* **2016**, *33*, 513–518. [CrossRef]
135. Evans, C.H.; Huard, J. Gene therapy approaches to regenerating the musculoskeletal system. *Nat. Rev. Rheumatol.* **2015**, *11*, 234. [CrossRef]
136. Mangraviti, A.; Tzeng, S.Y.; Kozielski, K.L.; Wang, Y.; Jin, Y.; Gullotti, D.; Pedone, M.; Buaron, N.; Liu, A.; Wilson, D.R. Polymeric nanoparticles for nonviral gene therapy extend brain tumor survival in vivo. *ACS Nano* **2015**, *9*, 1236–1249. [CrossRef]
137. Valero, A.; Post, J.N.; van Nieuwkasteele, J.W.; ter Braak, P.M.; Kruijer, W.; van den Berg, A. Gene transfer and protein dynamics in stem cells using single cell electroporation in a microfluidic device. *Lab Chip* **2008**, *8*, 62–67. [CrossRef]
138. Pérez, C.; Sanluis-Verdes, A.; Waisman, A.; Lombardi, A.; Rosero, G.; La Greca, A.; Bhansali, S.; Bourguignon, N.; Luzzani, C.; Pérez, M. Single cell transfection of human induced pluripotent stem cells using a droplet-based microfluidic system. *bioRxiv* **2020**. [CrossRef]
139. Zhan, Y.; Wang, J.; Bao, N.; Lu, C. Electroporation of cells in microfluidic droplets. *Anal. Chem.* **2009**, *81*, 2027–2031. [CrossRef]
140. Ho, Y.-P.; Grigsby, C.L.; Zhao, F.; Leong, K.W. Tuning physical properties of nanocomplexes through microfluidics-assisted confinement. *Nano Lett.* **2011**, *11*, 2178–2182. [CrossRef]
141. Grigsby, C.L.; Ho, Y.-P.; Lin, C.; Engbersen, J.F.; Leong, K.W. Microfluidic preparation of polymer-nucleic acid nanocomplexes improves nonviral gene transfer. *Sci. Rep.* **2013**, *3*, 3155. [CrossRef]
142. Yang, S.-M.; Yao, H.; Zhang, D.; Li, W.J.; Kung, H.-F.; Chen, S.-C. Droplet-based dielectrophoresis device for on-chip nanomedicine fabrication and improved gene delivery efficiency. *Microfluid. Nanofluid.* **2015**, *19*, 235–243. [CrossRef]
143. Hsieh, A.T.-H.; Hori, N.; Massoudi, R.; Pan, P.J.-H.; Sasaki, H.; Lin, Y.A.; Lee, A.P. Nonviral gene vector formation in monodispersed picolitre incubator for consistent gene delivery. *Lab Chip* **2009**, *9*, 2638–2643. [CrossRef]
144. Madrigal, J.L.; Stilhano, R.S.; Siltanen, C.; Tanaka, K.; Rezvani, S.N.; Morgan, R.P.; Revzin, A.; Han, S.W.; Silva, E.A. Microfluidic generation of alginate microgels for the controlled delivery of lentivectors. *J. Mater. Chem. B* **2016**, *4*, 6989–6999. [CrossRef] [PubMed]
145. Vizoso, F.J.; Eiro, N.; Cid, S.; Schneider, J.; Perez-Fernandez, R. Mesenchymal stem cell secretome: Toward cell-free therapeutic strategies in regenerative medicine. *Int. J. Mol. Sci.* **2017**, *18*, 1852. [CrossRef] [PubMed]
146. Yildirimer, L.; Thanh, N.T.; Seifalian, A.M. Skin regeneration scaffolds: A multimodal bottom-up approach. *Trends Biotechnol.* **2012**, *30*, 638–648. [CrossRef] [PubMed]

biosensors

Review

Droplet Manipulation under a Magnetic Field: A Review

Gui-Ping Zhu *, Qi-Yue Wang, Zhao-Kun Ma, Shi-Hua Wu and Yi-Pan Guo

Department of Aerospace Systems Engineering, School of Astronautics, Nanjing University of Aeronautics and Astronautics, Nanjing 210016, China; wqy2115@nuaa.edu.cn (Q.-Y.W.); zkk380@nuaa.edu.cn (Z.-K.M.); wushihuawushihua@outlook.com (S.-H.W.); guoyipan@nuaa.edu.cn (Y.-P.G.)

* Correspondence: zhuguiping@nuaa.edu.cn

Abstract: The magnetic manipulation of droplets is one of the emerging magnetofluidic technologies that integrate multiple disciplines, such as electromagnetics, fluid mechanics and so on. The directly driven droplets are mainly composed of ferrofluid or liquid metal. This kind of magnetically induced droplet manipulation provides a remote, wireless and programmable approach beneficial for research and engineering applications, such as drug synthesis, biochemistry, sample preparation in life sciences, biomedicine, tissue engineering, etc. Based on the significant growth in the study of magneto droplet handling achieved over the past decades, further and more profound explorations in this field gained impetus, raising concentrations on the construction of a comprehensive working mechanism and the commercialization of this technology. Current challenges faced are not limited to the design and fabrication of the magnetic field, the material, the acquisition of precise and stable droplet performance, other constraints in processing speed and so on. The rotational devices or systems could give rise to additional issues on bulky appearance, high cost, low reliability, etc. Various magnetically introduced droplet behaviors, such as deformation, displacement, rotation, levitation, splitting and fusion, are mainly introduced in this work, involving the basic theory, functions and working principles.

Keywords: microfluidics; magnetic field; magnetization; liquid actuation; droplet manipulation

Citation: Zhu, G.-P.; Wang, Q.-Y.; Ma, Z.-K.; Wu, S.-H.; Guo, Y.-P. Droplet Manipulation under a Magnetic Field: A Review. *Biosensors* **2022**, *12*, 156. https://doi.org/10.3390/bios12030156

Received: 31 January 2022
Accepted: 28 February 2022
Published: 2 March 2022

1. Introduction

In recent years, the manipulation of droplets and cells [1–3] has attracted much attention in biomedicine, chemistry and hydromechanics, especially microdroplets. The volume of microdroplets is usually on the microliter scale, which is low in production cost and minor in consumption. With high yield and purity ensured by the closed environment, which is almost free from external pollution, microdroplets become a preferable chemical reaction container [4,5], which promotes their wide application in synthesis [6], detection, delivery and packaging of various reagents, drugs and particles [7–9], culture, transportation, isolation and dissolution of cells and seeds [10], separation of plasma [11], etc. Microdroplets are typically mass generated and manipulated in microfluidic devices. The development of microelectromechanical systems (MEMS) technology enables the production of more powerful microfluidic devices, making the operation of microfluidic processes flexible. Meanwhile, with the development of computational fluid dynamics (CFD), the study of microdroplets is facilitated with numerical methods that contribute to robustness in high efficiency, low cost, underlying mechanism and so on.

Until now, microfluidic technology has been developed for a variety of droplet control methods coupled with light, sound, heat, electric field, magnetic field, pneumatics and mechanical devices to transport, sort, split, merge and deform droplets [12]. Among all the developed microfluidic strategies, the magnetic method is a prospering research area, which is very popular due to its remarkable advantages in remote and instant control of droplets. In addition, precise and localized manipulation of droplets is ensured by the programmable magnetic field distribution and thus the exerted force by adjusting the current of the electromagnetic coil or the position of the permanent magnet. Overall, the performance of the magnetic droplet manipulation is influenced by numerous factors, including liquid properties, surrounding medium, the interfacial characteristics between the immiscible fluids, the wetting behavior, the structure of the device, the configuration and tuning of the magnetic field and so on.

Various operations of droplets have been performed under a magnetic field, such as ferrofluid droplet stretching [13–17], deflecting [18,19], sorting [20,21], merging [14,22–25] and splitting [26–29], as well as the application in mixing chemical reagents [30], capsule synthesis [31], microlens [32], etc. Based on droplet manipulation, functional group-modified magnetic nanoparticles (MNPs) [33] dispersed in the carrier fluid are used for oil contamination separation [34], oil and gas processing [35], chemical extraction [36] and detection and extraction of heavy metal ions [37] under the magnetic field. With the development of magnetic manipulation of droplets and nanoparticles in molecular biology, MNPs further fulfilled the manipulation of the attached biological macromolecules, such as DNA [38,39], RNA [40,41], antibodies, proteins [42], enzymes [43,44] and even viruses [45]. Ferrofluid and the constituted MNPs are also widely used in drug delivery [46] and release [47], immunoassay [48], targeted therapy [49], biosensor [50], magnetic actuator [51–53], mechanical seal [54] and so on.

Ferrofluid is the most commonly used magnetic fluid, which is a colloidal dispersion synthesized by the suspension of MNPs (normally around 10 nanometers in diameter) in a carrier liquid [55], which exhibits the fluidity of liquid materials and the magnetic properties of solid materials. Thanks to paramagnetism, MNPs can be magnetically manipulated because of their large surface area and volume ratio. The precipitation and aggregation of MNPs are effectively prohibited by the thermal Brownian motion and surfactant ingredient in ferrofluid. Upon applying a magnetic field, ferrofluid is affected by the magnetic force exerted on individual MNPs and volumetric magnetic force determined by the field flux density, gradient, liquid susceptibility, droplet volume and so on. The approximated volumetric magnetic force is proposed for the relatively small volume of the manipulated droplets [56]. The magnetization is governed by the Langevin function in terms of magnetic susceptibility, which varies with the magnetic field [57]. Furthermore, the magnetic force on each particle is derived and presented for MNPs with shapes in regular spheres [58,59]. In addition, energy analysis in the magnetic material reveals the heat exchange or electromagnetic radiation in terms of the magnetic field characteristics, the magnetic susceptibility and the liquid volume [60]. With the magnetic material immersed in an instantly enhanced or weakened magnetic field, there will be generation or loss of heat applicable for cryogenic techniques exploring or temperature adjusting in ferrofluid. In numerical calculation, the interface of multiphase flow can be tracked by means of volume fraction method (VOF) [61], level set method (LSM) [62], phase field method (PFM), dynamic mesh method (DMM), etc. In microgravity, the force balance of the ferrofluid droplet is identified by magnetic force, resistance [63] and interfacial tension [64].

To sum up, magnetic tuning of liquid droplets in microscale provides a reliable approach for sample handling in numerous scientific pursuits. This work aims to provide a selected review of the progress of droplet manipulation in an external magnetic field. The accomplishment in this field is mainly summarized and categorized based on different basic processes discussed in the following sections, together with typical achievements and examples of their applications.

2. Droplet Manipulation in a Magnetic Field

There are many ways to control microdroplets, including optical [65], thermal [66], acoustic [67], electric [68], magnetic methods [69], etc. Among them, the magnetic approach has the advantages of remote control without sensitivity to pH and dielectric properties of the liquid medium. The high-throughput handling of droplets was achieved together with the extraction of magnetic materials due to the adsorption and magnetic actuation function of MNPs in ferrofluid [70,71]. The magnetic force on MNPs or ferrofluid droplets varies with the magnitude and direction of magnetic intensity. The regulation of magnetic control facilitates droplet processes, including generation, deformation, motion, transport, fusion and splitting, sorting and so on. Electromagnet needles [72] and permanent magnets [73] are both capable of manipulating magnetic droplet transport, fusion and dispersing. Moreover, Park et al. [74] flexibly used a permanent magnet to transport and merge magnetic droplets in a 3-dimensional space. With the development of MEMS technology, complex electronic components were integrated on microfluidic chips for droplet programming [75]. The magnetically induced manipulation is thus presented in this work according to the droplet process with analysis on liquid properties, channels and field configurations, etc.

2.1. Droplet Generation

Droplet generation is a prerequisite for microfluidic manipulation. Preferred due to high reliability and low cost, passive methods are widely explored for operating purely based on the channel configurations and liquid properties, while active methods are developed taking advantage of an external force field to overcome the issues of low efficiency and throughput. The active approach is implemented with microfluidic devices commonly coupled with magnetic fields, lasers, electric fields and ultrasonic waves to realize and regulate the generation of droplets [76] directly applicable in inkjet printing, metal 3D printing, capsule synthesis, etc.

In magnetofluidic droplet generation, a magnetic field raised by electromagnetic coils or permanent magnets is applied mainly to control the droplet generation rate, size and shape. The generation process, as well as the resulting droplet frequency, size and interval, are governed by the continuity equation and the Navier–Stokes equation with the volumetric force, including inertial, viscous, interfacial, gravitational and externally applied magnetic force [77]. For a physical understanding, the significance of the forces is usually analyzed by the nondimensionalized number [78]. In Table 1, the most commonly adopted dimensionless number is illustrated by Reynolds, Capillary, Weber, Bond and Magnetic Bond numbers, specifically for magnetic droplet generation. From the definition of the number, Re, Ca, We and Bo are calculated in terms of average fluid velocity (u), density (ρ), characteristic dimension (R, radius or contact radius), dynamic viscosity of the liquid (μ), gravity acceleration (g), interfacial tension (σ) and the volume mass difference between the continuous and dispersed phase ($\Delta\rho$). For the magnetic approach, Bm is determined by the permeability of the vacuum ($\mu_0 = 4\pi \times 10^{-7} NA^{-2}$), susceptibility of the magnetic fluid (χ), volume of the droplet (V) and the applied magnetic field intensity (H_0).

Table 1. Dimensionless number in magnetic droplet generation.

Dimensionless Number	Formula and Physical Description
Reynolds number	$\mathrm{Re} = \frac{\rho u R}{\mu} = \frac{Inertial\ force}{Viscous\ force}$
Capillary number	$\mathrm{Ca} = \frac{\mu u}{\sigma} = \frac{Viscous\ force}{Interfacial\ tension}$
Weber number	$We = \frac{\rho u^2 R}{\sigma} = \frac{Inertial\ force}{Interfacial\ tension}$
Bond number	$Bo = \frac{\Delta\rho g R^2}{\sigma} = \frac{Gravitional\ force}{Interfacial\ tension}$
Magnet Bond number	$Bm = \frac{\mu_0 \chi V^{1/3} H_0^2}{2\sigma} = \frac{magnetic\ force}{Interfacial\ tension}$

With different flow rate ratios, the droplet generation process is typically defined as slug flow, dripping flow and jetting flow with or without a magnetic field. The flow pattern and ferrofluid droplet generation can be illustrated in terms of the capillary number of the continuous phase and the Weber number of ferrofluids [78]. The low capillary number makes a slug flow when the continuous phase flow rate is relatively small. In this case, ferrofluid completely occupies the channel and leads to an increment in upstream pressure. As the flow rate of the continuous phase increases, a dripping flow pattern is formed with viscous force playing a prominent role in droplet breakup. With a continuous phase flow rate much larger than that of ferrofluid, the effect of capillary instability gives rise to the jetting flow regime. With an increment in magnetic strength, the flow tends to be a dripping pattern. The induced magnetic force acts on the head of the ferrofluid and thus decreases the length of the generated droplet. However, the magnetic field plays a negligible role in the transition between dripping and jetting flow [79].

Conventional straight channels or nozzles were commonly employed for droplet generation directly as shown in Figure 1a. By coupling with a permanent magnet, the tunability of the ferrofluid droplet generation was realized by the non-uniform magnetic field with strength high in the center and weak at both sides [80]. The generation rate of ferrofluid droplets was increased as the droplet stretched forward. Without external syringe pumping, Kahkeshani et al. [81] proposed the generation of ferrofluid droplets in a straight-channel under the driving of the gradient magnetic field, as shown in Figure 1b. In addition, a perpendicular uniform magnetic field was applied for the numerical study of ferrofluid droplet generation in a straight-channel microfluidic device [82]. The flow velocity increment and droplet size declination were investigated on a larger magnetic bond number.

Other typical microfluidic devices were proposed to control ferrofluid droplet generation in T-shaped [83–85] and cross-shaped [10,86,87] channel structures coupled with an electromagnetic field. For cross-shaped chips with a magnetic field applied along the flow field, the diameter of the ferrofluid droplets would become larger as the magnetic intensity gradually increases at a constant flow rate. On the contrary, higher magnetic intensity results in a smaller droplet size for the T-shaped chip. The process of droplet generation in the cross-shaped and T-shaped microfluidic chip is respectively shown in Figure 1c [88] and Figure 1d [84]. Alternatively, a permanent magnet was positioned above the T-shaped channel for generation of ferrofluid droplets with the reduction in generation frequency and increment in curvature radius attributed by the permanent magnetic field [85].

Except for the magnetic intensity, the shape and size of the generated droplets are related to the flow rate of the liquid sample, as well as the direction of the applied magnetic field. Generally, a larger flow rate and stronger magnetic field would accelerate the breaking up of the ferrofluid droplets, giving rise to a smaller volume. For a straight vertical nozzle structure, Fabian et al. [89] investigated the dripping of ferrofluid under a horizontal or vertical magnetic field, which contributed to larger or smaller ferrofluid droplets, respectively. Wu et al. [90] systematically studied ferrofluid droplet generation in a cross-shaped channel under a transverse (perpendicular to liquid flow) or a longitudinal (parallel to the liquid flow) magnetic field. In the transverse case, the generation rate was increased with the length of ferrofluid droplets shortened. In the longitudinal case, droplet generation slowed down with the length elongated, as shown in Figure 1e. For a T-shaped channel under a transverse magnetic field, the interval of the generated ferrofluid droplets is larger at a stronger field strength [91], Figure 1f.

Furthermore, programmability was implemented for non-magnetic water droplet generation in ferrofluids with rotating permanent magnets [92] and metal droplet generation under the action of Lorentz force formed by an electromagnetic field [93]. In brief, the size and shape of generated droplets, as well as the generation rate, are related to characteristic parameters, such as the flow rate, surface tension, magnetic permeability, the applied magnetic field, etc.

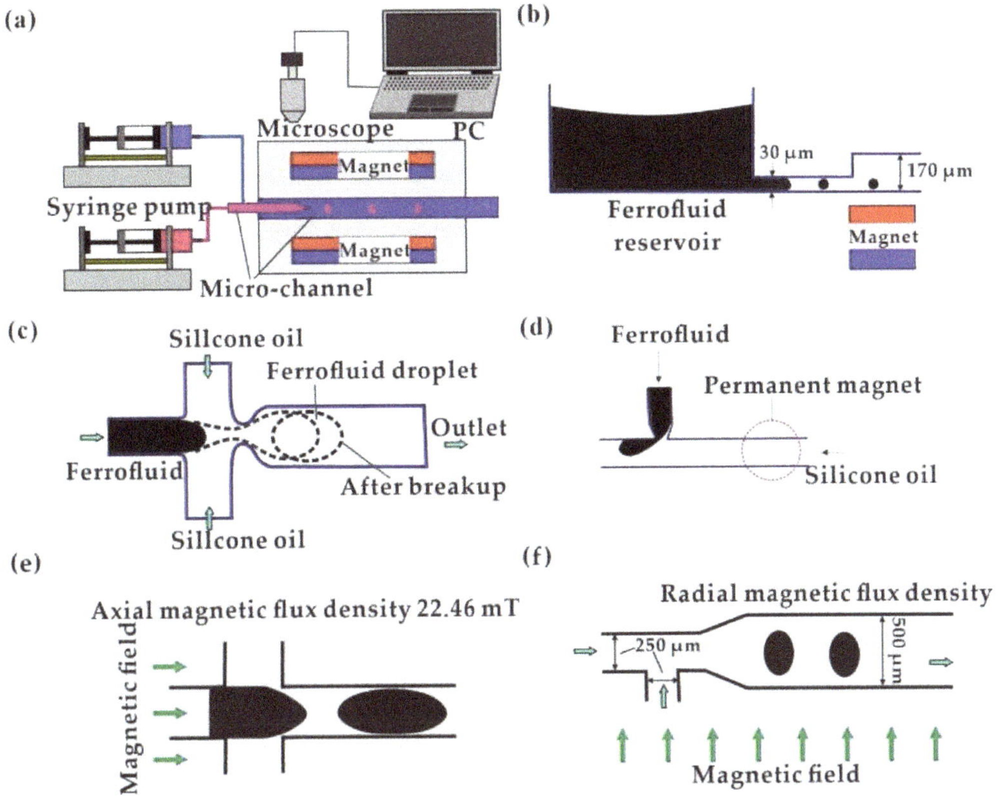

Figure 1. Ferrofluid droplet generation devices: (**a**) Straight channels with permanent magnetic field [80]; (**b**) Driven by a permanent magnet without a pump [81]; (**c**) In cross-shaped channel with a perpendicular uniform magnetic field [88]; (**d**) In T-shaped microfluidic chip with a permanent magnet [84]; (**e**) Horizontal magnetic field [90]; (**f**) Radial magnetic field [91].

2.2. Droplet Deformation

Once immersed in a magnetic field, magnetic forces on MNPs or magnetic liquid cause various deformations of the magnetofluidic droplets. The suspended ferrofluid droplet undergoes tensile deformation along the direction of the applied magnetic field [13–17]. Normally, the droplet appears to be stretched into a larger aspect ratio in a stronger magnetic field (Figure 2a) [15]. For sessile droplets, the vertically applied magnetic field forces the droplet to undergo deformation or even instability [94]. As shown in Figure 2b, Lee et al. [95] applied magnetic dots with a diameter of 0.5–0.95 times that of the sessile ferrofluid droplet to analyze the aspect ratio of the droplet in terms of magnetic field strength. In addition to profile analysis, wettability of the droplet [96] has been extensively studied in terms of contact angle variation [97–100]. A permanent magnet was placed under the sessile droplet, where the contact angle gradually decreased as the magnetic intensity increased [101]. Upon reaching a critical condition, the sessile droplet incepted and slid continuously with the moving permanent magnet, with the dynamic advancing and receding contact angles shown in Figure 2c. For the permanent magnet placed above, the droplet on a hydrophobic substrate was stretched and even split along the vertical direction with contact angle decreasing at increasing field strength, as shown in Figure 2d [102]. The shape deformation and the splitting process were determined by the combined effects of surface tension and magnetic force, which were also affected by nanoparticle concentration.

The direction of the droplet deformation is consistent with the magnetic force, which directs to a higher value of magnetic flux density. Ghaderid et al. numerically studied the influence of the vertical magnetic field on the droplet falling process [103]. The magnetic forces symmetrically distributed on the upper and lower interface, which can intuitively explain the suspended droplets' uniform tensile deformation to both ends. Haberad et al. [104] numerically calculated the magnetic field distribution around the droplet immersed in a uniform magnetic field and further verified the working mechanism. Due to magnetization, a strong gradient of magnetic intensity was generated at both ends of the droplet without significant fluctuations in the center area. For ferrofluid droplets in a magnetic field with non-negligible gravitational force, the droplets would have asymmetrical deformation during the falling process [105]. The shape of the droplet went through the initial oblate ellipsoid to a sphere, then to a long ellipsoid and finally to a teardrop shape, as shown in Figure 2e. In addition to the above-mentioned meniscus under a magnetic field only, Jackson et al. [106] proved the ability to acquire ferrofluid peaks with coupling of magnetic and electric fields, as shown in Figure 2f. Furthermore, sessile ferrofluid droplet deformation was studied under the control of the magnetic field in conjunction with gravity and sound [107].

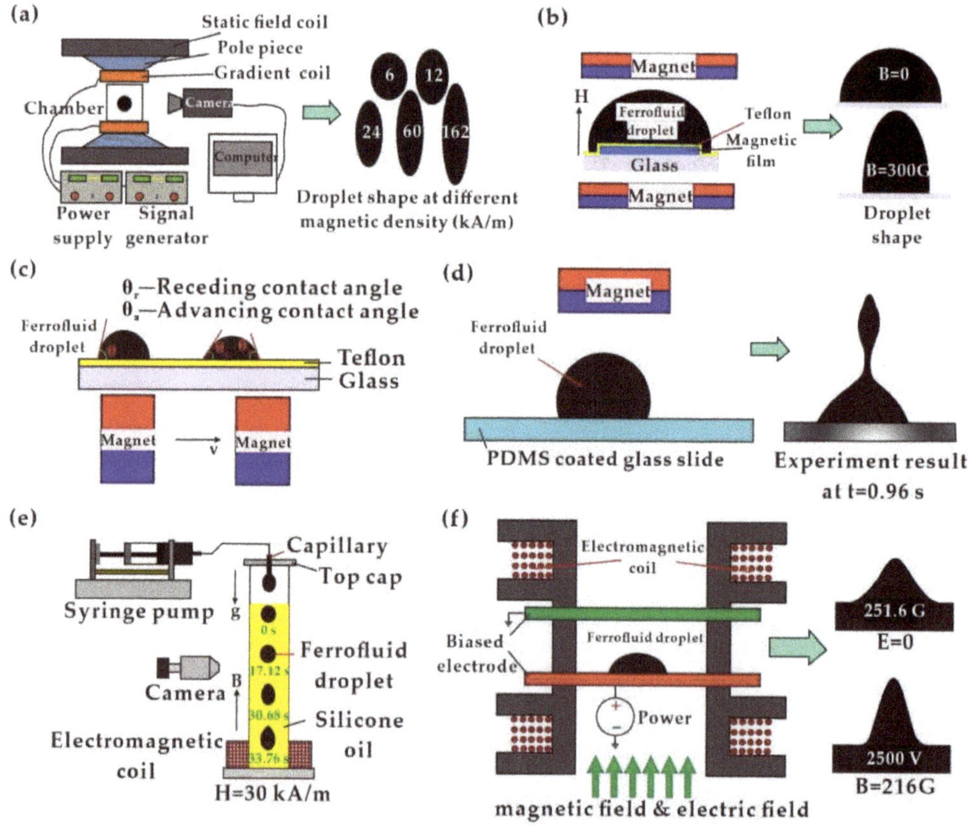

Figure 2. Suspended and sessile ferrofluid droplet deformation in a magnetic field: (**a**) Suspended ferrofluid droplet undergoes tensile deformation [15]; (**b**) Sessile ferrofluid droplet with magnetic dots [95]; (**c**) Sessile ferrofluid droplet above permanent magnet [101]; (**d**) Sessile ferrofluid droplet beneath permanent magnet [102]; (**e**) Asymmetrical deformation with gravitational force [105]; (**f**) Ferrofluid peak with coupling of magnetic and electric fields [107].

On the basis of magnetic droplet deformation and contact angle shifting in a magnetic field, various applications have been fulfilled, such as microfluidic actuators [52,108] and drive other transparent liquid via communicating vessels for making adaptive liquid microlens, grating, etc. [109]. With an electromagnetic field, the deformation of magnetic sessile droplets would lead to an adjustable droplet microlens by connecting droplets through a channel [110] or cavity [111]. The liquid properties of ferrofluid and the characterization of the external field play essential roles in the deformation of the magnetic droplet, and thus the performance of the lens. For miniaturization, a microcoil array [112] was adopted to actuate ferrofluid droplets for precise fabrication and easy integration.

2.3. Droplet Transportation

Magnetically induced droplet transportation enables reliable sample handling for bioassays, drug preparation, chemical reactions and so on. The magnetically actuated droplet motion can be regulated as predefined specific trajectories, such as rectilinear motion and rotational motion. The magnetic field configuration plays a vital role in the actuating and driving of magnetic droplets, including distribution, field strength and field gradient. The motion of ferrofluid droplets in a gradient magnetic field has been extensively investigated with the magnetic component performing as valve and actuator to pump other immiscible liquids. Normally, the pumping performance was tuned and optimized with a wise determination of channel configuration and magnetic properties. In a trapezoidal silicon microchannel, the driving of ferrofluid droplets was applied for liquid sampling in a magnetic field gradient generated by a permanent magnet with a stepping motor [113] and an array of electromagnets [114].

By applying only permanent magnets, Hatch et al. [115] proposed a droplet handling strategy for the pumping of immiscible continuous phases in a ring channel by manipulating ferrofluid plugs with rotating permanent magnets (Figure 3a). For a systematic understanding of the pumping behavior, the flow rate was studied with various chip dimensions and the speed of magnets directing the ferrofluid droplets. With minimal backpressure, the maximum flow rate and pressure head obtained were 45.8 μL/min and 1.2 kPa, respectively. In the originally adopted circular channel configuration, ferrofluid served as a rotating or sealing component in the pump with plunger and blade shape. With a similar design of rotary magnet driven by a motor, Fu et al. [116] proposed pumping of diamagnetic fluids by ferrofluid working as the plunger of the pump. The rotational motion of the ferrofluid droplets was realized [117] and further demonstrated for droplet-based PCR [118,119]. In conventional straight microchannels, multiple [120] ferrofluid droplets were introduced and tested for pumping of the DI water. In digital microfluidics, the permanent magnet was robust in droplet wetting properties shift, deformation, continuous sliding, and thus the fusion and split processes (Figure 3b) [73].

To avoid the pinching moment of the droplet under moving magnets, a magnetic coil was adopted in investigating various volumes of ferrofluid droplet (5 μL to 150 μL) transportation with a bio-compatible surfactant [121]. The stationary coil was beneficial in preventing the disengagement of the droplet during the sliding process, which possibly induced difficulties in predicting the exact trace line [101]. The platform was examined for its versatility in droplet transport both on a hydrophobic solid substrate and immersed in olive oil, with various magnetic flux densities, droplet volume, duty cycles and frequencies. For further precise position and manipulation of droplets at a distance from the magnets, an array of micro-coils were enormously developed and optimized for single or multiple droplet transportation in the demanded trajectory. Chakrabarty et al. [56] applied a gradient magnetic field generated by an embedded micro-electromagnetic coil to drive 0.5 μL ferrofluid droplets on the surfactant–water solution moving toward the center in a linear or predefined meandering path. A single planar squire coil formed a magnetic field with strength owning the maximum value at the center and dissipating sharply toward the periphery. The sequentially switched array of square spiral microelectromagnets facilitated the sophisticated regulation of droplet sliding paths (Figure 3c).

The superposition with a permanent magnetic field offers versatility in droplet actuation in an electromagnetic field. On a digital microfluidic platform, Beyzavi et al. [122] designed a magnetic field generated by two pairs of electromagnetic coils to restrain the ferrofluid droplets (0.5~3 mm in diameter) from moving in the horizontal direction, as shown in Figure 3d. The superposition with a pair of permanent magnetic fields was necessary for creating a single field maximum for droplet attraction. The additional permanent magnetic field also helped in stronger field strength, which induced a larger peak velocity of the droplet [123]. Probst et al. [124] sequentially powered four electromagnetic coils to realize the spiral motion of ferrofluid droplets. The magnetic remote control provided an approach for precise positioning, as well as steering along any desired 2-dimensional path of a single ferro fluid droplet. In addition to ferrofluid, droplet transportation was realized for magnetic liquid marbles by a permanent magnet [125] and EGaIn droplets by the generated Lorentz force [126].

An alternative option of magnetic actuation is a functioning substrate for working as a driving component. Inspired by deformable paramagnetic liquid substrate [127], Damodara et al. [128] utilized a permanent magnet to achieve droplet movement on the MNP-embedded PDMS chip, as shown in Figure 3e. Similarly, a superhydrophobic magnetic elastomer [129] was developed with surface depression under the action of a magnetic field to accomplish the motion of a non-magnetic droplet. Magnetic nano/micropillar arrays (MNA) [130] were proposed to transport droplets through pillar bending in a permanent magnetic field, as shown in Figure 3f. By coupling with gravitational force, Wang et al. [131] concluded that the reduction in height and shift in motion trace of the droplet rebound in a magnetic field.

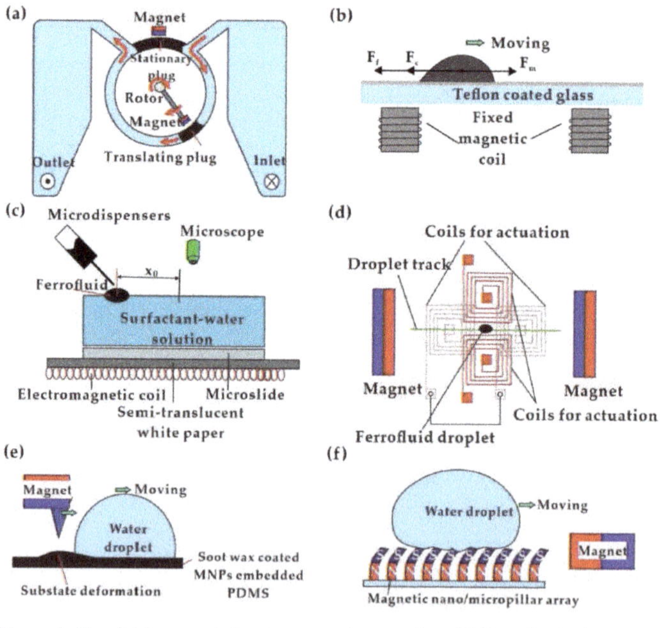

Figure 3. Droplet transportation in magnetic actuation: (**a**) Ferrofluid plugs with rotating permanent magnets [115]; (**b**) Ferrofluid droplet with sliding permanent magnets [73]; (**c**) Ferrofluid droplets with electromagnetic coils [56]; (**d**) Ferrofluid droplets with permanent and electromagnetic fields [122]; (**e**) Water droplet moving on substrate deformation [128]; (**f**) Water droplet moving based on magnetic nano/micropillar arrays [130].

2.4. Droplet Sorting

Droplets generated in all kinds of devices need to be sorted for an experiment or industrial production with specific requirements, such as speed, size, material and composition. According to the driving mechanism, droplet sorting is categorized into passive and active droplet ways [132]. The passive method sorts droplets by taking advantage of the physical or chemical properties of droplets, such as speed, interfacial tension [133], size [134], viscosity [135], etc. However, this sorting method has a shortage in delay and slow response. The active method mainly sorts droplets by utilizing an externally applied driving source, such as pneumatic actuator, ultrasonic, thermal, electric and magnetic force [3].

Magnetic active control is an instant and efficient way to sort ferrofluid droplets or magnetic particles. According to the difference in magnetism, Figure 4a shows a ferrofluid droplet sorting method using a permanent magnetic field [136]. During the sorting process, the droplet would be deflected to different extents, varying with the size of magnets placed under the microfluidic channel and the magnetic particle concentration in ferrofluid. A similar device was adopted to transport single or batch superparamagnetic droplets with their speed reaching 10 per second (Figure 4b) [18]. To sort droplets with magnetic beads, Teste et al. [21] developed a magnetic rail for steering the droplets along the designed track (Figure 4c). In addition, ferrofluid droplets were deflected to different extents in terms of flow speed or droplet size (Figure 4d) [19]. Hydrophobic ferrofluid was applicable in wrapping water droplets for indirectly sorting non-magnetic water droplets in a magnetic field [137]. Water droplets were deflected to flow out along three different passages by continuously moving the horizontally positioned permanent magnet (Figure 4e).

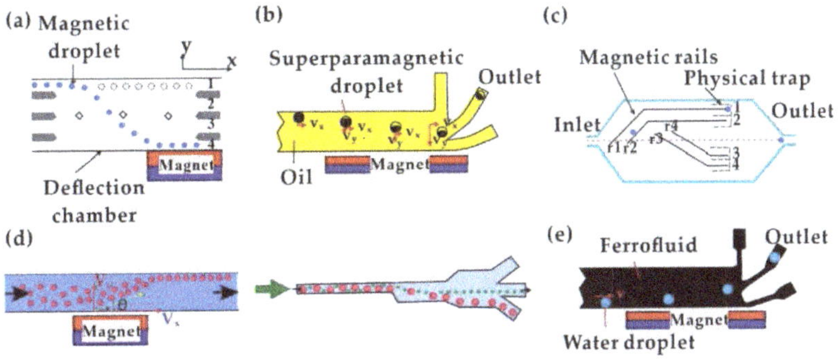

Figure 4. Different magnetic sorting methods: (**a**) Based on the difference in magnetism [136]; (**b**) Droplets deflected to the indicated outlet [18]; (**c**) Magnetic rails [21]; (**d**) Based on the difference in flow speed and droplet size [19]; (**e**) Deflect of water droplets in ferrofluid medium [137].

Magnetically induced droplet sorting is mainly applied in life science, bioengineering, medicine, diagnosis and medical testing, etc. By successfully encapsulating biological reagents or chemical samples in droplets, sorting of cells in microfluidic chips offered a promising way for disease detection and analysis [138]. With technology developed for labeling with superparamagnetic iron oxide nanoparticles, cell detection, loading, delivering and further sorting are ensured in all kinds of biomedical tasks [139]. In addition, Sung et al. [140] magnetically sorted droplets containing different microalgae cells with various densities to extract microalgae with high growth performance. Compared with conventional methods, magnetic droplet sorting is characterized by high throughput and purity, zero contact pollution, low sample consumption and cost, but a relatively slower response. Another fundamental drawback is the need for magnetic components for interaction with the externally applied magnetic field [141,142]. Moreover, the labeling of non-magnetic target droplets by magnetic material such as MNPs may give rise to the variation of physical and chemical properties, which could undermine the chemical/biological compatibility.

2.5. Droplet Coalescence and Splitting

The process of droplet coalescence or splitting in the microfluidic channel can be actively controlled by a magnetic field. In the microfluidic platform, magnetically induced droplet coalescence can be classified into the following two categories according to the working principle. Through controlling the flow moving direction and flow rate, Varma et al. [22] applied a 1000-mT uniform magnetic field to droplets in a cross-shaped channel. The speed of the droplets changed abruptly, resulting in the coalescence of two different types of droplets (Figure 5a). A permanent magnet was also functional for flowing droplets deflecting toward the magnet and merging into a larger droplet in the cross-shaped channel (Figure 5b) [23]. The non-uniform magnetic field was formed with a peak value of 50 mT. Similarly, a permanent magnet was placed in direct contact with the top surface of a Y-shaped channel to magnetically realize the assembling of double emulsion droplets near the magnet and then coalescence, as shown in Figure 5c [24].

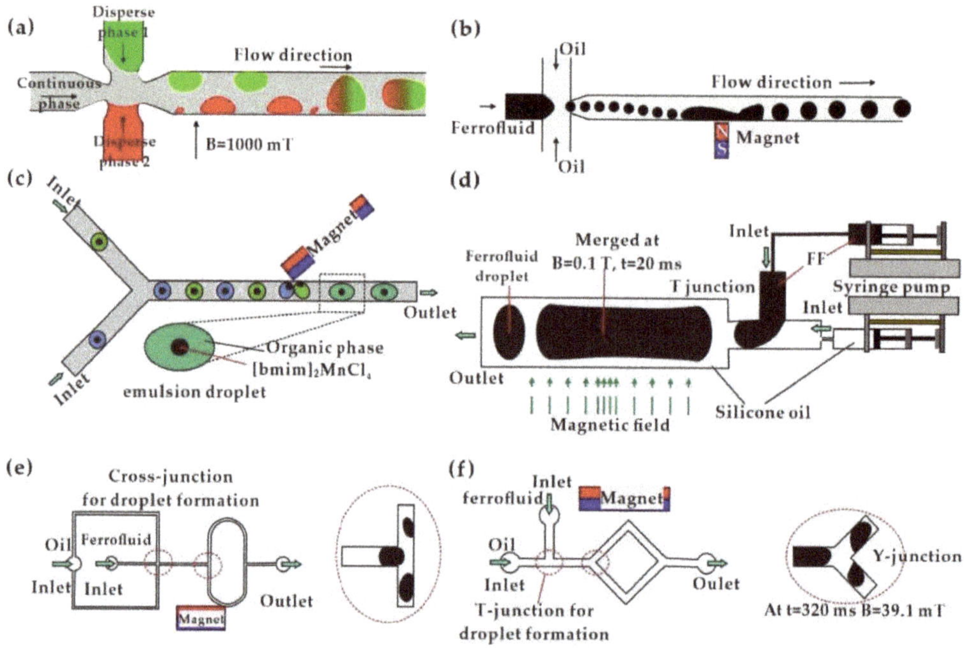

Figure 5. Droplet coalescence and splitting in a magnetic field: (**a**) Uniform magnetic field induced droplet merging in a cross-shaped channel [22]; (**b**) Permanent magnet induced droplet merging in a cross-shaped channel [23]; (**c**) Permanent magnet induced droplet merging in a Y-shaped channel [24]; (**d**) Droplet merging in coupling uniform and non-uniform magnetic fields [25]; (**e**) Droplet splitting in T-shaped channel [26]; (**f**) Droplet splitting in Y-shaped channel [27].

Alternatively, droplet coalescence would be accomplished by profile deformation in an external magnetic field. In a uniform magnetic field, Ghaffari et al. [14] studied the coalescence of two falling ferrofluid droplets based on the fundamental understanding of pendant droplet deformation in terms of field intensity, magnetic susceptibility, surface tension and droplet size (Figure 5d). The analysis was presented with droplet aspect ratio as a function of magnetic bond number and susceptibility. The numerical results, together with the verification by comparison with experimental data, indicated a magnetic approach to be beneficial in both coalescence and breaking of emulsion. In a T-shaped microchannel, the combination of uniform and non-uniform magnetic fields was utilized to explore multiple ferrofluid droplet actuation, inter-droplet space tuning, droplet deformation and merging [25].

In addition, the microfluidic droplets would be split when passing through the T-shaped [26,143–145], λ-shaped [146] and Y-shaped [27,147] channels. As shown in Figure 5e,f, when a permanent magnet is placed on one side of a T-shaped [26] or Y-shaped [27] channel, there would be an asymmetric splitting of the droplet with size and frequency programmable through manipulating the field. A stronger non-uniformly distributed magnetic field facilitates the non-breakup of droplets at a relatively small flow rate ratio between the continuous and dispersed phase. Chen et al. [28] proposed splitting of the magnetic droplet by adopting an orifice under the attraction of an electromagnetic field. The droplet characteristics, such as the sizes of the daughter droplets and the stretching lengths, were interpreted according to orifice diameter and local field strength. By droplet merging, quantitative methylation-specific PCR was realized in a microfluidic device with three identical and parallel lanes [148]. Droplets containing human cell suspension underwent bisulfite conversion and cell lysis through droplet coalescence. In addition, further merging with secondary droplets led to DNA binding to magnetic particles for further processing.

Besides the above-mentioned continuous microfluidic process, droplet coalescence and merging are also widely explored and analyzed in digital microfluidic devices with droplets deposited on an open platform. By inducing magnetic hydrodynamic instability, a large mother ferrofluid droplet on the magnet disc arrays was uniformly split into several small daughter droplets with the lattice structure affected by the field configuration [29]. Koh et al. [149] utilized a permanent magnet to drive the ferrofluid droplet fusion with the water droplet on the Teflon-coated substrate. The control of oil-based ferrofluid droplet transportation and fusion with a diamagnetic oil droplet was realized on superhydrophobic ZnO nanorod arrays by adopting an external magnetic field [150].

During the coalescence and splitting process, droplet sizes and flow rate can be precisely adjusted by the magnetic field to satisfy the requirement of further investigation. Fusion and splitting of microdroplets have broad applications in chemical reagent synthesis and dispersion, capsule synthesis and pharmaceutical detection, etc. By droplet splitting, Lehmann et al. [39] implemented the purification process by using a coil matrix to manipulate MNPs for extracting DNA from cell samples. In a permanent magnetic field with a peak at 400 mT, a magnetic Janus droplet was synthesized in the cross-shaped channel [151] and further applied for protein detection [152,153]. Furthermore, Alorabi et al. [31] successfully synthesized multi-layered capsules for drug delivery by controlling the magnetic droplets flowing into the polyelectrolytes under a magnetic field.

2.6. Droplet Levitation

Droplet levitation achievable under a magnetic field exhibits relatively stable equilibrium without media contact. Undoubtedly, the static magnetic field can levitate paramagnetic medium based on precise regulation. Both experimental and numerical results were presented for ferrofluid droplet levitation by an electromagnetic coil. The droplet dynamics were studied as a function of the applied magnetic field, including equilibrium shape and oscillation, as shown in Figure 6a [154]. It is worth noting that magnetized diamagnetic liquid and its levitation are of great importance in industrial and scientific fields. In terms of the propulsion working principle, the magnetic levitation of diamagnetic droplets was introduced with static and alternating magnetic fields individually.

The levitation of diamagnetic droplets offers the capability of investigating thermal characteristics, surface properties and hydrodynamics, sophisticated convection and so on. For levitation of diamagnetic liquids (such as water and alcohol), the field is required to have characteristic parameters large enough in both strength and gradient magnitude. Commonly, a strong static magnetic field generated by a superconducting coil was utilized to realize the suspension of diamagnetic liquid. In the preliminary work, Beaugnon et al. [155] used a super-static magnetic field of up to 20–27 T to levitate weakly diamagnetic liquids such as alcohol, water, acetone, etc. By being immersed in a magnetic field gradient, liquids were forced to lower field regions under the effect of a driving force stronger than gravity. The technique is applicable in biological fields, as most organic materials have a diamagnetic susceptibility almost in the same order of magnitude as water [156]. In addition, Liu et al. [157] used a 17 T static magnetic field to levitate a large water droplet with a few mm in diameter. The generated magnetic force follows $\mathbf{B}(z)\nabla\mathbf{B}(z) = -\mu_0\rho g/|\chi|$, where $B(z)$, μ_0, χ and $\mathbf{B}$ are the vacuum permeability, the magnetic flux density in the vertical direction, the magnetic susceptibility of the droplet and the externally applied magnetic flux density [158,159]. Therefore, the droplet is levitated by the generated repulsive magnetic force balancing the volumetric gravitational force. To fulfill the working mechanism on magnetism, Pang et al. [160] designed experiments for observing the levitation and interaction of pure water droplets in a superconducting magnetic field with strength up to 16.1 T. The suspension state significantly relied on the interaction between the superconductive magnetic field and magnetized water droplets.

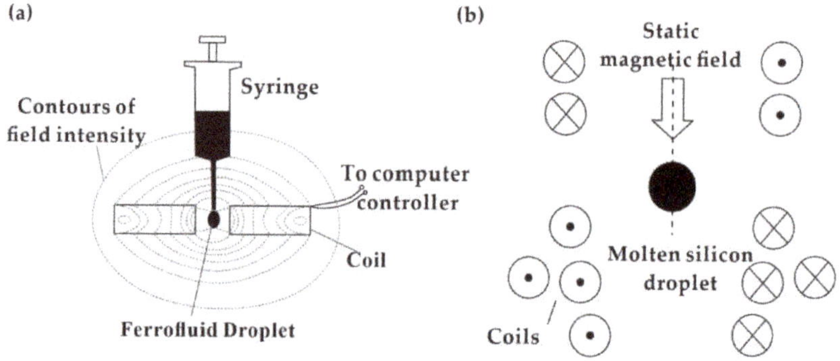

Figure 6. Magnetic levitation of droplets: (**a**) Ferrofluid droplet levitation and splitting in an electromagnetic field [154]; (**b**) Molten silicon droplets levitation in a coupled field [161].

In addition, an alternating magnetic field is adopted to levitate conductive liquids such as molten metal. Commonly, an AC coil is applied to generate the magnetic field and thus results in the Lorentz force as the driving source. It is worth mentioning that there will be generation of eddy current and heat in the droplet. The coupling of an AC coil with a static magnetic field was capable of levitating molten silicon droplets, as shown in Figure 6b [161]. The oscillation, equilibrium, deformation and temperature field of the droplet were systematically interpreted in terms of the static field strength ranging from 1–5 T and the Marangoni effect. The same group numerically studied the flow and temperature field distribution of molten copper droplets levitated by an electromagnetic approach [162]. The influence of the eddy effect was considered in the droplet and coils for taking into account the non-uniformity of current density. The eddy effect gave rise to higher magnetic force, velocity and unevenly distributed temperature in the levitation process. In microgravity, Bojarevics et al. [163] performed a numerical simulation on the melting of silicon droplets in an AC+DC magnetic field and presented the typical temperature and velocity field. The stable droplet suspension helped in studying heat transfer in microgravity environments [164], ensuring demanding chemical reactions without bacterial and virus pollution, smelting high-purity materials [165], promoting the fusion of incompatible alloys [166] and optimizing the welding effect [167].

3. Conclusions and Future Directions

Through the above analysis, we witnessed the development of magnetically induced droplet manipulation and its application prospects. As one of the most critical issues in the microfluidic area, magnetic control exhibits the most vital ability to wirelessly manipulate droplets constituted with biological reagents or chemical samples. In terms of fabrication and integration, magnetic manipulation is relatively easy to implement with advantages in locality, non-contact, strong anti-interference, etc. A variety of scientific works are presented in this review. The recent achievements and advances are categorized and discussed in terms of different processes, including droplet generation, deformation, transportation, sorting, fusion, splitting and levitation. Other specific functions are given by extracting magnetic particles in droplets, sorting droplets with magnetic differences, driving or transporting droplets, smelting metal, etc. Microfluidic devices coupled with wisely designed and regulated magnetic fields have unique advantages in biomedical and chemical applications, which can manipulate cells with natural magnetism, label non-magnetic target cells with MNPs, encapsulate cells or viruses, culture cells, sort cells, synthesize drug capsules, detect cancer and conduct targeted therapy. Overall, the magnetic control of droplets can be achieved by regulating fluidic properties, channel and magnetic field configuration to precisely tune the liquid wettability, droplet profile and size, droplet intervals, etc. The development of magnetic droplet manipulation can promote cross-disciplinary cooperation and exchange, which break the barrier between disciplines and provide novel vision in scientific research.

On the other hand, it is also worth mentioning that the defects of the magnetic approach are closely related to field configuration and material properties and need further excavation and improvement. The main characterization of the general performance in droplet handling includes response time and throughput, ease of fabrication, sample and device cost, system lifetime, accuracy and so on. Some of the interesting research directions and prospects are stated in the following section. Magnetic actuation is preferred in certain applications due to its advantages, such as easy fabrication and low cost. But to what extent magnetic droplet manipulation could contribute to the industry or biotechnology revolution would still depend on the response to the challenges. The integration of all components on the same platform demands the miniaturization of functional devices, such as pumps, valves, mixers, etc. The compromise would be necessary for the introduction of efficient active methods unfavorable in miniaturization and reliability, or passive methods with easy integration but less efficiency. We envision that the growing complexity of microfabrication, synthesis and assembly of composite materials would be required to support the adequate functions of the magnetic platform. For magnetic handling in biotechnology, the volume reduction of the macroscopic magnetic environment to microscopic cases in certain applications would be another considerable challenge, especially for droplet levitation with superconducting magnets. The molecules and MNPs in miniaturized systems also need to deal with the issue of adherence to boundaries due to the large surface/volume ratio or difficulty in entering microchannels due to the effect of capillary forces.

In addition, the experimental investigation requires further efforts on massive handling, instant tracking and monitoring. The manipulation of droplets would depend on technology for addressing information to identify and code a single droplet or a series of droplets. The addressing technology is foreseen to be a remote way realized by integrating components in the handling platform. The compatibility of magnetic handling with digital processing facilitates the programmability and diversity of droplet identification and treatment. Furthermore, a multi-functional droplet platform needs to be explored to provide complex, generic and robust handling. The application would offer integrated chemical, biological and medical functionality with specific strategies adopted. To conclude, on the basis of current accomplishments on the topic of droplet manipulation under magnetic field, this work summarized valuable theories and experimental behaviors, together with potential application prospects and challenges, in the hope of contributing to the construction of a sound and rigorous research structure that would be beneficial both for the understanding of its comprehensive working mechanism and for the future application and commercialization of this technology.

Author Contributions: Drafting, organization and proof reading of the manuscript, G.-P.Z.; Drafting of droplet generation and sorting, Q.-Y.W.; Drafting of droplet deformation and coalescence, Z.-K.M.; Drafting of droplet merging and transportation, S.-H.W.; Drafting of droplet levitation and some of the inserted images, Y.-P.G. All authors have read and agreed to the published version of the manuscript.

Funding: This work was funded by the National Natural Science Foundation of China (11702134, 11905105) and the Natural Science Foundation of Jiangsu Province (BK20160797, BK20180419).

Institutional Review Board Statement: Not applicable.

Informed Consent Statement: Not applicable.

Data Availability Statement: Not applicable.

Conflicts of Interest: The authors declare no conflict of interest. The funders had no role in the design of the study; in the collection, analyses, or interpretation of data; in the writing of the manuscript, or in the decision to publish the results.

References

1. Xiang, N.; Ni, Z.J.T. Electricity-free hand-held inertial microfluidic sorter for size-based cell sorting. *Talanta* **2021**, *235*, 122807. [CrossRef] [PubMed]
2. Ren, H.; Zhu, Z.; Xiang, N.; Wang, H.; Zhang, J.J.S.; Chemical, A.B. Multiplexed serpentine microchannels for high-throughput sorting of disseminated tumor cells from malignant pleural effusion. *Sens. Actuators B Chem.* **2021**, *337*, 129758. [CrossRef]
3. Xi, H.D.; Zheng, H.; Guo, W.; Ganan-Calvo, A.M.; Ai, Y.; Tsao, C.W.; Zhou, J.; Li, W.H.; Huang, Y.Y.; Nguyen, N.T.; et al. Active droplet sorting in microfluidics: A review. *Lab Chip* **2017**, *17*, 751–771. [CrossRef] [PubMed]
4. Haswell, S.J.; Middleton, R.J.; Sullivan, B.O.; Skelton, V.; Watts, P.; Styring, P. The application of micro reactors to synthetic chemistry. *Chem. Commun.* **2001**, *32*, 391–398. [CrossRef]
5. Ghazimirsaeed, E.; Madadelahi, M.; Dizani, M.; Shamloo, A. Secondary Flows, Mixing, and Chemical Reaction Analysis of Droplet-Based Flow inside Serpentine Microchannels with Different Cross Sections. *Langmuir* **2021**, *37*, 5118–5130. [CrossRef] [PubMed]
6. Hung, L.H.; Choi, K.M.; Tseng, W.Y.; Tan, Y.C.; Shea, K.J.; Lee, A.P. Alternating droplet generation and controlled dynamic droplet fusion in microfluidic device for CdS nanoparticle synthesis. *Lab Chip* **2006**, *6*, 174–178. [CrossRef] [PubMed]
7. Hu, H.; Eustace, D.; Merten, C.A. Efficient cell pairing in droplets using dual-color sorting. *Lab Chip* **2015**, *15*, 3989–3993. [CrossRef]
8. Wu, L.; Chen, D.; Yingsong, F.; Xiaojun, L.; Bi, F. Encapsulation of single cells on a microfluidic device integrating;droplet generation with fluorescence-activated droplet sorting. *Biomed. Microdevices* **2013**, *15*, 553–560. [CrossRef]
9. Yoon, S.; Kim, J.A.; Lee, S.H.; Kim, M.; Park, T.H. Droplet-based microfluidic system to form and separate multicellular spheroids using magnetic nanoparticles. *Lab Chip* **2013**, *13*, 1522–1528. [CrossRef]
10. Alegret, S. Integrated Analytical Systems. *Ind. Relat. J.* **2003**, *43*, 195–196.
11. Mielczarek, W.S.; Obaje, E.A.; Bachmann, T.T.; Kersaudy-Kerhoas, M. Microfluidic blood plasma separation for medical diagnostics: Is it worth it? *Lab Chip* **2016**, *16*, 3441–3448. [CrossRef] [PubMed]
12. Seemann, R.; Brinkmann, M.; Pfohl, T.; Herminghaus, S. Droplet based microfluidics. *Rep. Prog. Phys.* **2012**, *75*. [CrossRef] [PubMed]
13. Zhu, G.P.; Nguyen, N.T.; Ramanujan, R.V.; Huang, X.Y. Nonlinear deformation of a ferrofluid droplet in a uniform magnetic field. *Langmuir* **2011**, *27*, 14834–14841. [CrossRef] [PubMed]
14. Ghaffari, A.; Hashemabadi, S.H.; Bazmi, M. CFD simulation of equilibrium shape and coalescence of ferrofluid droplets subjected to uniform magnetic field. *Colloid Surf. A* **2015**, *481*, 186–198. [CrossRef]
15. Afkhami, S.; Tyler, A.J.; Renardy, Y.; Renardy, M.; Pierre, T.S.; Woodward, R.C.; Riffle, J.S. Deformation of a hydrophobic ferrofluid droplet suspended in a viscous medium under uniform magnetic fields. *J. Fluid Mech.* **2010**, *663*, 358–384. [CrossRef]
16. Capobianchi, P.; Lappa, M.; Oliveira, M.S.N. Deformation of a ferrofluid droplet in a simple shear flow under the effect of a constant magnetic field. *Comput. Fluids* **2018**, *173*, 313–323. [CrossRef]
17. Filali, Y.; Er-Riani, M.; El Jarroudi, M. The deformation of a ferrofluid drop under a uniform magnetic field. *Int. J. Non-Linear Mech.* **2018**, *99*, 173–181. [CrossRef]
18. Zhang, K.; Liang, Q.; Ma, S.; Mu, X.; Hu, P.; Wang, Y.; Luo, G. On-chip manipulation of continuous picoliter-volume superparamagnetic droplets using a magnetic force. *Lab Chip* **2009**, *9*, 2992. [CrossRef]
19. Zhang, J.; Yan, S.; Yuan, D.; Zhao, Q.; Tan, S.H.; Nguyen, N.T.; Li, W. A novel viscoelastic-based ferrofluid for continuous sheathless microfluidic separation of nonmagnetic microparticles. *Lab Chip* **2016**, *16*, 3947–3956. [CrossRef] [PubMed]
20. Lin, G.; Baraban, L.; Han, L.; Karnaushenko, D.; Makarov, D.; Cuniberti, G.; Schmidt, O.G. Magnetoresistive Emulsion Analyzer. *Sci. Rep.* **2013**, *3*, 2548. [CrossRef] [PubMed]

21. Teste, B.; Jamond, N.; Ferraro, D.; Viovy, J.-L.; Malaquin, L. Selective handling of droplets in a microfluidic device using magnetic rails. *Microfluid. Nanofluidics* **2015**, *19*, 141–153. [CrossRef]
22. Varma, V.B.; Ray, A.; Wang, Z.M.; Wang, Z.P.; Ramanujan, R.V. Droplet Merging on a Lab-on-a-Chip Platform by Uniform Magnetic Fields. *Sci. Rep.* **2016**, *6*, 37671. [CrossRef] [PubMed]
23. Ray, A.; Varma, V.B.; Jayaneel, P.J.; Sudharsan, N.M.; Wang, Z.P.; Ramanujan, R.V. On demand manipulation of ferrofluid droplets by magnetic fields. *Sens. Actuat. B-Chem.* **2017**, *242*, 760–768. [CrossRef]
24. Misuk, V.; Mai, A.; Giannopoulos, K.; Alobaid, F.; Epple, B.; Loewe, H. Micro magnetofluidics: Droplet manipulation of double emulsions based on paramagnetic ionic liquids. *Lab Chip* **2013**, *13*, 4542–4548. [CrossRef]
25. Ray, A.; Varma, V.B.; Wang, Z.; Wang, Z.; Jayaneel, P.J.; Sudharsan, N.M.; Ramanujan, R.V. Magnetic Droplet Merging by Hybrid Magnetic Fields. *IEEE Magn. Lett.* **2016**, *7*, 1–5. [CrossRef]
26. Wu, Y.; Fu, T.; Ma, Y.; Li, H.Z. Active control of ferrofluid droplet breakup dynamics in a microfluidic T-junction. *Microfluid. Nanofluidics* **2014**, *18*, 19–27. [CrossRef]
27. Li, H.; Wu, Y.; Wang, X.; Zhu, C.; Fu, T.; Ma, Y. Magnetofluidic control of the breakup of ferrofluid droplets in a microfluidic Y-junction. *RSC Adv.* **2016**, *6*, 778–785. [CrossRef]
28. Chen, C.-Y.; Chen, C.H.; Lee, W.F. Experiments on breakups of a magnetic fluid drop through a micro-orifice. *J. Magn. Magn. Mater.* **2009**, *321*, 3520–3525. [CrossRef]
29. Lee, C.P.; Tsai, H.Y.; Lai, M.F. Field evolution of self-assembled lattice structures of ferrofluid microdroplets on magnetic disc arrays. *Soft Matter* **2012**, *8*, 11537. [CrossRef]
30. Saroj, S.K.; Asfer, M.; Sunderka, A.; Panigrahi, P.K. Two-fluid mixing inside a sessile micro droplet using magnetic beads actuation. *Sens. Actuators A: Phys.* **2016**, *244*, 112–120. [CrossRef]
31. Alorabi, A.Q.; Tarn, M.D.; Gomez-Pastora, J.; Bringas, E.; Ortiz, I.; Paunov, V.N.; Pamme, N. On-chip polyelectrolyte coating onto magnetic droplets—Towards continuous flow assembly of drug delivery capsules. *Lab Chip* **2017**, *17*, 3785–3795. [CrossRef] [PubMed]
32. Cheng, H.-C.; Xu, S.; Liu, Y.; Levi, S.; Wu, S.-T. Adaptive mechanical-wetting lens actuated by ferrofluids. *Optics Commun.* **2011**, *284*, 2118–2121. [CrossRef]
33. Pamme, N. Magnetism and microfluidics. *Lab Chip* **2006**, *6*, 24–38. [CrossRef] [PubMed]
34. Ko, S.; Kim, E.S.; Park, S.; Daigle, H.; Milner, T.E.; Huh, C.; Bennetzen, M.V.; Geremia, G.A. Amine functionalized magnetic nanoparticles for removal of oil droplets from produced water and accelerated magnetic separation. *J. Nanoparticle Res.* **2017**, *19*, 132. [CrossRef]
35. Simonsen, G.; Strand, M.; Øye, G. Potential applications of magnetic nanoparticles within separation in the petroleum industry. *J. Pet. Sci. Eng.* **2018**, *165*, 488–495. [CrossRef]
36. Zarei, A.R.; Nedaei, M.; Ghorbanian, S.A. Ferrofluid of magnetic clay and menthol based deep eutectic solvent: Application in directly suspended droplet microextraction for enrichment of some emerging contaminant explosives in water and soil samples. *J. Chromatogr. A* **2018**, *1533*, 32–42. [CrossRef]
37. Gharehbaghi, M.; Farahani, M.D.; Shemirani, F. Dispersive magnetic solid phase extraction based on an ionic liquid ferrofluid. *Anal. Methods* **2014**, *6*, 9258–9266. [CrossRef]
38. Hung, P.-Y.; Jiang, P.-S.; Lee, E.-F.; Fan, S.-K.; Lu, Y.-W. Genomic DNA extraction from whole blood using a digital microfluidic (DMF) platform with magnetic beads. *Microsyst. Technol.* **2015**, *23*, 313–320. [CrossRef]
39. Lehmann, U.; Vandevyver, C.; Parashar, V.K.; Gijs, M.A. Droplet-based DNA purification in a magnetic lab-on-a-chip. *Angew. Chem. Int. Ed. Engl.* **2006**, *45*, 3062–3067. [CrossRef]
40. Ferraro, D.; Champ, J.; Teste, B.; Serra, M.; Malaquin, L.; Viovy, J.L.; Cremoux, P.D.; Descroix, S. Microfluidic platform combining droplets and magnetic tweezers: Application to HER2 expression in cancer diagnosis. *Sci. Rep.* **2016**, *6*, 25540. [CrossRef]
41. Shi, X.; Chen, C.H.; Gao, W.; Chao, S.H.; Meldrum, D.R. Parallel RNA extraction using magnetic beads and a droplet array. *Lab Chip* **2015**, *15*, 1059–1065. [CrossRef] [PubMed]
42. Kowalczyk, A.; Matysiak-Brynda, E.; Bystrzejewski, M.; Sutherland, D.S.; Stojek, Z.; Nowicka, A.M. Conformational control of human transferrin covalently anchored to carbon-coated iron nanoparticles in presence of a magnetic field. *Acta Biomater* **2016**, *45*, 367–374. [CrossRef] [PubMed]
43. Ostafe, R.; Prodanovic, R.; Lloyd Ung, W.; Weitz, D.A.; Fischer, R. A high-throughput cellulase screening system based on droplet microfluidics. *Biomicrofluidics* **2014**, *8*, 041102. [CrossRef] [PubMed]
44. Najah, M.; Calbrix, R.; Mahendra-Wijaya, I.P.; Beneyton, T.; Griffiths, A.D.; Drevelle, A. Droplet-based microfluidics platform for ultra-high-throughput bioprospecting of cellulolytic microorganisms. *Chem. Biol.* **2014**, *21*, 1722–1732. [CrossRef]
45. Pipper, J.; Inoue, M.; Ng, L.F.; Neuzil, P.; Zhang, Y.; Novak, L. Catching bird flu in a droplet. *Nat. Med.* **2007**, *13*, 1259. [CrossRef] [PubMed]
46. Agiotis, L.; Theodorakos, I.; Samothrakitis, S.; Papazoglou, S.; Zergioti, I.; Raptis, Y.S. Magnetic manipulation of superparamagnetic nanoparticles in a microfluidic system for drug delivery applications. *J. Magn. Magn. Mater.* **2016**, *401*, 956–964. [CrossRef]
47. Liu, Y.M.; Wu, W.; Ju, X.J.; Wang, W.; Xie, R.; Mou, C.L.; Zheng, W.C.; Liu, Z.; Chu, L.Y. Smart microcapsules for direction-specific burst release of hydrophobic drugs. *Rsc Adv.* **2014**, *4*, 46568–46575. [CrossRef]

48. Chen, P.; Huang, Y.Y.; Hoshino, K.; Zhang, J.X. Microscale magnetic field modulation for enhanced capture and distribution of rare circulating tumor cells. *Sci. Rep.* **2015**, *5*, 8745. [CrossRef]
49. Alexiou, C.; Arnold, W.; Klein, R.J.; Parak, F.G.; Hulin, P.; Bergemann, C.; Erhardt, W.; Wagenpfeil, S.; Lubbe, A.S. Locoregional cancer treatment with magnetic drug targeting. *Cancer Res.* **2000**, *60*, 6641–6648.
50. Lin, G.G.; Makarov, D.; Schmidt, O.G. Magnetic sensing platform technologies for biomedical applications. *Lab Chip* **2017**, *17*, 1884–1912. [CrossRef]
51. Leidong, M.; Hur, K. Towards ferrofluidics for μ-TAS and lab on-a-chip applications. *Nanotechnology* **2006**, *17*, S34.
52. Nethe, A.; Schöppe, T.; Stahlmann, H.-D. Ferrofluid driven actuator for a left ventricular assist device. *J. Magn. Magn. Mater.* **1999**, *201*, 423–426. [CrossRef]
53. Melikhov, Y.; Lee, S.J.; Jiles, D.C.; Schmidt, D.H.; Porter, M.D.; Shinar, R. Microelectromagnetic ferrofluid-based actuator. *J. Appl. Phys.* **2003**, *93*, 8438–8440. [CrossRef]
54. Volder, M.D.; Reynaerts, D. Development of a hybrid ferrofluid seal technology for miniature pneumatic and hydraulic actuators. *Sens. Actuators A Phys.* **2009**, *152*, 234–240. [CrossRef]
55. Torres-Diaz, I.; Rinaldi, C. Recent progress in ferrofluids research: Novel applications of magnetically controllable and tunable fluids. *Soft Matter* **2014**, *10*, 8584–8602. [CrossRef]
56. Chakrabarty, D.; Dutta, S.; Chakraborty, N.; Ganguly, R. Magnetically actuated transport of ferrofluid droplets over micro-coil array on a digital microfluidic platform. *Sens. Actuators B Chem.* **2016**, *236*, 367–377. [CrossRef]
57. Elmore, W.C. The magnetization of ferromagnetic colloids. *Phys. Rev.* **1938**, *54*, 1092–1095. [CrossRef]
58. Rosensweig, R.E. *Ferrohydrodynamics*; Cambridge University Press: New York, NY, USA, 1985.
59. Nacev, A.; Komaee, A.; Sarwar, A.; Probst, R. Towards Control of Magnetic Fluids in Patients: Directing Therapeutic Nanoparticles to Disease Locations. *Control Syst. IEEE* **2012**, *32*, 32–74. [CrossRef]
60. Hu, L.; Zhang, R.; Chen, Q. Synthesis and assembly of nanomaterials under magnetic fields. *Nanoscale* **2014**, *6*, 14064–14105. [CrossRef]
61. Hirt, C.W.; Nichols, B.D. Volume of fluid (VOF) method for the dynamics of free boundaries. *J. Comput. Phys.* **1981**, *39*, 201–225. [CrossRef]
62. Osher, S.; Fedkiw, R.P. Level Set Methods: An Overview and Some Recent Results. *J. Comput. Phys.* **2001**, *169*, 463–502. [CrossRef]
63. White, F.M.; Majdalani, J. *Viscous Fluid Flow*; McGraw-Hill: New York, NY, USA, 2006.
64. Brackbill, J.U.; Kothe, D.B.; Zemach, C. A continuum method for modeling surface tension. *J. Comput. Phys.* **1992**, *100*, 335–354. [CrossRef]
65. Dixit, S.; Faris, G. Optically Controlled Droplet Adhesion and Coalescence: A New and Versatile Microfluidic Technique. In *Biomedical Optics*; Optica Publishing Group: Washington, DC, USA, 2008; p. BTuF59.
66. Huang, S.; Yin, S.; Chen, F.; Luo, H.; Tang, Q.; Song, J. Directional transport of droplets on wettability patterns at high temperature. *Appl. Surf. Sci.* **2018**, *428*, 432–438. [CrossRef]
67. Laurent, M.; Ricard, J.-L.; Bahain, J.-J.; Voinchet, P.; Rousseau, L. Datation du site Paléolithique moyen de la Butte d'Arvigny (Moissy-Cramayel, Seine-et-Marne). *Comptes Rendus De L'académie Des Sci.-Ser. IIA-Earth Planet. Sci.* **2000**, *330*, 581–583. [CrossRef]
68. Nelson, W.C.; Kim, C.-J.C. Droplet Actuation by Electrowetting-on-Dielectric (EWOD): A Review. *J. Adhes. Sci. Technol.* **2012**, *26*, 1747–1771. [CrossRef]
69. Shin, D.J.; Wang, T.H. Magnetic droplet manipulation platforms for nucleic acid detection at the point of care. *Ann. Biomed. Eng.* **2014**, *42*, 2289–2302. [CrossRef]
70. Brouzes, E.; Kruse, T.; Kimmerling, R.; Strey, H.H. Rapid and continuous magnetic separation in droplet microfluidic devices. *Lab Chip* **2015**, *15*, 908–919. [CrossRef]
71. Zhang, Y.; Wang, T.H. Full-range magnetic manipulation of droplets via surface energy traps enables complex bioassays. *Adv. Mater.* **2013**, *25*, 2903–2908. [CrossRef]
72. Yang, C.; Ning, Y.; Ku, X.; Zhuang, G.; Li, G. Automatic magnetic manipulation of droplets on an open surface using a superhydrophobic electromagnet needle. *Sens. Actuators B Chem.* **2018**, *257*, 409–418. [CrossRef]
73. Long, Z.; Shetty, A.M.; Solomon, M.J.; Larson, R.G. Fundamentals of magnet-actuated droplet manipulation on an open hydrophobic surface. *Lab Chip* **2009**, *9*, 1567–1575. [CrossRef]
74. Park, Y.; Jeon, J.; Chung, S.K. Three-dimensional (3D) magnetic droplet manipulation for biomedical applications. In Proceedings of the 2018 IEEE Micro Electro Mechanical Systems (MEMS), Belfast, UK, 21–25 January 2018.
75. Yang, C.; Zhang, Z.; Li, G. Programmable droplet manipulation by combining a superhydrophobic magnetic film and an electromagnetic pillar array. *Sens. Actuators B Chem.* **2018**, *262*, 892–901. [CrossRef]
76. Chong, Z.Z.; Tan, S.H.; Ganan-Calvo, A.M.; Tor, S.B.; Loh, N.H.; Nguyen, N.T. Active droplet generation in microfluidics. *Lab Chip* **2016**, *16*, 35–58. [CrossRef]
77. Sattari, A.; Hanafizadeh, P.; Hoorfar, M.J.A.i.C.; Science, I. Multiphase flow in microfluidics: From droplets and bubbles to the encapsulated structures. *Adv. Colloid Interface Sci.* **2020**, *282*, 102208. [CrossRef] [PubMed]
78. Amirifar, L.; Besanjideh, M.; Nasiri, R.; Shamloo, A.; Nasrollahi, F.; de Barros, N.R.; Davoodi, E.; Erdem, A.; Mahmoodi, M.; Hosseini, V.; et al. Droplet-based microfluidics in biomedical applications. *Biofabrication* **2022**, *14*, 022001. [CrossRef] [PubMed]
79. Liang, D.; Ma, P.; Zhu, C.; Fu, T.; Luo, G.J.I.; Research, E.C. Manipulable Formation of Ferrofluid Droplets in Y-Shaped Flow-Focusing Microchannels. *Ind. Eng. Chem. Res.* **2019**, *58*, 19226–19238. [CrossRef]

80. Huang, J.-P.; Ge, X.-H.; Xu, J.-H.; Luo, G.-S. Controlled formation and coalescence of paramagnetic ionic liquid droplets under magnetic field in coaxial microfluidic devices. *Chem. Eng. Sci.* **2016**, *152*, 293–300. [CrossRef]
81. Kahkeshani, S.; Carlo, D.D. Drop formation using ferrofluids driven magnetically in a step emulsification device. *Lab Chip* **2016**, *16*, 2474–2480. [CrossRef] [PubMed]
82. Ghaderi, A.; Kayhani, M.H.; Nazari, M.; Fallah, K. Drop formation of ferrofluid at co-flowing microcahnnel under uniform magnetic field. *Eur. J. Mech. B-Fluid* **2018**, *67*, 87–96. [CrossRef]
83. Tan, S.H.; Nguyen, N.T. Generation and manipulation of monodispersed ferrofluid emulsions: The effect of a uniform magnetic field in flow-focusing and T-junction configurations. *Phys. Rev. E Stat. Nonlin Soft Matter Phys.* **2011**, *84*, 036317. [CrossRef]
84. Tan, S.-H.; Nguyen, N.-T.; Yobas, L.; Kang, T.G. Formation and manipulation of ferrofluid droplets at a microfluidicT-junction. *J. Micromech. Microeng.* **2010**, *20*, 045004. [CrossRef]
85. Zhang, Q.; Li, H.; Zhu, C.; Fu, T.; Ma, Y.; Li, H.Z. Micro-magnetofluidics of ferrofluid droplet formation in a T-junction. *Colloids Surf. A Physicochem. Eng. Asp.* **2018**, *537*, 572–579. [CrossRef]
86. Liu, J.; Yap, Y.F.; Nguyen, N.T. Numerical study of the formation process of ferrofluid droplets. *Phys. Fluids* **2011**, *23*, 395. [CrossRef]
87. Yan, Q.; Xuan, S.; Ruan, X.; Wu, J.; Gong, X. Magnetically controllable generation of ferrofluid droplets. *Microfluid. Nanofluidics* **2015**, *19*, 1377–1384. [CrossRef]
88. Liu, J.; Tan, S.-H.; Yap, Y.F.; Ng, M.Y.; Nguyen, N.-T. Numerical and experimental investigations of the formation process of ferrofluid droplets. *Microfluid. Nanofluidics* **2011**, *11*, 177–187. [CrossRef]
89. Fabian, M.; Burda, P.; Švikova, M.; Huňady, R. The Influence of magnetic field on the separation of droplets from ferrofluid jet. *J. Magn. Magn. Mater.* **2017**, *431*, 196–200. [CrossRef]
90. Wu, Y.; Fu, T.; Ma, Y.; Li, H.Z. Ferrofluid droplet formation and breakup dynamics in a microfluidic flow-focusing device. *Soft Matter* **2013**, *9*, 9792. [CrossRef]
91. Varma, V.B.; Ray, A.; Wang, Z.; Wang, Z.; Wu, R.; Jayaneel, P.J.; Sudharsan, N.M.; Ramanujan, R.V. Control of Ferrofluid Droplets in Microchannels by Uniform Magnetic Fields. *IEEE Magn. Lett.* **2016**, *7*, 1–5. [CrossRef]
92. Katsikis, G.; Breant, A.; Rinberg, A.; Prakash, M. Synchronous magnetic control of water droplets in bulk ferrofluid. *Soft Matter* **2018**, *14*, 681–692. [CrossRef] [PubMed]
93. Khoshmanesh, K.; Tang, S.Y.; Zhu, J.Y.; Schaefer, S.; Mitchell, A.; Kalantar-Zadeh, K.; Dickey, M.D. Liquid metal enabled microfluidics. *Lab Chip* **2017**, *17*, 974–993. [CrossRef] [PubMed]
94. Jamin, T.; Py, C.; Falcon, E. Instability of the origami of a ferrofluid drop in a magnetic field. *Phys. Rev. Lett.* **2011**, *107*, 204503. [CrossRef] [PubMed]
95. Lee, C.-P.; Yang, S.-T.; Wei, Z.-H. Field dependent shape variation of magnetic fluid droplets on magnetic dots. *J. Magn. Magn. Mater.* **2012**, *324*, 4133–4135. [CrossRef]
96. Latikka, M.; Backholm, M.; Timonen, J.V.I.; Ras, R.H.A. Wetting of ferrofluids: Phenomena and control. *Curr. Opin. Colloid Interface Sci.* **2018**, *36*, 118–129. [CrossRef]
97. Mats, L.; Young, R.; Gibson, G.T.T.; Oleschuk, R.D. Magnetic droplet actuation on natural (Colocasia leaf) and fluorinated silica nanoparticle superhydrophobic surfaces. *Sens. Actuators B Chem.* **2015**, *220*, 5–12. [CrossRef]
98. Berim, G.O.; Ruckenstein, E. Nanodrop of an Ising magnetic fluid on a solid surface. *Langmuir* **2011**, *27*, 8753–8760. [CrossRef] [PubMed]
99. Zhou, Q.; Ristenpart, W.D.; Stroeve, P. Magnetically induced decrease in droplet contact angle on nanostructured surfaces. *Langmuir* **2011**, *27*, 11747–11751. [CrossRef] [PubMed]
100. Saha, A.K.; Das, D.; Srivastava, R.; Panigrahi, P.K.; Muralidhar, K. Fluid Mechanics and Fluid Power–Contemporary Research. In Proceedings of the 5th International and 41st National Conference on FMFP 2014; Springer: Berlin/Heidelberg, Germany, 2017.
101. Nguyen, N.-T.; Zhu, G.; Chua, Y.-C.; Phan, V.-N.; Tan, S.-H. Magnetowetting and Sliding Motion of a Sessile Ferrofluid Droplet in the Presence of a Permanent Magnet. *Langmuir* **2010**, *26*, 12553–12559. [CrossRef] [PubMed]
102. Banerjee, U.; Sen, A. Shape evolution and splitting of ferrofluid droplets on a hydrophobic surface in the presence of a magnetic field. *Soft Matter* **2018**, *14*. [CrossRef] [PubMed]
103. Ghaderi, A.; Kayhani, M.H.; Nazari, M. Numerical investigation on falling ferrofluid droplet under uniform magnetic field. *Eur. J. Mech. -B/Fluids* **2018**, *72*, 1–11. [CrossRef]
104. Habera, M.; Hron, J. Modelling of a free-surface ferrofluid flow. *J. Magn. Magn. Mater.* **2017**, *431*, 157–160. [CrossRef]
105. Shi, D.; Bi, Q.; He, Y.; Zhou, R. Experimental investigation on falling ferrofluid droplets in vertical magnetic fields. *Exp. Therm. Fluid Sci.* **2014**, *54*, 313–320. [CrossRef]
106. Jackson, B.A.; Terhune, K.J.; King, L.B. Ionic liquid ferrofluid interface deformation and spray onset under electric and magnetic stresses. *Phys. Fluids* **2017**, *29*. [CrossRef]
107. Bashtovoi, V.; Reks, A.; Baev, A.; Mansoor, A.-J.T.M. Topological instability of a semi-bounded magnetic fluid drop under influence of magnetic and ultrasound fields. *J. Magn. Magn. Mater.* **2017**, *431*, 42–45. [CrossRef]
108. Oh, D.W.; Jin, J.S.; Choi, J.H.; Kim, H.Y.; Lee, J.S. A microfluidic chaotic mixer using ferrofluid. *J. Micromech. Microeng.* **2007**, *17*, 2077–2083. [CrossRef]
109. Lee, C.P.; Chen, Y.H.; Lai, M.F. Fabrication of Microlens Arrays by Utilizing Magnetic Hydrodynamic Instability of Ferrofluid Droplets. *IEEE Trans. Magn.* **2014**, *50*, 1–4. [CrossRef]

110. Xiao, W.; Hardt, S. An adaptive liquid microlens driven by a ferrofluidic transducer. *J. Micromech. Microeng.* **2010**, *20*, 055032. [CrossRef]
111. Malouin Jr, B.A.; Vogel, M.J.; Olles, J.D.; Cheng, L.; Hirsa, A.H. Electromagnetic liquid pistons for capillarity-based pumping. *Lab Chip* **2011**, *11*, 393–397. [CrossRef]
112. Schultheis, T.; Molella, L.S.; Reithmeier, E.; Rissing, L.; Hardt, S. Performance of an adaptive liquid microlens controlled by a microcoil actuator. *Microfluid. Nanofluidics* **2012**, *13*, 299–308. [CrossRef]
113. Ahn, J.; Oh, J.G.; Choi, B. A novel type of a microfluidic system using ferrofluids for an application of μ-tas. *Microsyst. Technol.* **2004**, *10*, 622–627. [CrossRef]
114. Greivell, N.E.; Hannaford, B. The design of a ferrofluid magnetic pipette. *Biomed. Eng. IEEE Trans.* **1997**, *44*, 129–135. [CrossRef]
115. Hatch, A.; Kamholz, A.E.; Holman, G.; Yager, P. A ferrofluidic magnetic micropump. *Microelectromech. Syst. J.* **2001**, *10*, 215–221. [CrossRef]
116. Lung-Ming, F.; Wei-Ching, F.; Ting-Fu, H.; Chia-Yen, L. A Magnetic Micropump Based on Ferrofluidic Actuation. *Int. J. Autom. Smart Technol.* **2014**, *4*, 77–82. [CrossRef]
117. Nguyen, N.T.; Chai, M.F. A Stepper Micropump for Ferrofluid Driven Microfluidic Systems. *Micro Nanosyst.* **2009**, *1*, 17–21. [CrossRef]
118. Ohashi, T.; Kuyama, H.; Hanafusa, N.; Togawa, Y. A simple device using magnetic transportation for droplet-based PCR. *Biomed. Microdevices* **2007**, *9*, 695–702. [CrossRef] [PubMed]
119. Lok, K.S.; Kwok, Y.C.; Lee, P.P.F.; Nguyen, N.-T. Ferrofluid plug as valve and actuator for whole-cell PCR on chip. *Sens. Actuators B Chem.* **2012**, *166–167*, 893–897. [CrossRef]
120. Hartshorne, H.; Backhouse, C.J.; Lee, W.E. Ferrofluid-based microchip pump and valve. *Sens. Actuators B Chem.* **2004**, *99*, 592–600. [CrossRef]
121. Bijarchi, M.A.; Favakeh, A.; Sedighi, E.; Shafii, M.B.J.S.; Physical, A.A. Ferrofluid droplet manipulation using an adjustable alternating magnetic field. *Sens. Actuators A Phys.* **2020**, *301*, 111753. [CrossRef]
122. Beyzavi, A.; Nguyen, N.-T. One-dimensional actuation of a ferrofluid droplet by planar microcoils. *J. Phys. D Appl. Phys.* **2009**, *42*, 015004. [CrossRef]
123. Nguyen, N.T.; Ng, K.M.; Huang, X.J.A.P.L. Manipulation of ferrofluid droplets using planar coils. *Appl. Phys. Lett.* **2006**, *89*, 648. [CrossRef]
124. Probst, R.; Lin, J.; Komaee, A.; Nacev, A.; Cummins, Z.; Shapiro, B. Planar Steering of a Single Ferrofluid Drop by Optimal Minimum Power Dynamic Feedback Control of Four Electromagnets at a Distance. *J. Magn. Magn. Mater.* **2011**, *323*, 885–896. [CrossRef]
125. Khaw, M.K.; Ooi, C.H.; Mohd-Yasin, F.; Nguyen, A.V.; Evans, G.M.; Nguyen, N.-T. Dynamic behaviour of a magnetically actuated floating liquid marble. *Microfluid. Nanofluidics* **2017**, *21*, 1–12. [CrossRef]
126. Shu, J.; Tang, S.Y.; Feng, Z.; Li, W.; Li, X.; Zhang, S. Unconventional locomotion of liquid metal droplets driven by magnetic fields. *Soft Matter* **2018**, *14*, 7113–7118. [CrossRef]
127. Vialetto, J.; Hayakawa, M.; Kavokine, N.; Takinoue, M.; Varanakkottu, S.N.; Rudiuk, S.; Anyfantakis, M.; Morel, M.; Baigl, D. Magnetic Actuation of Drops and Liquid Marbles Using a Deformable Paramagnetic Liquid Substrate. *Angew. Chem. Int. Ed. Engl.* **2017**, *56*, 16565–16570. [CrossRef] [PubMed]
128. Damodara, S.; Sen, A.K. Magnetic field assisted droplet manipulation on a soot-wax coated superhydrophobic surface of a PDMS-iron particle composite substrate. *Sens. Actuators B Chem.* **2017**, *239*, 816–823. [CrossRef]
129. Seo, K.S.; Wi, R.; Im, S.G.; Kim, D.H. A superhydrophobic magnetic elastomer actuator for droplet motion control. *Polym. Adv. Technol.* **2013**, *24*, 1075–1080. [CrossRef]
130. Lin, Y.; Hu, Z.; Zhang, M.; Xu, T.; Feng, S.; Jiang, L.; Zheng, Y. Magnetically Induced Low Adhesive Direction of Nano/Micropillar Arrays for Microdroplet Transport. *Adv. Funct. Mater.* **2018**. [CrossRef]
131. Wang, L.; Gao, C.; Hou, Y.; Zheng, Y.; Jiang, L. Magnetic field-guided directional rebound of a droplet on a superhydrophobic flexible needle surface. *J. Mater. Chem. A* **2016**, *4*, 18289–18293. [CrossRef]
132. Sajeesh, P.; Sen, A.K. Particle separation and sorting in microfluidic devices: A review. *Microfluid. Nanofluidics* **2013**, *17*, 1–52. [CrossRef]
133. Movafaghi, S.; Wang, W.; Metzger, A.; Williams, D.D.; Williams, J.D.; Kota, A.K. Tunable superomniphobic surfaces for sorting droplets by surface tension. *Lab Chip* **2016**, *16*, 3204–3209. [CrossRef] [PubMed]
134. Joensson, H.N.; Uhlen, M.; Svahn, H.A. Droplet size based separation by deterministic lateral displacement-separating droplets by cell-induced shrinking. *Lab Chip* **2011**, *11*, 1305–1310. [CrossRef]
135. Hatch, A.C.; Patel, A.; Beer, N.R.; Lee, A.P. Passive droplet sorting using viscoelastic flow focusing. *Lab Chip* **2013**, *13*, 1308–1315. [CrossRef]
136. Al-Hetlani, E.; Hatt, O.J.; Vojtíšek, M.; Tarn, M.D.; Iles, A.; Pamme, N.; Häfeli, U.; Schütt, W.; Zborowski, M. Sorting and Manipulation of Magnetic Droplets in Continuous Flow. *AIP Conf. Proc.* **2010**. [CrossRef]
137. Zhang, K.; Liang, Q.; Ai, X.; Hu, P.; Wang, Y.; Luo, G. On-demand microfluidic droplet manipulation using hydrophobic ferrofluid as a continuous-phase. *Lab Chip* **2011**, *11*, 1271–1275. [CrossRef] [PubMed]
138. Toner, M.; Irimia, D. Blood-on-a-chip. *Annu. Rev. Biomed. Eng.* **2005**, *7*, 77–103. [CrossRef] [PubMed]

139. Kolosnjaj-Tabi, J.; Wilhelm, C.; Clément, O.; Gazeau, F. Cell labeling with magnetic nanoparticles: Opportunity for magnetic cell imaging and cell manipulation. *J. Nanobiotechnol.* **2013**, *11*, 1–19. [CrossRef]
140. Sung, Y.J.; Kim, J.Y.H.; Choi, H.I.; Kwak, H.S.; Sim, S.J. Magnetophoretic sorting of microdroplets with different microalgal cell densities for rapid isolation of fast growing strains. *Sci. Rep.* **2017**, *7*, 10390. [CrossRef]
141. Borlido, L.; Azevedo, A.M.; Roque, A.C.; Aires-Barros, M.R. Magnetic separations in biotechnology. *Biotechnol. Adv.* **2013**, *31*, 1374–1385. [CrossRef]
142. He, J.; Huang, M.; Wang, D.; Zhang, Z.; Li, G. Magnetic separation techniques in sample preparation for biological analysis: A review. *J. Pharm. Biomed. Anal.* **2014**, *101*, 84–101. [CrossRef]
143. Aboutalebi, M.; Bijarchi, M.A.; Shafii, M.B.; Kazemzadeh Hannani, S. Numerical investigation on splitting of ferrofluid microdroplets in T-junctions using an asymmetric magnetic field with proposed correlation. *J. Magn. Magn. Mater.* **2018**, *447*, 139–149. [CrossRef]
144. Ma, R.; Fu, T.; Zhang, Q.; Zhu, C.; Ma, Y.; Li, H.Z. Breakup dynamics of ferrofluid droplet in a microfluidic T-junction. *J. Ind. Eng. Chem.* **2017**, *54*, 408–420. [CrossRef]
145. Leshansky, A.M.; Pismen, L.M. Breakup of drops in a microfluidic T junction. *Phys. Fluids* **2009**, *21*, 023303. [CrossRef]
146. Menetrier-Deremble, L.; Tabeling, P. Droplet breakup in microfluidic junctions of arbitrary angles. *Phys. Rev. E Stat. Nonlin Soft Matter Phys.* **2006**, *74*, 035303. [CrossRef] [PubMed]
147. Yamada, M.; Doi, S.; Maenaka, H.; Yasuda, M.; Seki, M. Hydrodynamic control of droplet division in bifurcating microchannel and its application to particle synthesis. *J. Colloid Interface Sci.* **2008**, *321*, 401–407. [CrossRef] [PubMed]
148. Stark, A.; Shin, D.J.; Wang, T.H.J.B.M. A sample-to-answer droplet magnetofluidic assay platform for quantitative methylation-specific PCR. *Biomed. Microdevices* **2018**, *20*, 31. [CrossRef]
149. Hang Koh, W.; Seng Lok, K.; Nguyen, N.-T. A Digital Micro Magnetofluidic Platform For Lab-on-a-Chip Applications. *J. Fluids Eng.* **2013**, *135*, 021302. [CrossRef]
150. Feng, H.; Xu, X.; Hao, W.; Du, Y.; Tian, D.; Jiang, L. Magnetic field actuated manipulation and transfer of oil droplets on a stable underwater superoleophobic surface. *Phys. Chem. Chem. Phys.* **2016**, *18*, 16202–16207. [CrossRef] [PubMed]
151. Varma, V.B.; Wu, R.G.; Wang, Z.P.; Ramanujan, R.V. Magnetic Janus particles synthesized using droplet micro-magnetofluidic techniques for protein detection. *Lab Chip* **2017**, *17*, 3514–3525. [CrossRef]
152. Nisisako, T. Recent advances in microfluidic production of Janus droplets and particles. *Curr. Opin. Colloid Interface Sci.* **2016**, *25*, 1–12. [CrossRef]
153. Lone, S.; Cheong, I.W. Fabrication of polymeric Janus particles by droplet microfluidics. *Rsc Adv.* **2014**, *4*, 13322–13333. [CrossRef]
154. Potts, H.E.; Barrett, R.K.; Diver, D.A. Dynamics of freely-suspended drops. *J. Phys. D Appl. Phys.* **2001**, *34*, 2529–2536. [CrossRef]
155. Beaugnon, E.; Tournier, R. Levitation of organic materials. *Nature* **1991**, *349*, 470. [CrossRef]
156. Simon, M.D.; Geim, A.K. Diamagnetic levitation: Flying frogs and floating magnets (invited). *J. Appl. Phys.* **2000**, *87*, 6200–6204. [CrossRef]
157. Liu, Y.; Zhu, D.M.; Strayer, D.M.; Israelsson, U.E. Magnetic levitation of large water droplets and mice. *Adv. Space Res.* **2010**, *45*, 208–213. [CrossRef]
158. Berry, M.V.; Geim, A.K. Of flying frogs and levitrons. *Eur. J. Phys.* **1997**, *18*, 307–313. [CrossRef]
159. Geim, A.K.; Simon, M.D.; Boamfa, M.I.; Heflinger, L.O. Magnet levitation at your fingertips. *Nature* **1999**, *400*, 323–324. [CrossRef]
160. Pang, X.F.; Zhong, L.S. The Suspension of Water Using a Superconductive Magnetic-Field and Its Features. *IEEE Trans. Appl. Supercond.* **2016**, *26*, 1–4. [CrossRef]
161. Feng, L.; Shi, W.-Y. The influence of Marangoni effect on flow and deformation of an electromagnetically levitated molten droplet under static magnetic fields. *Int. J. Heat Mass Transf.* **2016**, *101*, 629–636. [CrossRef]
162. Feng, L.; Shi, W.-Y. The Influence of Eddy Effect of Coils on Flow and Temperature Fields of Molten Droplet in Electromagnetic Levitation Device. *Metall. Mater. Trans. B* **2015**, *46*, 1895–1901. [CrossRef]
163. Bojarevics, V.; Hyers, R.W. Levitated Liquid Dynamics in Reduced Gravity and Gravity-Compensating Magnetic Fields. *JOM* **2012**, *64*, 1089–1096. [CrossRef]
164. Bojarevics, V.; Pericleous, K. Droplet Oscillations in High Gradient Static Magnetic Field. *Microgravity Sci. Technol.* **2008**, *21*, 119–122. [CrossRef]
165. Yu, J.; Koshikawa, N.; Arai, Y.; Yoda, S.; Saitou, H. Containerless solidification of oxide material using an electrostatic levitation furnace in microgravity. *J. Cryst. Growth* **2001**, *231*, 568–576. [CrossRef]
166. Kitahara, T.; Tanada, K.; Ueno, S.; Sugioka, K.-I.; Kubo, M.; Tsukada, T.; Uchikoshi, M.; Fukuyama, H. Effect of Static Magnetic Field on Recalescence and Surface Velocity Field in Electromagnetically Levitated Molten CuCo Droplet in Undercooled State. *Metall. Mater. Trans. B* **2015**, *46*, 2706–2712. [CrossRef]
167. Sun, Q.; Li, J.; Liu, Y.; Jiang, Y.; Kang, K.; Feng, J. Arc characteristics and droplet transfer process in CMT welding with a magnetic field. *J. Manuf. Processes* **2018**, *32*, 48–56. [CrossRef]

MDPI
St. Alban-Anlage 66
4052 Basel
Switzerland
Tel. +41 61 683 77 34
Fax +41 61 302 89 18
www.mdpi.com

Biosensors Editorial Office
E-mail: biosensors@mdpi.com
www.mdpi.com/journal/biosensors

www.ingramcontent.com/pod-product-compliance
Lightning Source LLC
Chambersburg PA
CBHW061809210926
51850CB00006B/653
* 9 7 8 3 0 3 6 5 7 1 2 6 3 *